# Human Development for Occupational and Physical Therapists

# Human Development for Occupational and Physical Therapists

## Margaret A. Short-DeGraff, Ph.D., O.T.R.

Research Fellow, Center for Child Study
Assistant Professor, Department of Psychology
Skidmore College
Saratoga Springs, New York

*with a chapter contributed by*

## Robert J. Palisano, Sc.D., P.T.

Assistant Professor
Department of Orthopedic Surgery and Rehabilitation
Programs in Physical Therapy
Hahnemann University
Philadelphia, Pennsylvania

WILLIAMS & WILKINS
Baltimore • Hong Kong • London • Sydney

*Editor:* John P. Butler
*Associate Editor:* Linda Napora
*Copy Editor:* Elia A. Flanegin
*Design:* JoAnne Janowiak
*Illustration Planning:* Wayne Hubbel
*Production:* Raymond E. Reter

Copyright © 1988
Williams & Wilkins
428 East Preston Street
Baltimore, MD 21202, U.S.A.

*Printed in the United States of America.*

**Library of Congress Cataloging in Publication Data**

Short-DeGraff, Margaret A.
  Human development for occupational and physical therapists.
  Includes bibliographies and index.
  1. Child development. I. Palisano, Robert J. II. Title. [DNLM: 1. Child Development. WS 105 S5597h]
RJ131.S5153 1988    612′.65    87-8106
ISBN-0-683-07703-1

88  89  90  91
10  9  8  7  6  5  4  3  2  1

The best thing for being sad, replied Merlyn, beginning to puff and blow, is to learn something. That is the only thing that never fails. . .Learn why the world wags and what wags it. That is the only thing which the mind can never exhaust, never alienate, never be tortured by, never fear or distrust, and never dream of regretting. Learning is the thing for you. Look at what a lot of things there are to learn. . ."

T.H. White
*The Once and Future King*

I write these essays primarily to aid my own quest to learn and understand as much as possible about nature in the short time alotted.

Stephen Jay Gould
*The Flamingo's Smile*

# Preface

This book was developed to fit several specific needs. I had been teaching developmental psychology for many years before beginning to teach similar topic material in an interdisciplinary program for physical and occupational therapists. I noticed that a need existed for a comprehensive text that not only covered normal developmental processes but also focused on specific aspects of development of importance to occupational and physical therapists. While many developmental psychology texts existed, they did not provide details in certain areas of interest to therapists. Similarly, in pediatrics, many clinical texts existed, but they did not cover *normal* developmental theories and processes. Thus, this text was designed to include information regarding developmental psychology (the study of normal behavioral change over time) while including more specific details about sensory processes, perception, motor and reflex development, neural development and plasticity, and the effects of early sensorimotor experiences.

To guarantee that the chapter on motor development addressed the needs and concerns of therapists, I enlisted the help of Bob Palisano, a physical therapist and clinical researcher. Bob and I had often discussed the need for a textbook that would be oriented to the scope and needs of students in both physical and occupational therapy. In both of these fields, there has been a growing focus on developmental theory and research. For example, the American Occupational Therapy Association included as one of its primary goals for 1990 the development of an integrated theory of human occupation. This text addresses these areas by providing an integrated view of human development, by including a comprehensive analysis of current developmental theories, and by synthesizing these theories with common, contemporary approaches to health and rehabilitation.

As an experimental psychologist who then became an occupational therapist and clinical researcher, I am concerned about students who feel the pressure to learn about, but still have little or no introduction to, research methods. This text was designed to provide numerous examples of research in a developmental context, to provide extensive references from across disciplines, and to view the research methods in terms of clinical problem solving. Case studies are included to illustrate that research methods enable us to look at how behavior changes over time, whether that change occurs as a result of normal development or therapeutic intervention. It is my hope that this approach will not only simplify the learning of research methods but also prompt students to be more comfortable as research consumers in general.

While this book incorporates and reviews the major developmental theories and research, two particular approaches, both of which focus on interactionism, are highlighted. This first approach, or theme, is that development occurs across the life-span and can best be understood by studying the whole person as a complex system in interaction with other people and objects in the environment. This is translated into what is termed an "ecological" or "ecological-systems approach" throughout the text. It is evident in some of the case studies as well as some of the models of intervention that are described in the text or in illustrations. The other approach is more specific and has emerged from a synthesis of research regarding the early experiences of animals and humans. This second line of thought recognizes the significance of self-produced movement or active involvement and feedback in facilitating neural (and developmental) change. Active involvement generalizes to the concept of reciprocity in social interaction and has become an important underlying concept for researchers in neurobiology and psychology as well as for educators, developmentalists, clinicans, and parents.

Finally, although I believe in the significance of studying development across the life-span, it was not possible to incorporate in this book the specific developmental changes that characterize adulthood and old age. I firmly believe that development and aging are synonymous, and my preference would be to study all stages of life together; however, this was just not possible given the scope and size of this text. For gerontology students, however, all of the developmental stages are introduced as are the developmental theories and research methods that apply to all stages of life.

This is a special book to me because it synthesizes information from many disciplines in which I have had experiences, ranging from neurobiology and ethology, to psychology and allied health education, and clinical work with people of all ages. This book is also special because it includes information, contributions (photographs, illustrations, research), and ideas that are important to me. Writing this has been an experience that has promoted my own development and resulted in my understanding how much more there is to learn. Development changes us and moves us on.

# Acknowledgments

Writing this book has been a pleasure, due in large part to the people who have worked with me on this project. Publishing a book is a multistep process that involves the input and assistance of many different people. I am grateful to them for helping to make what could have been a difficult task instead a memorable, pleasurable event—what a friend described as "an obvious labor of love."

The idea for this book was a pipe dream 10 years ago when fellow graduate student and therapist Pat Paradies suggested I enter the field of occupational therapy. She had the foresight to see the growing emphasis on research and theory in the field and suggested that someone with a doctorate in experimental psychology might be able to contribute to occupational therapy as well as fulfill the multiple dreams of combining clinical work, teaching, and conducting research.

Nearly 10 years later, another friend and occupational therapist, Hilda Versluys, actually provided the impetus for this text's publication with Williams & Wilkins. I am grateful for Hilda's confidence and for John Butler's subsequent spirit of cooperation in conceptualizing this project. At a lecture I recently attended, a publisher explained that some kinds of editing can not only change authors' words but also amend the spirit of their work. I feel as if the Williams & Wilkins staff have respected the spirit, but still enhanced the quality, of my work. My thanks go to John Butler, Linda Napora, Ray Reter, and my copy editor, Elia Flanegin—I have not met her but feel that we know each other.

The opportunity to actually sit down and write this book was afforded by a postdoctoral research fellowship I received from the Center for Child Study in the Education Department at Skidmore College. The purpose of the fellowship was to provide freedom from numerous extraneous academic responsibilities so that academic research goals could be pursued. This purpose was achieved, as the final chapter of this book was completed and mailed out on the day the fellowship ended! I am grateful to Karen Diamond and to Bill LeFurgy for their vision of this fellowship.

At no other institution have I received such widespread and congenial support across programs than at Skidmore College. I specifically want to thank the Psychology Department's Chair, Dave Burrows, and Acting Chair, Joan Douglas, as well as Donna Evans, former Chair of the Education

Department. The Director of the Center for Child Study, Karen Diamond, played many important roles in production of this book. Her trusting supervisory style promoted my responsibility and productivity. In addition, Karen contributed photographs, edited, gave critical feedback regarding several chapters, and also provided resources and collegial stimulation that enhanced my research.

The confidence to take on such a large project has developed gradually and is partly due to previous successful collaborative research with Ken Ottenbacher and Paul Watson. I am grateful for the opportunity to have worked with these talented individuals and to have learned from them both. In addition, I am appreciative of Ken's editorial comments on the research methods chapter. The actual approach to this chapter was prompted by a discussion with Mary Ann Foley. Her suggestion for viewing research methods in terms of a problem-solving continuum provided the framework for chapter 4.

There are many others whom I also want to thank for their contributions to this text. These include:
—The Arde Bulova Foundation and the Ehrmann Foundation, which provided support through the Center for Child Study at Skidmore College.
—Bob Palisano who gave encouragement as well as feedback on numerous chapters and who willingly contributed one of the chapters in this text.
—Sandy Jung Vrem, Hilary, and Jamie.
—Judy Pelletier, who enabled me to obtain photographs courtesy of the Neonatal Intensive Care Unit, University of Connecticut Health Center's John Dempsey Hospital, Farmington, Connecticut.
—Gregory Kriss, photographer, who took most of the photographs from the John Dempsey Hospital.
—The March of Dimes Birth Defects Foundation.
—Ruth M. Wimmer of Wimmer-Ferguson Child Products for the photograph of the Infant Stim-Mobile.
—The New Yorker, United Features Syndicate, and Dave Strickler of Strix Pix for photographs or cartoons.
—and the many family members, colleagues, and neighbors who tolerated my intrusions and picture-taking.

It is important for me to point out that many ideas in this text were prompted by thoughts and questions raised by students during the almost 15 years I have been teaching courses in human development. I cannot help but be influenced by their ideas and by information gained from various texts I have used as well as other's research. I have tried to acknowledge those formal influences thoroughly with references throughout the text. My personal views and optimism about development (aging) and about the continuity of life come not from books but from direct experiences with my grandparents and my mother and father. As an example, in between his gardening, woodworking, birding, and other crafts, my father created all the original line illustrations for this text. My understanding of the applications of the ecological approach and the real meaning and limitations of occupational and physical therapy are also not drawn from textbooks.

Instead, they have been gradually acquired by listening to clients in home care and by everyday life with my husband, Skip DeGraff. He has demonstrated to me and to many others that with motivation, compassion for others, and a sense of humor, disabilities do not impede, but instead challenge and enhance development. Finally, I need to acknowledge the influence of Mucky and Chan. For more than the ten years this book has been occupying the back or the front of my mind, they have provided companionship, comfort, continuity, and diversion. While the text was being written, they added occasional keystrokes (which kept me alert during proofreading) and kept the manuscript warm while I was out. Aldous Huxley is attributed with saying, "If you want to write, keep cats." I believe it.

To readers of this book: I would encourage you, if you have feedback, to let me hear from you. Your thoughts and ideas are important.

Peggy Short-DeGraff
*September 1987*

# Contents

# 1

## Introduction to the Study of Human Development

# NED, JIMMY, MARIE, AND JESSICA

Ned is 14 years old. About two years ago he started to take an interest in science, especially in astronomy and chemistry. He used to do science experiments with a small chemistry set in the basement of his parents' house, and his mother has supported him in these interests over the past couple of years. In fact, she plans to give him a small telescope for his birthday. Ned likes to listen to music, and often stays up late in his room, with a small reading light, listening to the radio and reading science fiction. At school he participates in sports, especially baseball and track. While he is not an excellent athlete, he is small and quick and has a keen sense of competition.

Ned's interests are quite different from those of his siblings: Jimmy who is 9; Marie, 6; and Jessica, the baby. Although Ned has always been cooperative and helpful at home, this past year has been difficult for him as well as for his mother and new stepfather. While not a star student, Ned has consistently achieved grades in the B and C range. Lately, however, his grades have slipped, and disciplinary problems have occurred. Last month Ned was temporarily suspended from school because of an incident in the school parking lot. The events are not clear, but some students were found in and around a car that had empty and opened beer cans on the dashboard and in the backseat. School officials are not sure who was drinking and which students were involved, as no one had been in possession at the time they were discovered. Seven students, four boys and three girls, were either inside or around the car at the time the incident occurred, and Ned was one of the seven. Since none of the students would indicate what was going on, who was involved, who purchased the beer, or where it was obtained, all seven were disciplined by the school. Ned's parents were called in, and the parents and involved students had to meet with several school officials on repeated occasions.

Ned's mother, Jan, is not quite sure what is going on with her eldest son. She feels that somehow everything has turned around since he started going to the large, impersonal junior/senior high school. Things seemed so much simpler when he attended the small, local neighborhood school where Jimmy and Marie now attend. There, Ned had the same teacher all day, and the teacher and Ned's parents often kept close track of his progress, his activities, and his circle of friends. Now, however, Ned is in a much larger school. His circle of friends has widened to include people his mother does not know, and since he moves to different classrooms during the day, he has a wide variety of teachers. No one really seems to have a handle on his progress or his activities during the day.

Lately Ned has been spending less time at home. He often grabs a snack after school and claims he is "just going out" with his friends. Ned's parents are concerned. He talks to them infrequently, and they feel that whenever they try to get close to him, he moves farther away. This is so different from the way he used to be that they feel that this must be some kind of "typical adolescent stage" he is going through. Although she has not clearly identified her feelings, Jan feels angry with Ned. She just recently gave birth to Jessica, and she feels pulled by the attention she wants to give her new husband, the baby, and the other children. Ned has always been a big help to her, and she was depending upon him to help out when the baby came. Instead, he has disappointed her, and she feels confused and a little betrayed. Just when she felt that she needed a little extra help, her most dependable child has become unpredictable and moody. And although things are really tight financially, she has saved money every month to make payments on the telescope for Ned. She knew she could not afford it, but when she placed the order months ago, she really wanted to give him something special for all his help. But now, Ned acts like he does not care about astronomy or, for that matter, about anything or anyone.

Ned senses his mother's feelings and he, too, feels betrayed. He thinks that maybe she does not want him or need him as she used to. He used to be the "man of the house." Everyone told him so, and although he felt lonely and depressed when his father moved out several years ago, he did not show his feelings. His father kept telling him to be tough and to help his mother with Jimmy and Marie, who were much younger then. Recently, however, it seems as if everything has changed. Life was easier before, when he went to the other school. He knew everyone and everyone knew him. He was one of the oldest students, and he knew all the rules and how to bend them a little to meet his needs. Now, however, he is faced with so many choices. Everyone keeps asking him what he wants to do, what his interests are, what courses he wants to take, what curriculum he needs to meet his vocational goals. Vocational goals, indeed! Ned does not know what he is interested in. Some days it is one thing, and other days it is something else. Some days he just wishes everyone would leave him alone. He has so many teachers, and they all seem to have different demands and expectations. It is as if the rules keep changing, at school and at home, and no one really cares about him or about what he is doing. They even have unstructured free periods during school, and no one really cares whether he goes to study hall or leaves the school. He figures that if they really cared, they would make sure he studied.

Just like his parents, Ned's old friends seem to have drifted apart. With so many new people in school, and with students going to so many different classes, his old gang has broken up. There are some new students in the class he has before free period, and they keep asking him to leave school with them instead of studying. While Ned knows his mother would not approve of them, Ned is intrigued by these other guys who seem to like and accept him. They are interesting, they do things he has never heard of, and he is curious. They know things he does not, and already just by listening to them he has learned some things about sex and about what is happening to his body.

In addition to all the changes at school, at home, and with his friends, Ned's body also seems to be changing. Ned finds himself increasingly interested in and confused by girls his own age. He finds he cannot sleep at night sometimes, and his sexual arousal scares and mystifies him. He is afraid that everyone knows when he gets aroused, and that seems to be happening all of the time now. His brother Jimmy teases him about combing his hair, and Marie keeps asking him when he is going to kiss a girl and does he want to practice on her dolls. Sometimes he feels he could hit them both. His stepfather, who thinks he knows just about everything in the world, keeps telling him to take it easy. He tells Ned that they ought to go out together some weekend, maybe camping or hiking, but then he never comes up with a serious date or time to actually go. The guy tousles Ned's hair, and tries to hug him a little and roughhouse. Ned just wishes everyone would leave him, his hair, and his personal life alone! He feels uncomfortable, strange, confused, and out of control.

Everything seems to be changing too fast, and Ned blames his parents for changing it all. Two years ago, the family had seemed to be coping with the divorce. Now, it is as if no one can understand him—or has time to really care about him. Everything has changed at home. His new stepfather seems to be taking over, and if it is not him, it is the baby making demands on everyone. His father, who lives across town, now has a steady girlfriend who has a son who is Ned's age. His father, it seems, when Ned really needs his attention, is going places with "another son." Ned resents this and feels that he does not really belong anywhere. He wishes he could go back in time, maybe to when he was Jimmy's age. Then his parents would be together, he would have a teacher who cared, and he could spend his days playing with the toys and games he gave to Jimmy.

# Jimmy

Jimmy is 9 years old. He loves comic books, television, football, and drawing pictures of cars and spaceships. Jimmy has a small group of friends from his neighborhood. They all play football or tag or build forts and spaceships after school and belong to the same scout troop. All his playmates are other boys his age, and they share most of the same interests. Jimmy tries to be like his big brother Ned, but finds Ned leaving him out more and more. Jimmy cannot fully understand why Ned spends more time with other guys and with girls. Jimmy does not really care for school that much. He does not like to sit at his desk all of the time, but he likes art and physical education. He has the same classroom teacher for most of his classes, and he walks to and from school with his friends. Sometimes he starts out with his sister Marie, but usually, they all meet up with their own groups of friends on the way. Then, he and his friends throw snowballs, kick leaves, or race each other.

One of Jimmy's favorite things is to go to his grandparents' house after school. He often nags his mother to let him go and stay there for the afternoon and for supper. They have a big playroom and a lot of television channels. He keeps some of his toys there and many of his comic books. His grandmother makes him cookies or popcorn, and he likes to curl up and snack and watch television. He especially likes the time by himself and the attention of his grandparents. He gets tired, sometimes, of Ned yelling at him, that is, when Ned is home; he also gets tired of the baby. He cannot seem to do anything right when it comes to her. Mostly, he is yelled at for making too much noise when it is the baby that has turned the house upside down. What confuses Jimmy the most is why his mother expects him to do certain things, such as take care of Marie or check on the baby, but then refuses to let him do other things he knows he can handle, such as lighting the barbecue or driving their snowmobile by himself. It does not seem fair. So he likes to, as he explains, "hang out at his grandparents where he doesn't get any grief."

When Ned was around the house more often, he and Jimmy used to play a lot together in the backyard or in the basement. Jimmy would help Ned with his experiments, or they would build forts together. Now, Ned is not home much, and with the baby always crying, Jimmy's mother keeps telling him to "look after Marie," or to "take Marie playing with you." Doesn't she understand that Marie is no fun to play with? Marie ignores the rules to all the games he likes, she likes things he does not, she gets hurt easily, she cries all the time, and she tattles on him whenever he does any little thing. After all, Marie is 6, and Jimmy is 9.

## Marie and Jessica

Marie likes to do things with both of her big brothers, but mostly with Jimmy because lately Ned seems to be always snapping at her. She likes it when Jimmy's friends come over. Then she goes outside with them. They climb trees and play ball, and often her mother makes them let her play too. Sometimes though, they ride their bikes faster and farther than she is allowed to go, or they play keep-away and will not give her the ball. They get angry with her because she "messes up" their games. All their games have these silly rules, which she does not understand. Why can't they just play without rules like she and her friends do? She would play with her own friends, but her mother does not like her to go to their houses after school because they live too far for Marie to walk alone. And now, with the baby, it is too difficult for her mother to drive her.

Marie likes school. She has a teacher who reminds her of her grandmother who lives far away. She likes learning new things at school, and she thinks it is fun to read. She wants to write like Jimmy and Ned, and she wants to read bigger books like those they have. She likes plays and puppet shows at school and thinks she will be an actress when she grows up. What Marie really likes to do sometimes is to take care of Jessica, the baby. It makes her feel important, but she cannot understand why everyone keeps telling her not to drop the baby or to pay attention to what she is doing. She knows what she is doing, and Jessica is fun—at least until she starts crying. Then, if Marie is with the baby, everyone blames her for waking her up or for not holding her right. Jessica is 5 months old and seems to get all the attention lately. She cries all the time and even wakes everyone up in the middle of the night; yet when Marie plays her piano game or her computer toy during the day, her mother gets cross with her for making too much noise. When things are quiet, Marie likes to hold the baby, but then she loses interest and wants to do something else. Then it seems everyone gets mad at her for fidgeting. They call her "wiggle worm" and tell her she will never go to her mother's studio to get her hair cut unless she can learn to sit still.

Marie likes to go with Jimmy to visit her grandparents, but she does not go often because Jimmy will not take her and she is too young to go by herself. She cannot understand why Jimmy does not take her; he keeps telling her he is going to Gramma's so he won't have to stay around "all the little babies." She is not a baby, and she is not at all like Jessica, who cannot talk, walk, get dressed, or go to the bathroom. Why, Jessica cannot even feed herself! Marie knows she is not a baby, yet everyone keeps telling her not to behave like one. For example, she tries to sit next to her mother and Jessica on the couch, and she tries to crawl into her mother's lap. But her new stepfather keeps telling her she's a big girl now and that Jessica is the baby. So Marie goes to her room and plays with her dolls. What she really wants is a puppy to play with, but her mother says she cannot handle that now, not with the baby and all. Everyone tells Marie to wait until summer. Maybe in the summer, when things settle down, then they can talk about a dog. In the summer, the baby will be older, and maybe Ned will have calmed down too. Then, maybe Gramma will take Jessica for a couple of days, and all of them, Mom, Dad, Ned, Jimmy, and Marie, will go camping together, like old times.

# Developmental Differences Among the Children

Ned, Jimmy, Marie, and Jessica all belong to the same family, yet each one is distinctly different from the other. Ned, for example, thinks, acts, feels, interacts, talks, and, overall, behaves differently than his siblings. While Ned has things in common with Jimmy because Jimmy is 9 years old and is also a male, Ned is still very different. He is experiencing things inside of his body and in his social and intellectual life that are common for a 14-year-old boy but that are vastly different from what a 9- or a 6-year-old experiences. Jimmy and Marie are only 3 years apart in age, but Jimmy also thinks and feels differently than does Marie. This is not just because of their sex differences but also because Jimmy, at age 9, knows things that Marie cannot understand, and he possesses skills that Marie has yet to master. Marie, on the other hand, is vastly different from Jessica, the baby. Despite the fact that they are both females, there is a world of difference between Marie's skills and Jessica's dependence. As a young child, Marie has passed through the demanding physical dependency of infancy, and even though she is still quite immature, she has many basic motor, language, and emotional abilities that she will now refine during the rest of her childhood. Jessica, in many ways, is the most different from all of the other children; this is partly because she is an infant, partly because of her parentage, and because she has most of her life ahead of her. She will acquire skills and abilities and go through unique experiences as the fourth child in a family whose siblings are all very different in traits and in age.

# DIMENSIONS OF SCIENTIFIC STUDY OF HUMAN DEVELOPMENT

The many potential reasons for the differences between individuals, such as Ned, Jimmy, Marie, and Jessica, are examined in the scientific study of human development, which is the topic of this book. The study of human development examines how people such as Ned, his siblings, his parents, and his grandparents change over time. In order to explore all the different dimensions of the scientific study of human development, this book examines major areas corresponding to the following five sections:

1. Issues in Development
2. Research and Theory
3. Prenatal Development and Birth
4. Sensory, Perceptual, and Motor Development
5. Cognitive-Intellectual and Social-Emotional Development

## Purposes of the Book

This book has two primary purposes. One is to explore the many potential differences that exist between people of varying ages ranging from birth through adolescence. This involves the scientific study of how people change over time, i.e., the study of human development. Thus, the book explores different stages of life, e.g., adolescence, childhood, and infancy while also looking at the different areas of life in which an individual experiences change, e.g., social, emotional, motor, and intellectual. The other purpose of the book is to introduce the scientific study of human development to occupational and physical therapists. Both of these clinical fields share a common concern about how and why people of all ages change, as a result of therapeutic intervention and recovery *and* as a result of normal developmental processes. While many other books in these fields have addressed how people change as a result of intervention and recovery, this book addresses change as a result of normal development.

## Themes

Two fundamental themes implicit throughout this book are that people undergo change throughout their lives and that all aspects of one's life affect each other. Thus the book is geared toward looking at all aspects of an individual's life, i.e., at the "whole" person, and at how change occurs throughout life, i.e., development across the life span. Although a major theme of this book is that people change throughout their lives, it is not possible, with the amount of space needed to cover human development and its significance to physical and occupational therapy, to include the complete life span. Thus, this text ends its discussion with adolescence, not because human development ends there but because it is a logical stopping point for the scope of this book.

## Organization

This book is organized into five sections. Sections 1 and 2 are included because professional organizations and leaders in physical and occupational therapy have stated the need for students to become familiar with research and theory about human behavior and its change over time. Sections 4 and 5 examine the many developmental areas of the total person. Included are

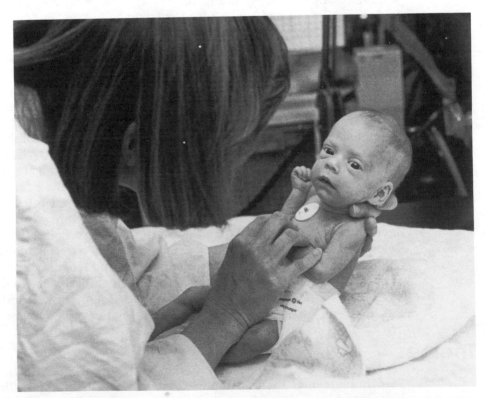

**Figure 1.1.** The topic of this book, the scientific study of human development, looks at how people change from conception throughout the life span.

different trends in sensing, moving, thinking, and feeling that occur during infancy and childhood. These chapters also explore the differences between infants (such as Jessica), young children (such as Marie), and older children (such as Jimmy). In contrast, section 3 looks at a specific age-related division, the prenatal period. Included in this section are the foundations for infant and childhood development that arise before birth. Prenatal development and environmental and genetic risk conditions are examined as well as factors affecting pregnancy and birth.

The topic of this text, the scientific study of human development, consists of specific approaches and methods that identify the field as a science. These "scientific methods," as they are called, which are discussed in section 2, enable researchers in development to make formal studies of behavior as it changes over time. Thus, developmental researchers, who may study individuals such as Ned and his family, use specific methods for conducting their studies. They then offer speculations as to how and why Ned and Jessica, or other people of differing ages, behave the way they do. These scientific speculations are called "theories" and are also covered in section 2. In this book a considerable amount of space is dedicated to these theories. As section 1 explains, the study of development and knowledge of developmental theories are important for clinicians such as physical and occupational therapists.

For any discipline (or for any individual), knowledge of its history illuminates its current status. The scientific study of human development has its own history, which, in ways, parallels many changes that also occurred in applied clinical fields such as pediatrics. Sections 1 and 2 look at the history and at the many dimensions of the science of human development and explore their applications to the health care fields. As leaders in both professions of physical and occupational therapy have recently indicated, study of developmental theories and research is important and may provide a bridge for furthering continued understanding of clinical change.

This introductory chapter started with a description of four siblings: Ned, Jimmy, Marie, and Jessica. These youngsters were described for several reasons. One was to illustrate how vastly different individuals can be. Despite the fact that they all belong to the same family and are growing up in the same home, each of these siblings is distinctly different from the others. In part this is due to their different ages and levels of development. In part this is also due to each youngster's unique combination of genetic background and life experiences. The three oldest children, Ned, Marie, and Jimmy, have the same parents and grandparents. Thus although each has his or her own uniqueness, each also shares some traits with the others because of similar genetic backgrounds. In contrast, the baby, Jessica, has a father and a set of grandparents who differ from those of the other youngsters. Because of her age, status in the family, the age of her parents and siblings, and because of her special genetic heritage, Jessica will be an individual quite different from her parents and from her brothers and sister.

Human development occurs as a result of a special blend between what we inherit and what we experience throughout our lives. The next chapter will explore this issue with regard to the interaction of genes and environment as a basis for understanding how and why people change over time.

# SECTION

# 1

## ISSUES IN DEVELOPMENT

# Section 1

Section 1 includes chapters 2 and 3, which introduce basic terms and concepts as well as historical approaches to the study of human development. Chapter 2 focuses on an issue of historical interest that continues to influence research and thinking in the contemporary study of human development—the nature-nurture controversy. This controversy is discussed as it relates to the development of intelligence and as it affects clinical attitudes toward intervention and change. Chapter 3 examines the history of scientific study of development as well as parallel changes in the field of pediatrics. The important applications from the topic of human development to clinical fields such as occupational and physical therapy are pointed out. Developmental stages, developmental tasks, developmental therapy, and developmental approaches to health and rehabilitation are examples of some of the numerous applications that are explored.

# 2

---

# Understanding Behavior and the Mechanisms of Change

---

There are many different beliefs about what causes behavior to change over time. The purpose of this chapter is to examine the major reasons for studying behavior and to examine the basic mechanisms and the controversies surrounding these mechanisms that control development—behavior as it changes over time.

# WHY STUDY BEHAVIOR?

Students of human development want to determine how and why people behave the way they do. The goals of psychology are to describe, explain, and predict human behavior (1). While such goals sound simple, in reality, they are not easy to attain. Describing behavior is considerably easier than either explaining or predicting it, and the fields of psychology and of occupational and physical therapy have made enormous advances in accumulating descriptive data regarding large varieties of human behavior.

# Description

In clinical fields it is important to be able to adequately and objectively describe behaviors that are observed. Written and oral descriptions of behavior are the means for recording observations for future reference and communicating those observations to others. Behavioral descriptions become essential parts of professional reports and records, enabling therapists to have accurate profiles of patients' past and current performance levels.

Explaining behavior, on the other hand, is more difficult than describing it. See the following example.

---

*Describing vs. Explaining Behavior*

A 3-year-old white male named Scott is sitting in a manual wheelchair that is pushed up to a table in a school classroom. The child is sitting at the table by himself, his knees and hips are flexed, and his feet are resting on the footrests of the wheelchair.

On the table in front of Scott is a small wooden stool with 10 different colored pegs inserted snugly into 10 holes in the top of the stool. While using a cylindrical grasp to hold a small wooden hammer in his right hand, Scott is hammering the pegs so that they are flush with the top of the stool. His left hand is gripping and stabilizing the stool, and while he works, he closely watches what he is doing. He is able to hit a peg with almost every hammer strike, and he has been working at this task for about 5 minutes.

Scott's head and trunk are in alignment, and he maintains stable balance while he hammers. He has no trunk supports attached to his wheelchair. Occasionally his tongue protrudes about ¼ inch from the right side of his mouth as he concentrates on the task.

This is a fairly complete description of Scott's overt behavior. Can you *explain* what Scott is doing?

---

## Explanation

While descriptions of a single behavior can vary from specific to general, they tend to be fairly consistent. Explanations, however, frequently vary in content. For example, one person may explain that Scott is playing, or more specifically, that he is engaging in solitary play. Another explanation might be that he is exercising his right upper extremity while maintaining trunk stability and head control. Someone who knows him better might explain that he is showing off in front of other students in the room, or maybe, if this is all that he does for hours at a time, that he is engaging in repetitive, stereotyped behavior. A therapist might explain that Scott is developing eye-hand coordination, and another might explain how he is developing specific muscle groups in his right shoulder and arm.

Explanations of behavior clearly vary depending upon one's point of view. Individuals with different training and orientations may easily agree about behavioral descriptions, whereas they may *explain* the same behavior in entirely different terms.

## Prediction

Even more difficult than explaining behavior is being able to predict it. While many of us feel that we can predict some of our own or our friends' behaviors, we are still often surprised by the idiosyncratic responses that people make to what we thought were simple, predictable circumstances. Predicting Scott's specific behaviors over the next 4–5 hours would be nearly impossible. Human behavior is so rich and variable as well as unexpected that even general predictions are often very inaccurate.

Our understanding of normal human behavioral functions is slowly growing but it is still limited. Even more challenging than understanding normal function is understanding what happens when functions go wrong. Clinical fields such as occupational and physical therapy focus on ameliorating human dysfunction, so it is important for us to know about basic normal function before focusing on all of the things that may go wrong. One of the most important goals in science and medicine is to understand the nature and causes and the response of the human body and brain to sickness, disease, and trauma.

Continued study and research into human behavior will direct us toward the goals of explaining and predicting human behavior, but it is possible that some behaviors will never be completely understood (adequately explained) and that predicting human behavior will always remain a challenge.

# MECHANISMS OF BEHAVIORAL CHANGE

There are different processes that cause behavior to change over time. Four of these basic mechanisms are:

Growth
Development
Maturation
Learning

Even though we informally use some of these terms interchangeably, technically, each refers to a different type of process; students of human development must understand the differences between these processes. This is, in part, the subject of *developmental psychology*, the study of behavior as it changes over time.

Numerous and long-standing controversies have arisen regarding the relative influences of these different processes on human behavior, and ongoing controversies about the definitions of these terms exist. In this chapter, these controversies will be explored, and some of the basic definitions of these processes will be examined. What is important for the student is to understand the concepts represented by these terms, the effects of these processes on human functions, the controversies surrounding these terms, and the implications of these controversies on the philosophy and practice of occupational and physical therapy.

# Definitions of Growth and Development

Growth, from a biological perspective, is a proliferation of cells. It is illustrated by the increase in size, or "additive accumulation" (2) that occurs in the skeleton and muscles of children's trunk and limbs as they progress from infancy to adolescence.

The term *development* is often used to refer to all of the changes that occur in one's life from conception to death. More specific definitions vary. For example, some theories regard life as an initial expansion followed by reduction or decline that terminates in death. Development, according to this view, is the initial process of expansion, whereas aging refers to the phase of decline. This definition is controversial because life is not always viewed as a time of progression followed by decline. Others see life as a continual progression, with aging viewed as a process, like development, that occurs throughout, not just at the end of life (3).

A more general definition of development is that it is a change in form. Humans grow and also change in form during life. Thus, a child's limbs elongate and also change qualitatively. Consider how different are the forearms of a 3-year-old child compared with either those of an adult or a 2-day-old infant. With age, the arms grow and also change in form, undergoing structural changes that make them firmer and more muscular. "Development" is used to describe this process of changing form.

It is important to recognize that many structures increase in size (grow) and change in form (develop) as an individual passes through life. The terms growth and development do not apply *only* to physical structures. For example, enormous changes occur in human language, which not only grows but also develops over time. The vocabulary of a 1-year-old child is initially very small, but it *grows* with time. Language also *develops*, starting with single words or word combinations and increasing in complexity to a predictably structured but enormously complex means of oral and written communication.

Many other aspects of human behavior or functions grow and develop over time. All these changes—social, emotional, physical, cognitive, and more—are the subjects of this book.

# Nature-Nurture Controversy

Historically, philosophers and scientists have taken opposing views regarding the types of forces behind behavioral change. Some groups have emphasized that behavior changes primarily as a result of internal forces (maturation, genetics, or heredity), and others have attributed those influences to primarily external factors (environment, experience, or learning). Currently, most scientists agree that *both* internal and external forces contribute to changes in behavior over time (the *interactionist* position), and disagreement today tends to be one of emphasis. The controversy, however, has had considerable historical impact on the field of human development and continues to resurface in contemporary work. The nature-nurture debate will be examined in detail in this chapter because it is fundamental to the study of human development.

The traditional debate regarding the relative effects of inheritance or learning on development is referred to as the *nature-nurture controversy*. Many scientists and philosophers have contended that behavior is either primarily inherited (*nature*) or acquired (*nurture*). The resulting controversy over these effects takes many names including *maturation vs. learning, nativism vs. empiricism, innate vs. acquired, preformed vs. epigenetic, heredity vs. environment* (4) or *biology vs. experience* (5).

## Philosophical and Historical Origins of Nature vs. Nurture

This debate has its origins in early religious and philosophical views regarding the nature of humans and their ideas. For example, early religious views that believed in original sin supported the position that the infant comes into the world already possessing certain tendencies and traits (a *nativist* view).

Philosophical foundations of the nature-nurture debate are generally traced to the works of Jean-Jacques Rousseau (1712–1778), a Swiss-French philosopher, and to a British philosopher, John Locke (1632–1704), respectively. The significance of inheritance and of experience on behavior were recognized prior to the time of these men, however.

Descartes (1596–1650), for example, wrote of the existence of innate ideas, and Plato (427–347 BC) believed that learning involved drawing out knowledge that was already in existence in the soul at birth. Plato did not, however, espouse the view of innate ideas (nativism), since he contended that knowledge is acquired through effort and hard work (6). In his *Republic*,

Plato emphasizes the importance of education and clearly acknowledges the impact of early experiences on development. Cornford (7, p 68) cites Plato: "And the beginning, as you know, is always the most important part, especially in dealing with anything young and tender. That is the time when the character is being moulded and easily takes any impress one may wish to stamp on it."

Both Rousseau and Locke believed that infants possessed unformed minds and that early experiences had significant consequences upon later behavior. Both philosophers believed that the education of children was a serious issue, but their philosophies diverged. Rousseau (8), in his book *Emile*, indicated that children possess an innate goodness that should not be spoiled by exposure to punitive teachers. Children, according to Rousseau, should be exposed to nature, which will guide them in a natural progression through the stages of life.

### Nativism and Empiricism

Nativists, such as Rousseau, believe that humans are born with a form of built-in knowledge that must merely unfold or be accessed through life (8). The nativist approach is in contrast to the *empiricist* approach of Locke, a member of the British School of Empiricism (5). Locke is most commonly associated with the concept of the *tabula rasa*, or clean slate. This concept views the infant as naive at birth, like a slate that starts out clean and is gradually etched with information through time. Locke claimed that through interaction with the world, the child learns gradually through repeated associations formed by practice and drill. This belief in the significance of association is characteristic of the empiricist point of view, which emphasizes the role of experience, or nurture, on human development.

## The Nature Side: Genetics and Maturation

In the context of the nature-nurture controversy, the terms, "maturation," "genetics," and "inheritance" are used interchangeably to refer to innate contributions to development. These terms are not synonymous. *Maturation* is frequently defined as the emergence or unfolding of genetic potential (9) and is illustrated by the onset of motor abilities such as sitting, crawling, and walking. To fully understand the concept of maturation, one must understand some elementary genetics. *Genetics* is the field of biology that studies inheritance and variation. *Inheritance* refers to the transmission of characters across generations.

Understanding genetics requires a familiarity with the basic units of life, which are cells. Living organisms consist of microscopic structures called *cells*. Within organisms are many different kinds of cells, and these are specialized to carry out specific functions that maintain the organism. Thus, brain cells, liver cells, skin cells all contain even smaller elements which direct them to function in their own specific ways. The actual director of each cell is the *nucleus* (plural *nuclei*), which itself consists of smaller substances and structures.

### Chromosomes, DNA, and Genes

Despite the fact that cells have specialized functions, they all contain the same hereditary substance within their nuclei. This material is called *deoxyribonucleic acid*, or *DNA*. In human cells, the DNA is concentrated and organized in specific long, thin patterns within the nucleus. These organized structures of DNA are termed *chromosomes* (10), and certain segments along DNA strands are called *genes*, of which there may be thousands on one DNA fiber (11).

Chromosomes tend to group in pairs, thus DNA is typically depicted as two long strands of material, wrapped around each other in what is popularly referred to as the *double helix* (12). Nobel prize winners Watson and Crick came up with the now popular model of the DNA molecule as a double helix that can be compared to a twisted or spiral ladder with two parallel sides and central rungs (13). This organization is significant, for it is this structure which contributes to DNA's capacity for duplication.

DNA is a large molecule that comprises many different kinds of smaller molecules. Some smaller molecules form the parallel sides of the "ladder" structure, and other molecules, called *bases*, form the "rungs". These smaller molecules have different chemical structures, resulting in four different bases in the DNA molecule. The actual order of these bases within DNA is what provides the basic program for the cell and ultimately for function of the organism (11, 13).

### Reproduction

"One of the fundamental properties of cells and organisms is their ability to reproduce. All cells are, in principle, capable of giving rise to a new generation of cells by undergoing a replication process" (11, p 42). The actual structure of DNA simplifies this process. Basically what happens during replication is that the double-helix structure breaks in half in the center of each rung. The halves of the helix then realign with other molecular structures identical to their previous mates, resulting in two identical double helixes. Thus, simple cell division involves a duplication of the cell nucleus and its chromosomes to form two new daughter cells, identical to each other and to the original parent cell (Fig. 2.1).

The human body consists of many kinds of cells. Most cells are body, or *soma*, cells. The second, and smaller category of cells are the *sex cells*, also termed *germ cells* or *gametes*. These are the *sperm cells* in males and *egg cells* (also called *ova*, or singular *ovum*) in females.

Soma cells and germ cells replicate in different ways. Soma cells reproduce through the process of *mitosis* (Fig. 2.1). This is the means by which identical genetic material contained in the chromosomes is distributed to two new nuclei (referred to as "daughter nuclei") (10). The process of mitosis results in daughter nuclei that have the same amount and the same kind of chromosomes as is found in the parent cells.

Cells of different organisms contain different amounts of DNA. The nuclei of human soma cells contain a specific amount of DNA organized in

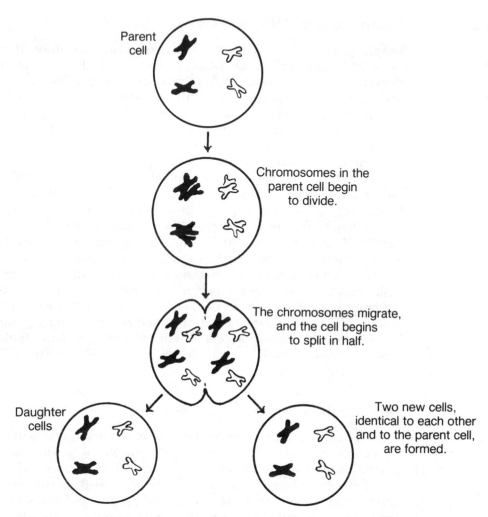

Parent
cell

Chromosomes in the
parent cell begin
to divide.

The chromosomes migrate,
and the cell begins
to split in half.

Daughter
cells

Two new cells,
identical to each other
and to the parent cell,
are formed.

**Figure 2.1.** Mitosis, known as "cell division," duplicates the original parent cell and produces two identical daughter cells. (Human chromosomes actually contain 46 chromosomes, but for illustrative purposes, only four are shown here.)

46 chromosomes (Fig. 2.2), which are arranged in a double-helix pattern of 23 pairs. Thus, when each soma cell replaces itself by mitosis, it creates two daughter cells also containing 46 (or 23 pairs of) chromosomes. In contrast, human germ cells, which are responsible for carrying the genetic material of the individual from one generation to another, undergo an additional level of cell division.

## Fertilization and Meiosis

During the process of *fertilization*, a new organism is formed by the union of a sperm from the male and an egg from the female parent. Each human

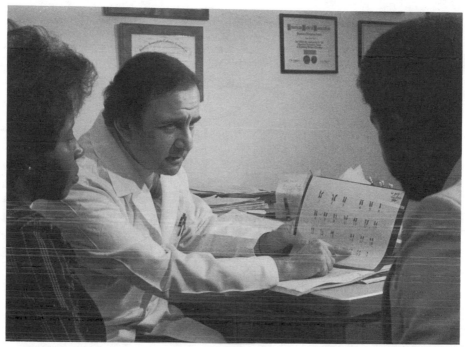

**Figure 2.2.** Human chromosomes. The normal cell contains 46 chromosomes arranged in 23 pairs. Here a health professional discusses one spouse's chromosomal makeup. Such genetic counseling is advised for families that may be at risk for passing on inherited disorders to their offspring (see Chapter 7). (Courtesy of March of Dimes Birth Defects Foundation.)

sperm and egg, when united, contains the necessary hereditary material to direct development of a new organism. Human cells contain a consistent number of chromosomes because of a process that accommodates the union of a sperm and egg without increasing the chromosomal material at every new generation. This system is a specialized form of cell division known as *meiosis.*

Germ cells result from meiosis (Fig. 2.3), which, in effect, is one cell division beyond mitosis. In mitosis, the 23 pairs of chromosomes replicate themselves, resulting in cells containing 23 pairs that are identical to the parent cell. In meiosis, while the operation is more complicated than noted, the result is that the 23 pairs of chromosomes split up and stay halved, This is essential if the accurate quantity of chromosomal material is to be transferred from one generation to another. If the chromosomal material did not divide during meiosis, then each new generation would end up with twice the needed chromosomes. For example, parents would start with 46 chromosomes. The offspring would then end up with 2 × 46, or 192 chromosomes, and the grandchildren would end up with 2 × 192 chromosomes, etc.

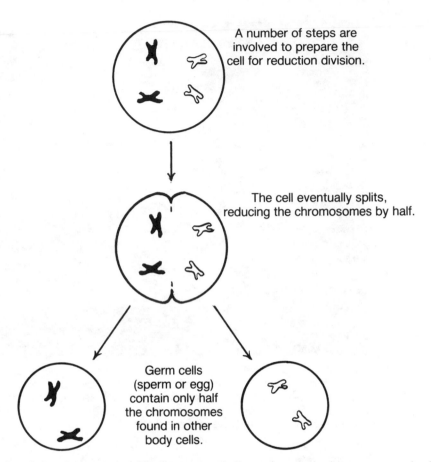

A number of steps are involved to prepare the cell for reduction division.

The cell eventually splits, reducing the chromosomes by half.

Germ cells (sperm or egg) contain only half the chromosomes found in other body cells.

**Figure 2.3.** An oversimplified representation of meiosis, known as reduction division, which halves the genetic material to produce germ cells (sperm or eggs). The germ cells end up with only 23 of the original 46 chromosomes. This is necessary so that at fertilization, the union of the sperm and egg produces a new human with the normal number of 46 chromosomes. (For illustrative purposes, only four chromosomes are shown here.).

The process of meiosis results in each sperm and egg cell containing 23 chromosomes. Thus, fertilization, the union of the sperm and egg, gives the new human the appropriate number of chromosomes, 23 from its mother and 23 from its father. The structure that results from the union of an egg and a sperm is called a *zygote*. This structure, itself, undergoes replication many, many more times and eventually results in a complex, mysterious, special, individual human being, which will be carried in a protective environment in its mother's uterus for approximately 9 months until birth. This process is described in more detail in the section on prenatal development.

Thus, DNA regulates inheritance. It is present in every cell of every organism, and it is through DNA that a genetic program is transmitted by every organism to its offspring. "DNA carries the instructions both for its own reproduction and for the production of the entire chemical machinery of the organism" (11, p 40).

### Genetics and Maturation

During development, DNA regulates the emergence of physical traits and the growth and variability of each new organism. The biological study of this process is *genetics*. The actual emergence of each organism's genetic potential is *maturation*. Maturation is the expression of an individual's own timetable; thus, a baby's first tooth, an adolescent's initial surge of sex hormones, and an adult's first gray hairs are all maturational in nature.

Maturation includes the processes of growth and differentiation. *Growth* was earlier defined as additive accumulation, whereas *differentiation* is "the process by which the growing organism forms qualitatively different and specialized parts and processes" such as eyes, legs, and language (2, p 47).

What exactly is inherited and how much inheritance contributes to behavior as it changes over the life span is under continuous examination and debate, the topic of the nature-nurture debate.

# The Nurture Side: Environment and Learning

On the nurture side of the controversy, the terms "environment," "learning," and "experience" are used interchangeably to refer to external, as opposed to innate influences. *Experience* has been defined as the effect of environment (or milieu) on maturational processes (2). Learning is a complex phenomenon, which is defined simply here. (Learning and maturational theories, their relationship to development, and their application to occupational therapy and physical therapy are explored later in this book.) From an empiricist point of view, *learning* is defined as, "the establishment of new relationships—bonds or connections—between units that were not previously associated" (14, p 100).

Acknowledging the nature-nurture controversy forces students of human development to dichotomize terms such as "learning" and "maturation" that can and should be viewed as complementary. For example, maturation has been defined as an internal force, with environment as a strictly external force. As Hilgard and Bower (14) point out, this is really a false dichotomy:

> Growth is learning's chief competitor as a modifier of behavior. If a behavior sequence matures through regular stages irrespective of intervening practice, the behavior is said to develop through maturation and not through learning. If training procedures do not speed up or modify the behavior, such procedures are not causally important and the changes do not classify as learning. Many activities are not . . . clear-cut, but develop through a complex interplay of maturation and learning. A convenient illustration is the development of language in the child. The child does not learn to talk until old enough, but the langauge which he learns is that which he hears. In such cases it is an experimental problem to isolate the effects of maturation and of learning. The ambiguity in such cases is one of fact, not of definition (14, p 4).

# Interactionism

Most authorities in the field of development believe that the nature-nurture controversy is old news and should be dropped as a topic of study. Lorenz (15) contends that it is indeed a false dichotomy conveniently used by scientists to defend their theoretical positions during the early years of American psychology. The recommended position is an interactionist approach that acknowledges that heredity and experience complement one another, blending together in the production of behavioral change.

## Polarized Views of Interactionism

Interactionism does not have a simple definition. There are varying levels of interactionistic approaches. For example, some theorists accept very polarized definitions of maturation and learning but still recognize that an interaction must occur between them. For example, a polaristic definition of learning is that it is "an environmentally produced change in behavior." The problem with this definition is that it excludes the organism. We know from common sense that when environmental forces cause an organism to change, the process of responding entails a complex interaction between the organism and its environment. It makes little sense to talk of environmental pressures while ignoring the organism's multiple responses to them.

Equally extreme is defining maturational behaviors as those that occur "irrespective of" or "despite" environment. Behavior cannot be expressed except in some kind of context. Thus, it also makes little sense to speak of behaviors occurring *independent* of environment. Despite what appears to be a common sense interactional approach, some scientists and lay people still tend to separate learning from maturation. Such approaches typically emphasize one process over the other, thus perpetuating a nature vs. nurture dualism.

A similar approach is to ascribe certain percentages (i.e., how much) that nature or nurture contributes to specific behaviors:

> Laymen are apt to think of the distinction in competitive terms and have been wont to ask themselves what fractions nature and nurture respectively contribute to a certain character difference—say, a difference of intelligence. We can only proffer a general rule about attempts to attach exact percentages to the two contributions: the more confident and dogmatic the attempt, the more likely it is to be wrong—or, more often, wrongheaded (16, p 194).

As will be pointed out later in this chapter, this tendency to assign percentages to nature and to nurture is characteristic not only of lay people but also of educators and scientists. For example, some psychologists have attempted "to demonstrate that intelligence is 75 to 80 percent genetically controlled and only influenced to the tune of 20 to 25 percent by education and nurture" (16, p 195).

Schnierla (17, p 286) aptly points out that scientists "tend to evade the question of behavioral development by relying on implicit or outright assumptions of innate behavior as distinct from learned behavior. . . . Behavior develops and changes . . . yet ideologies directed at interpreting it are too often cast in rigid forms."

### Holistic, Complementary Views of Interaction

A more integrated level of interactionist thinking is holistic in form, recognizing the existence of internal and external pressures within *both* processes of maturation and learning. A synthesis of both heredity and learning is recognized with each being, "indispensable to the attainment of normal patterns of behavior" (18). Schnierla provides definitions characteristic of such an approach.

> [I suggest that the factors contributing to an organism's development] be redefined objectively so as to exclude implications that a nature-nurture dichotomy exists. Accordingly, I suggest that *maturation* be redefined as the contributions to development of growth and of tissue differentiation, together with their organic and functional trace effects surviving from earlier development; and that experience be defined as the contributions to development of the effects of stimulation from all available sources (external and internal) including their functional trace effects surviving from earlier development . . . These are objective and heuristic definitions and not subjective ones based on any assumed dichotomy of the innate and acquired. The developmental contributions of the two complexes, maturation and experience must be viewed as *fused* (i.e., as inseparably coalesced) at all stages in the ontogenesis of any organism. This holistic theory conceptualizes all process of progressive organization in consecutive early stages of development as fused, coalescing maturational and experiential functions (17, pp 288–289).

While Schnierla's holistic approach is not simple, it does demonstrate how developmental processes need to be viewed as complementary and interactive. It is important to keep this view in mind when reading about the specific areas of development (e.g., motor, perceptual, and cognitive development) that are discussed in the major sections of this textbook.

A basic tenet of this book is that human behavior results from complex interactions within the organism and between the organism and its environment. The concept of interactionism is an important one in the fields of human development, occupational therapy, and physical therapy. Thus, while holistic definitions and explanations may be cumbersome, they must be kept in mind, especially when encountering more popular definitions of maturation and learning. As the rest of this chapter illustrates, however, interactionism is not always promoted. Choosing one side of the nature-nurture debate instead of interactionism has profound implications for therapists and others working in human development fields and studying human behavior.

# SIGNIFICANCE OF NATURE AND NURTURE TO OCCUPATIONAL AND PHYSICAL THERAPY

Recognizing the interaction of nature and nurture is pivotal in health-related fields that provide services for individuals with special needs.

Therapy involves a combination of a wide variety of external forces along with each patient's own internal resources. For example, therapists (external forces ourselves) design clinical and home environments that enable our patients to maximize the use of their own intrinsic functional capacities. We use environmental factors such as ourselves, other people, animals, activities, adaptive devices, or environmental modifications to alter or ameliorate patients' functions. These environmentally produced changes (one way of defining learning) as well as the patient's own personal history, culture, age, and state of readiness to perform (heredity, maturation, and recovery of function) are pivotal to effective therapeutic intervention.

A core philosophy of any clinical practice field is that human behaviors and functions can be changed. Disbelieving this eliminates the need for intervention (e.g., the need for service professions) and also leads to other profound political and ethical consequences. One example of this is found in the treatment of the elderly.

A common misconception is that elderly people cannot benefit from therapy and, as such, should not be availed of rehabilitation intervention. Only lately have attitudes changed, with increasing recognition that many elderly people show beneficial improvements with occupational and other therapies. Many previously institutionalized or hospitalized elderly people are now benefiting from various therapies that enable them to remain productive, independently functioning contributors to our society. The old adage "You can't teach an old dog new tricks" is entirely fallacious. Elderly people are capable of learning and changing positively over time.

# THE NATURE AND NURTURE OF INTELLIGENCE

Probably nowhere is the nature-nurture debate more controversial than in the study of intelligence. A detailed analysis of this controversy is beyond the scope of this book, in fact entire books (e.g., 19–22) have been devoted to this topic. However, this area is introduced here because it serves as an excellent illustration.

Intellectual development is only one small area in the study of human behavior, but it provides an excellent example for understanding contemporary issues and for demonstrating the connection between occupational and physical therapy and human development. First, the study of intelligence serves as an illustration of how the nature-nurture debate still pervades contemporary thinking. Second, it illustrates how attitudes and beliefs regarding one aspect of behavior can have multiple carryover effects. Professionals' attitudes about the nature and nurture of intelligence have had profound educational, social, legal, and clinical consequences.

"There is perhaps no issue in the history of science that presents such a complex mingling of conceptual, methodological, psychological, ethical, political, and sociological questions as the controversy over whether intelligence has a substantial genetic component" (19, p xi). The basis of the controversy deals with two related issues. One issue deals with whether intelligence is genetically determined (i.e., nature), and if so, to what degree.

The second issue has to do with whether intelligence (even if it is genetically determined) can be altered by experience (i.e., nurture), and if so, to what degree. Fundamental disagreement over definitions of intelligence confuses the issue even more. This problem is discussed in more detail in the chapter on "cognition and intelligence," and it will be raised again later in this chapter. For the purposes of this discussion, intelligence is simply defined as a score on an intelligence test. This figure is referred to as an *intelligence quotient*, or IQ, score.

## Genetics and Twin Studies

Those who believe that intelligence is primarily a result of genetics point to experiments involving twins. These experimental data provide some rationale for the notion that intelligence has a genetic basis (for a summary, see Ref. 22). These experiments, sometimes referred to as "twin studies," are designed to compare the IQs of children with varying genetic and environmental backgrounds. To understand these studies, it is necessary to introduce some basic material regarding the genetics of twins.

Since siblings (brothers or sisters) inherit genetic material from their parents, these individuals are expected to be more similar to one another than are unrelated individuals. The most striking example of this is found in twins.

**Figure 2.4.** Twins, who have identical genetic backgrounds but different experiences, are often studied to determine the relative contributions of nature and nurture on behavior.

There are two different types of twins, identical and fraternal, which are produced through two different processes occurring at fertilization. Identical twins (*monozygotic*) are the result of a single (mono) zygote (fertilized egg) splitting in two, yielding two separate organisms with the *exact same* genetic material (Fig. 2.4). Fraternal twins (*dizygotic*) occur when two separate eggs are fertilized by two separate sperms, resulting in two zygotes, each with *different* genetic material. Fraternal twins are like siblings except that they share the same fertilization, birth dates, and prenatal conditions.

## Phenotype and Genotype

The physical expression of one's genetic makeup is termed *phenotype* and is defined as "the organism's observable characteristics as expressed within a given environment" (23, p 33). *Genotype* refers to one's actual genetic organization. Identical twins look exactly alike because they have the same genetic makeup, i.e., they have the same phenotype and possess identical genotypes. Scientists argue that if for some reason identical twins end up looking, or behaving, differently, then those differences are environmentally determined.

## Twin Studies and IQ

A variety of studies have examined the effects of genetic or environmental forces on IQ, and the results of these studies are under constant scrutiny because they are controversial and subject to multiple levels of interpretation (e.g., 19–22). Basically, twin studies support the notion of the inheritability of IQ. For example, the IQs of identical twins are more similar than those of fraternal twins (24, 25). Also, the IQs of biological mothers and their children (who share genetic material) are more similar than the IQs of children and their adoptive mothers (who do not share genetic material) (26).

While these data seem straightforward in support of the heritability of IQ, the issue is not that simple. The conclusions from these studies are subject to constant debate and reinterpretation (e.g., 19–22). For example, it is widely accepted (but still controversial among some researchers) that environmental stimulation can alter IQ (27, 28), including the IQs of adopted children (29).

In refuting some of the twin studies, scientists argue that the data can be reinterpreted to support an environmental position rather than pointing solely to the genetic regulation of IQ. For example, identical twins look and act alike and are more likely to be treated similarly. Their comparable IQs may not be due to genetics but to similar environmental circumstances. Thus, fraternal twins, with different phenotypes, are more likely to be treated differently, which accounts for the fact that their IQs are not as similar as those of identical twins.

# Why the Genetics of Intelligence is so Controversial

Common misconceptions occur regarding the mechanisms that regulate human behavior. Scientists and lay public often exhibit prejudices about the relative contributions provided by nature or nurture. Such prejudices "have had a profound influence upon politicians and through them upon legislation" (16, p 195). For example, early in this century, the results of IQ tests were used in political campaigns promoting the restriction of certain immigrants into the United States

## IQ and Immigration Restrictions

Intelligence tests were once administered from 1916 to the early 1920s to recent immigrants from southeastern Europe and to other immigrants previously admitted into the United States from England and northern Europe. The test data revealed a superior performance of the latter group and, some believed furthermore, the genetic inferiority of the southeastern European immigrants. Despite the fact that these data were biased and had been misinterpreted, some politicians demanded the placement of immigration quotas on what were mistakenly viewed as "intellectually inferior" nationalities.

The discrepancies in the IQ scores of different immigrant groups have been subject to multiple reinterpretations and is now more clearly understood after extensive analysis of the nature of IQ tests and the circumstances in the United States at the time the immigrants were tested (e.g., 19–21). First of all, the new immigrants who had scored poorly had been tested upon arrival to the United States. Many of these people had just arrived from long and arduous voyages. Some were confused and fearful. Some had never taken a test nor held a pencil before in their lives (20). The IQ scores of this new group were compared with those of individuals who had already set up residence here. Thus, the differences between these groups may have reflected variations in comprehension of American values and culture as well as levels of anxiety, both of which may affect IQ test performance. Compounding the social and political fears that somehow the "gene pool" of the United States would be undermined by admitting "inferior" races or "discards" from other nations was the additional social, political, and economic issue—fear of competition from cheap labor provided by new immigrants (19).

While the immigration issue was highly controversial in 1922–1923, it died down quickly (19, p 2). Why, then, is it brought up for discussion in a textbook written well over 50 years later? The answer, in part, has to do with a highly controversial article written by Arthur Jensen in 1969 (30). The article, "How Much Can We Boost IQ and Scholastic Achievement?", brought up once again the nature-nurture issue by emphasizing the impact of the genetics (and politics) of IQ.

## Not Nature or Nurture but *How* the Two Work Together

Twelve years before Jensen's article (30) was published, Anne Anastasi (31), President of the American Psychological Association in 1957, pointed this out:

> Two or three decades ago, the so-called heredity-environment question was the center of lively controversy. Today, on the other hand, many psychologists look upon it as a dead issue. It is now generally conceded that both heredity and environmental factors enter into all behavior. The reacting organism is a product of its genes and its past environment, while present environment provides the immediate stimulus for current behavior (31, p 197).

Anastasi further noted that studies had failed in their attempts to point to either heredity or environment or to assign proportional contributions of these to specific behaviors. This is her suggestion:

> Perhaps we have simply been asking the wrong questions. The traditional questions about heredity and environment may be intrinsically unanswerable. Psychologists began by asking *which* type of factor, heredity or environment, is responsible for individual differences in a given trait. Later, they tried to discover *how much* of the variance was attributable to heredity and how much to environment. It is the primary contention of this paper that a more fruitful approach is to be found in the question "*How*"? (31, p 197).

The utility of the question "how" is central to the concerns of the practitioner as well as the social reformer and the student of human development. Determining *how* behavior changes over time addresses the fundamental goals for the scientific study of development because it provides a basis for explanation and prediction. If we understand how behavior changes over time, then we can develop effective intervention strategies designed to alter development that may be interrupted or delayed.

Unfortunately, Anastasi's recommendations were not universally heeded. While some scientists have been concentrating on determining how genes and environment interact in behavioral regulation, others (such as Jensen) continue to assign relative percentages to nurture's or to nature's control of such behaviors as intelligence. The result was a widespread controversy that occurred in the 1960s and is still being debated two decades later. Familiarity with this controversy will illustrate to the therapist the importance of thinking about development in interactive terms. Also, it will illustrate how the nature-nurture debate still provides a focus for developmental and therapeutic intervention.

## Racial Differences in IQ

In the 1960s large-scale intervention programs were developed to help infants and children who exhibited delays in various aspects of their development. This *early-intervention* movement was spurred partly by governmental programs that were directed toward providing services for culturally disadvantaged children as well as children with physical disabilities. It was assumed that children in the former group exhibited delays in development because their environments were not sufficiently stimulating.

The early-intervention programs were designed to provide extraneous stimulation not present in their culturally deprived environments.

Part of the rationale for the onset of these programs was the conviction on the part of educators, therapists, and psychologists that environmental factors were partly responsible for racial differences obtained on IQ tests in the United States. These test differences consistently indicate an advantage for white over black Americans (21). Jensen (30) had pointed this out in his 1969 article. In it, he summarized data from twin and other studies of the heritability of IQ and concluded that intelligence is primarily biologically determined, i.e., about 80%. The reason for consistently observed racial differences in IQ, Jensen concluded, is due to innate superiority of whites over blacks.

This issue is obviously controversial in many fields. Jensen's (30) conclusions, and their social and political ramifications, have been compared to the errant conclusions drawn about early American immigrants except that Jensen's conclusions had even greater impact. His 1969 paper was immediately met with criticism, and it is still hotly debated (e.g., 20, 21).

While Jensen is correct in reporting racial differences in IQ, the reasons and the implications for these differences are not clear. Jensen and others have claimed that racial differences in IQ are attributed to genetics, but other noted scientists and educators have pointed to other equally plausible explanations. For example, some scientists point out that IQ tests are culturally biased. It is this kind of bias, they claim, that accounted for the immigrant controversy and accounts now for racial IQ differences.

This contemporary quagmire illustrates the extent to which nature and nurture are separated and debated in one small area of human development. At present, research in psychology and in educational testing is particularly centered around determining the actual nature of human intelligence and what is measured by IQ tests. For example, psychologist Howard Gardner (23, p 18) reports, "Much of the information probed for in intelligence tests reflects knowledge gained from living in a specific social and educational milieu . . . In contrast, intelligence tests rarely assess skill in assimilating new information or in solving new problems . . . Moreover, the intelligence test reveals little about an individual's potential for further growth." IQ tests have traditionally been constructed by white middle-class Americans, and the tests are assumed to naturally assess the information judged important to this group.

A further edge for whites taking IQ tests is that there seems to be different cultural emphasis placed on test-taking and educational achievement. White children, who are brought up to respect and compete in an academic environment, may indeed have an edge over other children who lack the motivation, who do not pay attention in school, and who do not strive to perform well on IQ—or any other—tests.

A separate but interrelated issue has to do with equal social and economic opportunities for blacks and whites. Blacks have traditionally been exposed to greater degrees of poverty over longer periods of time and have had fewer overall opportunities than white Americans. Many sociol-

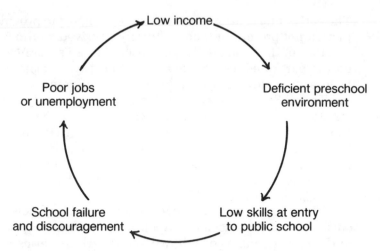

**Figure 2.5.**  Chain of assumptions supporting the theory, widely espoused during the 1960s, that preschool education would help reduce poverty by striking at its roots. (From Gallager JJ, Haskins R, Farran EC: Poverty and public policy for children. In Brazelton TB, Vaughan VC (eds): *The Family: Setting Priorities.* New York, Science and Medicine, 1979, p 240.) (Reproduced with permission of publisher)

ogists, psychologists, and politicians have pointed out that constant poverty leads to poor nutrition, poor health, and inadequate housing. These factors result in reduced stimulation, poor school attendance, reduced motivation and attention, and overall fewer opportunities (32–34). Any or all of these factors subsequently lead to inferior performance on educational tests (Fig. 2.5).

Thus, IQ racial differences may not reflect *innate* black-white differences but rather the cumulative effects of oppressed and inferior *environments* and cultural biases to the disadvantage of blacks.

## Compensatory Education and the War on Poverty

The controversy introduced by Jensen (30) went beyond a resurrection of the nature-nurture debate (which many thought had been put to rest). Jensen's claims stirred up a wide-ranging controversy about the effectiveness of compensatory programs designed to narrow racial and socioeconomic inequalities.

Jensen's 1969 article (30) was timely. Remarkable social and political reforms had been initiated in the United States under the legislation of the Economic Opportunity Act of 1964. Programs that were part of the War on Poverty, as these reforms were popularly referred to, attempted to reduce the disparity between the environments and opportunities of children brought up in poverty and those given relative social, educational, nutritional, and medical advantages of being born and raised in America's middle and upper socioeconomic classes.

### Cycle of Poverty

Government-supported welfare and community-based compensatory education programs were developed to give low-income infants and children a head start with nutrition, education, and health care before they entered public school. The most noted program, started in 1965 and still in operation 20 years later, is Head Start, a preschool program that provides health, nutrition, education, and parent involvement for low-income children in communities across the United States. The premise of Head Start and the War on Poverty was that compensatory educational, health, and other programs could make up for reduced opportunities that resulted from what was termed "the cycle of poverty" (Fig. 2.6).

There is no one starting point in the cycle of poverty; it is continuous. From a developmental perspective, however, the starting point is in infancy. At this time, poverty is responsible for environmental insufficiency and low level or inappropriate infant and childhood stimulation. For the preschool

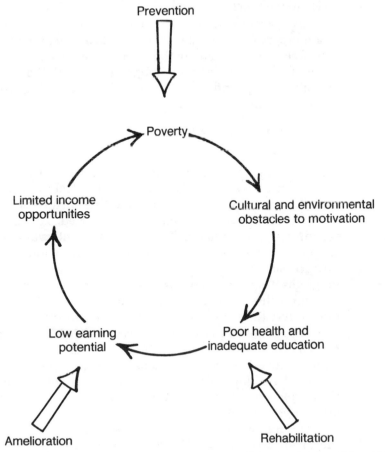

**Figure 2.6.** Attacking the poverty cycle. (Modified from Moynihan DP: The professors and the poor. In Moynihan DP (ed): *On Understanding Poverty.* New York, Basic Books, 1969, p 9.)

child, this results in inadequate skills for entry to public school. Continued exposure to poverty results in poor health, inadequate nutrition, and reduced motivation to attend and participate in scholastic activities throughout childhood and adolescence. This, in turn, leads to failure in school, discouragement, and ultimately to reduced employment potential as an adult. The result of unemployment or poor job skills is a lifetime of continued poverty and limited opportunities (32, 34).

The cycle of poverty is particularly relevant to human development and human service fields. Educators, therapists, psychologists, sociologists, etc. were consulted regarding how to interrupt this vicious cycle, and programs were aimed at all ages of individuals across the life span. As Moynihan (32) noted, *prevention* programs were geared toward the elimination of poverty by focusing on youngsters and parental attitudes and involvement. *Rehabilitation* programs focused on adults in their work years, and *amelioration* programs were aimed at aged individuals, physically and mentally disabled people, or those for whom prevention and rehabilitation were ineffective (Fig. 2.6).

There were wide-ranging expectations for these environmental stimulation programs. It was believed that if the cycle of poverty were broken, it would reduce or eliminate the environmental causes of social, educational, economic (and intellectual), inferiority (32–34), thereby equating racial and disability groups and, most importantly, eliminating further need for welfare and other similar programs.

## Implications for the Success or Failure of Compensatory Programs

The success or failure of these programs obviously had wide-ranging academic, social, economic, clinical, and political implications. If the compensatory programs were effective, then the nurture or environmental side of the nature-nurture dichotomy was supported. This would provide a rationale for continued financial investment in intervention programs designed to help those in poverty break out of the vicious cycle. Support for the nurture side of the controversy would be a vote for welfare and other programs designed to give assistance to social (or racial) groups in need.

If, on the other hand, the initial compensatory programs were found ineffective, then the data would be used to support those, such as Jensen (30), who believed in the genetic basis of IQ. They, in turn, could argue that people in poverty were not there because of reduced opportunities or environmental deficits but because of innate, potentially unalterable genetic causes. Thus, Jensen's (30) claims in 1969 were so important and timely. His article was published just four years after Head Start and other antipoverty programs had been in effect.

The initial data from studies examining these programs revealed that they were not effective. IQ scores of the graduates of the first Head Start programs did not seem significantly improved to eliminate racial differences. These initial findings, some claimed, supported Jensen's (30) conclusions and disappointed the academic and social reformers who were counting on obtaining support for compensatory programs. Jensen concluded that compensatory education would be ineffective because intelligence, being 80% genetically determined, cannot be appreciably altered. The lack of differences obtained upon initial examination of the Head Start data, Jensen and others claimed, supported the notion of innate racial differences in IQ.

If Jensen were correct, then these data might be used to provide a rationale for discontinuing compensatory programs that involved thousands of teachers and therapists and even more children. Fortunately, others realized that alternative interpretations were possible.

### Wide-ranging Impact of the Nature and Nurture of Intelligence

Some scientists pointed out that even if compensatory programs were found ineffective in reducing racial differences in IQ, as Jensen predicted, these findings had multiple interpretations (e.g., 34). For example, Stephen Jay Gould (35, p 247), an anthropologist, points out that one possibility is that "we didn't spend enough money, we didn't make sufficiently creative efforts, or (and this makes an established leader jittery) we cannot solve these problems without a fundamental social and economic transformation of society."

It is hard to believe that in the 1970s and 1980s, the nature-nurture controversy regarding one aspect of human development would have such wide-ranging and serious consequences. Scientists and practitioners from numerous fields jumped on this issue and began to debate whether compensatory programs were effective and whether, ultimately, intelligence was immutable. For some reason, despite the advice of Anastasi and many other scientists over the years, interactionism was not adopted; in regard to intelligence, nature-nurture was an either/or choice. Scientists Medawar and Medawar (16) point out the profound political and philosophical implications of taking sides in the nature vs. nurture debate:

> There is a strong political coloration to the inclination to attribute inequalities of intellectual performance to nature rather than to nurture. For if the poor and needy have become so by reason of their genes, there is nothing very much we can do about it: their poverty and inadequacy are not of our doing, and we have no moral compulsion to take social steps to remedy their condition. If, on the other hand, like dedicated modern Marxists we categorically deny that there are inborn differences among human capabilities so that a human being is only what his environment and upbringing can make of him, then in a just society it is an obligation upon the state to provide for the education and upbringing of its citizens (p 196).

[As a postscript, it should be noted that subsequent studies of Head Start and other early intervention programs indicated positive effects, many of which were long-range and unable to be assessed until years later (e.g., 36). This issue will be examined in detail in chapter 9, where the consequences of early intervention and infant stimulation will be investigated.]

# IQ, Nature-Nurture, and Intervention

The nature-nurture debate on intelligence demonstrates that preconceived notions about which mechanism actually causes behavior may result in inaccurate and negative generalizations. For example, a significant consequence of believing that nature is the main contributor to intelligence (or to any other behavior) is that most individuals associate inheritance with immutability. That is, people tend to believe that if some characteristic or behavior is genetically determined, it cannot be changed (20). This is not so. As the eminent anthropologist Gould (35, p 245) points out, "[I]nheritability carries no implications of inevitability or of immutable entities beyond the reach of environmental influence. Eyeglasses correct a variety of inherited problems in vision; insulin can check diabetes."

The genetics or mutability of behavior extends beyond the issue of IQ to all aspects of human development as well as having important implications for the health service fields. Physical and occupational therapists work with a large variety of patients. Some patients have deficits caused by inherited genetic diseases or due to some genetic anomaly (see chapter 7). Others have problems traced directly to environmental factors (e.g., irradiation sickness, lung cancer from smoking, amputation from an accident). While we may refer to "genetic" or "nongenetic" problems, we realize that genes and environment interact in complex ways.

For example, genetic conditions may be ameliorated or aggravated by the environment. Genetically caused auditory impairments may be functionally or actually eliminated by hearing aids or surgery. In contrast, someone with a slight genetic tendency toward deafness may become completely deaf because of continued exposure to industrial noise. We also know that individuals' interactions with their environments are affected by genetic conditions. Males and females behave differently as a result of inherited sex characteristics. For example, males do not give birth and cannot nurse infants. Females typically do not become bald nor do they typically develop facial hair and deep voices in adolescence. People with inherited lung or joint problems may choose to live in environments that minimize the effects of their diseases. Others with inherited disabling diseases may use wheelchairs for mobility, which in turn, alters their daily life activities and their access to different environments. People with certain inherited skin pigments or body shape or size may interact differently depending upon environmental contexts.

It is, indeed, difficult to sort out where genetics or environment stops and the other starts. An excellent example of this is mental retardation. As we have noted, intelligence is a result of a complex interplay between environment and genetics. Mental retardation is associated with subaverage intellectual functions and deficits in adaptive behavior (37). The causes of mental retardation may initially be genetic or environmental, or there may be "aggravating environmental factors [that] interact with organic factors in such a complex fashion that the developing overlay of problems blocks out forever the primary cause" (37, p 463). Despite the primary cause of mental retardation, intervention is offered to improve as much as possible each individual's intellectual functions and adaptive skills.

The primary treatment orientation of human service fields is to provide intervention that will help patients achieve their highest rehabilitative, educational, or developmental potential. What we most need to learn about our patients is not limited to the causes of their deficits but extends to determining the most effective methods of prevention and intervention. Our patients' heredity, culture, and immediate and past environment all interact to effect their current levels of performance. As Anastasi (31) pointed out, research needs to be aimed at understanding *how* organism-environment interactions work together. Knowing this will enable us to eventually prevent deleterious conditions from occurring or to offer the most effective interventions if and when they do occur.

Whether conditions have a genetic basis should not prevent us from environmental intervention attempts to provide opportunities for therapy or for education. As Gould (35, p 246) points out, "I do not claim that intelligence, however defined, has no genetic basis—I regard it as trivially true, uninteresting, and unimportant that it does. The expression of any trait represents a complex interaction of heredity and environment. Our job is simply to provide the best environmental situation for the realization of valued potential in all individuals."

# NATURE-NURTURE DEBATE AS A BASIS FOR STUDYING DEVELOPMENT

While some claim that by now it is irrelevant, the nature-nurture debate continues to resurface in academic debate and in popular work (e.g., 35, 38, 39). For example, Canadian author Robertson Davies (40, p 231) states this in his 1985 novel *What's Bred in the Bone*: "Nature and nurture are inextricable; only scientists and psychologists could think otherwise, and we know all about them, don't we?". Despite such cautions from popular writers such as Davies (40), or from scientists such as Gould (20, 36), Schnierla (17), Lorenz (15), or Anastasi (31) (to name a few), some scientists and lay people still continue to estimate the degree to which heredity or environment contribute to specific behaviors. In 1981, for example, Eysenck and Eysenck (41), reviewed the literature relating to interindividual differences in personality traits and concluded that 70–75% of the differences were due to hereditary factors (see Ref. 42).

Students and clinicians need to be familiar with such views and with the assumptions and potential biases underlying them. Extreme or polarized views of nature or nurture may have, as we have seen in regard to intelligence, "important philosophical implications and major political consequences" (35, p 247) that may affect our attitudes and approaches to treatment and education.

One of the tenets of this textbook is that the nature-nurture controversy serves as a foundation for many of the basic concepts and research in the field of human development. Understanding the roots of the debate provides a basis for organizing and categorizing developmental theory and research, and this will be fully illustrated in later chapters that examine specific aspects of development. In addition, understanding how the nature-nurture issue has led to academic, philosophical, and political debate is an important lesson for physical and occupational therapy students who will move on to assume a variety of roles during their careers.

Occupational therapists and physical therapists take on a wide variety of professional roles. Not only do they provide treatment, but many also assume other roles of leadership in their communities and in national affairs. Some clinicians become educators, researchers, or administrators; and most participate in program development and decision making regarding their occupational therapy or physical therapy departments. These departments interact closely with other health care and human service departments and agencies. Many therapists become directly or indirectly involved in community programs and local health-policy decision making, or they participate in policy-making at state or national levels (e.g., 43).

In addition, therapists are constantly interacting with patients and their families, acting as representatives for existing health care attitudes, and often advocating for change. Health care issues transcend personal and local scope. Thus, it is important for future therapists to recognize that academic issues and biases often find their way into everyday life and into policy-making. This is especially true with regard to issues related to human development. As the IQ controversy demonstrates, academic assumptions often have social, economic, and political consequences. Our beliefs and our expectations are important, as Gould (20, p 65) points out: "Expectation is a powerful guide to action."

One of Gould's (20, p 28) claims regarding the mismeasure of intelligence equally applies to therapeutic intervention: " We pass through this world but once. Few tragedies can be more extensive than the stunting of life, few injustices deeper than the denial of an opportunity to strive or even to hope, by a limit imposed from without, but falsely identified as lying within."

Occupational and physical therapists must recognize that it is a common tendency for humans to think in terms of dichotomies rather than interactions. Yet interactionism is one of the most important concepts in treatment and development. The important view for clinicians involves understanding that in every organism, internal and external forces interact in unique and often unpredictable combinations to produce behavior. The goal of therapy is not to identify limiting factors but to touch on special interactions between the therapist, the environment, and the patient so that they all work together to enhance and maximize each person's special potential for change.

## References

1. Mussen PH, Conger JJ, Kagan J: *Child Development and Personality*, ed 3. New York, Harper & Row, 1969.
2. Michel GF, Moore CL: *Biological Perspectives in Developmental Psychology*. Monterey, CA, Brooks/Cole, 1978.

3. Timiras PS: Biological perspectives on aging. *American Scientist* 60:605–613, 1978.
4. Lerner RM: *Concepts and Theories of Human Development.* Reading, MA, Addison-Wesley, 1976.
5. Kagan J: *The Nature of the Child.* New York, Basic Books, 1984.
6. Watson RI: *The Great Psychologists from Aristotle to Freud.* Philadelphia, JB Lippincott, 1963.
7. Cornford FM: *The Republic of Plato.* New York, Oxford University Press, 1966.
8. Gardner H: *Developmental Psychology. An Introduction.* Boston, Little, Brown & Co, 1978.
9. Ripple RE, Biehler RF, Jaquish GA: *Human Development.* Boston, Houghton Mifflin, 1982.
10. Levine RP: *Genetics.* New York, Holt, Rinehart & Winston, 1966.
11. Luria SE, Gould SJ, Singer S: *A View of Life.* Menlo Park, CA, Benjamin/Cummings, 1981.
12. Watson JD: *The Double Helix.* New York, Signet, 1968.
13. Lampton C: *DNA and the Creation of New Life.* New York, Arco, 1983.
14. Hilgard ER, Bower GH: *Theories of Learning,* ed 3. New York, Appleton-Century-Crofts, 1966.
15. Lorenz KZ: *The Foundations of Ethology.* New York, Simon & Schuster, 1981.
16. Medawar PB, Medawar JS: *Aristotle to Zoos. A Philosophical Dictionary of Biology.* Cambridge, MA, Harvard University Press, 1983.
17. Schnierla TC: Behavioral development and comparative psychology. *Q Rev Biol* 41:283–302, 1966.
18. Greenough WT: Preface. In *The Nature and Nurture of Behavior. Developmental Psychobiology.* San Francisco, WH Freeman, 1973.
19. Block NJ, Dworkin G (eds): *The IQ Controversy.* New York, Random House, 1976.
20. Gould SJ: *The Mismeasure of Man.* New York, WW Norton, 1981.
21. Kamin LJ: *The Science and Politics of IQ.* Potomac, MD, Lawrence Erlbaum, 1974.
22. Jencks C, Smith M, Acland H, Bane MJ, Cohen D, Gintis H, Heyns B, Michelson S: *Inequality. A Reassessment of the Effect of Family and Schooling in America.* New York, Basic Books, 1972.
23. Gardner H: *Frames of Mind. The Theory of Multiple Intelligences.* New York, Basic Books, 1983.
24. McCall RB: Intelligence quotient pattern over age: comparison among siblings and parent-child pairs. *Science* 170: 644–648, 1970.
25. Wilson RS: Twins and siblings: concordance for school-age mental development. *Child Dev* 48:211–216, 1977.
26. Munsinger H: The adopted child's IQ: a critical review. *Psychol Rev* 82:623–659, 1975.
27. McVicker Hunt J: Environmental programming to foster competence and prevent mental retardation in infancy. In Walsh RN, Greenough WT (eds): *Environments as Therapy for Brain Dysfunction.* New York, Plenum Press, 1976.
28. Richardson SA: The influence of severe malnutrition in infancy on the intelligence of children at school age: an ecological perspective. In Walsh RN, Greenough WT (eds): *Environments as Therapy for Brain Dysfunction.* New York, Plenum Press, 1976.
29. Skodak M, Skeels H: A final follow-up study of children in adoptive homes. *J Genet Psychol* 75:85–125, 1949.
30. Jensen AR: How much can we boost I.Q. and scholastic achievement? *Harvard Educ Rev* 39:1–123, 1969.
31. Anastasi A: Heredity, environment, and the question "How?" *Psychol Rev* 65: 197–208, 1958.
32. Moynihan DP: The professors and the poor. In Moynihan DP (ed): *On Understanding Poverty. Perspectives from the Social Sciences.* New York, Basic Books, 1969.
33. Laosa LM: Social policies toward children of diverse ethnic, racial, and language groups in the United States. In Stevenson JW, Siegel AE (eds): *Child Development Research and Social Policy.* Chicago, University of Chicago, 1984, vol 1.
34. Gallagher JJ, Haskins R, Farran DC: Poverty and public policy for children. In TB Brazelton, VC Vaughan (eds): *The Family: Setting Priorities.* New York, Science & Medicine, 1979, pp 239–269.
35. Gould SJ: Racist arguments and IQ. In Gould SJ: *Ever Since Darwin.* New York, WW Norton, 1977.

36. Brown B: Head Start. How research changed public policy. *Young Children* July: 9–13, 1985.
37. Dunham JR, Dunham CS: Mental retardation. In Goldenson RM (ed): *Disability and Rehabilitation Handbook*. New York, McGraw-Hill, 1978.
38. Rensberger B: Margaret Mead. The nature-nurture debate. I. *Science '83* Apr: 28–37, 1983.
39. Rensberger B: On becoming human. The nature-nurture debate. II. *Science '83* Apr: 38–42, 1983.
40. Davies R: *What's Bred in the Bone*. New York, Viking, 1985.
41. Eysenck H, Eysenck M: *Mindwatching*. London, Michael Joseph, 1981.
42. Magnusson D, Allen VL: An interactional perspective for human development. In Magnusson D, Allen VL (eds): *Human Development—An Interactional Perspective*. New York, Academic Press, 1983, pp 3–31.
43. Moersch MS: Occupational therapy and public policy. *Am J Occup Ther* 40: 202–205, 1986.

# 3

# Historical and Contemporary Perspectives of Development

A consistent theme throughout this text is interactionism. Both fields of occupational therapy and physical therapy are necessarily concerned with forces that are within and outside of people and that interact in special ways to create unique, individual behavior. In chapter 2, the concept of interactionism was discussed primarily in regard to the synthesis of these external, environmental, and internal, hereditary forces in the regulation of behavior.

Interactionism also applies to other perspectives of development in addition to nature and nurture. For example, while we tend to think that academic fields are independent or "pure", they are, in fact, subject to multiple interacting influences. This is particularly true of the fields of human development and those with an ongoing, dynamic interaction between clinicians and clients. Such fields are so diverse that they are constantly affected by the research and thinking that goes on in other academic and clinical areas, such as sociology, philosophy, anatomy, or medicine.

The last chapter's examination of the nature and nurture of IQ illustrates how one specialized area of academic interest is subject to and also has influenced other social, economic, and political concerns (Fig. 3.1). This seems to be characteristic of science. We know, for example, that the ideas of the great scientists Copernicus, Newton, and Darwin were subject to widespread religious and social criticism (1). Noted anthropologist Gould illustrates the interaction between science and social forces in the following quote:

> Science must be understood as a social phenomenon . . . Science, since people must do it, is a socially embedded activity. It progresses by hunch, vision, and intuition. Much of its change through time does not record a closer approach to absolute truth but the alteration of cultural contexts that influence it so strongly. Facts are not pure and sullied bits of information; culture also influences what we see and how we see it. Theories, morever are not inexorable inductions from facts. The most creative theories are often imaginative visions imposed upon facts; the source of imagination is also strongly cultural. (2, pp. 21–22)

Thus, it is not surprising that such a wide-ranging field as the study of human development has reflected enormous changes as a result of multiple social reforms over the years (Fig. 3.1).

To accurately understand contemporary views of human development, we must necessarily look at how our thinking in this field has changed over time. The scope and direction of developmental research has changed, as has our approach to the field and to the various stages of human development. Contemporary thinking is shaped from past biases and beliefs. Thus, we must look back in time to see where we have been in order to determine how we have gotten to where we are now.

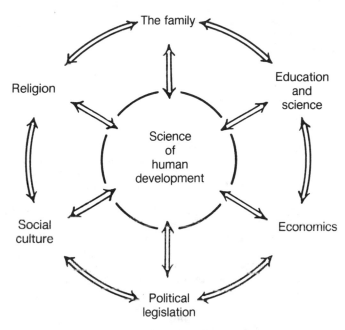

**Figure 3.1.** Interactionism. Our understanding of human development is altered by multiple, interacting social forces.

# CHANGING CONCEPTS OF CHILDHOOD AND FAMILY

Children in American society today are relegated to positions of extreme importance. This was not so in the past. As cultural historian Phillipe Aries (3) has noted, societal views of children have radically changed over the centuries. Aries analyzed literature and art over four centuries and described how differently children's roles were depicted in medieval times compared with our present view.

> Nowadays our society depends, and knows that it depends, on the success of its educational system ... New sciences such as psycho-analysis, pediatrics and psychology devote themselves to the problems of childhood, and their findings are transmitted to parents by way of a mass of popular literature. Our world is obsessed by the physical, moral and sexual problems of childhood. (3, p. 411)

## Children in Medieval Times

This preoccupation with childhood, Aries points out, was unknown to medieval civilization. As soon as he or she had been weaned, or shortly thereafter, the child was treated as an adult. Except for size, there were no noticeable differences between children and adults in art and literature. Children were depicted as miniature adults, who dressed as adults, engaged in adult games and rituals, and for most intents and purposes, were integrated into the adult world (3).

# Factors Affecting the Emerging Concept of Childhood

After the 17th century, perspectives of children changed. A new view of childhood emerged, reflecting massive changes in the nature of the family, society, and in the world at large:

> Important economic, political, social, and religious movements (growth of the towns in the thirteenth and fourteenth centuries, the discovery of new lands in the fifteenth and sixteenth centuries, the intellectual Renaissance of the fifteenth century, the religious Reformation of the sixteenth century, and the American and French Revolution of the eighteenth century) all combined to reduce the influence of the Church, loosen the social fabric, and provide fresh opportunities for those in lower social and economic positions . . . .
>
> These diverse factors brought about a decisive alteration of social life. With greater mobility, extended families were less likely to remain together, and family units became small and more private. Street life declined; personal interactions were by and large restricted to those in similar occupations or social positions. And the child became increasingly central in the family's view. Children had a better chance than before of surviving to adulthood, and they could now aid their family and improve its prospects. By acquiring skills and knowledge as children, including the mastery of reading and writing, individuals could rise rapidly in the society (4, p. 141).

No longer was it so widely accepted that one's place in society was preordained. Instead, because of religious, social, economic, and intellectual changes in the world, individuals could move through social strata by virtue of accomplishments, accumulated goods, or education. Education shifted in focus from religious training in the Middle Ages for a few children, mostly boys, to increased study for more children, including girls, in the 16th, 17th, and 18th centuries (4).

# Family Roles

Children's roles in family life, and family life itself, experienced widespread change. Religious literature in the 16th and 17th centuries taught parents that "they were spiritual guardians, that they were responsible before God for the souls, and indeed the bodies too, of their children" (3, p. 412). Families were given new roles in the education of their children:

> This new concern about education would gradually install itself in the heart of society and transform it from top to bottom. The family ceased to be simply an institution for the transmission of a name and an estate—it assumed a moral and spiritual function, it moulded bodies and souls. The care expended on children inspired new feelings, a new emotional attitude, to which the iconography of the seventeenth century gave brilliant and insistent expression: the modern concept of the family. Parents were no longer content with setting up only a few of their children and neglecting the others. The ethics of the time ordered them to give all their children, and not just the eldest—and in the late seventeenth century even the girls—a training for life. (3, p. 413)

# PHILOSOPHICAL VIEWS OF CHILDHOOD

Strong philosophical views of children and child rearing were espoused during the 17th and 18th centuries. As discussed in the previous chapter, John Locke and Jean-Jacques Rousseau are associated with opposing philosophical views that influenced the field of human development for centuries. Their philosophies formed the basis for views of child rearing and development as well as the basis for the nature-nurture controversy.

Rousseau's liberal approach to child rearing and his belief in the innate goodness of childhood were in direct opposition to the trend of rigid, disciplined training of children in the 1770s. At this time, Puritanical religious views concurred with Rousseau's concept of innate characteristics but attributed to children the concept of original sin. Children were perceived as innately evil, requiring instruction to direct them away from the natural, sinful course of life.

British empiricist Locke did not accept the nativistic concept of innate characteristics, but his views were consistent with the times. His emphasis on practice and drill was consistent with the contemporary views of structured rigid training designed to mold children's characters and shape their impressionable minds. On the other hand, Rousseau's more humanitarian view, purported in his book, *Emile*, emphasized leaving children alone to develop naturally, on their own, without strict disciplinary forces to direct them. Rousseau contended that children are separate and unique individuals, whose talents need to be nourished but not coerced (4).

The 1700s and 1800s witnessed an expanded interest in child development in religious and philosophical fields. Yet, despite their increasingly important roles in society and in family life, children in the 18th and 19th centuries still did not share the position of importance they now possess, especially in the United States. For example, during the Industrial Revolution in the early 1800s, young children provided easy, cheap sources of labor. Many children, because of their size, were ideally suited for factory positions that most adults could not perform, and children were commonly exploited, working long hours in severe conditions. Children certainly did not share the social and legal protection afforded them nowadays. The prominent position enjoyed by children in contemporary American society is relatively recent, primarily affected by the scientific and medical advances begun around 1900.

# SCIENTIFIC AND MEDICAL PERSPECTIVES OF CHILDHOOD

In the field of medicine, perspectives about children paralleled those of art and literature. Specialized training in the medical care of children did not exist before 1880 because, just as Aries notes regarding art and literature, children were perceived and treated as miniature adults. Then, around

1900, with the high infant mortality rates (approximately 25% for infants under age 2) physicians became increasingly concerned with the special needs of children (5). Thus, at the same time that the specialty of pediatrics was emerging in the field of medicine, so also was a science of child development beginning in the United States.

Despite their parallel growth, pediatrics and child development have for the most part developed separately. As Butler and colleagues note:

> Except in research on infants, where a promising pattern of collaboration has evolved, there are few good examples of substantially shared research and teaching between these disciplines. There also are few institutional structures by which developmentalists can contribute routinely to medical education; likewise, most students in child development are poorly exposed to medical topics or topics in the delivery of health care. This is a problem caused by organizational and professional boundaries and the constraints of time for training rather than the disdain of one profession for the other. (6, p. 172)

One marked exception to the separation between child development and pediatrics is in the fields of occupational and physical therapy, where students study normal human developmental processes and also receive specialized training in pediatrics. While these areas are commonly not addressed in one text, they fit together naturally in a text for occupational and physical therapy because of their dual concerns with academic and clinical study.

## Development of Child Study

The end of the 19th and the beginning of the 20th century marked the onset of child study as a scientific discipline in the United States. G. Stanley Hall is often credited with beginning the systematic study of children at the turn of this century, but as Gardner points out, Hall's work, "was preceded by numerous scientific events which shaped the way for developmental research. The great scientists such as Galileo and Newton in the 1700s and Darwin in the 1800s had determined the course of scientific inquiry. Methods derived from these individuals provided a background for the application of scientific principles in the study of child development" (4, p 149).

Darwin's evolutionary theories of the mid-1800s set the stage for the scientific analysis of developmental changes. His work has been considered by some developmental experts as "the single most vital force in the establishment of child psychology as a scientific discipline" (7, p 13). Darwin is often also noted for his specific contribution to the field of child development—the diary of his son's early life. In one sense, such diaries, or "baby biographies" as they are called, legitimized child study. Although diaries are recognized as biased and selective means for collecting information, these early biographies pointed out some essential issues in the field of child development and provided an impetus for more systematic and more detailed examinations of future generations of children (7).

# Early Approaches to the Study of Children

As the study of children became a legitimate field of investigation, scientists adopted different strategies of scientific inquiry. These early scientists' methods, as well as their goals for research, were varied. For example, in the early 1900s, G. Stanley Hall investigated large groups of children, used questionnaires for obtaining data, and applied Darwinian theory to interpret behavior. Hall is widely acclaimed in developmental psychology for initiating some of the first empirical studies of children in the United States. In addition, he is recognized as "the father of the psychology of adolescence" because he was the first to scientifically examine adolescence as a separate life stage. He also examined later stages of life, though his book, *Adolescence* (8), which was published in 1904, is more widely recognized than his book on aging, *Senescence* (9), published in 1922 (10).

Another psychologist in the early 1900s, John B. Watson, initiated work that was to change the shape of American psychology. Watson's focus was on overt, observable behaviors. His method of investigation involved conducting experiments in which he would introduce specific stimuli to infants and children and then observe their overt behavioral responses. In contrast to Hall's widespread use of questionnaires and the baby biographies preceding them, Watson's methods were viewed as even more systematic and objective. Such methods enabled scientists to objectively examine the behaviors of infants and children who could not be assessed with questionnaire approaches. This opened up a whole arena of infant and early childhood investigation.

Arnold Gesell adopted a different strategy for infant study. In the 1920s, he began observing or photographing infants' behaviors. His work resulted in detailed timetables listing the age of onset of various developmental milestones in motor, social, and adaptive behavior. Gesell's work is still widely consulted by therapists and researchers because it provides age-referenced data useful in developmental evaluations.

The type of work conducted by Gesell is characteristic of the research that occurred in the early 1900s. Study in child development focused on description and the determination of age trends for the appearance of such behaviors as intellectual or motor skills.

From the 1930s and 1940s to the 1960s and 1970s there was a progressive shift away from describing behavior toward analyzing more complex and more abstract developmental processes such as socialization and personality development. In addition, research approaches moved away from simple data collection and description toward formulating and testing theories designed to explain behavior (7, 11, 12). The research methods we have mentioned here will be discussed in greater detail in the next chapter of this book.

The current focus of human development is in areas of increasingly greater specialization. Sophisticated technologies have enabled social and clinical researchers to focus on areas of study that at one time were inconceivable. Microelectrodes, computer technology, and other instrumentation (Fig. 3.2) enable scientists to examine very discrete events such as responses of single brain cells or simple muscular reactions. Computer analyses enable researchers to examine complex behavioral processes and look at the collected effects of entire systems such as muscle groups or human interactions.

Current study in human development is focusing on all ages of life. Partly as a result of our increasingly sophisticated technology, Americans are living longer, and developmental research has shifted some of its focus toward middle and old age. In American society in the 1980s, younger infants, older adults, and disabled individuals who years ago may not have lived, are now surviving. Medical, psychological, and allied health research is looking at the extremes as well as at the more commonly studied areas of the developmental continuum. It is hard to believe that at one time, the scientific study of infants and newborns or the study of the elderly did not exist as it does today.

Germane to the fields of both physical and occupational therapy is the increased investigation across many fields such as allied health, education,

**Figure 3.2.**   Computer technology and electronics have enabled scientists and medical researchers to examine very discrete behavioral responses. Here a psychologist is measuring and recording various psychophysiological responses of an individual located in another room.

psychology, and medicine regarding the development of the handicapped. Social and educational efforts to "mainstream" or integrate children and adults with disabilities into schools and the community have also been examined. Also under scrutiny are the multiple interactive effects of disabilities on the personality and social relations of the disabled person as well as society's responses to these individuals.

It is estimated that by the year 2000, nearly 20% of our population will be over the age of 65. The sheer numbers of people alone will necessitate increased scientific, social, and medical attention to this population. Many of these elderly individuals will require some form of assistance due to health-related dysfunctions. Although it is impossible to predict the consequences of this developmental age shift in the American population, it seems very likely that the scientific study of issues related to aging and handicapping conditions will continue to expand.

# LIFE STAGES VS. LIFE-SPAN APPROACHES

"Before the twentieth century ... early life began with infancy and was followed by a period of childhood that lasted until around puberty, which occurred several years later than it does today. After puberty, most young men and women simply entered some form of apprenticeship for the adult world" (13, p 4). When childhood became a recognized stage of life, developmental study focused primarily on that stage alone.

Traditional views of development such as that held by Hall (9), paralleled the biological or reproductive phases of life (Fig. 3.3). These phases are characterized by growth, stability, and decline (14). Human psychosocial processes, i.e., social interactions and participation in activities, have been compared to these phases (15). Developmental research for the most part focused on the beginning phase (growth-expansion) and tended to disregard the phase of contraction.

In contrast, contempory research examines many different stages of development and recognizes that change occurs throughout the life span. There are two different but complementary views of human development (16). One views life as a continuous, integrated passage from, what is commonly referred to as "womb to tomb"; the other view recognizes discrete developmental stages. This latter view is age specific; it recognizes certain time periods, each with particular characteristics separating them from one another.

The actual markers delineating one stage from another may be biological, social, or even cultural, and some of the markers may be arbitrary. For example, the prenatal period is clearly circumscribed by the biological events of conception and birth, whereas the stages of adolescence or adulthood are more difficult to define.

The reference to life stages does not imply that development is a series of radical transitions in moving from one stage to another. In fact, probably one of the major rationales for separating human development into discrete

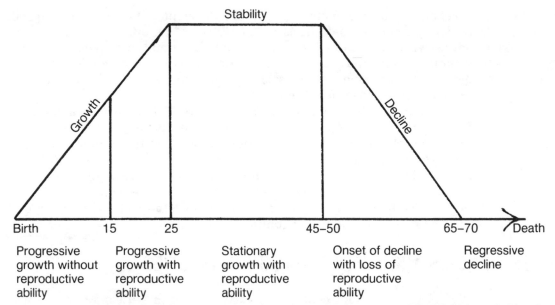

Figure 3.3.   Life cycle based on reproductive process. (Modified from Buhler C: The general structure of the human life cycle. And Buhler C: The developmental structure of goal setting in group and individual studies. In Buhler C, Massavik F (eds): *The Course of Human Life. A Study in the Humanistic Perspective.* New York, Springer, 1968, pp. 14 and 44.)

stages is for convenience of communication. It is easy to catalog information chronologically and categorically.

## Life-span View of Development

The more integrated perspective of development is called the *life-span view.* This regards life as a developmental continuum, not a progressive series of separate stages. The inverted-U perspective of development as expansion, stability, and decline (Fig. 3.3) is rejected by the life-span approach in favor of developmental change throughout, rather than just at the beginning, of life (11). Some researchers in human development may combine approaches, using a traditional approach to life's stages while also stressing the significance of life-span research. This is illustrated by the following quote from psychologist, N. Bayley, in 1963 (17, p. 5). (Italics are inserted to illustrate the traditional cumulation-plateau-decline view of development.)

The study of change in persons over time is recognized as an important aspect of psychological research and theory, concerning both children [*development*] and the elderly [*decline*]. There is less heed paid, however, to the evidence that the processes of change over the lifespan are continuous even during the relatively *stable* adult years. Such evidence indicates that any segment of the lifespan should be considered appropriate for the investigation of psychological change in relation to age. Furthermore, these investigations may appropriately be directed toward practically all major fields of psychology: experimental, learning, personality, emotions, social, clinical, educational, and very obviously in such areas as intelligence and motivation.

In the field of human development, focus on the life span is relatively new, although the approach has been around for nearly two centuries. Baltes (18) points out that this approach can be traced to work in the early 1800s. The approach then lay dormant until it resurfaced early in this century. As an example, Lerner and Hultsch (19) provide a quote from a 1927 publication by Hollingworth that emphasizes a life-span approach. Hollingworth (20, p 57) states: "The infant does not merely start out from birth to become a school child or to attain voting age; his aim is far more complex than that. He sets out, from the moment of conception rather than from the moment of birth, and his goal is the production of an old man or woman; indeed, the ultimate goal of development is death."

Although some researchers were looking at life-span issues in the 1920s, the major focus of developmental study at this time was the early stages of life. As we will examine in chapters 5 and 6, many prominent theories did not continue beyond childhood or adolescence, when it was felt that development had reached its peak. Thus, although the life-span view has been around for some time, it has not been emphasized until the last 20 years. This may be due, in part, to the incompatability of the life-span view with what were the significant views of the world at that time (20).

In the last 10 to 20 years, specific research from a life-span view has tended to focus on selected areas of development such as the biological changes associated with aging. Research has tended to be experimental and psychometric, i.e., using certain psychological tests to measure performance skills. Examinations of cognition (thinking and reasoning) and emotional and social processes throughout life have tended to lag behind these other areas of research (12, 21).

Looft's (12) analysis of the evolution of developmental psychology points out that the life-span view has been emphasized more by specialists in aging rather than in child psychology. He interprets this as evidence that psychologists who study aging (psychogerontologists) are more likely to take retrospective views of development, whereas "[i]t is extremely rare for specialists in child development to take a prospective view of their subjects (at least not beyond adolescence)" (12, p. 200).

Looft's (12) conclusions, published in 1972, have lately been contradicted, as witnessed by the growing numbers of life-span development texts, even by authors traditionally associated with the field of child development (e.g., 22). We should expect that the life-span view will grow even more and supplant the traditional primary focus on the early phases of the life cycle.

This textbook regards the life stage and the life-span views of development as complementary. The student is challenged, thus, to keep in mind interacting perspectives of development. Life stages provide simple means for cataloging developmental research, but life, in general, is characterized by continuity rather than abrupt, choppy changes over time.

# Developmental Stages and Tasks

With the increase in specialization in our society, it is common to speak of increasingly smaller divisions in nearly every area of study. This is also true of human development. Instead of the historical view of a brief division between childhood and adulthood, we now speak of substages such as early, middle, and late childhood and early, middle, and late adulthood.

Stages of development are divided both chronologically, i.e., according to age, and according to specific sets of behaviors that are characteristic of each age. Underlying this division of development into concrete stages is the recognition that certain developmental "responsibilities" must be accomplished before one progresses from one stage to another. These responsibilities or behaviors, acquired during a specific age period, are called *developmental tasks*.

## Developmental Tasks

Educator R. J. Havighurst (23) is one of the fundamental sources regarding developmental tasks, which he describes as

> . . . things a person must learn if he is to be judged and to judge himself to be a reasonably happy and successful person. A developmental task is a task which arises at or about a certain period in the life of the individual, successful achievement of which leads to his happiness and to success with later tasks, while failure leads to unhappiness in the individual, disapproval by the society, and difficulty with later tasks. (23, p. 2)

Havighurst's (23) description of the origin of the developmental task concept is reminiscent of the nativism-empiricism of Rousseau and Locke in their approach to the education and treatment of children. Havighurst reports that developmental tasks stand midway between two opposing theories of education. One theory (reminiscent of Rousseau), is what Havighurst terms the *theory of freedom*. This assumes that the best approach to childhood education is to leave children alone, where they are free to explore and develop on their own. The other theory (reminiscent of Locke) is the *theory of constraint*, which assumes that a child must learn how to become a worthy, responsible adult via discipline and restraints imposed by society.

Some developmental tasks characteristic of certain developmental stages will be listed in the following pages. Before these are listed, however, the difference between ages and stages of development must be noted. Some specialists divide developmental stages according to discrete time periods, e.g., infancy: birth to 18 months; early childhood: 2–5 years; late childhood: 6–12 years (Table 3.1). While this is an appealing approach because it simplifies information, it is also misleading. It tends to reinforce the notion of life as divided into discrete, separate phases, when in fact age-related divisions are rather arbitrary and may vary from source to source.

Table 3.1
Comparison of Age-Related Stages of Life

| Chronological Phases of Life[a] | Age | Age | Life Stage or Period of Development[b] |
|---|---|---|---|
| Germinal, Embryonic, Fetal | before birth | conception to birth | prenatal period |
| Infancy | 0–2 | 0–3 | infancy |
| Early childhood | 3–5 | 3–5 | preschool years |
| Middle childhood | 6–8 | 6–12, 13 | middle childhood |
| Late childhood | 9–11 | 12, 13–17, 18 | early adolescence |
| Middle or late adolescence | 16–18 | 17, 18–early 20s | later adolescence |
| Late adolescence or emerging adulthood | 19–22 | early 20s–middle 30s | early adulthood |
| Early adulthood | 23–43 | | |
| Middle adulthood | 43–64 | middle 30s–middle 60s | middle adulthood |
| Late adulthood | 65+ | middle 60s–death | later adulthood |

[a] Data from Kaluger G, Kaluger MF: *Human Development. The Span of Life*. St. Louis, CV Mosby, 1974.
[b] Data from Schiamberg LB, Smith KU: *Human Development*. New York, Mac-Millan, 1982.

## Prenatal Stage of Development

The first stage of development is the period from conception to birth, which is known as the *prenatal period* and will be covered in more detail in chapter 7. The prenatal stage of development is the easiest to categorize because it has finite biological markers to denote its beginning and end. In addition, as chapter 7 will discuss, it can be subdivided into different stages depending upon specific major prenatal events.

### Prenatal Research

Most developmental research examining this stage of life has tended to focus on prenatal genetic and physiological processes. Research has explored the consequences of different maternal diseases, trauma, and drugs, on the developing organism. Current research is looking at prenatal social-emotional and motor characteristics such as fetal responses to parents' voices, early predictors of later developmental problems, and genetic engineering.

As a result of advances in medical technology in the past 10 years, infants, who at one time could only be studied in the womb, are now surviving. These infants were once collectively called premature but are now divided according to age or weight: "preterm" refers to infants born early, whereas small for small for gestational age (SGA) refers to infants with a birth weight less than the 10th percentile for their age (24) (see "Preterm or Low Birthweight Infants," chapter 7). Some of these premature and SGA infants require hospitalization in neonatal intensive care units (NICUs), and clinical researchers have been examining these infants to determine the long-term consequences of being born extremely small or early, the consequences of hospitalization at this age, and the consequences of separation from the family at birth.

In addition, occupational and physical therapists are particularly interested in the motor development and adaptive behaviors of healthy premature infants. Such information will help those therapists working with infants who are ill or who have special needs to develop more effective evaluation and treatment approaches based on the therapists' understanding the principles of normal development.

Other medical research directed at the prenatal period is concerned with understanding what factors may contribute to complications during pregnancy or delivery. Prospective parents and health professionals are interested in developing simple and successful fertilization methods for partners with fertility problems. Even before fertilization occurs, couples are seeking genetic counseling regarding problems and diseases that may be passed from one generation to another.

Only since the 1970s have a variety of diagnostic tests (see chapter 7) been available for medical personnel to determine whether genetic or physiological problems exist prenatally. If so, parents may be counseled regarding the consequences of terminating or continuing the pregnancy. The study of prenatal development is becoming increasingly technical and sophisticated, and we are benefiting from new knowledge gained from these medical advances. It is reasonable to assume that our understanding of this stage of life will continue to grow during the rest of this century.

### Prenatal Developmental Tasks

The developmental tasks of this period can be examined from the perspective of the mother (as a host organism—or hostess) and the developing fetus. They do indeed share similar if not identical environments so that the actions and health of one have a strong chance of affecting the other. While it is probably not completely accurate to speak of developmental tasks for the fetus, there are certain expectations for "its" behavior. For example, parents (and relatives) expect the fetus to begin moving at a certain time during pregnancy. With new devices for determining the sex of the child before birth, some parents have even named the fetus and call its name and refer to it by name during pregnancy. Parents commonly talk to the fetus and respond to it when it moves or changes position.

Current research indicates that the emotional attachment that occurs between parents and children may start even before birth (e.g., 25). This attachment process, which we will discuss in greater detail in chapters 7 and 12, is based upon a reciprocal interaction between the parents and the fetus. This indicates responsibility for both the parents and the fetus to act and to respond to one another.

In fact, the birth process can be viewed as an interactive process between the mother and the fetus. Milani-Comparetti (26) has suggested that in some deliveries where complications occur, the fetus may be deficient in the motor skills necessary for facilitating or participating in its own delivery. Thus, elementary forms of *responsibility* for the prenatal organism involve rhythmic or reactive motor activity as rudimentary but essential developmental tasks.

## Infancy

The actual period of infancy spans from birth to 18 months or 2 years of age. Infancy is the time when the organism's basic needs are expressed, and motor, social, cognitive (thinking), and language skills are shaped. The typical infant's needs include such activities as being comforted, cleaned, sheltered, and fed; those needs are normally clearly expressed by a variety of vocalizations and facial and body movements, such as crying, fidgeting, cooing, or smiling.

Infancy is often regarded as a time of dependency, especially when seen in contrast to the emergent skills of childhood. This is not to say that the infant is passive or helpless. As we will see in subsequent chapters, the motor skills, inquisitiveness, and vocalizing evident during this stage of life are preparatory for the mobility, rapid learning, and extended socializing respectively observed in childhood.

### Neonatology and the Scientific Study of Infants

The scientific study of infancy is relatively new in the United States. The medical treatment of children (pediatrics) as distinctly different from adults, began only at the turn of this century (5), and systematic research investigating infant learning, for example, has proliferated in the United States only since the 1960s (27). Presently, there exists an enormous amount of literature regarding development, medical conditions, and medical complications during the first 1–2 years of life.

Since the 1970s, the growth and sophistication of intensive care units designed to maintain premature or SGA infants has resulted in the expansion of the medical subspecialty of *neonatology*, which is the study of newborns. The *neonatal* (*neo*, new; *natal*, birth) *period* is a brief period of hours or days during which the infant stabilizes from the trauma of birth. Most infants stabilize quickly, but those who have medical problems, complications associated with prematurity, or are very small often require specialized technical and medical care. These neonates are treated by medical specialists in addition to occupational and physical therapists and other staff who work with the infants and their families.

### Developmental Tasks of Neonates and Infants

The developmental tasks of the neonatal period include breathing without aid of a respirator, crying, eliminating, sucking, swallowing, and maintaining body temperature. Some premature, SGA, ill, or disabled neonates will require assistance with these normal developmental tasks. These infants may be placed briefly, or for long periods of time, in enclosed, protected "isolettes" in special hospital units where their breathing, food and fluid intake, and body temperatures are monitored and maintained (Fig. 3.4).

The developmental tasks of the normal infant are diverse. Only a few examples are listed here because they will be examined in greater detail in later chapters of this book. Some developmental tasks include motor skills such as rolling or lifting up the head; visual skills such as looking and following objects with the eyes; social-emotional skills such as looking at a caregiver and smiling. These motor, visual, and social tasks are coordinated with an infant's other emotional, social, auditory, language, and sensory skills such as laughing, reaching, turning the head in a certain direction, crying, babbling, and imitating a diverse array of sounds, movements, and facial expressions.

## Childhood

The transition from infancy to childhood, at approximately the second year of life, is characterized by the attainment of such skills as walking and talking. Such skills afford the child a greater degree of freedom than was exhibited in infancy. Childhood, is therefore, often considered a time of independence compared with infancy. The infant is viewed as *acquiring the motor skills for walking*, whereas the child *walks*. The infant is in the *process of developing language*; whereas the child *talks*.

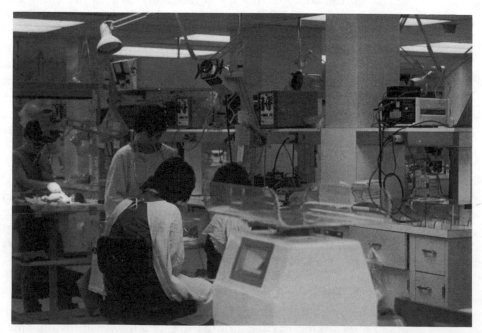

**Figure 3.4.**   The technology of neonatal intensive care nurseries is keeping alive preterm and small infants who, only years ago, would not have survived. (Courtesy of March of Dimes, Birth Defects Foundation.)

As a time period, childhood is longer than infancy. Thus, it is often subdivided on the basis of predominant developmental tasks. For example, the first stage, "early childhood", from age 2 or 3 to 5, may also be called the "preschool period." This time period is characterized by the acquisition of skills preliminary to attending school, e.g., dressing independently, socializing, following instructions, perhaps learning to hold a crayon and scribble or letter one's name.

The developmental tasks of middle and later childhood include incorporating masculine or feminine sex roles into a self-image; socializing (playing) with peers; developing a moral sense that involves understanding and following rules; and being able to operate independently of parents or other adults (23).

## Adolescence

Recognition of adolescence as a stage of life is relatively new. Not until childhood was recognized could a transitional stage between childhood and adulthood be considered. Although adolescence was introduced in the early 1900s as an area of scientific study (8), research at this time still focused primarily on childhood. Through the years, adolescence has been increasingly recognized as a life stage because of cultural, social, and economic changes in the United States. As adolescence became more widely recognized, scientific study of this age group increased.

### Adolescence and Culture

This stage of development, probably more than any other, provides an illustration of how life stages are influenced by culture. While our society typically views the stage of adolescence as a time of turmoil and adjustment, this is not the case for all cultures (28). Schell and Hall (29) explain how interacting social factors can alter life stages:

> The way in which people in a society view the life span depends largely on its social and economic system. If the preparation for adult roles is gradual and continuous from early childhood and if the necessary technology can be acquired by apprenticeship, then adulthood is likely to begin shortly after a person reaches reproductive maturity. On the other hand, if full participation in the economic system depends on years of technical education, a period of adolescence is likely to be recognized (p. 9).

In some societies, when children become physically mature, they mate, bear and raise children, and participate in adult social tasks. In the United States, especially with our focus on higher education and specialized training for technical jobs, adolescence is prolonged.

There is controversy regarding when adolescence ends and when maturity starts. For example, legislation regarding this issue is clearly inconsistent. The legal age for draftee status in the Armed Services, the voting age, the age for obtaining a driver's license, the age when one graduates from high school, and the age for legally purchasing alcoholic beverages all vary. Some 12- and 13-year-old adolescents have given birth to children and may hold jobs to support their families, which are tasks clearly associated with maturity. Thus, specific ages associated with the onset and termination of adolescence are largely based on culture or arbitrarily determined.

### Developmental Tasks and Adolescence

The stage of adolescence includes such tasks as adapting to the physiological changes underlying the body's preparation for fertility and childbearing. In American and some other societies, typical developmental tasks of adolescence involve separating from the family, establishing an independent and separate sense of self, socializing with members of the opposite sex preliminary to childbearing and child rearing, and coping with responsibilities of acquiring job skills. As Havighurst (23) states:

> Some developmental tasks are practically universal and invariable from one culture to another. But other tasks are found only in certain societies, or they are peculiarly defined by the culture of the society. The tasks that are most completely based upon biological maturation, such as learning to walk, show the smallest cultural variation. Others, and especially those that grow principally out of social demands on the individual, show great variation among various cultures. The task of selecting and preparing for an occupation, for example, is a very simple one in a primitive society where there is little or no division of labor and everyone has practically the same occupation, whereas it is one of the most complex and worrisome tasks of middle-class adolescents in America. (p 37–39)

Lately, research has been directed at adolescent behaviors that have considerable social impact, e.g., adolescents' roles in politics, juvenile delinquency, the incidence and cause of dropping out of high school, drug use and abuse, family-adolescent conflict, and sexuality and adolescent pregnancy. The study of this age group will tell us more than specific information about adolescents, it will also direct us toward a better understanding of American culture.

### Youth

In the 1960s, Kenneth Keniston (13) proposed that a separate stage of life was emerging. He claimed that because of the massive social and political changes that had occurred in American society in the 1960s, a separate stage of development was appearing between adolescence and adulthood. This stage, called "youth," accommodated not all individuals but those who had achieved developmental tasks of adolescence and yet had still not assumed the responsibilities typically associated with maturity and adulthood.

The typical youth, according to Keniston, had a clear sense of self, was comfortable with sexual roles and relationships, but had made the decision to "drop out", i.e., not be a productive member of adult society. Typical of the counter-culture of the times, youths experimented with religious, consciousness-altering, or social group alternatives prevalent during the 1960s. Keniston recognized that these individuals are not searching for a newly established sense of self, as do adolescents, nor are they experimenting with newfound sexual roles as do adolescents. Instead, Keniston claims, youths are characterized by their experimentation with life and with making the decision not to participate in the tasks typically associated with maturity, such as marriage, child rearing, and vocational productivity (13).

Some researchers in the field of human development do not recognize youth as a separate life stage, while others acknowledge that the American trend toward higher education and technical training are cause for a separate life stage that bridges the developmental tasks generally associated with adolescence and adulthood.

It is interesting to note that, just as the stage of childhood emerged during times of religious, philosophical, economic, and scientific revolution, so also has Keniston's concept of youth. During the 1960s, there was considerable social and political unrest in the United States. Student groups banded together with community peace movements, protested the war in Vietnam, and supported a variety of human rights movements.

The first generation to grow up under the influence of the nuclear bomb, this age group, more than any other since, participated in active and passive strategies to change political, social, and economic traditions. While they were not in traditional, recognized positions of political power, these individuals participated in and supported civil rights legislation and the War on Poverty, which was discussed in chapter 2. Debate and unrest on college campuses focused also on other religious and social issues of considerable impact.

Technology in the 1960s was literally opening new frontiers. Just as the discovery of new lands and the scientific revolution of the Middle Ages led to changes in the family and the status of childhood, the space program of the 1960, as well as medical and electronic technologies led to other changes in the American way of life. Families are different now than they were in the 1940s and 1950s. Education, occupational specializations, the economy, and global communication have affected how we view life and development. Keniston's concept of youth is another example of how multiple interacting forces can alter our concept of developmental stages.

## Adulthood

Maturity or adulthood is typically regarded as the time of decision making and productive contribution to society through child rearing, maintaining an occupation, and maintaining a role in the community (e.g., 23). This stage of development may span 50 years, and in comparison to earlier life stages, has until recently been relatively neglected as a focus of research. As Havighurst (30, p 5) reports: "Developmental studies of infancy, childhood, and adolescence originated in the nineteenth century and were flourishing by 1940, when studies of adulthood and aging were just getting started."

## Research on Adulthood

Reviewing the research on adulthood is essentially the same as generalizing about life-span research. It is interesting to note that most textbooks on infancy and childhood include or imply the concept "development" in their title. Studies of adulthood have typically been associated with aging. Now, with a greater focus on life-span approaches, development and aging are viewed as the same process, and life-span models of development have emerged.

Much of the past research on adulthood investigated psychomotor behaviors (i.e., motor skills), intelligence, and memory. The data were used to generalize about the effects of aging on the systems underlying these abilities. Contrary to research with children, efforts were primarily empirical and did not attempt to produce or test theories of (adult) development (30).

## Adulthood and Developmental Tasks

The traditional relationship of the life cycle to reproductivity (Fig. 3.3) forms the basis for the substages of adulthood. For example, Havighurst (23) lists these as the developmental tasks associated with early adulthood: selecting a mate, starting a family, rearing children, and providing a home environment for that family by holding a job, keeping a home, and participating in civic affairs. The developmental tasks of middle adulthood include preparing older children to make the transition to adulthood, adjusting to physiological changes associated with aging, developing leisure time skills (for which there is more time available due to reduced family demands), and resuming a spousal role where parents can relate as mates rather than as fellow parents (23).

Now that research in the field of adulthood has virtually exploded in the past 10 years, new models of adulthood and aging are regularly being proposed. Some of these data come from studies that were started 20 or 30 years ago and have been following the course of specific adults' development over time (e.g., 31). The research of Bernice Neugarten (32) indicates that middle age is not a time of loss or decline but instead a time of certainty when individuals take on what she terms "an executive role toward life."

It is difficult to predict what will occur in developmental research regarding adulthood over the next decade. Common stereotypes regarding aging are being altered by contemporary reports of fitness and health associated with middle adulthood (e.g., 33). Current academic research is examining all aspects of adult development (social, emotional, intellectual) and interrelating these findings to all stages of the life span. New information is constantly updating existing views of this long stage of life.

## Expanding Stage of Late Adulthood

This substage of adulthood is important to examine separately because of its significance to the fields of physical and occupational therapy. Late adulthood or old age encompasses the time from about age 65 until death.

This stage of life has particularly been neglected as an area of scientific investigation because, until this century, large numbers of people did not survive to their 70s and 80s.

In 1900 the average life expectancy was 47 years compared with present life expectancies in the 70s. In 1900, only 4% of the population was 65 and older, whereas now people in the over-75 and over-85 age groups are the fastest growing age segments in the United States (34, 35). Individuals once negatively classified as "elderly" are in positions of authority and responsibility in our country. Ronald Reagan, for example, at 75, has been the oldest President of the United States, and despite some health problems, he emanates an image of vitality and vigor.

While old age is commonly stereotyped as a time of decline and withdrawal, it is increasingly recognized as a time of productivity and activity. With retirement an option rather than an obligation, many older Americans are remaining with their jobs or in some advisory capacities throughout their lives. Other elderly individuals are taking advantage of the increased leisure time and diminished family responsibilities and are developing new hobbies and skills.

There is probably no other time when health is such a significant issue as during late adulthood. Activities can be curtailed when health is compromised by the cumulative effects of disease, malnutrition, or normal wear and tear throughout life. In addition, with our increasingly sophisticated medical prevention and technology, elderly individuals, who years ago would not have survived, are maintaining fairly independent lifestyles because of orthopedic, pharmaceutical, or surgical advances.

Many other elderly individuals require some form of health-related assistance. Due to hospitalization costs and discharge regulations, many people are discharged from hospitals, as what is commonly termed, "sicker and quicker." This places a responsibility on families and spouses for providing regular daily assistance for spouses or parents whose sensory, physiological, or skeletal functions may be impaired. Just at a time when American society has become increasingly mobile and the extended family is spread out, elderly family members are living longer and needing assistance from family that may be unavailable to help.

Alternative forms of residences and different kinds of nursing care have been developing to take care of this growing population. Springing up in many cities and towns are independent-living apartments or centers that may be associated with nursing facilities or linked by communication devices to centralized assistance. Community and private agencies employ home care providers who are trained to give specialized services to people in their homes. Many of these programs (such as Nursing Homes Without Walls) have been developed to meet the growing needs of the elderly population. Other agencies, such as the Visiting Nurses' Association, have been in operation for many years but are finding they must shift their orientation to an increasingly elderly caseload.

Physical and occupational therapists' roles in such agencies involve providing direct care to patients or consulting with nursing staff regarding a variety of factors such as client's mobility with walkers or wheelchairs in the home; making the home safe and functional for individuals with sensory impairments such as blindness or deafness; and providing assistive devices for clients who want to function as independently as possible in daily living tasks, such as bathing, meal preparations, and dressing.

While there is a tendency to think that most elderly people are frail, sick, and institutionalized, this is an inaccurate stereotype. Only 5% of the elderly are institutionalized at one time, with the other 95% living in the community with family, friends, or by themselves (34). Many elderly people take advantage of innovative programs (e.g., convenient transportation to community centers for socialization or to shopping centers) that enable them to continue to participate as much as they can in community life. Some agencies provide free or low-cost meals (e.g., Meals on Wheels) which are delivered to people's homes. This guarantees that a person who may be inactive and tend toward malnutrition is provided at least one nutritious meal each day.

Other community-based programs as well as technological advances reduce the loneliness that is common among elderly individuals who have lost family, friends, and spouses. For example, the telephone has provided an excellent socialization outlet for many whose mobility is reduced.

### Developmental Tasks of Late Adulthood

An important developmental task of this life stage is adjusting to losses typical of this age. These losses occur within individuals as well as in their relationships with others. For example, with age comes the loss of physical attributes associated with youth. Even in early adulthood individuals may have to adjust to baldness, graying hair, and skin, skeletal, muscular, and sensory changes that accompany aging. These changes accumulate especially in old age so that one can see physical changes and feel progressive effects of declining strength and stamina.

Other social changes have to do with the loss of family, spouse, and friends through death. In addition, physical conditions or retirement force the loss of occupation, i.e., purpose, activities, and vocational and avocational skills. Some elderly individuals seek new living arrangements and must adjust to the subsequent change in friends and environment. Others, Havighurst (23) points out, must adjust to the effects of reduced income as a result of retirement.

While old age is a time of adjustment and loss, many people, however, maintain interactions with family and friends, substituting new friends and grandchildren for family members they have lost. Many keep up active, healthy lifestyles. Clearly, for most, late adulthood is not a time of progressive decline as was traditionally viewed many years ago.

Zubin (36) concludes this about the future of the growing field of *gerontology* (the study of aging):

As one views the year 2000 with reference to the problems of the aged, it becomes quite clear that a larger and larger proportion of voters, consumers, and men and women of leisure, as well as workers, will be drawn from the age group beyond 65 ... It will be as great if not greater than the shift from rural to urban centers, and its impact on education, industry, health, and business can only be guessed. How to prepare to serve such a tremendous population is a crucial issue. Certainly such areas as play, recreation, and continuing education will play an important part. (36, p. 11)

Zubin's (36) summary points to two important issues for readers of this text. One issue is the recurrent theme of interactionism. Zubin points out how the increased size of one age group, the elderly, will have multiple social effects, i.e., on education, industry, health, and business. The second issue has to do with the significant role of clinicians working with this age group. Occupational and physical therapists will play a role not only in treatment of elderly people who are sick or disabled but also in social issues relating to community care and activities, increased leisure time, and the concern about values and respected human occupation, function, and productivity throughout the life span.

# DEVELOPMENTAL TASKS AND THERAPY

The developmental tasks described in "Developmental Stages and Tasks" vary for different ages as well as for different cultures. They are not always universal and are therefore subject to numerous, interacting influences. For example, Havighurst recognized that task development involves an interaction between individuals and society, that it implies an active level of involvement on the part of the person striving to accomplish a task, and that it is a concept that brings together different, interacting disciplines (23): "A developmental task is midway between an individual need and a societal demand. It assumes an active learner interacting with an active social environment. Accordingly it is a useful concept for students who would relate human development and behavior to the problems and processes of education" (23, p. vi). In addition, it is a useful concept for relating human development to both physical and occupational therapy.

Task development is a major focus in clinical practice, however, it has been more widely promoted in occupational therapy. As occupational therapist, Barbara S. Banus (37, p. 15) reports, Havighurst's (23) concept of development task does not differ from the concept of "occupational performance;" for example, one such definition of occupation is, "learned and developmental patterns of behavior which are the prerequisite foundations of self-care, work, and play/leisure skills."

Banus points out: "Labels used to describe the tools or dimensional stimuli of the developmental therapist include: activities, objects, media, tasks, play, equipment, and toys. *Tasks*, however, is the most accurate term because it implies doing which is central to occupational therapy" (37, p. 219). A common definition cited for "occupation" is the "goal-directed use of time, energy, interest, and attention (38)." Thus, occupations may include activities or tasks, both of which, as Banus (37) points out, are the tools of the occupational therapist.

At times, the terms "activities" and "tasks" are used synonymously, but in developmental literature, the term "task" has a special meaning. This concept of a developmental task as Havighurst (23, p vi) has defined it as "mid-way between an individual need and a societal demand," implies an individual and social *expectation* of successful performance. As Kaluger and Kaluger (39, p 5) report, "developmental tasks are functions that individuals *must do* (italics mine) to achieve a certain level of maturity that is expected at each stage of life."

In contrast to the concept of a "task", an "activity" is more general. Activities, like tasks, imply action and the process of doing, but we do not commonly associate with activities the expectation of achievement. People do activities in the process of keeping occupied, whereas people *are expected* to do tasks. Thus, tasks are generally associated with some end product or accomplishment, i.e., we speak of *accomplishing* a task and *doing or performing* an activity.

## Task and Activity Analysis

*Task analysis* is the examination and reduction of specific behaviors, skills, or abilities into the component behaviors necessary for performing them. *Developmental task analysis* involves two steps: recognizing that one's performance (skill level) of specific tasks and one's interest in tasks changes over time, and then determining what tasks are associated with or appropriate for a certain age or level of development. Thus, task analysis refers to examining an expected behavior or function and determining what skills or steps are essential for its successful achievement. Developmental task analysis examines those tasks in terms of society's expectation of the age at which they should be accomplished.

In contrast, activity analysis determines all the characteristics of activities, how to modify them for specific pathologies, and the nature and the extent of all the skills required for productive use of those activities in treatment. Developmental task analysis as well as activity analysis are fundamental to occupational therapy and physical therapy because of the professions' focus on people's skills, actions, and productive occupations over time.

## Developmental Approaches in Therapy

Clark (40) has suggested that human development through occupation become a guiding philosophical approach for the profession of occupational therapy. Her proposal, which is based upon an analysis of the history and existing theories in her field, is testimony to the overriding significance of the concept of development as a basis for occupational theory and practice. In fact, many physical and occupational therapy theories and treatment approaches are centered around the concept of development. While many of these will be discussed later in chapters 5 and 6, two are introduced here. These are described here because of their interdisciplinary focus and their integration of developmental terms and concepts within a framework for treatment.

Both approaches have been introduced by occupational therapists, but, as will be described, the approaches are broad enough to incorporate a variety of clinical disciplines. For example, occupational therapist Lela Llorens (41) has applied a developmental stage perspective to the concept of patient rehabilitation. This approach may be useful for any clinical practice field dealing with ill or injured individuals.

## Developmental Theory for Health and Rehabilitation

According to Llorens (41), growth and development are chronological and directed at the achievement of mastery in various skills, abilities, and relationships. Therapy is aimed at the individual who, as a result of disease, injury, or environmental or personal deficits, experiences trauma that interferes with the capacity to continue the normal progression through life (41).

Based on various sources, including Havighurst (23), Llorens (41) delineates specific developmental tasks that are appropriate for the following stages of life: infancy to age 2, ages 2–3, ages 3–6, ages 6–11, adolescence, young adulthood, adulthood, and maturity. In addition, she notes specific areas of human development of particular concern to occupational therapists as well as other rehabilitation professionals. The areas and their definitions are listed in Table 3.2.

**Table 3.2**
**Areas of Development of Concern to Therapists Working with the "Whole Person"[a]**

| Area | Definition |
|------|-----------|
| Neurophysiological | Dealing with nervous system and its control of body functions |
| Physical or physical-motor | Observable, coordinated physical behavior |
| Psychodynamic | Interactive psychological forces operating within a person |
| Psychosocial | Behavior that integrates psychodynamic and social behavior |
| Social language | Verbal and nonverbal communication |
| Sociocultural | Social interaction behaviors dictated by and learned from culture |
| Activities of daily living | Routine tasks of everyday living |
| Intellectual | Thinking and reasoning, interactions with environmental stimulation |

[a] Modified from Llorens, LA: *Application of a Developmental Theory for Health and Rehabilitation.* Rockville, MD, American Occupational Therapy Association, 1976.

According to Llorens (41), some kind of trauma may interrupt functions in any or all of the areas of development listed in Table 3.2. As a result, what may occur are gaps *between* these different areas of development as well as gaps in the chronological progression through life. Thus, an individual who experiences some trauma may become what we commonly refer to as "setback". The term "setback" is particularly descriptive in the context of Llorens' approach. Setback literally means moving backwards or interrupting normal progress (in this case, through life).

Therapy, according to Llorens (41) enables patients to return, as much as possible, to their previous level of functional performance. It accomplishes this by providing a bridge between the gaps that separate patients from their expected level of function and the actual functional level that has resulted from trauma. The therapy bridge, Llorens points out, links up the various areas of development (Table 3.2) or links the individual chronologically with the stage of development previously attained prior to the setback.

## Developmental Therapy

Banus (37, 42) has applied Llorens' work in an approach to pediatric treatment. In the 1971 edition of the text, *The Developmental Therapist*, Banus (42, p 251) describes the mastery of developmental tasks approach (MDTA) which, she points out, is a direct application of Llorens' ideas: "It is characterized by helping the child master tasks appropriate to his age level . . . until parity is achieved between the child's chronological age and his level of functioning . . . It is applicable to all areas of dysfunction in children—indeed, to dysfunction of all individuals according to Llorens."

Banus' approach, however, is directed at children who have not developed to the level of function expected for their age. Examples of such children are those who are born with wide-ranging disabilities that interfere with their normal course of motor, social, emotional, and intellectual development. These individuals are developmentally delayed.

### Developmental Delays

The term "developmental delay" is frequently used to refer to immature performance in many areas of development, e.g., all those listed by Llorens (41) as significant to therapy (Table 3.2). Developmental delay, however, can also apply to one or a few specific areas of function. For example, Banus (37) has illustrated how a child exposed to environmental deprivation that leads to malnutrition and reduced sensory stimulation results in delays in the many areas of development illustrated in Figure 3.5.

Nelson (43, p 69) defines developmental delay as "an ability or a function that would be normal for a younger child but that is immature for an older child." He points out that it can be in one or in a combination of areas and that "the presence of a developmental delay usually implies the failure to develop in a normal progression beyond the delay" (p 69). Treatment for these patients is aimed at *habilitation* or at *rehabilitation*, and treatment is termed "developmental therapy."

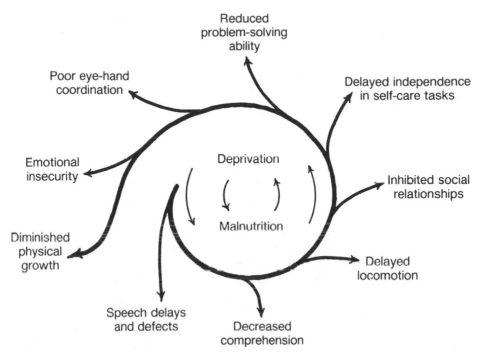

**Figure 3.5.** The spin-off effect of deprivation and malnutrition that together produced multiple developmental and growth delays. (From Banus BS: Development and occupation. In Banus GS, Kent CA, Norton YS, Sukiennicki DR, Becker ML (eds): *The Developmental Therapist*, ed 2. Thorofare, NJ, Charles B Slack, 1979, p. 5. Reproduced with permission from publisher.)

### Developmental Therapist

A developmental therapist, according to Banus (37) is

> ... anyone who analyzes the child's development, recognizes the discrepancies between his development and the norm, plans an appropriate program of maintenance, correction or prevention, and provides therapy for the child. It might be a child psychologist, early childhood educator, physical therapist, occupational therapist, pediatrician, special education teacher or speech pathologist. Different developmental therapists focus on different aspects of development. It is the emphasis on occupational performance that differentiates the occupational therapist from the others. (37, p. 14–15)

Tyler and Chandler (44) point out that in the field of pediatrics, occupational and physical therapists' roles have merged such that they are both collectively referred to as "developmental therapists." These individuals are concerned with various aspects of children's development, and they work in interaction with other treatment specialists with similar concerns:

> The developmental therapist is a member of an interdisciplinary team concerned primarily with a child's strengths and problem areas in gross and fine motor development and sensory skills. The evaluation and treatment of motor movements overlaps with the child's total development and, therefore, the therapist works closely with other team members. For example, in the manage-

ment of a child with a feeding problem, the therapists may contact the nutritionist, nurse, social worker, dentist, or educator. If there is also a speech problem, the communication specialist and psychologist may provide assistance. Information regarding a child's social, emotional, and physical growth is important in any remediation program. The total team assessment of the child provides a priority list of rehabilitation goals that should reflect the parents' and child's concerns. (44, p. 171)

## Summary of Llorens' and Banus' Approaches

The use of the term *rehabilitation* is significant in Llorens' (41) developmental theory for health and rehabilitation. "Rehabilitation" refers to the treatment and training of a patient so that he or she "may attain the maximal potential for living physically, psychologically, socially, and vocationally" (45, p. 1). The prefix *re* means "again", thus the term "rehabilitation" literally refers to individuals receiving treatment to gain back skills they once had. *Habilitation* refers to treatment aimed at individuals who are furthering their development for the first time. It is not uncommon, however, for professionals to use "rehabilitation" as a general term that encompasses the concept of habilitation.

Llorens' (41) and Banus' (37, 42) approaches are similar, differing primarily in emphasis regarding certain aspects of the life span. Llorens focuses on development across the life span. In fact, she reports that the developmental frame of reference in occupational therapy refers to "recognizing, motivating, and facilitating growth and development throughout the life span in accordance with the individual's ability to progress" (46, p. 8). Banus' emphasis is on pediatrics and the developmental stages up to and including adolescence. In a general sense, Llorens' is a rehabilitation approach, while Banus' approach is habilitative in nature.

Banus, however, has modified the concept of the mastery of the developmental task approach (MDTA). In 1979, in an updated version of the 1971 text, Banus (37, p. 360) refers not to MDTA, but to mastery of occupational performance skills (MOPS). She defines MOPS identically to MDTA except for substituting the term "occupational performance skill" for "developmental task," i.e., "helping the child master occupational performance skills appropriate to his age level . . . until parity is achieved between the child's chronological age and his level of functioning" (Fig. 3.6).

The MOPS approach, Banus (37) points out, is applicable to all areas of dysfunction *in children*. Since her text is pediatric in nature, the MOPS approach has been applied primarily to early development. However, consistent with Llorens' theory and with Havighurst's application of developmental tasks across the life span, Banus MDTA (or MOPS) approach can also be applied across different age groups, i.e., across the life span.

Using Llorens' and Banus' approaches provides a developmental reference point for therapists. Thus, by modifying Banus' terminology, we can define the roles of occupational therapists and physical therapists as, facilitating the mastery of occupational and physical functional and performance skills to a level that is developmentally appropriate for individuals throughout life. This is illustrated in the following section.

**Figure 3.6.** As a result of developmental therapy and adaptive equipment, individuals with disabilities are often able to perform developmental tasks that are appropriate for their age.

# Developmental Frame of Reference for Therapy

The fundamental clinical roles for occupational and physical therapists are evaluation and treatment. Evaluation determines the patient's existing level of function and the extent of complications due to dysfunction. Based on the results of the evaluation, treatment is geared to the patient's special needs. Some of these needs are developmental in nature and can be incorporated into an integrated developmental approach toward treatment.

For example, Llorens' (41) and Banus' (37, 42) approaches examine treatment and evaluation from a developmental orientation, termed a *frame of reference*. Frames of reference are basic assumptions about a certain content area (47), and have historically been studied more in occupational than in physical therapy (e.g., 46, 48). Since treatment fields are so diverse, it is natural for many different frames of reference to be used in clinical practice. As an example, the developmental frame of reference takes into consideration each patient's level of development prior to trauma, the level of development as a result of trauma, and the design of treatment that is geared toward meeting the specific developmental needs of each patient.

The following case example illustrates how therapists take into consideration the specific developmental and individual needs of two patients who are widely different in ages and in interests. Both patients have been experiencing severe psychosocial (psychological and social) or emotional disturbances that have interfered with their abilities to function normally in everyday life.

### A Case Study

Two different patients have been referred to a psychiatric hospital that is well-known for its interdisciplinary and innovative focus on treatment. Both patients lately have been experiencing emotional disorders that interfere with their abilities to function in everyday life. Both patients have been fairly inactive for the past several months but prior to that time were otherwise functioning normally in everyday living situations.

Both patients have withdrawn from other people; they tend not to regularly get out of bed or get dressed and have engaged in few regular daily activities. Out of concern for them, the patients' families have encouraged them to enter this hospital for treatment. Both patients have agreed.

Upon admission to the hospital, these patients have agreed to the hospital's criteria that they get up at a certain time each day, get dressed, attend meals in a separate dining facility with other residents, and participate in a variety of therapies. The treatment and evaluation team has collectively addressed the patient's problems and concurred that each requires a variety of approaches to treatment that is aimed at reducing their current depression and social isolation.

### Evaluation

During the initial evaluation of the patients, therapists consulted with both to determine their past history and current interests, hobbies, skills, level of activity, general daily routine, level of fitness, and current living situation. One patient is an athletic 13-year-old female named Gail, who is of average height but considerably below average weight for her age. She lives at home with her parents and older brother and younger sister. The other patient is a 50-year-old, very obese male named Joe, who lives alone at home and whose wife recently died.

An analysis of interests revealed that Gail has participated in a variety of individual and team sports at school and at summer camp. Her family has just recently moved, however, and she has lately been spending most of her time at home. She does not belong to any community groups, and school has not started because of summer vacation. She has heard that a lot of teenagers who are her age and live in her neighborhood attend a community recreation program where they swim and play tennis. She can do neither. She feels lonely, left out, and very anxious that she will not fit in with her peers when school starts.

The older male patient, Joe, has no hobbies. He has been accustomed to spending most of his time either at work or at home interacting with his wife, watching television, or napping. His wife recently died, and he has increased time on his hands as well as increased home responsibilities for which he is unprepared. His wife used to manage the household by shopping, planning and preparing the meals, and doing housekeeping, laundry, and other chores.

Both patients are of normal intelligence, but both have mild health problems associated with weight and nutrition. Gail had been very active but now spends most of her time in her room or watching television. Joe has normally worked a regular 40-hour week as an engineer, but lately he finds it hard to get to work. His employer is concerned about his frequent lateness and absenteeism. While Gail has only recently developed an inactive, solitary lifestyle, Joe has been sedentary in his job and in his spare time for most of his life.

### Treatment

Treatment goals for these patients need to address their current sedentary lifestyles, their reduced social involvement and emotional withdrawal characteristic of depression, additional nutritional and weight complications, and poor management of time and occupation. The overall treatment goals for both patients are essentially identical, but the treatment activities that are selected are very different for each one because of their different developmental levels and interests. Different kinds of treatment activities will need to be selected for their appropriateness to Joe's and Gail's age, sex, physical condition, habits, previous lifestyle, and personal goals for the future.

### Gail

Gail's interest in swimming and in other sports in addition to her age and physical condition (confirmed by medical reports in the hospital and referral for participation in any activities) make her appropriate for the hospital's swim program and for participation in some of the team sport and recreation activities that involve other adolescents in the hospital. Consistent with the interdisciplinary focus of the hospital, these activities are offered under the combined supervision of recreational, occupational, and physical therapy staff.

For Gail, appropriate introduction to the groups and a gradual involvement in sports will help to increase her socialization with peers, provide success experiences for her (e.g., learning how to swim), and provide her with daily exercise that will attack her temporary, current sedentary habits. Gradual carryover into some community activities will help her to meet other people her age in the community and help her to replace old friends whom she lost when she moved. The gradual introduction into community activities will serve as a bridge for her for when she leaves the hospital.

Exercise and physical activity will also help to improve Gail's appetite and help her develop a more positive attitude toward herself. Since Gail also exhibits dietary problems associated with low body weight, she is a candidate for a youth cooking and nutrition group that is conducted collectively by the activities, nutritional, and occupational therapy staff. In addition, she participates in a counseling group whose nature is primarily discussion and exploration of personal concerns and goals.

*Joe*

In conjunction with the medical and nutrition staff, Joe is starting to attack his problem of obesity. In a combined program supervised by various therapies, Joe participates in a walking program designed to gradually increase stamina and physical activity in individuals who have had habitual sedentary lifestyles. This program is attended by other patients of various ages, and there is time during and after exercise for socialization. In addition, this program includes discussion groups led by various therapists who address time management, the benefits of exercise, and the interaction of lifestyle, exercise, and diet.

Joe also has started exploring with therapists some people-oriented avocational interests that he can pursue when he returns home. One possibility he may be interested in is a local nature-study group in his community. Participation in the group, he feels, will take him out of his home, involve him socially with other people, and provide a fairly flexible challenge in terms of demands for physical activity.

In regard to home-care skills, Joe has started to participate in a meal planning and preparation class with other patients his own age. Here, several goals are addressed at once, i.e., how to go about planning, shopping for, and actually preparing balanced meals that take into account specific caloric requirements. This class also addresses other home care skills; various therapists, nutritionists, and social service workers lead small-group sessions to discuss time management, how to do actual skills (vacuum, operate a clothes washer and dryer) required for home care, how to plan meals, how to locate help in the community, and how to supervise and manage help in the home for those who want it.

*Summary*

Both Joe and Gail have similar needs. Both have activity, nutritional, emotional, and social problems. Both have recently experienced losses in their social lives, and both have recently withdrawn more and more into themselves. Despite these similar problems, however, treatment for each is individualized because of their different developmental needs. Good communication between team members has taken into consideration the individual interests and the developmental tasks of each patient. As a result, specially designed programs are specifically geared toward both Joe's and Gail's eventual rehabilitation and reintegration of active, healthy lifestyles in their community.

Understanding how to select specific activities or modalities and how to adapt them and make them appropriate for individual patients at specific functional and developmental levels are essential, fundamental clinical skills in occupational and physical therapy. Developmental skills and abilities and how they change over time, how they characterize certain life stages, and the research analyzing them is the subject matter of much of this text. This is not, however, a textbook on treatment and clinical skill development. This book does provide basic concepts that physical and occupational therapy students can later adapt to models of clinical treatment, which they will obtain from other textbooks and coursework.

# INTERACTION OF DEVELOPMENTAL STAGES WITH CONTEMPORARY LIFE-SPAN VIEWS

The developmental-stage approach is fundamental to the study of human development and to clinical work. For example, the goals of study in human development are to understand, explain, and predict how behavior changes over time. These goals are most commonly addressed by determining general principles of behavior. This is done by finding average abilities shared by most individuals at specific ages. This results in categories of age-related behaviors, which are the developmental stages and tasks that have been discussed so far. These developmental stages also form a basis for Havighurst's (23) developmental task approach and Lloren's clinical theory of health and rehabilitation (41).

While a developmental frame of reference and developmental stages are important for the clinician, there are limitations to holding this as the sole point of view. The problem with thinking *only* in terms of stages is that it may lead to the expectation that all people represent "the average person". This approach tends to group all people together and to neglect the concept of individual differences. A well-rounded study of human development acknowledges averages while also recognizing differences between people.

## Individual Differences in Evaluation and Treatment

Physical and occupational therapists encounter an enormous variety of individuals at all ages of life. As illustrated in the case examples in this chapter, two primary roles for therapists in the clinic are evaluation and treatment. Evaluation involves comparing each patient's individual performance with some group-related average, called a *norm*. These averages are obtained from tests that are called *norm referenced*.

A key factor in developmental evaluation involves understanding what each patient can and cannot do in terms of occupational performance and physical function for his or her age. This requires an understanding of average skills and abilities for all age groups, e.g., knowing the normal age that most children can independently tie their shoes, or the normal age that most adolescents reach sexual maturity.

While we all know that there is an average age for acquiring these skills, we also know that individual children show considerable variability in the actual time when they accomplish such skills. This is the concept of individual differences. A skilled evaluator refers to norm-referenced tests while *also* recognizing that individuals show considerable variability in their own rates of development.

A well-rounded clinical evaluation takes into account each patient's unique deficits and strengths. Treatment is then aimed at reducing or eliminating those deficits while taking advantage of the strengths. Treatment is based on understanding norm references for setting goals but also appreciating each patient's individual variation from those goals. As in the cases of Joe and Gail, treatment takes into consideration patients' existing performance levels, the expected developmental tasks for their ages, and their own personal skills, histories, and desires. This latter category may deviate from developmental tasks expected of them, and the astute therapist will take this into account.

Some treatment strategies are also based on developmental approaches, where it is recognized that patients with similar diseases may go through similar stages of recovery. However, at the same time that therapists acknowledge similar patterns of recovery, they also recognize the individual expectations and goals of each patient. Treatment plans are geared toward age-related goals (developmental tasks) while also considering personalized, individual patterns of life and culture.

## Individuals as Special and Unique

Havighurst's (23) concept of developmental tasks acknowledges the significance of culture on task attainment. Several factors that affect our own and our patients' attitudes, approaches, and motivation toward changing behavior over time are listed in Table 3.3. Human beings are variable, and although psychologists develop general principles of behavior, they also recognize that all people, at all times, will not obey those principles. People differ in skills, abilities, appearance, culture, history, genetics, home environment, religion, child-rearing practices, nutritional habits, and interests. People vary within families and across cultures and nations; they also vary in relation to such factors as exposure to diseases and environmental hazards and susceptibility to accidents and trauma.

Thus, different but complementary views of life are necessary. Developmental stages and developmental tasks, which are based on norms, are

Table 3.3
**Factors Causing Individual Variations in Rate of Development**[a]

Innate sex differences (i.e., male or female)
Glandular function or dysfunction
Adequacy of nutrition
Genetic endowment
Rate of intellectual development
Health status
Exposure to fresh air or pollutants
Birth order of child in family in relation to siblings
Level of motivation or drive
Parents' attitudes: encouragement or interest

[a] Adapted from Kaluger G, Kaluger MF: *Human Development. The Span of Life.* St. Louis, CV Mosby, 1974.

useful for reference points, especially in performing evaluations and in setting goals for treatment. But in everyday dealings with people and especially during clinical practice, we must recognize and respect each individual's special history, skills, and interests. Related to this is each person's desire to be unique and his or her right to be treated that way.

### General Principles of Development

Individual differences must also be considered when referring to general principles of development. These general principles are usually used to describe the processes of early growth and development and include such descriptors as "cumulative, sequential, orderly, predictable, directional, and progressive" (4, 15, 29, 49). Such terms reflect a view of life as an orderly, predictable accumulation of skills leading to a progression in one direction (upward) from each stage to the next in the hierarchical sequence of life (Fig. 3.7).

These principles, it should be noted, fit within conventional views of life as a progressive series of stages. Contemporary views of development integrate the stage perspectives with new approaches that emphasize the interaction of organisms and their environments across the life span. The remainder of this chapter will discuss how these new interactional and ecological views are changing our perspectives of human development and also how well these views overlap with our clinical views of human occupation.

## Models of Development

A model is a symbolic representation used to explain an idea (47). Figure 3.3 represents a typical, historical model of the life cycle. In this traditional model, the first two decades, referred to as "development," are characterized by cumulative expansion. Later life, aging, is literally downhill, represented by the mirror image of development, which is decline.

Achenbach (50) points out that because of this type of model of the life cycle, aging has typically been posed as a mirror image of development. Thus, the study of development has typically looked at increments in behavior, such as the infant obtaining motor skills, the child acquiring self-care skills, and the adolescent acquiring an independent sense of identity.

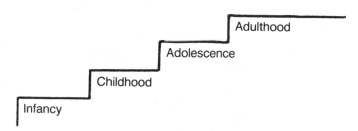

**Figure 3.7.** The life cycle as progressive, sequential, and cumulative.

In contrast, common models of adult behavior have inaccurately but traditionally been decremental in nature, i.e., depicting the changes that accompany aging in the direction of slowness, reduced behavior, reduced acquisition of skills, less performance, and reduced independence (51).

As scientists are finding, this is an overgeneralized and inaccurate profile of the life span. In promoting an interactional view of development, Magnussen and Allen (51) explain how such general, unidirectional descriptions are inaccurate. In contrast, the life-span view

> does not imply any particular direction in the changes we are looking at, nor does it suggest that people improve with age or that they degenerate. It is far too early in the history of adult psychology to offer precise descriptions of how aging and maturity progress. But it is probably safe to say that simple, one-directional models of development have not been supported by research. People don't always appear to become more mature, complex, discriminating, and intelligent with age, nor is there any simple model of degeneration. (51, p. 15)

Thus the model of development/aging as mirror images characterized by expansion/decline is not a representative appraisal of the actual functional changes that occur during life. Emerging life-span views of development are altering these traditional orientations toward development and aging. Now, rather than referring to opposing (mirror image) processes, the concepts of aging and development are often used synonymously to refer to one process—change across the life span.

## Contemporary Models of Interaction in Human Development

Life-span models recognize that many different approaches must be used to accurately portray the multitude of operations taking place during human development. Baltes, (21) for example, has proposed a multidirectional model (Fig. 3.8) to account for the many levels of human behavioral change over time. Here, different skills and developmental tasks are represented by the varying curves. They illustrate that different skills may start and peak at different times. Some may not peak at all; they may, for example, plateau, gradually increase and never decline, or gradually decline.

Such multidirectional, life-span models of development acknowledge the many interacting systems within people and the variability of the onset and duration of the development of each of these systems over time. Baltes (21, 52) refers to this as *intraindividual variability*. This concept should not be confused with the concept of individual differences, which has to do with variations that occur in rates of development *between* people. Intraindividual differences deal with varying degrees of change in the many different systems *within* individuals.

For example, Figure 3.3 illustrates how only one aspect of development, reproductivity, was generalized to portray the entire life cycle. Life-span researchers now recognize that many other systems and abilities have their own cycles that may or may not parallel reproductivity. Speech and lan-

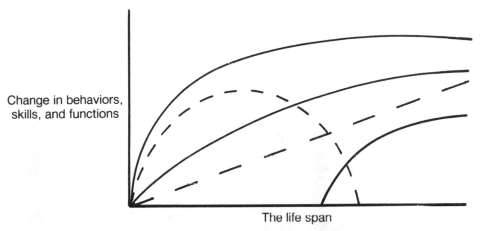

**Figure 3.8.** Over the course of the life span, different abilities accelerate, peak, and possibly decline at different rates. (Modified from Baltes PB: Life-span developmental psychology: some converging observations on history and theory. In Baltes PB, Brim OG (eds): *Life-Span Development and Behavior.* New York, Academic Press, 1979, p 264.

guage skills, prehension, gross motor abilities, interpersonal interactions, or moral reasoning may all express their own developmental cycles independent of the reproductive life cycle.

## Contemporary Systems Models of Development

Baltes' recognition of different developmental systems is not unique. Systems' views, which are based on systems theory that will be described in chapter 6, are common to a science of human development as well as clinical practice. For example, a systems model of human occupation offered by Kielhofner and Burke (53) depicts many different, ongoing levels of interaction within individuals. These levels are perceived in constant flux and interaction with other systems inside and outside of the person (Fig. 3.9). In this model, the internal subsystems of the person are shown in interaction with the external environment. Kielhofner and Burke note that, all together, these interacting internal and environmental systems result in occupational behavior, and the components of this system work together and change over time during the course of human development. "Because the subsystems are part of an open system, they are not static structures determining behavior. Rather they are patterns of organization manifest in a living system and they change throughout the history of that system's existence" (53, p. 35).

Interactional views in the field of human development have elaborated the systems' view. For example, Magnussen and Allen (51) report on the variety of systems that must be examined to understand human development:

Within the person-environment system itself, the person is, of course, one

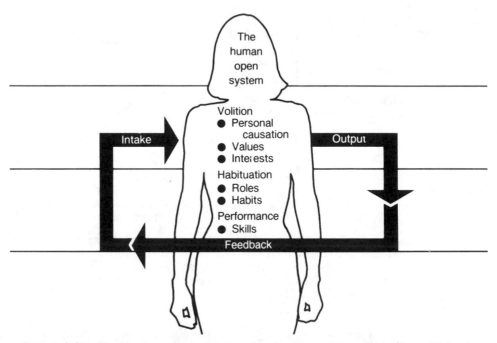

**Figure 3.9.** The open system representing human occupation. (From Kielhofner G, Burke JP: Components and determinants of human occupation. In Kielhofner G (ed): *A Model of Human Occupation. Theory and Application.* Baltimore, MD, William & Wilkins, 1985, p 35. Reprinted with permission of publisher.)

system. The person is also composed of a set of subsystems that operate at another level of organization that is: (*a*) a mediating system (including world conceptions and self-conception systems); (*b*) a biological-physiological system; and (*c*) an action-reaction system ... Beyond the level of the person-environment interaction are other larger and more encompassing systems as well, such as the family, the small group, the community, and so on. Thus, other open systems exist both above and below the level of the person-environment interaction. (51, p. 8)

The study of human development is, indeed, a study of multiple interactions.

### Contemporary Ecological Views of Development

Most of these interactional views of development are also *ecological* in nature, i.e., they refer to changing organism-environmental relations over time. They view the person and the environment as "being interconnected in a dynamic process of mutual influence and change" (52, p 8). One widely recognized and respected ecological approach to development, which is discussed further in later chapters, has been proposed by Bronfenbrenner (54).

Horowitz (55) has successfully developed a three-dimensional model (Fig. 3.10) to account for organism-environment interchange. She explains how organism-environmental interactions must be taken into account when

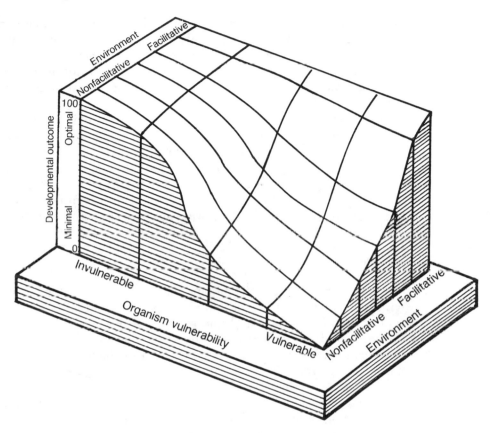

**Figure 3.10.** A three-dimensional model of organism environment relationships and developmental outcome. (From Horowitz FD: Toward a model of early infant development. In Brown CC (ed): *Infants at Risk: Assessment and Intervention.* Skillman, NJ, Johnson & Johnson Baby Products Co, 1981, p. 32. Reproduced with permission from publisher.)

assessing or predicting an infant's potential for continued development (i.e., *developmental outcome*):

> An evaluation of only the infant or only the environment will never net us an understanding of human development, which is probably the most complex phenomenon on this planet. An accurate model of development must include variables associated with the organism as well as variables associated with the environment, and it should depict developmental outcome as resulting from interactions within and across these domains.

The three dimensions accounted for in the model in Figure 3.10 are (*a*) the organism's vulnerability, (*b*) the environment, and (*c*) the outcome of the child's development over time. This model equally applies to development across the life span. For example, the organism can be classified according to the continuum of vulnerability-invulnerability (Fig. 3.10). The environment can be classified according to the dimension of facilitative-nonfacilitative; and developmental outcome can be classified according to the continuum of minimal-optimal. Optimal development is promoted in an organism that is invulnerable and exposed to a facilitative environment, whereas developmental outcome is progressively diminished as the organism becomes increasingly vulnerable and the environment increasingly nonfacilitative.

What makes Horowitz' model dimensional and flexible is the recognition that some nonvulnerable organisms can have good developmental outcome even though they are exposed to nonfacilitative environments. Likewise, some vulnerable organisms can have good outcome when exposed to facilitative environments. Such a model reinforces the contemporary recognition that human development occurs as a result of a complex interaction between multiple internal and external systems changing over time.

# CHAPTER SUMMARY

Contemporary models of development account for the wide array of abilities within humans and the fact that abilities vary from person to person. This reinforces the image of multiple systems within people and the fact that the study of development is a study of interaction (i.e., of forces within an individual and between an individual and his or her environment). Thus, while developmental stages and developmental tasks are important as guidelines for normal development, they must be considered flexibly as they tend to portray development linearly rather than interactively.

In addition, while developmental stages are useful, they must be interpreted within a context. That context first of all recognizes the wide variability of individual differences in human functions and development; secondly, it recognizes the significance of interacting systems in development. Life, it is found, cannot be characterized by a simple linear progression upward nor a simple expansion and decline. Instead, life consists of an ongoing series of advances and retreats, times of growth and regression that may be explored in different systems of an individual's development as well as, cumulatively, across time (56). While we do not yet know about all of these changes during life, especially those occurring beyond adolescence, our understanding of life-span development is expanding rapidly. The actual methods used to obtain this information and to explore the enormous developmental changes during life are examined in the next chapter.

## References

1. Boorstin DJ: *The Discoverers. A History of Man's Search to Know His World and Himself.* New York, Vintage, 1985.
2. Gould SJ: *The Mismeasure of Man.* New York, WW Norton, 1981.
3. Aries P: *Centuries of Childhood.* New York, Vintage, 1962.
4. Gardner H: *Developmental Psychology.* Boston, Little, Brown & Co, 1978.
5. Yogman MW: Pediatric concerns in the perinatal period: the need for interdisciplinary research. In Bloom K (ed): *Prospective Issues in Infancy Research.* Norwood, NJ, Erlbaum, 1981.

6. Butler JA, Starfield B, Stenmark S: Child health policy. In Stevenson HW, Siegel AE (eds): *Child Development Research and Social Policy.* Chicago, University of Chicago, 1984, vol 1 pp 110–188.
7. Mussen PH, Conger JJ, Kagan J: *Child Development and Personality,* ed 5. New York, Harper & Row, 1979.
8. Hall GS: *Adolescence. Its Psychology and its Relations to Physiology, Anthropology, Sociology, Sex, Crime, Religion and Education.* New York, Appleton, 1904, vols 1 and 2.
9. Hall GS: *Senescence: The Last Half of Life* New York, Appleton, 1922.
10. Watson RI: *The Great Psychologists. From Aristotle to Freud.* Philadelphia, JB Lippincott, 1963.
11. Lerner RM: *Concepts and Theories of Human Development.* Reading, MA, Addison-Wesley, 1976.
12. Looft WR: The evolution of developmental psychology. *Hum Dev* 15: 187–201, 1972.
13. Keniston K: *Youth and Dissent.* New York, Harcourt-Brace Jovanovich, 1971.
14. Buhler C: The general structure of the human life cycle. In Buhler C, Massarick F (eds): *The Course of Human Life.* New York, Springer, 1968, ch 1.
15. Kimmel DC: *Adulthood and Aging.* New York, John Wiley, 1974.
16. Labouvie EW: Issues in life-span development. In Wolman BB (ed): *Handbook of Developmental Psychology.* Englewood Cliffs, NJ, Prentice-Hall, 1982.
17. Bayley N: The life span as a frame of reference in psychological research. In Charles DC, Looft WR (eds): *Readings in Psychological Development Through Life.* New York, Holt, Rinehart & Winston, 1973, pp 5–17.
18. Baltes PB: Life-span developmental psychology: some converging observations on history and theory. In Baltes PB, Brim OG, (eds): *Life-span Development and Behavior.* New York, Academic Press, 1979, vol 2.
19. Hollingworth HL (cited by Lerner): *Mental Growth and Decline: A Survey of Developmental Pscyhology.* New York, Appleton, 1927.
20. Lerner RM, Hultsch DF: *Human Development. A Life-Span Perspective.* New York, McGraw-Hill, 1983.
21. Baltes PB, Willis SL: Toward psychological theories of aging and development. In Birren JE, Schaie KW (eds): *Handbook of the Psychology of Aging.* New York, Van Nostrand Reinhold, 1977, pp 128–154.
22. Mussen PH, Conger JJ, Kagan J, Geiwitz J: *Psychological Development: A Life-Span Approach.* New York, Harper & Row, 1979.
23. Havighurst RJ: *Developmental Tasks and Education,* ed 3. New York, David McKay, 1972.
24. Schanzenbacher KE: Diagnostic problems in pediatrics. In Clark PN, Allen AS (eds) *Occupational Therapy for Children.* St. Louis, CV Mosby, 1985, pp 78–113.
25. Klaus MH, Kennell JH: *Parent-Infant Bonding,* ed 2. St. Louis, CV Mosby, 1982.
26. Milani-Comparetti A: Pattern analysis of normal and abnormal development: The fetus, the newborn, the child. In Slaton DS (ed): *Development of Movement in Infancy.* Chapel Hill, NC, University of North Carolina, Division of Physical Therapy, 1981.
27. Denenberg VH: Learning. In Denenberg VH (ed): *The Development of Behavior.* Stamford, CT, Sinauer Associates, 1972, p 195.
28. Mead M: *Coming of Age in Samoa.* New York, American Museum of Natural History, 1973.
29. Schell RE, Hall E: *Developmental Psychology Today,* ed 2. New York, Random House, 1979.
30. Havighurst RJ: History of Developmental Psychology: socialization and personality development through the life span. In Baltes PB, Schaie KW (eds): *Life-Span Developmental Psychology. Personality and Socialization.* New York, Academic Press, 1973, pp 3–24.
31. Vaillant G: *Adaptation to Life.* Boston, Little Brown, 1977.
32. Neugarten BL: Adult personality: a developmental view. In Charles DC, Looft WR (eds): *Readings in Psychological Development Through Life.* New York, Holt, Rinehart, & Winston, 1973, pp 356–366.
33. Fonda J: *Women Coming of Age.* New York, Simon and Schuster, 1984.
34. Butler RN, Lewis MI: *Aging and Mental Health.* St. Louis, CV Mosby, 1982.
35. Botwinick J: *Aging and Behavior,* ed 2. New York, Springer, 1978.
36. Zubin J: Foundations of gerontology: history, training and methodology. In Eisdorfer C,

Powell Lawton M (eds): *The Psychology of Adult Development and Aging*. Washington, DC, American Psychological Association, 1973, pp 3–10.

37. Banus BS: Development and occupation. In Banus BS, Kent CA, Norton YS, Sukiennicki DR, Becker ML (eds): *The Developmental Therapist*, ed 2. Thorofare, NJ, Charles B Slack, 1979, pp 3–20.

38. American Journal of Occupational Therapy: Occupational therapy: its definition and functions. *Am J Occup Ther*, 26: 204–205, 1972.

39. Kaluger G, Kaluger MF: *Human Development. The Span of Life*. St. Louis, CV Mosby, 1974.

40. Clark PN: Human development through occupation: A philosophy and conceptual model for practice, part 2. *Am J Occup Ther* 33: 577–585, 1979.

41. Llorens LA: *Application of a Developmental Theory for Health and Rehabilitation*. Rockville, MD, American Occupational Therapy Association, 1976.

42. Banus BS: *The Developmental Therapist*. Thorofare, NJ, Charles B Slack, 1971.

43. Nelson DL: *Children with Autism and Other Pervasive Disorders of Development and Behavior: Therapy Through Activities*. Thorofare, NJ, Charles B Slack, 1984.

44. Tyler NB, Chandler LS: The developmental therapists: the occupational therapist and physical therapist. In Allen KE, Hold VA, Schiefelbusch RL (eds): *Early Intervention—A Team Approach*. Baltimore, University Park Press, 1978, pp 169–198.

45. Krusen FH: The scope of physical medicine and rehabilitation. In Krusen FH (ed): *Handbook of Physical Medicine and Rehabilitation*, ed 2. Philadelphia, WB Saunders, 1971, pp 1–13.

46. Llorens, LA: Theoretical conceptualizations of occupational therapy: 1960–1982. *Occup Ther Mental Health* 4(2): 1–14, 1984.

47. Reed KL: *Models of Practice in Occupational Therapy*. Baltimore, Williams & Wilkins, 1984.

48. Mosey AC: *Three Frames of Reference for Mental Health*. Thorofare, NJ, Charles B Slack, 1970.

49. Daub MM: Human development. In Hopkins HL, Smith HD (eds): *Willard and Spackman's Occupational Therapy*, ed 5. Philadelphia, JB Lippincott, 1978, pp 29–57.

50. Achenbach TM: *Research in Developmental Psychology: Concepts, Strategies, Methods*. New York, Free Press, 1978.

51. Magnusson D, Allen VL: An interactional perspective for human development. In Magnusson D, Allen VL (eds): *Human Development: An Interactional Perspective*. New York, Academic Press, 1983, pp 3–31.

52. Baltes PB, Goulet LR: Status and issues of a life-span developmental psychology. In Goulet LR, Baltes PB (eds): *Life-Span Developmental Psychology. Research and Theory*. New York, Academic Press, 1970, pp 3–21.

53. Kielhofner G, Burke JP: Components and determinants of human occupation. In Kielhofner G (ed): *A Model of Human Occupation. Theory and Application*. Baltimore, Williams & Wilkins, 1985, pp 3–36.

54. Bronfenbrenner U: *The Ecology of Human Development*. Cambridge, MA, Harvard University Press, 1979.

55. Horowitz FD: Toward a model of early infant development. In Brown CC (ed): *Infants at Risk: Assessment and Intervention An Update for Health Care Professionals and Parents*. Skillman, NJ, Johnson & Johnson, 1981.

56. Bever TG (ed): *Regressions in Mental Development: Basic Phenomena and Theories*. Hillsdale, NJ, Lawrence Erlbaum, 1982.

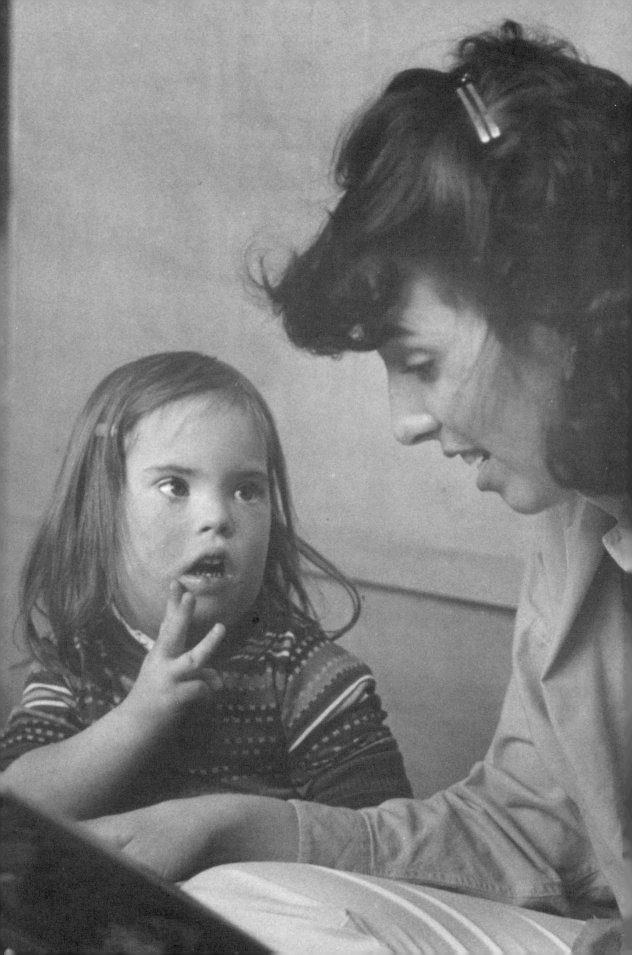

# 2

# RESEARCH AND THEORY

# Section 2

Section 2 covers an area that is receiving considerable interest and attention in the fields of physical and occupational therapy—theory and research. In chapter 4, basic research methods, approaches, and data-gathering instruments used in the study of development are described. Research is approached as structured problem solving, and the various research methods are fully illustrated as they are applied in a clinic setting. Chapters 5 and 6 contain in-depth discussions of traditional and contemporary theoretical approaches to the study of behavior as it changes over time. Each theory is examined in terms of its application and in terms of parallel concepts in the fields of physical and occupational therapy. Of particular interest to many therapists will be the discussions of the recently evolving holistic and systems approaches in clinical and developmental study as well as an evaluation of recapitulation theory, which forms the basis for sensorimotor approaches to treatment in both physical and occupational therapy professions.

# 4

# Reasons and Methods for Developmental Study

The last chapter pointed out how our knowledge of the different stages of life has changed over the past century. For example, conventional models of development portrayed adulthood as a time when skills and abilities peaked, leveled off, and then declined with age. Now, increased research into aging and adulthood has forced us to generate new, more dynamic models that accommodate developmental change over the life span. Similarly, traditional views of infancy have changed over the years, as we acquire more information about the active, participatory role that infants play in family interactions.

Part of the reason for the gradual changes in our understanding of human capabilities is due to the fact that over the years scientific investigators have used varied approaches in their studies. They are now looking at different aspects of development, and they are now asking different questions than they did 50, or even 20, years ago.

These approaches, the questions asked about development, and the methods used to ask these questions are the topics of this chapter. Research, which is an area of increasing importance in both fields of occupational therapy and physical therapy, is discussed here. This chapter will look at the different reasons for studying development, the purpose of scientific problem solving, and different approaches or methods that are used to ask questions in the field of development.

# REASONS FOR STUDYING HUMAN DEVELOPMENT

The purposes for studying human development are to describe, explain, and predict how human behavior changes over time. As was pointed out previously in this book, early in this century, scientists investigating development concentrated primarily on description. Therefore, since research focused primarily on the first two decades of life, most descriptive accounts were of childhood.

As more and more descriptive data were collected, scientists gradually began to offer explanations and make predictions about behavior. Models and theories were constructed and were tested for their adequacy in explaining developmental processes (1).

Models are used for offering explanations (2). They are symbolic representations of ideas, like the various models of the life cycle and of human occupation that were described in the last chapter. Theories are broader in scope than models. Theories organize large bodies of information, and they are used to generate predictions. For example, in the next chapter, we will discuss the major theories that have organized the field of human development. One of these theories, the theory of cognitive development, explains how individuals obtain intellectual capacities and how those abilities change from infancy to adulthood. This theory not only offers explanations about cognitive development, but it makes predictions about how development occurs over time.

The current role of research in cognitive development is to test these predictions and thereby to determine the accuracy of the theory. Theories are not fact. They organize information, and they change. Research tests theories and offers revisions when the theories must be changed.

As time passed in the study of human development, the reasons for scientific investigation changed. Once many behaviors had been described in sufficient detail, scientists began to generate models and theories to explain or predict behavior. This represented a shift from normative to systematic reasons for studying development.

# Description

Normative study in development has to do with gathering information that describes average (or normal) performance for certain age groups. The information that is collected is called *normative data*. These data are typically obtained by observing or testing a very large group of subjects and recording the age at which each subject accomplishes the specific skill(s) of interest.

---

### Gathering Normative Data

Suppose you are interested in knowing when children typically start to walk and can finally walk with only one hand used for support. There are several methods you can use to find out when this skill develops. You can, for example, observe many infants closely until they start pulling up, start to walk with both hands supported, and then eventually walk with only one hand supported. Another, easier method is to survey a group of parents, asking them to record when their infant first starts to walk with only one hand for support.

To conduct a good normative study, you should collect data from a very large group of subjects. But for this study and for illustration purposes, we will look at 16, which is a manageable number of subjects. You decide that it will be easier for you to survey parents than to sit and wait for infants to begin walking. So, you ask 16 parents to record the age of their infants when they began to walk and were supported by only one hand.

After surveying each parent, you end up with 16 replies that you list on a piece of paper (Table 4.1) organized chronologically, according to the month each infant walked while supported with only one hand. From examining your list in Table 4.1, you conclude that, at least from the families you sampled in your study, the average age the parents reported their infants' walking with only one hand supported was 12 months.

How did you come to this conclusion? There are two ways. First of all, you probably just "eyeballed" the data, i.e., you looked at your list in Table 4.1 and saw that age 12 months was the most frequent response and that it was located right in the center of the table, with an even number of data points on each side. Another way to confirm your conclusion is to actually compute an average, i.e., add up all of the months (Total = 192) and divide by 16, the number of subjects you sampled. The result is 12 months.

Given the fact that you only sampled 16 subjects, visual examination was a simple method for looking at the data you collected and coming to some

**Table 4.1**
**Ages of Children When Parents Reported Their Walking with One Hand Supported**[a]

| Subject No. (Child) | Age (in months) |
|---|---|
| 1 | 10½ |
| 2 | 11 |
| 3 | 11 |
| 4 | 11½ |
| 5 | 11½ |
| 6 | 11½ |
| 7 | 12 |
| 8 | 12 |
| 9 | 12 |
| 10 | 12 |
| 11 | 12½ |
| 12 | 12½ |
| 13 | 12½ |
| 14 | 13 |
| 15 | 13 |
| 16 | 13½ |
| | Total months = 192 |
| | Average = 12 |

[a] Data from example discussed in text.

**Table 4.2**
**Frequency of Children Whose Parents Reported Their Walking with One Hand Supported**[a]

| Frequency | Age (in months) |
|---|---|
| ☇ | 10½ |
| ☇☇ | 11 |
| ☇☇☇ | 11½ |
| ☇☇☇☇ | 12 |
| ☇☇☇ | 12½ |
| ☇☇ | 13 |
| ☇ | 13½ |

[a] Data from Table 4.1, example discussed in text.

conclusion. However, for your average to be truly meaningful as normative data, it should be representative of all infants.

*Frequency Distribution*

When you collect data about hundreds of subjects, it becomes a lot less easy to accurately visually inspect your results. So you need other ways to organize and condense your data in order to make sense of it yourself and to explain it to others. One way to organize the information in Table 4.1 is to count the number of infants in each age category. For example, one youngster walked with support at 10½ months, two walked at 11 months, three at 11½ months, four at 12 months, etc. If this were to be tabulated, it would provide a distribution of the number, or frequency, of responses for each age category. This results in a *frequency distribution* such as that depicted in Table 4.2.

Frequency distributions are easy to graph, and graphs provide easy means for summarizing information and for communicating it to others. For example, if you were to take the left column in Table 4.2 and turn it 90° to the left, as we have done in Figure 4.1, it would end up as a graphed frequency distribution. This can easily be made into a bar graph, as in Figure 4.1, by making columns for each age category. Visual inspection of the bar graph illustrates that the average age when your sample of infants began to walk with one hand supported is 12 months.

## Normal Curve

Normative data are often described by graphs such as that in Figure 4.1. If you were to draw a smooth line over the frequencies listed on the graph, you would end up with a curve that in this case is *symmetrical* (i.e., equal on both sides) and gradually slopes down on each side (Fig. 4.2). This is similar to what is called a *bell-shaped/normal curve*, and is the typical shape for normative data because most scores cluster around the central point (*the norm*) with fewer and fewer scores tapering off at each end. Figure 4.2 illustrates, just as your data from Table 4.1 indicated, that the average is at the center of the curve with other scores above and below the average.

We know from practice and from theory that the scores would be distributed as a normal curve if we were to sample the age that *every* infant walked supported by only one hand. This is characteristic of data that are obtained from an entire population and is typical for all traits, e.g., IQ, height, weight, motor performance, or grip strength.

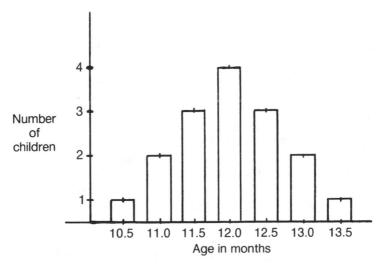

**Figure 4.1.**   A bar graph based on data presented in Tables 4.1 and 4.2, illustrating that the children discussed in the text started walking with one hand supported between 10½ and 13½ months of age. The average age when the children started to walk was 12 months.

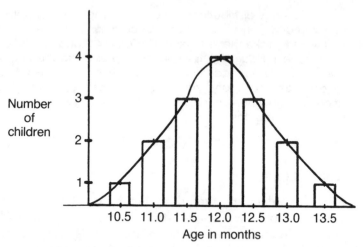

**Figure 4.2.** If a line is drawn through the bar graph in Figure 4.1, the result is a symmetrical curve that resembles the normal or bell-shaped curve in Figure 4.3.

## Descriptive Statistics

With a normal curve, the centralmost scores represent the *norm*, or average, and are the most common in the frequency distribution. Scores that fall in the middle of a normal curve are referred to as *measures of central tendency*. These are statistics that are used for describing data.

Descriptive statistics include three measures of central tendency. One of these measures is the average, also called the norm or the *mean*. The *median* is the most central data point, i.e., the one occurring right in the middle of the frequency distribution. The *mode* is the most frequently occurring data point.

In our study of the 16 youngsters, the mean, median, and mode are all the same—12 months. In other data collections and distributions, however, these measures may vary. For a further discussion of these terms, the student may want to consult an elementary statistics or research book.

Descriptive statistics and normative data are used in many fields other than human development. We encounter descriptive statistics in everyday living. For example, weather reports give us the average rainfall or snowfall or the average temperature for certain cities and towns at every day of the year. Census reports tell us the average age or life expectancy of individuals, and economic forecasts give us median incomes for U.S. citizens. Insurance tables show us the average life expectancy for adults in the United States, and advertisers make common use of descriptive statistics in monitoring the numbers and characteristics of consumers who use their products.

Similarly, developmental specialists can tell us the average age that children will sit up, roll over, ride a bicycle, tie their shoes, kick a ball, and walk up and down stairs without holding onto a railing or someone else's hand for support. All this information is available because scientists have concentrated on gathering descriptive data regarding human performance skills at many different ages.

## Gesell's Normative Data

Such information is valuable in the field of human development. As has been repeatedly stated in this book, the first goal in understanding behavior is describing when and under what circumstances it occurs. Thus, scientists in the field of human development initially gathered normative data regarding an enormous complex of human behaviors. One of these scientists, who was discussed in the last chapter, was Arnold Gesell. He observed and filmed large numbers of children of various ages as they performed a large variety of tasks. Then, Gesell and his colleagues assembled the findings in charts much like the one introduced in Table 4.1. Such charts provide age-related categories of behavior that illustrate the average age that most children can perform specific behaviors as well as the normal developmental sequence of behaviors.

An example of some behavioral items from one of Gesell's charts, which was published in 1938 (3), is included in Table 4.3. Gesell's chart differs somewhat from Table 4.1. Instead of listing the number of children performing one behavior, Gesell listed in the right-hand columns the percent of children who accurately passed each of many behavioral items on which they were tested.

## Age-Referenced Data for Evaluations

One of the significant uses of normative data is for evaluation purposes. Evaluations are administered by therapists when we want to determine whether an individual is functioning within normal limits, and if not, how much the individual deviates from what is considered normal. Develop-

**Table 4.3**
**Behavior Norms for Standing and Walking Behavior[a]**

| Behavior | Age In Weeks | | | | | | | | | | | | | | |
|---|---|---|---|---|---|---|---|---|---|---|---|---|---|---|---|
| | 4 | 6 | 8 | 12 | 16 | 20 | 24 | 28 | 32 | 36 | 40 | 44 | 48 | 52 | 56 |
| Head sags | 73 | 52 | 36 | 4 | 0 | | | | | | | | | | |
| Head erect (momentarily) | 30 | 14 | 8 | 0 | 0 | | | | | | | | | | |
| Head steadily erect | 3 | 14 | 24 | 48 | 62 | | | | | | | | | | |
| Supports no weight | 77 | 62 | 41 | 43 | 32 | 28 | 15 | 14 | 3 | 0 | 3 | | | | |
| Supports entire weight | 0 | 0 | 4 | 8 | 2 | 28 | 28 | 44 | 55 | 91 | 94 | 97 | 97 | 98 | 100 |
| Stands on toes | 3 | 10 | 26 | 38 | 38 | 28 | 31 | 31 | 50 | 18 | 29 | 6 | 7 | | |
| Lifts foot | 23 | 10 | 41 | 60 | 59 | 34 | 41 | 31 | 28 | 28 | 40 | 52 | 86 | 81 | 88 |
| Hand-supported balance inadequate | | | | | | | | | | 34 | 64 | 45 | 38 | 31 | 12 |
| Stands independently | | | | | | | | | | | 9 | 10 | 22 | 33 | 68 |
| Walks using support | | | | | | | | | | 3 | 6 | 16 | 47 | 63 | 80 |
| Walks independently | | | | | | | | | | | | 3 | 3 | 26 | 44 |

[a] Adapted from Gesell A, Thompson H: *The Psychology of Early Growth Including Norms of Infant Behavior and A Method of Genetic Analysis.* New York, Macmillan, 1938.

mental scales such as Gesell's give us norm references, that is they enable us to refer to a standard of comparison for evaluation purposes.

If, for example, we know that 63% of infants walk using support at 52 weeks of age, and 80% walk using support at 56 weeks of age (Table 4.3), then we can compare other children with that norm. Thus, if we assess a child and discover that the child is not walking with support by 96 weeks (2 years) of age, there may be cause for concern. But, suppose we see a child who is not walking with support by 62 weeks of age. Should we be concerned? How do we measure how much one can deviate from the norms and still be considered to be developing within normal limits? To answer this question we need some method of gauging how far one can deviate from the norm and still be considered average.

## Measures of Variability

Graphs or tables of data such as those in Figure 4.1 and Tables 4.1 and 4.2 show us the centralmost scores in addition to showing us how far those scores deviate from the average. These latter scores are termed *measures of variability*. These include the *range*, the *standard deviation*, and the *variance*. The range is an index of how far apart, or dispersed the scores are. A range is easily determined by subtracting the smallest score from the largest score. For example, of the 16 infants whose walking we assessed (Table 4.1), the lowest score was 10½ months and the highest was 13½ months. Thus, the range is 3 months.

The standard deviation and the variance measure how much the scores tend, on the average, to deviate from the mean. Logically, this could be determined by counting how much each score deviates from the norm, adding those scores, and obtaining an average deviation score. There are problems with this method, however. For example, as the second column in Table 4.4 illustrates, if you were to measure the average deviation of the scores in Table 4.1, you would end up with a score of 0. This is the case with a normal distribution, because the scores less than the mean will always offset the scores greater than the mean.

To accommodate this problem in determining the average deviation from the mean, we square the deviation scores first, then add them up, and then compute an average. This is the average square deviation, also called the variance. The standard deviation is the square root of the variance.

Knowing the standard deviation enables us to determine how much scores differ from the measures of central tendency. For example, Figure 4.3 illustrates a normal distribution curve and the percentage of scores that fall within 1, 2, 3, and 4 standard deviations from the mean. This illustration is useful in enabling us to judge how close or how far away a particular score is from average. Knowing that enables us to evaluate whether a subject or a patient's performance is normal, and if not, how much it deviates from normal.

Another use of age-reference data such as Gesell's (Table 4.3) is to illustrate the normal progression of development. Such information is useful

Table 4.4
Computing the Average Deviation of Scores from the Mean[a]

| Score − Mean | Deviation | Deviation Squared |
|---|---|---|
| 10½  −  12 | −1½ | 2¼ |
| 11   −  12 | −1 | 1 |
| 11   −  12 | −1 | 1 |
| 11½  −  12 | −½ | ¼ |
| 11½  −  12 | −½ | ¼ |
| 11½  −  12 | −½ | ¼ |
| 12   −  12 | 0 | |
| 12   −  12 | 0 | |
| 12   −  12 | 0 | |
| 12   −  12 | 0 | |
| 12½  −  12 | +½ | ¼ |
| 12½  −  12 | +½ | ¼ |
| 12½  −  12 | +½ | ¼ |
| 13   −  12 | +1 | 1 |
| 13   −  12 | +1 | 1 |
| 13½  −  12 | +1½ | 2¼ |
| | Total = 0 | Total = 10.00 (variance) |
| | | $\sqrt{10.00} = 3.162$ (standard deviation) |

[a] Data from Table 4.1.

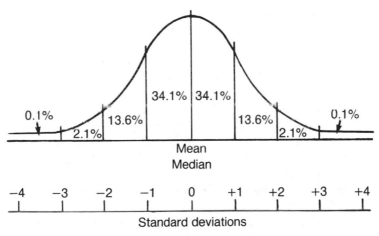

Figure 4.3. Percent of area and standard deviations associated with a normal curve.

to teachers, therapists, parents, and psychologists who work with children and who want to know the normal progression of skill acquisition, i.e., that most infants normally roll over before they sit up, that they sit up before they stand, and that they pull to standing and walk with support before they start ambulating independently. Such information has come from detailed descriptive studies of infant and child development.

## Normative Goals in Physical Therapy and Occupational Therapy Evaluation

Normative aims were common in developmental psychology early in this century, but then the focus of study shifted toward explanation and theory construction (1). In both fields of occupational and physical therapy, normative data are still pursued for many functional skills that are of specific interest to our professions. While we may use normative data that already has been collected, such as that obtained by Gesell and others, we also require age-referenced data for many additional occupational and physical performance skills. This is illustrated by Rose's claim regarding the need for observation and description in physical therapy:

> Historically, the body of knowledge in most science-based disciplines has been developed by initially focusing investigative efforts on observing and describing phenomena of interest. Subsequent to the initial observation and description, refinements were suggested, and eventually, consensus was achieved regarding specific observations. Finally, the observed phenomena were arranged according to systematic rules into classes or groups. I believe we should adopt this model, which has been effective in other disciplines, and commit ourselves to the objectives of observing, describing, and classifying pathokinesiological phenomena that exist in the patients we serve. (4, p. 379).

Some therapists have spent years developing evaluation instruments and obtaining normative data from large groups of subjects that have been sampled from across the United States. Occupational therapist Lucy J. Miller, with the help of many other therapists, devised an instrument to test preschool children's performance skills. The resulting test called the Miller Assessment for Preschoolers (5) was published in 1982, and it includes normative data obtained on a large group of children who were equally represented for sex, age, region of the country, and socioeconomic status (6). Therapists and educators can use this test as a means for determining whether children are functioning appropriately for their age or whether they need further assessment of various developmental skills. Many other therapists have been examining, on smaller scales, age-related changes that occur relative to certain reflexes, prehension skills, grip strength, types of balance, coordination, and many sensory or sensory-motor abilities (e.g., 7–18).

Some normative data used by occupational therapists are obtained by observation and with photography, just as Gesell did. However, instrumentation has become so sophisticated in the past two decades that therapists are using other strategies for precisely measuring a variety of skills and abilities (e.g., 19–23). Along with each new instrument comes a need for more precise and detailed norms.

A goniometer (Fig. 4.4), which is used to measure joint range of motion, is an example of this new instrumentation. There are changes in range of motion that accompany the muscular changes typically seen in the growing and developing neonate and infant. Most range of motion norms, however, have been obtained on adults and cannot be used by pediatric specialists

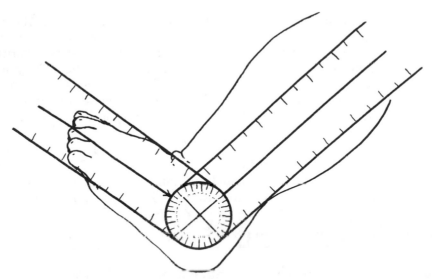

**Figure 4.4.** Goniometers, used to measure joint range of motion, have also been used to obtain norms of neonatal flexibility.

because it is not useful to compare adults', infants', and children's range of motion. Thus, some therapists (e.g., 10) have been collecting norms for range of motion for a variety of movements that are typically examined in neonates and infants. Such a collection of norms may be helpful for therapists who are assessing infants who may have neuromuscular or joint problems that would alter their joint range of motion.

Lately, many test instruments in medicine and in allied health fields have become more sophisticated and technical. Many of these are automated to reduce error in measurement by providing a standard mechanism for making recordings. An example of this is the electric goniometer, which is an automated version that provides readouts of the range of motion as the joint is moved (e.g., 21–23). As these devices become more sophisticated and precise, the norms that are used for comparison purposes in evaluations must also reflect this increase in precision. Thus, as long as new instruments are devised and revised and as new tests are introduced, normative data need to be gathered.

## Explanation and Prediction

Once a large body of descriptive data have become available, then the focus of research can naturally shift away from description and move toward explanation and prediction. This occurred in developmental psychology in the 1930s and 1940s. The descriptive studies of Gesell and others were supplanted by studies with a different orientation. From the mid-1940s on, research in human development became more theoretical and explanatory. Rather than describing the actual behaviors that changed over time, scientists became more systematic. They began to generate theories to account for the various mechanisms and processes of growth and development.

Scientists with *systematic* aims want to test and verify theories. As previously explained, theories organize information. They are constructed, proposed, and disseminated through publications and conference presentations, and then scientifically tested for their accuracy. Theories are tested by experimentation. In fields such as physical therapy, occupational therapy, and human development that are so diverse, many theories are required to organize the relevant information. These various theories will be examined in detail in the next two chapters. The purpose at present is to examine how our knowledge of human development changed as a result of the different kinds of studies that comprise developmental research.

# WHAT IS RESEARCH?

Many people tend to be intimidated by the concept of research, associating it with advanced, esoteric investigation that has no application to the real world. In fact, research has had widespread application in the study and practice of occupational and physical therapy and in the field of human development. This will become much more obvious as we explore the diversity of developmental studies introduced in this textbook.

What exactly is research? Although there are many definitions and many classifications for research, it can be simply defined as "structured problem solving" (24). What determines the actual structure that is used in problem solving varies greatly.

## Classifications of Research

Research may be classified according to several dimensions. Vasta (25) points out that classification schemes may be based upon the purpose of the research, the types of methods used, or the types of subjects that are studied. No one system can include all of the possible combinations of these methods. As Helmstadter (24, p. 27) states, "Almost every author will have worked out his own unique system of classification. It will help if the reader will keep in mind the fact that taxonomies, like theories, are best thought of not as right or wrong but rather as useful or useless for different particular purposes."

Tables 4.5 and 4.6 illustrate two different classification schemes for research. In Table 4.6, various research divisions are arranged along a continuum. According to Helmstadter (24), one division is based on the applicability (or purpose) of the research. This ranges from *basic* to *applied research*. Basic, also called "pure" or "fundamental", research is that which is conducted for its own sake. It is problem solving based on curiosity that is performed in order to gain some basic information, whereas applied research is conducted in order to obtain practical solutions (or applications) to specific problems.

**Table 4.5**
**Nine Basic Methods of Research**[a]

| Method | Purpose |
| --- | --- |
| Historical | To reconstruct the past objectively and accurately, often in relation to the tenability of an hypothesis. |
| Descriptive | To describe systematically a situation or area of interest factually and accurately. |
| Developmental | To investigate patterns and sequences of growth and/or change as a function of time. |
| Case and Field | To study intensively the background, current status, and environmental interactions of a given social unit: an individual, group, institution, or community. |
| Correlational | To investigate the extent to which variations in one factor correspond with variations in one or more factors based on correlation coefficients. |
| Causal-Comparative or "ex post facto" | To investigate possible cause-and-effect relationships by observing some existing consequence and searching back through the data for plausible causal factors. |
| True Experimental | To investigate possible cause-and-effect relationships by exposing one or more experimental groups to one or more treatment conditions and comparing the results with one or more control groups not receiving the treatment (random assignment being essential). |
| Quasi-Experimental | To approximate the conditions of the true experiment in a setting that does not allow the control and/or manipulation of all relevant variables. Researchers must clearly understand what compromises exist in the internal and external validity of their design and proceed within these limitations. |
| Action | To develop new skills or new approaches and to solve problems with direct application to the classroom or other applied setting. |

[a] From Isaac S, Michaels WB: *Handbook in Research and Evaluation.* San Diego, CA, EdITS, 1981, p 42. Reprinted with permission of publisher.

The purpose of *descriptive research* is to describe some behavior, groups of behaviors, or some situation. *Predictive research*, on the other hand, anticipates the future by making predictions about some behavior or circumstance. Descriptive research can also be contrasted with *manipulative research*. In descriptive research, the investigator may actually wait for some environmental change to occur, whereas in manipulative research the investigator deliberately and systematically produces some environmental manipulation and observes its effects on behavior (25).

The next two classifications in Table 4.6, location of study and degree

Table 4.6
Classifications of Research[a]

| Breadth of Appli-cation | Level of Outcome | Location of Study | Degree of Control |
|---|---|---|---|
| Basic | Descriptive | Library | Historical |
| ↓ | ↓ | ↓ | ↓ |
| | | | Descriptive |
| | | Field | ↓ |
| | | | Observational |
| | | ↓ | ↓ |
| Applied | Predictive | Laboratory | Experimental |

[a] Adapted from Helmstadter GC: *Research Concepts in Human Behavior.* New York, Meredith, 1970, p 28.

of control, are interrelated. For example, library research, named so because it is conducted in the library, is associated with very little control. Library research tends to be historical in nature, as the investigator gathers information from previous work. Thus, library or historical researchers tend not to exert control over their research topic but are more controlled by the nature of and the themes that emerge from their investigations. The primary control over the data is through the process of rejecting what the researcher determines irrelevant or by selecting relevant information to include in the final report (24).

*Field research* also takes its name from its location. It is conducted in natural settings ("in the field") as opposed to the laboratory. Field research is naturalistic, i.e., primarily descriptive or observational so as to interfere as little as possible with the natural conditions of the subjects under study. Thus, the amount of control exerted by the investigator tends to be limited to selecting the time and place for making observations (24). This is in contrast to *laboratory* (also termed "manipulative") research where the investigator may exert a considerable amount of control. In laboratories, environmental conditions can be completely regulated in systematic fashion. Factors ranging from diet to temperature, even day-night cycles, can be manipulated according to the goals of the investigator. Laboratory researchers tend to use the most systematic research (i.e., experiments) to solve their problems.

Research is also classified according to the degree of structure that is used during the problem-solving process. These varied structures are also called *research methods.*

# RESEARCH METHODS IN HUMAN DEVELOPMENT

Scientists studying human development have used a variety of research methods. These may vary depending upon the investigators' goals, purposes, resources, and backgrounds. In the history of developmental psychology, researchers' methods changed over time as their interests and purposes changed. Early in this century, scientists such as Gesell *described* large varieties of behavior and *observed* large groups of children in a variety of settings. As the goals of study in human development changed, research methods became more *systematic*, with *experimental* methods more widely used.

# Why Are Research Methods Important to Physical Therapists and Occupational Therapists?

The methods that are commonly used in the study of human development are introduced and described in this chapter. This information is important to clinicians and researchers for several reasons. In the field of developmental psychology, these methods have been widely examined, and researchers found that the types of methods that were used affected the nature of the data that were gathered. This, in turn, often affected the conclusions that were drawn about particular topics. Thus, knowledge of the methods of problem solving often helps an interested research "consumer" to evaluate and understand the limitations and generalizations of final research conclusions. This is important for therapists who are ongoing consumers of research reports about various patient assessments and treatment strategies. As Bohannon and LeVeau (26, p. 45) have noted, "Research is, after all, an ever-growing source of objective information that can serve as an evolutionary source for increasing professional effectiveness."

## Clinical Reasoning and Problem Solving

There are many additional reasons why therapists should become familiar with research methods. Research is a form of problem solving, and there is a diverse array of problems for which physical and occupational therapy practitioners seek solutions. Payton (27), Rogers and Masagatani (28), and Steger (29) note that an important component of clinical practice is problem solving to determine patient's problems and to determine a course of treatment. Rogers and Masagatani (28) note, however, that practitioners are often unable to specify the steps that are involved in clinical reasoning, and Steger (29, p. 52) claims that occupational therapy students "are taught the concepts and principles that they habitually use in the reasoning process, but ... are rarely asked to explain how these concepts and principles are organized ... to arrive at clinical solutions."

Research methods provide an organizational structure that can be used in the clinical reasoning process. Thus, occupational and physical therapists need to be introduced to and become familiar with the varieties and applications of these methods. If they do not gain this knowledge, then, as Steger (29, p. 52) explains, "they may not use or articulate the reasoning process with maximum effectiveness." Solid reasoning skills will help therapists to become more effective in their varied roles in assessment and in treatment.

A report from Rogers and Masagatani (28, p. 196) illustrates the varieties of stages of problem solving necessary for clinical reasoning. Assessment, they note, is the first component of clinical reasoning and, "consists of problem solving under conditions of uncertainty." They note that during the assessment process, therapists go through several stages in order to determine each patient's problems, goals, and prescriptions:

First, the therapist obtains some preliminary information about each patient. This may involve talking with other professionals, reading a report or notes from a referral form, or making a preliminary observation of the patient. Second, based on the information received, the therapist then selects an assessment tool for more formal evaluation of the patient's occupational and physical abilities. Third, the therapist formulates and implements an assessment plan while the patient's problems are clearly defined. Finally, the therapist determines the treatment objectives and goes back and determines whether the selected assessment tool was effective for delineating the problems of the particular patient under study (28).

If the therapist determines that the assessment device was ineffective, then the process may be started all over again. In addition to assessment, the treatment process also poses a different set of problems for clinical practitioners. For example, therapists must decide the kind of treatment that is appropriate for each patient's needs. Then therapy schedules must be organized around a particular therapeutic regimen that meets the needs and schedules of the patient, the therapist, the departmental program, and the institution (28).

A significant problem in the human service field is assessing treatment programs and determining under what conditions and with what patient populations they are most effective. This is a concern of physical and occupational therapists (e.g., 4, 26, 30–40) as well as of other practitioners who follow the progress of patients. As will be illustrated later in this chapter, the process of assessing treatment effectiveness may employ any of the research methods described next.

## Research Focus in Occupational and Physcial Therapy

Research is increasingly being emphasized in both professions of physical and occupational therapy (4, 26, 30–40). Rogers (34, p. 3) reports that research in occupational therapy "has been seen as a vehicle for improving intervention, developing the scientific basis of practice, and establishing occupational therapy as a full profession." Regarding physical therapy, Bohannon and LeVeau (26, p. 45) note: "The heart of a human service profession, such as physical therapy, is the objective of bettering the state of the client. The attainment of this objective requires a priori that the professional be familiar with interventions that are likely to succeed if applied with discernment. Although such interventions may be based on knowledge gained from a number of sources, knowledge that is research-based should be of particular interest to the clinician."

Physical therapy and occupational therapy educators are concerned that students be introduced to research methods and learn how to evaluate research in the context of their clinical courses (26, 35). Becoming familiar with the research methods used in other disciplines, such as developmental psychology, is a step toward addressing some of those concerns.

Finally, historical examinations of the field of developmental psychology note that knowledge in that field changed as the research methods changed. Understanding this trend has enabled researchers to critically evaluate information gained through the use of different methodologies. This same type of process may possibly occur (or has occurred) in clinical fields. For example, Ottenbacher and Short (41) have reported that research methods in occupational therapy seem to have shifted from descriptive to quasi-experimental. Similar shifts also occurred in the field of developmental psychology. Understanding parallels between the growth of our clinical fields and that of other more established academic fields may illuminate the development of health care professions.

# Research Methods

These are the three primary research methods that are commonly introduced in developmental psychology: observational, clinical, and experimental. The following section introduces these methods, discusses the pros and cons of the methods, and describes their applications in gaining knowledge about human development or occupational and physical performance.

## Observation

Observation is the simplest method for studying behavior. It involves looking at and then recording behaviors as they occur naturally in the settings of everyday life (Fig. 4.5). Direct observation involves no prior arrangement between the observer and the subject under investigation, and as such is rarely used in developmental psychology or in clinic settings where some form of intervention normally occurs between the investigator (the therapist) and the subject. Direct observation is typically used in areas where the observer is distanced from the subject under investigation, such as in some sociological studies or in astronomy (42).

As more structure is added, the observational method becomes more systematic. For example, this method includes an unstructured approach called "diary description," and a more systematic approach called "time sampling." Diary description is the oldest method of gathering data in the field of child development. Diaries of children's progress date back to the 1700s, and the most notable diarist was Charles Darwin, who chronicled the early development of his son (25).

### Diary Descriptions

Sometimes called "baby biographies," diary descriptions involve recording what behaviors the writer sees as important. Certainly new behaviors as they emerge are noted, just as parents, nowadays, note when their baby

**Figure 4.5.** The observational method involves looking at and then recording behaviors as they occur naturally in the settings of everyday life. In this photograph, the observer is using earphones that signal the onset and offset of discrete time periods that are used for making observations and recordings. This is the observational method's most sophisticated procedure called *time sampling*.

first talks, first eats solid food, first acquires and then loses a tooth, etc. A diary description can be useful because it is often a very personal, detailed account of one child's many behaviors. As such, it can serve as an excellent illustration of the continuity of behavior (42,43).

### Advantages and Disadvantages of Diary Descriptions

Diary descriptions can be useful for therapists who are learning how to improve their observational skills. Effective practitioners have excellent observational skills that enable them to obtain important clinical information about their patients' reactions during assessment and their progress and responses to treatment. Refined observational skills are obtained with practice; however, novice observers can sharpen their skills by learning some of the basics. For example, students who are initially introduced to the clinic and are asked to keep records of patients' performance are frequently unsure of which behaviors to observe and which records to keep. Fearing that they will miss some relevant details, students often record all of the behaviors they see. Later, after supervisory feedback and with experience, students or novice observers gradually learn how to condense their observations into a meaningful report that can be kept as a part of the patient's records. However, until students become adept at observation, they might benefit from initially making diary descriptions of patients' behaviors.

One of the disadvantages to diary descriptions is that it is a very time-consuming method and results in a large and unwieldy quantity of information that is difficult to communicate to others and to store. Oftentimes, what is recorded in the diary is a reflection of the bias of the observer. An additional disadvantage is that recordings and final interpretations may be biased, unreliable, or inaccurate (42). Thus, other observational methods may be adopted to enable the observer to abbreviate and to objectify the observation process. For example, the therapist or researcher may want to structure a specific amount of time for observations or to focus on certain specific behaviors to be observed.

## Time Sampling

Time sampling is the most advanced and most frequently used subcategory of the observational method (25). It involves recording very specific, previously determined (samples) of behavior within a previously determined time span. This type of method is useful in obtaining information about normal growth and development as well as about patients' performance during therapy (e.g., 44). In time sampling, a specific structure is followed. First of all, the amount (duration) of time that is allotted for making observations and for recording behavior must be determined. In addition, the behaviors of interest must be predetermined and clearly defined. Then, all that is needed is a clock or a watch and a checklist of the behaviors of interest. What follows is an example of how a time-sampling method could be used to obtain practical information in the clinic.

### Time Sampling in the Clinic

A therapist wants to see how many (and for how long) eight of his adolescent clients will participate in a cooperative, social group task. These clients have been attending an after-school program designed to help them learn productive use of their time. All of the students are on probation because of drug-related difficulties. (It is common for such individuals to have shortened attention spans, to tend to be impatient, and to have difficulty delaying gratification. They also may or may not have immature or uncooperative social behaviors.)

Ed, the occupational therapist, is working with this group two afternoons/week as part of a rehabilitative program aimed at helping them develop good occupational (productive time use) skills. Today, the activity on which the students are working is a huge collage which is to be used for a wall decoration. The activity is set up to take 60 minutes, and Ed wants a measureable way of recording whether and how much each youngster actually participates in the task.

As already noted, there are certain procedures that one goes through in setting up a time-sampling method. For example, Ed must define specifically what behaviors are of interest to him and how long and how frequent will be his observation time periods.

He decides that "participation" will be defined as "talking with group members about making the collage or actually working on (gluing, cutting, touching, talking about, or assembling materials related to) the collage." Sitting in the room, walking around the room, singing or talking about other events while not in proximity to the project do *not* qualify as participation. Talking about other events while actually working on the project does qualify as participation, as Ed realizes that many people socialize while also being occupied with a cooperative task.

Ed decides to make a 30-second observation at the beginning of each 3-minute interval. This will enable him to make observations while also participating in the activity. At the scheduled time for making observations, he will look from one client to another, in a particular order, for a total of 30 seconds. Then he can record whether or not each student was participating in the activity. At the end of 60 minutes, Ed will have data that indicate whether, and how often, each client participated in the group task. This objective, quantifiable information will be a profile of the students' participation and can form the basis for subsequent decisions regarding treatment plans or counseling.

## Advantages and Disadvantages of Observational Method

As a general category, the observational method has been widely used and has produced important information in the study of human development. It is a naturalistic approach that takes into consideration the organism and its relationship to the environment. Wright (42) described in 1960 how widely this method had been used and reported numerous applications of this method in child development. He claimed that the widespread use of the observational method would enable scientists to obtain a natural history of the child. Additionally, it provides norms for describing behavior, it can be clinically useful, and it is particularly beneficial in providing information about behaviors that are not easily examined with other methods, such as the experimental method that will be discussed later.

The observational process, however, is limited and is often replaced by other procedures that provide more structure or control, such as the clinical or experimental methods. Thus, the observational method is not as commonly used as it was in the 1920s and 1930s, which Hutt and Hutt (45) term the "heyday of observational study." In general, the observational method went out of vogue because it did not provide the amount of control that psychologists wanted to employ in developing a scientific study of human behavior. Observation is criticized because it is unstructured and time consuming. In many cases, the investigator must wait until the behaviors of interest occur, whereas in experimentation, the investigator can set up and manipulate situations at convenient times. In addition, it is generaly assumed that the observational method is used to generate normative or descriptive data but that the experimental method is necessary for systematic study.

Hutt and Hutt (45) claim that the observational method, just like any other, is useful for generating hypotheses and should be considered complementary to the other research approaches discussed in this chapter. Further, they point out that the lack of control that is characteristic of the observational method is what makes it particularly attractive. It is useful for assessing organisms' adjustment to physical and social environments such as during exploration or play, for examining uncooperative subjects such as young children or some psychiatric patients, and for assessing patient's clinical adjustments. In addition, Ottenbacher (40) notes that observation is particularly useful for assessing such behaviors as activities of daily living (ADL) skills, which are best examined in circumstances that are akin to the patient's home environment and daily routine. Readers interested in greater applications of the observational method in occupational therapy may consult Ottenbacher's text on evaluating clinical change. In it he provides details regarding the use of various categories of the observational method for assessing client changes in response to treatment.

## Clinical or Case Study Method

The clinical method involves a one-to-one interaction with the subject or patient under investigation (Fig. 4.6). This usually includes an in-depth interview that takes place over a period of time. The interview can be either very structured or informal. The clinical method is one of three (the other two are *observation* and *experimentation*) that are most commonly cited as the primary methods in the field of child development. Piaget, the famous cognitive theorist, is attributed with the widespread application of this method in the study of human development.

### *Piaget's Use of the Clinical Method*

Piaget observed and recorded how his own children reacted to a large variety of situations that he set up deliberately to test their reasoning and problem-solving skills. The following is an excerpt from one of Piaget's descriptions of his daughter, Jacqueline.

> Observation 182.—At [1 year, 8 months] Jacqueline has an ivory plate in front of her, pierced by holes of 1–2 mm in diameter. She watches me put the point of a pencil in one of the holes. The pencil remains stuck vertically there and Jacqueline laughs. She grasps the pencil and repeats the operation. Then I hold out another pencil to her but with the unsharpened end directed toward the plate. Jacqueline grasps it but does not turn it over and tries to introduce this end (the pencil is 5 mm in diameter) into each of the three holes in succession. She keeps this up for quite a while even returning to the smallest holes. On this occasion we make three kinds of observations:
> 1. When I return the first pencil to Jacqueline she puts it in the hole correctly at once. When I hand it to her upside down, she turns it over even before making an attempt, thus revealing that she is very capable of understanding the conditions for putting it in ... But if I offer [the second pencil] to her upside down, she does not turn it over and recommences wishing to put it in by the unsharpened end. This behavior pattern remained absolutely constant during thirty attempts, that is to say, Jacqueline never turned the second pencil over whereas she always directed the first one correctly ...

**Figure 4.6.**    The clinical method involves a one to one interaction between the clinician and the client. (Courtesy of March of Dimes Birth Defects Foundation)

2. Several times Jacqueline, seeing the second pencil will not go in, tries to put it in the same hole as the first one. Hence, not only does she try to put it in by the unsharpened end but also she wants to put it into a hole which is already filled by the other pencil . . .

3. At about the thirtieth attempt, Jacqueline suddenly changes methods. She turns the second pencil over as she does the first and no longer tries a single time to put it in by the wrong end. . . . (46, pp. 339–340)

Piaget continues to analyze Jacqueline's problem-solving abilities within a context of his theory of cognitive development. He has effectively used the clinical method to set up situations so that he could systematically examine his children's reasoning abilities, to make and record observations about these skills, and later to use the observations as data in formulating and testing his theory of development. (His theory will be examined in greater depth in chapters 5 and 11.)

### The Case Study Method

The case study method involves an in-depth analysis of an individual's behavior, and as such, is, by definition, the same as the clinical method. Rothney (47, p. 3) reports that "case studies of children were originally devised for use in mental hygiene or social work clinics. To the extent that clinical approaches were organized to portray an individual as a total functioning personality, they reflect what we refer to as a case study." However, the clinical approach, or clinical method is often also used, as is usual in occupational or physical therapy, to refer to the investigation of maladjustment or therapeutic intervention. For example, Sigmund Freud used this method to describe many of his patients. Freud, who is discussed at greath length in chapter 6, made important contributions to the field of developmental psychology and gave us significant insights into adult and child behavior.

Many investigators do not associate the case study solely with the clinical description of patients. Case studies are conducted for various purposes, and their contents will reflect that purpose. For example, a physical therapist, an occupational therapist, and a teacher may be requested to provide case study reports about a student. While all three professionals may include some of the same general biographical information, the focus of each professional will be different. The occupational therapist may be concerned with reporting on the child's functional performance skills; the physical therapist may be concerned with motor performance, physical function, strength and mobility; and the teacher would most likely focus on school grades, attendance, and behavior in the classroom. Rothney reports:

> Case studies have been written to serve many purposes. There are case reports about presidential candidates which appear in book form as election times approach and which tend to extoll the virtues or rationalize the shortcomings of the candidate. Case reports used to show the effectiveness of certain kinds of remedial treatments when they are applied to persons or to show how some therapy is utilized are fairly common. Case studies of individuals describing their struggles for success in a particular cultural setting are employed in studies of societies, and some case studies are simply literary creations by their authors. (47, p. 3)

This book includes several brief case studies of patients, such as those of Joe and Gail in chapter 3 and Jennifer in this chapter. While these are very brief, they illustrate the general form of the case-study method. Lucci (48) has assembled 41 clinical reports in her text, *Occupational Therapy Case Studies*, and others (e.g., 37, 49–51) have either reported on or demonstrated the usefulness of this method in physical therapy or occupational therapy. Ottenbacher (40) reports:

> In the rehabititation fields, case studies are typically conducted to determine the background, environment and characteristics of individuals with a handicapping condition. The primary purpose of a case study is to determine the factors, and the relationships among a set of variables which have resulted in the current disability or status of the person under study. In other words, the purpose of most case studies is to determine what a person's status is. The case study methodology does not allow the therapist to establish functional relationships between the various factors being described. (p. 92)

Establishing functional relationships between the variables under investigation is best done with the experimental method.

## Experimental Method

In contrast to the other methods discussed previously, the experimental method is the most systematic, manipulated, and controlled. The experimental method involves making a statement or prediction and then systematically testing it. The prediction is called a *hypothesis*, which is defined as, a statement of a relationship between variables. The term, "variable" refers to anything that varies, or that is free to change.

In *The Game of Science*, McCain and Segal (52 p. 105) cite numerous everyday examples of variables, such as a person's weight or the number of times a person blinks in an hour. "The color of cars, the make of cars, the number of cylinders in a car, how far the car goes on a gallon of gasoline, how much the car cost—all are variables . . . Whenever an entity or event is described, every dimension by which it could be described can be considered a variable, including where it is and its environmental conditions."

In the experimental method, the investigator wants to control as much as possible all of the variables except for the one(s) being studied. The hypothesis, as was stated, is a statement of a relationship between variables. The experimenter wants to explore this relationship without the interference of any other extraneous factors.

The variables that make up the hypothesis are termed "independent" and "dependent". The *independent variable* is free to be manipulated by the experimenter, whereas the *dependent variable* is the one that may or may not change as a result of experimental manipulations. The independent variable is typically the stimulus or intervention measure, whereas the dependent variable is the outcome measure or response. A convenient way to recognize the dependent variable is to remember that it varies *depending* upon the experimenter's manipulations. Variables that may interfere with the study and that the experimenter wants to control are referred to as *extraneous*, or nuisance, *variables*. Typically, the dependent variable is some measure of response on the part of experimental subjects. The subjects of an experiment are the organisms that are under investigation.

---

*Sample Hypothesis*

Assume that your hypothesis is that physical exercise results in improved self-concept in obese men. Can you identify the independent and dependent variables and subjects in this hypothesis?

The hypothesis is a statement of a relationship between variables. In this case, two variables exist: (*a*) the level of exercise and (*b*) self-concept. Obese men are the subjects of the experiment. To identify the independent and dependent variables, remember that the independent variable is the stimulus measure, the one that is selected by the experimenter to manipulate and to investigate as a possible cause or influence. In this case, the independent variable is the presence or amount of exercise.

The hypothesis states that exercise will have some kind of effect on a dependent variable. The response measure that depends upon the degree of exercise is self-concept. If this were an actual study, the investigator would want to control as much as possible for any other factors, that might interact with an exercise program and after self-concept, such as health, smoking, diet, medications, and other factors. These other factors, which the experimenter wants to rule out and does so through control measures, are the extraneous variables.

## Operational Definitions

In the experimental method in particular, all operations that take place must be clearly defined. Doing this enables the experimenter to have the greatest control over the variables and to most clearly communicate the method and the findings of the study to others. A clear explanation of the study is accomplished by using very specific definitions of the variables. The term *operational definitions* is derived from the fact that these definitions refer to the operations used by the experimenter to obtain them.

In order to test the hypothesis proposed above, one must specify operational definitions of both independent and dependent variables before initiating the study. One such definition of the independent variable in this hypothesis is an index or measure of exercise level, e.g., a specific description of the exercise program such as maintaining a target heart rate for 20 minutes 3 times/wk for 3 months. The operational definition of the dependent variable would be a measurable assessment of self-concept, such as a score from a self-concept test.

## Specifying the Subject Pool

In addition to operationalizing the variables under study, subjects must also be clearly recruited and described. Experimental subjects are selected from a much larger group of individuals displaying the same characteristics. All the subjects in this larger group are called the *population*, and the particular group selected for one study is termed the *sample*. In the previous example, the population would include all obese men who meet certain criteria based on their reported heights and weight. The sample for this particular experiment might come from a specific clinic or institution.

When they conduct experiments, researchers and clinicians are careful to use specific procedures for selecting their samples. Such procedures prevent them from introducing bias into the selection process. Experimenters avoid this kind of bias by selecting subjects at *random*. That means that each subject has an equal chance of being selected. *Random selection* is easily accomplished by drawing names out of a hat or by referring to tables set up specifically to generate random numbers. Many clinicians, however, are prevented from random-selection procedures because most clinical subjects have similar problems or are referred by a particular agency or group of physicians. Many patients seen by one therapist often have similar backgrounds, and they generally reside in a certain geographical region. Thus, clinical samples are often biased from the beginning and may or may not be representative of some larger population.

Thus, researchers must exercise caution so that they do not overgeneralize the results of studies that are based on narrow subject samples. Specifying the types of subjects and their method of selection enables other investigators to determine the extent to which the findings of a study have been generalized. In addition, statistics help the clinical researcher to draw conclusions or inferences (hence the term "inferential statistics," which are used in experimentation) about the conclusiveness and generalizability of experimental findings.

## Control

The major reason for selecting the experimental method over other research methods is that it enables the investigator to exert the greatest amount of control over the variables under study. The independent variable is frequently manipulated in order to produce some effect, and the extraneous variables are controlled by establishing separate treatment and control conditions.

For example, in the study of exercise and self concept, the investigator may look at the self-concepts of obese men who have exercised for three months compared with a control group of similar men who have not exercised. The advantage of this procedure is that the control group provides a comparison measure for drawing conclusions about the effects of the independent variable.

## Applications and Limitations of the Experimental Method

The experimental method has been extremely useful in the study of human development because of its utility in identifying, isolating, and testing variables. It has enabled researchers to identify specific factors that are involved in various developmental processes. Experimentation enables researchers to systematically test theories and to come to testable, logical conclusions about the relationships of specific developmental or causative factors. The systematic nature and the practical application of experimentation are clearly illustrated in a quote from an interview with Dr. Jonas Salk, the originator of the serum used to combat polio:

> I asked [Salk] how, in the midst of all the naysayers, he persuaded himself that he could solve the tragedy of polio.
>
> "My ego played no role in it," he said. "I didn't think I was the person appointed to do this. I was simply granted the opportunity to help. We do not all see the world in the same way. There are those of us who see it in terms of solvable problems. If you have a problem that can be solved, then it will be solved."
>
> I asked him if he used to go to work every day thinking that was the day that polio would be defeated.
>
> "It doesn't work that way," he said. "You're not on a golf course. You don't

say to yourself, 'Today's the day I'm going to break par.' What you do is have a continuing dialogue with nature. You ask questions, in the form of experiments. And you get answers. Yes or no. Yes or no. Yes or no. And then you use those answers to ask your next questions, and you keep doing it until you have the ultimate answer." (53, p. 29)

The psychologist James Watson, whom we will discuss in the next chapter, used experimental strategies to investigate the emotional development of infants. When others such as G. Stanley Hall were using questionnaires to test subjects, Watson demonstrated the utility of experimentation for communicating with subjects who could not be assessed with other methods. Experimentation lends itself to use with nonhuman subjects and with infants or nonverbal humans who cannot be easily tested with other means such as interviews or paper-and-pencil tests. In addition, the experimental method provides a mechanism for systematically comparing different species as well as different age groups of subjects (from rats to humans).

The experimental method also has been increasingly used over descriptive methodologies in the field of occupational therapy (41). Many scientists equate the experimental method with rigor and objectivity, factors often cited as essential ingredients for a "truly" scientific approach. Thus, this method is often promoted within health professions that are concerned with developing sound scientific groundwork for their fields.

There are limitations to the experimental method, however. When compared with observation, for example, the experimental method is contrived and unnatural. Experiments are controlled and fixed and are therefore not generally useful for the study of spontaneous behavior. In addition, experiments relay on such concepts as randomization and the use of large subject pools for drawing statistical inferences about results. Short-DeGraff and Ottenbacher (39) have pointed out that strict experimental methods are not useful in clinical fields where subject pools cannot be selected at random and where subject samples are often small.

Ottenbacher's (40) recent text on evaluating clinical change, discusses the drawbacks to experimental designs that have been used traditionally with large numbers of subjects. He points out that other single-system strategies can be adapted easily to clinical investigation. Such strategies require fewer subjects and also provide a systematic and structured alternative to clinical problem solving. (For greater detail about single-subject designs, the reader is referred to Ottenbacher's useful text.)

# Problem Solving in Occupational and Physical Therapy

Burkitt (37) describes how the clinician views research as problem solving:

All research is based on an essential human characteristic, the need to ask questions and to find answers . . . The real motivation is recognition of the need

for the physiotherapist to be a problem solver and therefore to have the knowledge and skills to know:
—That a question needs to be asked.
—What is the appropriate question.
—Ways to test our possible answers.
—Methods to evaluate the outcome. (p. 187)

The previous sections of this chapter introduced the three basic methods that are traditionally associated with research in the scientific study of human development. This section will illustrate how these three methods can be applied in a clinical context. Research is structured problem-solving which uses certain methods to find answers to "scientific" questions. These research methods fit on a continuum of structure that can be graded for different purposes. In the clinic, evaluation presents a problem that must be solved. The following discussions are based on the premise that the continuum of structure found in research methods can be applied to clinical problem-solving, where the concern is whether patients demonstrate improvement as a result of therapy.

The last chapter introduced the concept of developmental delay. In this section, therapists' approaches to a young girl who is developmentally delayed will be examined.

---

### Application of Research Problem Solving Methods to Assess Treatment Effectiveness

Jennifer is a 5-year-old whose parents took her 2 years ago to a community screening program that offers assessment services to any preschool age children residing in their county. Jennifer's parents were concerned about their daughter's slow development, so they took her to this community program to have Jennifer examined by a variety of professionals. A *screening* is a preliminary form of evaluation that is used to determine if further professional assessment is required.

The professionals who screened Jennifer recommended that she be further evaluated, especially in regard to her motor skills, which seemed to be lagging. At the time of evaluation, Jennifer's *chronological age* (age from birth) was 36 months of age.

The results of the evaluations by physical and occupational therapists indicated that Jennifer was functioning at the 18-month level in her gross motor and fine motor/adaptive skills. The team recommended that Jennifer be enrolled in an early intervention program designed to provide sensory, intellectual, social, and motor activities for children who are delayed in these areas. Jennifer has been attending this program and has been receiving a variety of services for 2 years. She attends the program with other children 3 days a week during the regular school schedule and receives physical and occupational therapy services 3 times per week.

Jennifer's therapists, teachers, and her family are naturally interested in determining whether she has made any progress during the time she has been attending this program. Jennifer's therapists, Elaine and Joyce, must solve the problem of determining how to assess Jennifer's progress over the past 2 years.

There are a variety of approaches that Elaine and Joyce might use in solving this problem, and each approach consists of a slightly different amount or type of structure. A relatively unstructured approach for Elaine would be to casually observe Jennifer and make some general conclusions about her response to treatment. Such conclusions might include such comments as "Jennifer seems to be getting better. She seems to like therapy, and it seems to be good for her."

This, however, is a very subjective and biased approach. One that is slightly more structured involves making more systematic observations about Jennifer's performance and coming to some conclusion based on what the therapist sees.

### Observation

Elaine and Joyce might each make observations of Jennifer's performance on some motor tasks performed during therapy. They could then report to the parents: "Jennifer seems to be making progress. She is able to go up and down stairs, she can draw circles, and she now puts on her jacket, shoes, and socks. When she entered the program, she did not perform these skills."

The actual observational process for collecting information about Jennifer can be loosely or highly structured. For example, Elaine and Joyce may set up specific times and specific behaviors to be observed, as in time sampling. Such systematic observations could serve as a record of Jennifer's progress over time and could be charted to indicate the degree of change she has made on certain skills since she started with the program.

### Case Study

Most therapists, like Joyce and Elaine, periodically reevaluate their client's performance. This is a naturally occurring part of treatment. Many therapists are required by state or federal legislation or by institutional criteria to periodically reassess their patients and to report on the progress indicated by such assessment. Each new assessment of a patient's skills and abilities becomes a part of the patient's record as an in-depth report.

This in-depth approach to problem solving (case study) could be used easily by Jennifer's therapists as part of their clinical responsibilities. For example, they could keep a chart of Jennifer's performance from the time of initial entry to the program and periodically thereafter. A summary report of the findings over time would comprise a case study of Jennifer's response to therapy.

### Experimentation

Both Joyce and Elaine are experienced therapists who recognize that Jennifer's performance skills might change merely as a result of maturation. Thus, they may want to manipulate their problem-solving approach to "tease out" any effects of treatment while also taking into consideration the effect of maturation. One way to structure such an approach, would be to compare Jennifer's progress during therapy with an equal (control) period of time when Jennifer was not receiving treatment, for example, during summer vacation. Elaine or Joyce could test Jennifer's motor skills on March 1 and again on June

1, during which time she is attending therapy. The amount of gain Jennifer made would be recorded and compared with the amount of gain she makes from June 1 to September 1, a control period when she is not receiving treatment. If Jennifer made greater gains during treatment than during an equal period of no treatment, then her therapists could report that treatment was effective.

Elaine and Joyce, however, have some concerns about this approach. Many children go through periods of growth spurts. What would happen if they were to administer the assessments when Jennifer was experiencing a growth spurt? Or suppose Jennifer was sick when she was tested? One possible solution to such problems is to systematically administer an assessment to Jennifer at different times when she receives therapy and when she does not. Then, hopefully, some of the problems of extraneous variables such as sickness or growth spurts might be washed out by multiple assessment. Then, Jennifer's gains during several treatment and control conditions can be compared.

Elaine and Joyce also recognize the importance of specifying the variables they are examining. Thus, they must clearly delineate the type, duration, and frequency of therapy (i.e., the independent variable) as well as describing clearly the assessment tools used to measure developmental changes over time (i.e., the dependent variable).

Jennifer's parents are interested in Jennifer's progress in motor and adaptive skills, and her therapists want to measure these skills to determine if they change as a result of treatment. Elaine and Joyce, however, are also interested in determining if their treatment programs are effective with other patients. Thus, they could set up an experiment to test the effects of treatment on a group of subjects.

To exercise the most control and to rule out the effects of any extraneous variables, they should have two groups of children that are fairly similar in terms of their ages, their interactions with their families, and their developmental problems. Then, the children should be assigned without any particular bias to two different groups. In one group, the children are administered therapy, while the no-therapy group serves as a control for such extraneous variables as sickness, growth spurts, summer weather and exercise.

Elaine and Joyce would administer developmental tests to all of the children before, perhaps during, and at the end of the experiment. Then they could compare the average gain of the children receiving treatment with the average gain of the children belonging to the control group. While Jennifer's parents may not be interested in such a study because it is not personal enough for Jennifer's specific case, such an approach would give Elaine and Joyce the most systematic and most conclusive generalizations about the overall effectiveness of their treatment programs.

# Steps in Problem Solving

Defining research as "structured" problem solving marks it as one kind of problem solving that must differ from other kinds, e.g., guessing. The actual "structure" includes different methods as well as specific steps that are followed during the problem-solving process. These steps are (a) recognizing and stating the problem, (b) determining a method for solving the problem, (c) gathering data relative to the problem, and (d) coming to some conclusion (24). The steps involved in structured problem solving are particularly relevant to the clinical reasoning process.

For example, in the previous case, Jennifer's therapists had to recognize and clearly state the problem, i.e., determining whether Jennifer was responding to treatment. Next, they recognized that a variety of different approaches could be used to collect data and solve the problem. These different approaches can be graded according to the actual degree of structure they impose on the problem-solving process (research). Thus, in the case of Jennifer, the structure ranged from casual observation to more systematic observation, then to a case study, and finally to experimentation.

Which kind of structure a therapist selects depends upon many factors: the nature and the purpose of the problem; the nature of the caseload (i.e., the kind of, the age of, the number of, and the type of patients being treated as well as the reasons for their requiring treatment); the treatment setting (hospital, clinic, school, home); the therapist's clinical and educational training, philosophical background, and approach to treatment (frame of reference); and the therapist's ongoing clinical experiences and resources.

## Other Problem-Solving Methods

Research methods other than observational, clinical, and experimental have been recognized. As Helmstadter (24, p. 5) points out, "Research is the activity of solving problems which leads to new knowledge *using methods of inquiry which are currently accepted as adequate by scholars in the field*" (italics mine). Different disciplines have their own particular orientations, and therefore they tend to use certain methodological approaches. Anthropologists working in the field, use different methods than do geneticists working in the laboratory. Methods vary, and no one method is proper or correct. The actual orientation of one's profession and one's frame of reference provides a framework for evaluating the usefulness and applications of specific research methods.

### Ethological Method

An example of an approach that emerged from a specific discipline is the ethological method. *Ethology,* which is associated with zoology, a branch of biology, studies organisms in their natural habitats. This field originated with Austrian physician and biologist, Konrad Lorenz. As will be explored in the next chapter, Lorenz and other ethologists have a particular orientation toward behavior. They believe in the evolutionary continuity between animals and humans and examine organisms in relation to the adaptations they have made to their environments during evolution (25, 54).

As methods, ethological and observational approaches do not differ. Operationally, they are identical—they both involve the study of organisms in their natural environments. However, each is associated with a specific orientation. The observational method, for example, is more commonly associated with American psychology and its tendency toward intervention and manipulation, whereas the ethological method is European in origin and has biological traditions. This separation between the fields is not as common today, although some still view these fields from such historical perspectives.

A schism has existed traditionally between the American-oriented *comparative psychology* [which is the study of the similarities and differences between the behaviors of different animals (including humans)] and the European-oriented field of ethology. American psychologists have frequently used the observational method in seminatural or nonnatural settings in a "combined experimental-observational" approach. Thus, while they may use observational methods for obtaining data, their settings may still be very structured and manipulated.

In such settings, investigators may be located behind a one-way mirror where they observe specifically, manipulated circumstances to which a subject responds. For example, they may observe the interchanges between two children in a room set up like a kindergarten classroom. The investigators may actually manipulate characteristics of the environment to determine the children's reactions, e.g., introduce an unfamiliar child or objects, or introduce an unfamiliar adult. In this case, the method for obtaining data is observation, but it does not fit the rigid definition of "observation in a naturalistic setting." Ethologists, on the other hand, do tend to observe organisms in their natural habitats. Many study animals in the wild, and those that examine human behavior tend not to intervene. However, ethologists may return to the laboratory to test some hypothesis that was generated from work conducted in the field (25).

While their methods may appear identical, ethologists are discriminated from other psychologists and biologists by their evolutionary orientation. Ethologists are particularly concerned with describing and classifying behaviors before conducting more complex analyses. Ethologists are also interested in finding relationships between behaviors and the environment where the species under investigation became adapted during evolution (45, 54, 55). Certain typical behavior patterns, often considered innate, such as maternal behaviors, dominance and aggression, or sexual behaviors, are of particular concern to ethologists.

## Applications of Ethological Method to Study of Human Development

Some researchers claim that ethology's influence has significantly diminished (56). Blurton Jones (55) points out that many may associate ethology with discarded concepts and therefore see it as outdated. On the other hand, Hinde (54) has promoted the synthesis of comparative psychology and ethology, and Blurton Jones (55) notes numerous applications and fundamental contributions of ethological theory and methodology.

When referring to ethology, one must distinguish between theory (which is discussed in the next chapter) and method. The ethological method has influenced developmental psychology and many other fields. It has provided information about the similarities and differences between animal and human behaviors and clearly demonstrated the continuity of some behaviors across species. Ethologists have provided detailed, careful descriptions and classifications of behaviors as well as pertinent information about cross-cultural studies of humans.

Michel and Moore (56) claim that the ethological method cannot be adequately applied to human behavior because of the difficulty of determining "natural" human behavior and "natural" or "field" settings for humans. On the other hand, diverse ethological studies of humans exist (e.g., 57), and the method of detailed behavioral description as well as many of the descriptions generated from ethology can be used in other disciplines.

The fields of ethology and occupational and physical therapy focus on adaptation, thus giving them a similar orientation toward behavior. Ethologists examine functional behaviors in regard to their adaptive significance in specific environments over time, and physical and occupational therapists are concerned with adapting environments to promote maximum functional performance for patients. Ethologists, for example, have examined human prehension and children's play behaviors (57), topics of practical and theoretical significance to many clinicians.

Therapists may apply or modify different ethological strategies for investigating adaptive responses. Practitioners obtaining normative data may benefit from ethological approaches to describing and classifying specific adaptive, functional behaviors. Although we may not be observing our patients' behaviors "in the wild", we promote adaptive and functional behaviors in our patients so they can be productively occupied in their own natural habitats, such as their homes and work places.

The ethological focus on behaviors associated with reproduction has contributed considerably to the knowledge of human development. Ethological studies of the attachment process between parents and offspring, the socialization of offspring, and the rituals associated with mating and parental care have been widely cited in the scientific study of human development. Such information has provided essential insights into human behavior and has resulted in practical information used in the clinic. For example, knowledge of the attachment process between parents and offspring has been used with benefit by clinicians who work with sick or premature infants who must be separated from their families at birth (e.g., 58).

The ethological method, therefore, has provided important, useful information for therapists and developmental psychologists. As a method, however, it is often judged as too circumscribed because of its association with a particular evolutionary orientation. The method of naturalistic observation, however, to which the ethological method is in some ways comparable, has recently become emphasized under a different name. The ecological method may in fact become more widely used and replace the more circumscribed ethological approach.

## Ecological Method

The ecological method also evolved out of a special area of interest. Operationally, it is the same as the observational or ethological methods, but there are differences in the orientations of these three. These differences may be subtle, but they do affect the topics and nature of investigations conducted by researchers aligned with one particular method. For example, the ethological approach grew out of zoology, the study of animals. Ethology is a new field, originating in the 1940s in Europe, and it assumes an evolutionary continuity between living organisms. In contrast, the observational method was used widely in the study of child development in the United States during the first half of the 20th century; although it is often associated with comparative psychology, ecology, like ethology, is actually a branch of biology. It is the study of the interrelationship between organisms and their environments.

The ecological approach in psychology was widely used by individuals at the Midwest Psychological Field Station at the University of Kansas in the 1950s (45), but an ecological emphasis in psychology has only become evident since the 1960s and 1970s (59). Presently, interest and research in ecology has virtually exploded. As Walsh (60 p. 2) reports, "More books have been published in this area in the last five years than in the preceding five decades."

The ecological method takes into account the nature of the organism, the environment, and the interactions between them. It emphasizes contexts and therefore interferes as little as possible with the behavioral transactions of the subject under study. As such, it is identical to naturalistic observation or the ethological approach. The ethological approach, however, tends to focus on evolutionary themes in the relation between the organism and the environment; and the observational approach is traditionally associated with the introduction of various structures for exerting control (e.g., time sampling).

## Ecological Method in Human Development and in Physical and Occupational Therapy

As noted in the previous chapter, developmental psychologists and clinicians have become increasingly concerned with identifying the nature of human's interactions with their environments. The focus of the ecological method meets that need by taking into consideration complex phenomena such as reciprocal or interdependent relationships.

That ecological thinking has affected the study of human development is clearly illustrated in the study of bonding or attachment. *Attachment* and *bonding* are reciprocal processes whereby infants and caretakers become attached to one another. At one time in developmental psychology, the study of attachment focused primarily on the infant alone and examined those characteristics that facilitated or inhibited the infant's attachment to his or her mother. There was a tendency to speak of attachment as an "event" rather than a "process". In time, however, researchers began to examine the interaction between the infant and the mother and to characterize the interchange between them. As thinking became more ecologically oriented, i.e., aimed at investigating reciprocal relations and contexts, the scope of research became broader (58, 61). Attachment and bonding research expanded to investigations of ongoing systems of reciprocal interactions between infants and their environment, including family members (mother, father, siblings, other caretakers, etc.) and opportunities for perceptual, motor, and other learning experiences (see chapter 12).

Thus, the growing ecological orientation in thinking and research has altered traditional concepts in human development and pointed investigators to new, relevant variables. Because of the ecological approach, current research in human development tends more than in the past to specify and operationally define fundamental characteristics of environmental and subject variables. Walsh (60) has provided a partial list of the variety of dimensions of environmental stimuli (summarized in Table 4.7). To illustrate the application of such information to occupational therapy and to human development, consider just one dimension from that list—the *temporal* (time-related) aspect of organism-environment interchange.

As occupational therapist Kielhofner (62, p. 235) has stated, "Time is the inescapable boundary for human existence and activity." Developmental psychology is the study of how organisms change *over time*. Temporal-related ecological investigations have to do with determining if certain environment-organism interactions have pronounced or limited effects because they occur at circumscribed times during that organism's life. (For a

Table 4.7
Dimensions of Environmental Stimuli[a]

| Dimensions | Explanation |
| --- | --- |
| Proximal-distal | Whether stimulation directly or indirectly affects organism |
| Social-nonsocial | Whether stimulation relates to groups, sex, population or to inanimate environment |
| Afferent-reafferent | Whether stimulation derives from the environment or in part from the organism |
| Sensory modality | Whether stimulation acts through one or more senses—and which ones |
| Intensity | Strength of stimulation |
| Complexity | Range, variety, number, dynamics of stimuli in environment |
| Temporal | Timing, periodicity, duration of stimulation |
| Learning | Prior exposure and meaning of stimulation |
| Reinforcement | Stimuli that modify behavior |
| Environment-organism system | Interactions between environmental stimuli and subsystems of the organism |

[a] Adapted from Walsh R: *Towards an Ecology of Brain*. New York, SP Medical and Scientific Book, pp 17–18, 1981.

further discussion of this topic, refer to the sections "Critical Periods" in Chapters 6 and 9.) Kielhofner has described how temporal adaptation was an important theme in the early development of occupational therapy. He suggests that clinicians reconsider the significant impact of time-related issues in everyday life (62). He also describes the importance of temporal adaptation to human occupational development (62, 63).

Temporal issues are particularly relevant in clinical practice, where such questions may arise: Is there, an optimum time for treating certain patients, such as those who have just experienced a stroke or hand surgery or an infant with some birth trauma (58)? Do patients of certain ages benefit differentially from therapy administered at varying times posttrauma? For example, should the times for administering postoperative range-of-motion therapy for youngsters be different than for elderly patients? Our knowledge of the tissue changes that occur with age certainly indicates that this is an important variable to investigate, but our research base has yet to provide a clear answer.

Clinical researchers Johnson and Deitz (64) illustrate how time can be used as a dependent variable in a study examining mother-child interactions. These authors reported the significance of the findings that mothers of handicapped children spend greater amounts of time during physical-care activities than do mothers of nondisabled children. The increasing time demands on the parents of handicapped children may have effects on interpersonal stress, family conflict, and parent's perception of the quality of their life. A separate but related issue is the exploration of the effects of time demands on the child (or adult) with a disability that severely interrupts performance of daily living activities. As Kielhofner (62) explains, early in this century, the purposeful use of time was considered an important dimension in occupational therapy practice.

Ecological approaches to human development may help to restore temporal issues as a continued clinical concern. The ecological method includes examining environmental contexts and determining their influences upon and response to participating organisms. This is an important aspect of therapy which also involves modifying environments to promote maximum patient function. Thus, inasmuch as the ecological method specifies the nature and characteristics of environmentally-induced change (e.g. see 65), it provides an important methodological resource for occupational and physical therapists.

Despite the fact that it lacks the control obtained with the experimental method, the ecological method has been designed for studying organism-environmental interactions which, as was pointed out in the last chapter, are highlighted in contemporary work. Since its introduction is recent, the ecological method should find extensive application to research examining numerous varieties of interactional circumstances that are important to developmental psychologists and clinicians. Most therapists are concerned with the nature of their patient's interactions with their environments and are also interested in delineating the significant factors involved in effective patient:therapist contact and interchange. The ecological method can address those issues.

In *Towards an Ecology of Brain*, Walsh (60, p. 2) points out that interest in ecology has become more popular, in part because "of the current awakening to some of the deleterious effects of pollution and dwindling resources. Physical and chemical pollutants have been increasingly recognized as exacting a toll on both physical and psychological well-being." This issue impacts on human development as well as on productive human occupation and health. As clinical fields become more concerned about the effects of environmental contexts on human physical and mental health, the ecological approach will be increasingly relevant.

Some contemporary researchers including astronauts, astronomers, other scientists, physicians, and philosophers note that the notion of environmental context extends beyond immediate realms (homes, neighborhoods, communities) to incorporate the entire earth, its atmosphere, and all its resources. The ecological method accomodates examination of the effects of global pollution, global politics, and global warfare on human behavior, development, and health. In *Staying Alive. The Psychology of Human Survival*, Walsh (66) points out that scientists and mental health professionals must recognize and begin to focus on the effects of global forces on human development and mental health. As long as researchers remain concerned with such global issues (and it appears that they must), the ecological method will be increasingly used in many fields of study.

# GATHERING DATA

The steps in the problem-solving process include (a) stating the problem or setting up the hypothesis, (b) developing a method, (c) gathering data, and (d) coming to a conclusion (24). This applies to research as well as to cinical problem solving. With the case of therapists Elaine and Joyce, the problem was determining whether Jennifer's treatment program would be effective in promoting motor development. Several examples of different methods that a therapist could use in investigating this relationship were described in the previous sections. The general categories of these methods were listed as observational, case study, and experimental. The ethological and ecological methods were also introduced.

So far, however, only the first two stages of the problem-solving process have been considered: (a) stating the problem and (b) developing a method. The next step, that is covered in this section, is the process of gathering or collecting data. In a structured approach to problem solving, any investigator must have some way of objectively measuring the dependent variable under study. In the case of Jennifer, some measure of motor skills might be used. Thus, to determine if her skills changed over time, she could be given the assessment upon initial entrance to the program and then later after participating in treatment. This would yield two different data points, one from each administration of the test, and the difference between these scores would reflect Jennifer's change over time.

The actual test instrument that is used to gauge Jennifer's progress will affect the type of data that are obtained. For example, the therapists may administer some comprehensive measure of motor skills and obtain a single score indicative of Jennifer's developmental level. As alternatives, a very comprehensive test of many different kinds of skills could be used to obtain many different scores, or some simple performance measure could be used. This might include the amount of time required to complete a task (e.g., cutting out a uniform-size square with a pair of scissors) or a measure of precision in completing a task (e.g., walking on a balance beam without falling and without outside support). Obviously, comparing two different scissor cutouts will provide data that differ from numerical scores on a test. Thus, it is important in research to be aware of and to specify the procedure used in data collection. The various kinds of devices used to collect clinical data are called *assessment* or *research tools, measurement devices,* or *evaluation instruments.*

# Categories of Measuring Devices

Many kinds of tools and devices can be used to gauge behavior (e.g., 67). These are given a variety of names such as measurement, test, or evaluation instruments, or assessment tools. They may be used to evaluate performance in the clinic, or in conjunction with research methods, to study behavior as it changes over time. Since there are so many instruments, they can be classified in a variety of ways. They can be classified according to their purpose or the specialized discipline of use, according to their structure, their type, or their location of use (Table 4.8).

The purpose of the following sections is to briefly introduce the categories of some commonly used research or evaluation instruments that have been applied in developmental research. The student should be aware that there are enormous numbers of evaluation instruments used in occupational therapy, physical therapy, and in the study of development. Only a few will be cited here, and the interested student can explore others by consulting specific clinical textbooks and journals.

### Structure of Evaluations

Evaluation instruments will vary according to the goals of the researcher or clinician. Obviously, a field researcher using an observational method will use a differently structured instrument for collecting data than will a laboratory technician assaying blood chemistry or a therapist testing a child's balance. Banus (68) distinguishes between informal, formal, and

Table 4.8
Categories of Measurement Instruments

| Location of Use | Discipline | Structure | Type[a] | Purpose (to assess) |
|---|---|---|---|---|
| Clinic | Allied health | Informal | Survey | Reading/education |
| Laboratory | Psychology | Formal | Interview | Achievement |
| Home | Social Work | Stanardized | Questionnaire | Cognitive level/intel- |
| School | Education | Norm referenced | Scale | ligence |
| Field | Biology | | Projective | Perceptual-motor |
| | Medicine | | Developmental | Manual dexterity |
| | Neurology | | Inventories | Reflexes |
| | Orthopedics | | Opinion polls | Daily living skills |
| | Pediatrics | | Observation | Range of motion |
| | Psychiatry | | Self-report | Muscle testing |
| | | | Sociometric | Feeding/oral motor |
| | | | Demographic | Moods/anxiety |
| | | | | Attitudes |
| | | | | Prehension |
| | | | | Environment |
| | | | | Self-concept |
| | | | | Motor skills |
| | | | | Personality |

[a] Data from Isaac S, Michael WB: *Handbook in Research and Evaluation*, ed 2. San Diego, CA, EdITS, 1981, p 149.

standardized evaluation, with informal being the least, and standardized the most, structured. Banus describes *informal evaluation* as a casual, unstructured observation of behavior that is typically used preliminary to selecting other suitable tests. Thus, a therapist may use an informal assessment when first meeting a client. Based on the information obtained at that time, the therapist might then follow up with a formal or standardized test for more structured data collection.

*Formal evaluations* are more structured, organized, and planned; they are frequently set up by one or by a group of therapists for assessing specific behaviors or patient populations for whom no other test exists. Thus, formal tests are often highly specific to an institution or a particular clinic (68, 69). An example of a formal test is a clinic-designed checklist of the functional kitchen skills of upper-extremity amputees who have recently been fitted with *prostheses* (artifical limbs), or a brief clinical endurance test administered to lower-extremity amputees when they first start to ambulate.

In contrast, *standardized tests* are very structured in form and in administration. They are typically developed by examining a large group of subjects and obtaining normative data regarding test performance. Standardized tests are so named because they are highly structured and because administrators must follow certain "standards" or specific procedures in order to accurately use and apply them. The applicability of the normative data used with such tests depends upon the evaluator's following specific instructions and methods of test administration and scoring. Thus, test instructions for administration, scoring, and interpretation of standardized evaluations are explicit and must be followed. In some cases, formal training must be obtained.

The purpose of *norm-referenced tests* is to compare how one person performs relative to others. Most widely known intelligence tests are standardized and norm-referenced as are some of the tests used in physical therapy or occupational therapy, e.g., the Miller Assessment for Preschoolers (5) and the Southern California Sensory Integration Tests (70), recently replaced by the Sensory Integration and Praxis Tests (71). All require familiarity with test construction and with the types of subjects for whom the test is designed as well as very specific training in methods of test administration and interpretation.

## Types of Research and Evaluation Instruments

In addition to classification according to the dimension of structure, evaluation instruments can also be classified according to type. Different types include surveys, interviews, questionnaires, and projective tests.

### Surveys

Isaac and Michaels (67, p. 128) report that surveys "are the most widely used technique in education and the behavioral sciences for the collection of data." Surveys are useful in the study of development because they can be used to gather data relevant to description or for examination of trends across time. Surveys are subdivided into interviews or questionnaires, both of which have been used to study adolescents and adults as well as to obtain information about children from their parents (67, 72).

*Interviews.* Interviews can be fairly unstructured, open-ended conversations or can be structured with specific, systematically introduced questions. Interviews are typically conducted as a one-to-one interaction with an individual, but they may be administered in a group format. The advantages of interviews are that they can be very flexible and can provide additional nonverbal information, such as gestures or tone of voice, as the subject responds. The disadvantages are that the process of interviewing itself can bias the data. The interviewer's reactions to the respondent may subtly affect the interviewee. Thus, in many cases, it is recommended that an interviewer be trained and skilled in specific techniques. Interviews are often expensive and time consuming, and it is often difficult to summarize all the findings obtained with them (67, 72).

*Questionnaires.* Questionnaires are essentially written forms of interviews. Thus, they are more structured and provide less opportunity for interaction. The written series of questions can be mailed or read directly to individuals. Compared with interviews, which are often open-ended and do not focus the answers, questionnaires frequently include a series or continuum of answers (e.g., multiple-choice responses; yes or no; strongly agree, agree, neutral, disagree, strongly disagree) that force the respondent to make a selection.

Questionnaires are widely used because they are inexpensive and can be easily administered to large groups; they can be very simple or complex and can be administered to individuals by themselves rather than requiring a time-consuming interaction between an investigator and subject. Since most questionnaires are mailed out, they can also be used to ensure the anonymity of subjects. The disadvantages of mailed questionnaires are that subjects often do not return them, and they are subject to misinterpretation and even deception. With mailed questionnaires, there is no way to confirm who completed and returned the form, whether the person understood the questions, or whether the person responded honestly and accurately. For clinicians, there is concern about the inability to administer questionnaires to patients who cannot read or who do not possess the motor abilities to make accurate responses (67, 73).

## Scales

Measuring devices that assign numbers or symbols for rating individuals or their behavior are called "scales", and they are commonly used in attitude assessment. Common dimensions along which attitudes are scaled include: "agree-disagree" or "like-dislike" (67). As was noted in regard to questionnaires, scales may be included in a questionnaire format, such as:

What is your opinion of ice cream? (Circle one)

1—Like        2—Dislike        3—No opinion

Rating scales may also be used in conjunction with the observational method; for example, by clinicians who are observing specific subjects and rating their reactions. Clinical dimensions that may be rated include: participate – not participate; happy – sad; functional – nonfunctional; muscle tone: high – low.

## Projective Tests

Projective tests provide subjects with a set of ambiguous stimuli which they must interpret. In so doing, the subjects "project" their thoughts or feelings, and the investigator analyzes and interprets the response. Such tests are particularly useful in mental health fields that want to tap people's "real" or "deep" feelings. Projective tests are based on the assumption that subjects will reveal their basic needs, fears, and defenses through perceptions and symbolism that emerge in their responses (74). For example, with the Thematic Apperception Test (75), a patient is shown a series of pictures (typically of people of different ages performing everyday activities) and asked to describe what is going on in the picture. It is assumed that the response will reveal some basic information about the thought processes and mental organization (as well as imagination, intellectual abilities, or language) of the patient.

When compared with other instruments such as questionnaires, projectives are less subject to deception by the respondent. On the other hand, the interpretation of the response may be highly biased by the examiner, who must be careful not to project his or her emotions during interpretation. A simple projective test that may be administered in an informal or structured format, or for clinic or research use, requires subjects to either draw a person or to draw a picture of themselves (e.g., 76–78). For example, a teacher or clinician may ask a child to draw a person and then have the child talk about and explain the drawing. This is useful in assessing writing, coordination, and perceptual skills as well as the child's emotions that are projected into the drawing. The child may draw a happy or a sad person and then describe the person engaged in some specific activity that is particularly relevant to the child's current personal life.

Some patients with neurological damage make very disorganized drawings, revealing clinical information that might not emerge during a yes/no test or during a conversation. One of the assumptions about projectives is that people actually project themselves into the picture. Thus, patients might draw themselves as they are or as they really want to be. This assumption is corroborated by findings that patients with physical handicaps (e.g., an amputation or paralysis) often draw a person with corresponding handicaps. For example, a patient whose body is paralyzed on one entire side (i.e., hemiplegic) may draw only half a person. In contrast, other patients may draw a person as they would like to be, either hopefully reconstructing their own self-image or perhaps experiencing an unhealthy denial of their problems. Patients with severe emotional disorders may draw people as they feel—torn apart, distorted, without heads, with large heads and probing, dark eyes, or with unusual sex organs. Thus, such drawings can inform therapists about patients' integration of physical deformities into their body images, or the drawings can be useful as a point of departure for an informative, revealing conversation that rounds out a clinical profile of an otherwise noncommunicative individual.

Some therapists use projective drawings not only for measurement but also for effective interaction and treatment. Children who have been sexually abused or who have fatal diseases often express their fears and emotions more effectively through drawing than through conversation alone. Thus, while projectives may be subjective, they offer a unique and useful tool for researchers and clinicians.

### Screening Tests

Screenings are used as preliminary means for determining whether or not an individual requires further assessment. The Miller Assessment for Preschoolers (5) or the Denver Developmental Screening Test (79), for example, may be administered to all children entering a regular preschool program. The purpose of administration is to screen out those children who show delays in several or in a specific area of testing. Those children with obvious delays are then referred for more comprehensive testing.

## Location of Instrument Use

Another way of classifying instruments is according to where they are used. Some instruments, such as the Tennessee Self-Concept Scale (80), have different versions for clinical and for research use. Some evaluations may be useful in the clinic but too cumbersome for itinerant therapists who are visting homes. For example, electronic instrumentation used to measure range of motion may be useful and necessary for precision readings for research, but it may be unwieldy, too time consuming, expensive, fragile, and impractical for everyday clinic or home use.

Some test instruments may be designed specifically for certain locations. Some evaluations have been structured deliberately for use in a classroom group setting (e.g., college entrance tests), whereas other tests of daily-living skills must be administered in a kitchen, laundry, or bathroom setting. Some checklists or surveys are designed specifically for use in the field by naturalists who remain detached from their subjects, e.g., for observing and recording migration patterns or ritualistic mating and maternal behaviors of sea mammals or birds. In contrast, comprehensive muscle testing of patients requires considerable one-to-one interaction with a patient in a private location in the clinic or home.

## Disciplines Associated with Measurements

In addition to classifying measurement devices by their location, they can also be divided according to the primary discipline for which they were designed. Obviously, ethologists' means for collecting data will differ from those of neurologists or cardiac surgeons. Although some devices are specific to certain disciplines, there is considerable overlap in the clinical use of tests. Although there may be problems with this procedure, especially when using standardized tests, occupational or physical therapists sometimes use parts of neurological reflex assessments and parts of psychiatric or psychological tests. This is particularly true in areas where assessments specific to one's discipline do not exist (e.g., 81).

Some evaluation instruments are designed to cover a wide range of behaviors and may therefore be used by diverse disciplines. An example of this is the Denver Developmental Screening Test (79), a broad-based screening instrument designed to examine performance of children from birth to age 6 in the following general areas of development: personal-social, fine motor–adaptive, language, and gross motor. Other instruments focus on areas of concern to specialists, or they require such specific training for administration and interpretation that they are not (or cannot be) used by other professions. Medical laboratory tests (e.g., blood tests, genetic analyses, x-rays) must be conducted by someone with specific training and expertise to conduct the test, operate the equipment, and interpret the results.

## Purpose of Measuring Instruments

Each measurement instrument is created for a specific purpose, and these instruments vary considerably depending on their purpose. Some tests are specifically designed for group administration in a classroom, e.g., some intelligence, achievement, reading, or college entrance examinations. Other evaluations, especially of clinical abilities, require a close one-to-one inter-action between the evaluator and subject. A very large number of tests exists, and reviewing them here is beyond the scope of this chapter. Some examples of different test types are listed in Table 4.8.

## Characteristics of Test Instruments

Two characteristics of measuring instruments can give an indication of their efficiency and accuracy: a test's *reliability* and *validity*. The reliability of a test refers to its accuracy of measurement; i.e., is it internally consistent, and is it consistent over time? "Internal consistency" refers to the similarity of all the test items in one assessment in measuring the same factor. "Consistency over time" refers to the ability of an evaluator to obtain similar scores with repeated administration of the same test. Thus, on a reliable assessment of motor development, which normally should not change in a few days, scores should be consistent from one day to the next.

Reliability is assessed using a statistic called a *correlation coefficient*. This is an index of agreement between items, and it can range from −1 (indicating perfect disagreement) to 0 (indicating no agreement or disagree-ment, i.e., random) to +1 (indicating perfect agreement) (82). To test the internal consistency of a reading test, for example, one could correlate half of the test items (i.e., statistically compare them) to the remaining half. This is called "split-half reliability," and if the test is reliable, the test halves should display a high positive correlation (i.e., close to +1). To test stability over time, one could conduct a test-retest study and similarly obtain a correlation coefficient. An example of this would be using a goniometer to gauge joint range of motion (ROM). In normal individuals, ROM should not change dramatically from one day to the next. Thus, to determine a student's accuracy in judging ROM, one may obtain a reliability coefficient.

*Validity* refers to the degree to which a measuring instrument assesses what it claims to. This has been an important issue in clinical fields where there are a variety of skills or traits that are difficult to assess. Muscle tone, for example, is very difficult to accurately and objectively measure. What makes it difficult to judge the validity of new methods designed to gauge muscle tone is that there is very little with which to compare them. For example, suppose you wanted to develop a new test of fine motor abilities for 4-year-olds. You could test your tools' validity by comparing the scores obtained on your assessment with the scores obtained on many other already proven assessments of fine-motor ability. If the tests were comparable, then you might claim that your new tool was valid. However, as with muscle tone, if there is no really powerful method in existence, then it is difficult to make comparisons with new instruments.

In the field of intelligence testing, which will be explored later in Chapter 11, validity is a common, controversial issue. Since intelligence is so difficult to define, valid intelligence tests are subject to discussion. For example, Howard Gardner (83), who claims that there are many different levels of intelligence, might question the validity of the more common intelligence tests used in high schools and colleges. Gardner, in fact, claims that there are linguistic, musical, perceptual, mathematical, somatic, and social forms of intelligence, all of which would be difficult to validly measure with a simple paper-and-pencil test.

# Measuring Development

Developmental psychology studies behavior *as it changes over time*. Thus, data collection revelant to developmental issues must take into consideration the concept of time. This is approached in three ways: by using developmental assessments or by taking a longitudinal or cross-sectional approach in research methods.

## Developmental Instruments

Many measuring devices are designed specifically to assess developmental level. These are referred to as *developmental instruments* or *developmental evaluations*. These are generally broad-based, norm-referenced tests that may include motor, social, language, cognitive, and other skills. They are based on developmental milestones, stages of development, or developmental tasks (as discussed in chapter 3). Thus, a child's performance on one of these assessments is compared to norms for children of his or her age. For example, the Peabody Developmental Motor Scales (84) provides the clinician with measures of gross and fine-motor abilities, whereas the Bayley Scales of Infant Development provide measures of motor and of mental abilities (85).

Developmental tests are primarily created for the welfare of children, i.e., they are used for discriminating children who deviate from the norm. These are children who may have motor handicaps, or mental retardation or who may be at risk for certain developmental deficits. The value of determining if such children are behaving appropriately for their age is that once they are discriminated, then needed treatment services can be justified and provided for their welfare (86).

In addition to using developmental tests for comparing a child's performance with some age-referenced (or time-related) norms, investigators must use appropriate approaches for incorporating time into their research. This may involve the longitudinal or cross-sectional approaches to research.

## Longitudinal and Cross-Sectional Approaches

Researchers in human development are interested in behavior as it changes over time. The most obvious approach for taking time into consideration is to examine how certain variables change in an individual or a group of individuals over a designated time span. That period may be weeks, months, or years, depending on the nature of the investigation. Thus, such an approach could conceivably be used to examine one person's entire lifetime.

This approach is *longitudinal* in nature; it involves examining the same individual or group of individuals over the course of some time period.

The longitudinal approach contrasts with the *cross-sectional* approach, which involves testing different people (or groups of people) at different points in time. For example, if you wanted to see how manual dexterity changes over time, you might want to assess separate groups of 1-, 5-, 20-, 40- and 60-year-old-individuals. This would contrast with a longitudinal study, which would start with a group of 1-year-olds and measure them periodically for 60 years.

## Longitudinal and Cross-Sectional Evaluation

The terms "longitudinal" and "cross-sectional" are also used in reference to developmental evaluations. Banus (68), refers to the *chronological* approach to evaluation. This is also called *serial evaluation*, or the periodic reassessment of an individual over time. This is synonymous with the longitudinal approach, except that in research, investigations often involve groups; whereas in the clinic, we tend to refer to serial assessments of individuals. *Horizontal* evaluations look at a variety of performance areas in one individual, e.g., adaptive, motor, cognitive, social-emotional. Since horizontal evaluations look *across* areas (or sections) of development at one point in time, they correspond to cross-sectional approaches.

Developmental evaluation takes into consideration both longitudinal and horizontal approaches. This enables the clinician to obtain a profile of how the *whole* person is functioning at one point in time and to see how the patient is progressing in all areas of development over time.

## Advantages and Disadvantages of Longitudinal and Cross-Sectional Research

Advantages and disadvantages exist for both the longitudinal and cross-sectional approaches in research. The obvious disadvantage of the longitudinal approach is that it is time consuming. Clearly, if you studied one person for 40 years, it would take your entire career to conduct your study. The longitudinal approach, however, provides information that another approach cannot. For example, it is particularly useful for examining how personality traits change over time or for exploring the continuity of behavior. Thus, it is particularly useful for examining such behaviors as intelligence, dependency, behavioral problems, or the effects of early experience on later development (43).

Cross-sectional studies, because they use groups of different people, cannot provide information relevant to continuity or stability of behavioral patterns. But the cross-sectional approach is immediate and therefore time-effective and cost-effective. With the longitudinal approach, subjects are often lost. They may die, move, change their names and addresses, or voluntarily drop out of the study. It takes time and money to keep in contract with subjects from longitudinal research. On the other hand, cross-sectional studies are conducted once, so the subject pool is obtained, sampled, and the study is completed.

There are disadvantages, however, with using just the cross-sectional approach, and this has been demonstrated in the field of human development. Although some longitudinal studies were started early in this century, obviously their results have only lately been forthcoming. For example, the longitudinal study of the Fels Institute in Yellow Springs, Ohio was started in 1929. It has been characterized by Achenbach (86, p. 2) as "one of the lengthiest and most comprehensive longitudinal studies ever attempted." However, it was not until 1959 that 30 years of data collection could be assessed and that teams of researchers could "piece together patterns and relationships that would expose the nature of behavioral development."

The accumulation of longitudinal data had some significant effects on existing knowledge about human development. As was pointed out in the last chapter, conventional, decremental models of the life cycle portrayed middle and old age as times of plateau or decline. Such models were derived partly from cross-sectional research that erroneously indicated that IQ, and in fact most human abilities, progressively declined from adolescence on (87).

Unfortunately, the results of studies examining developmental changes in IQ have been biased and often misinterpreted. As Achenbach (86, p. 225) notes, "Longitudinal studies in which the same adults were tested at repeated intervals have shown that IQ test performance continues to increase well into the middle adult years where cross-sectional studies had indicated declines." Botwinick (87) notes that cross-sectional and longitudinal studies of intelligence do show similar effects if they are cautiously analyzed. As indicated in chapter 2, intelligence and many other developmental processes are not simple, linear variables but complex, multivariable phenomena which research methods and studies must take into consideration. This is most effectively addressed by obtaining developmental information through the use of different approaches and methods.

### Cohort Effects and the Cross-Sequential Approach

A primary disadvantage of the cross-sectional approach is that it can be biased by cohort effects. A *cohort* is an age group, and cohort effects result when different age groups are affected by the historical periods during which they live(d) (Fig. 4.7). Suppose that we wanted to study whether attitudes change with age. Therefore we survey groups of 20-, 35-, 50-, 65-, and 80-year-olds regarding their attitudes about war. Then suppose we find that attitudes about war tend to be more positive in the older age groups than in the younger ones. As a result of this study, we may conclude that attitudes become more positive over the lifetime.

In contrast, suppose that we survey the same groups about attitudes toward sex. In this case, we may come up with opposite results. The problem with these studies is that, by using a cross-sectional approach, we may be sampling attitudes that are a result of the time period during which indi-

**Figure 4.7.** Cohort effects occur when subjects of different ages are affected by the trends and events of the time periods when they grew up. (© 1986 United Feature Syndicate, Inc.)

viduals grew up. Thus, rather than obtaining real developmental trends, we were obtaining cohort effects. It is quite possible that the oldest groups responded more positively about war because they experienced favorable national and sentimental feelings about the First and Second World Wars. The younger age groups may have reacted less favorably because their generations were reared under the threat of nuclear war and because their experience of war was associated with the bitter, national discord occurring during the Viet Nam era. Similar generational explanations could be used to explain the different findings regarding sexual attitudes. In fact, the groups may not differ, but the older generations may have learned that sex is a topic that should not be openly discussed.

Attempts have been made to eliminate cohort effects by using parts of both the cross-sectional and longitudinal approaches. The *cross-sequential* approach involves repeatedly studying one age group, while adding more groups that are the same age as the original one. Thus, in 1950 a sample of ten 25-year-olds are investigated. In 1960, the original sample is studied as well as a new sample of ten 25-year-olds. In 1970, an additional sample of 25-year-olds are added to the first two groups, and so on depending upon the resources of the investigator (86). The advantages of this approach is that the original group can be used for assessing continuity or stability of behavior while at the same time comparing different age groups.

# CHAPTER SUMMARY

As pointed out at the beginning of this chapter, methods and approaches for studying development have been changing over time. These changes have been either in content or in emphasis. Initially, investigations were primarily normative and descriptive in scope, but gradually systematic, theoretical studies became more common.

Researchers and clinicians adopt certain methods for specific purposes. For many different reasons, some methods tend to be in vogue for a period of time and widely used by a certain profession. When changes occur in society or in the profession, different methods may then be adopted. This was evident in the field of psychology when the scientific aims shifted from normative to systematic, and research methods changed in emphasis from observational to experimental and are now more ecological in scope. The methods one uses may affect the nature of the data that are collected as well as the overall conclusions that are generated from the study. Research consumers must exercise caution in overgeneralizing the results from any one study. Just as contexts are increasingly emphasized in human development, contexts must be considered when judging research conclusions.

At present, the ecological method is increasingly emphasized in academic as well as in clinical fields. As the next two chapters will discuss, studies of organism-environmental interaction are increasingly found in clinical theory and research, and there tends to be an association between certain research methods and theories used. The ecological method is associated with systems theory, the experimental method with learning theory, and the ethological method with ethological theory. Thus, to round out a complete investigation of human development and research methods, it is essential to become familiar with the major theories in the field. This is the topic of the next two chapters.

## References

1. Looft WR: The evolution of developmental psychology. *Hum Dev* 15:187–201, 1972.
2. Reed KL: *Models of Practice in Occupational Therapy.* Baltimore, Williams & Wilkins, 1984.
3. Gesell A, Thompson H: *The Psychology of Early Growth. Including Norms of Infant Behavior and A Method of Genetic Analysis.* New York, Macmillan, 1938.
4. Rose SJ: Description and classification—The cornerstones of pathokinesiological research. *Phys Therap* 66:379–381, 1986.
5. Miller LJ: *Miller Assessment for Preschoolers (Manual).* Littleton, CO, Foundation for Knowledge in Development, 1982.
6. Banus BS: The Miller Assessment for Preschoolers (MAP): An introduction and review. *Am J Occup Therap* 37:333–340, 1983.
7. Ager DL, Olivett BL, Johnson CL: Grasp and pinch strength in children 5 to 12 years old. *Am J Occup Therap* 38:107–113, 1984.
8. Bowman OJ, Katz B: Hand strength and prone extension in right-dominant, 6 to 9 year olds. *Am J Occup Therap* 38:367–376, 1984.
9. Crowe T, Deitz JC, Siegner CG: Postrotary nystagmus response of normal four-year-old children. *Phys Occup Therap Pediatr* 4:19–28, 1984.
10. Drews JE, Vraciu JK, Pellino G: Range of motion of the joints of the lower extremities of newborns. *Phys Occup Therap Pediatr* 4:49–62, 1984.
11. Erhardt RP: *Developmental Hand Dysfunction: Theory-Assessment-Treatment.* Laurel, MD, Ramsco, 1982.
12. Fike ML, Rousseau E: Measurement of adult hand strength: a comparison of two instruments. *Occup Therap J Res* 2:43–49, 1982.
13. Gregory-Flock JL, Yerxa EJ: Standardization of the prone extension test on children ages 4 through 8. *Am J Occup Therap* 38:187–194, 1984.
14. Williams M, Tomberlin JA, Robertson KJ: Muscle force curves of school children. In *Growth and Development. An Anthology.* Washington DC, American Physical Therapy Association, 1975.
15. Mathiowetz V, Rogers SL, Dowe-Keval M, Donahoe L, Rennells C: The Purdue Pegboard: Norms for 14 to 19 years olds. *Am J Occup Therap* 40:174–179, 1986.
16. Lefkof MB: Trunk flexion in healthy children aged 3 to 7 years. *Phys Ther* 66:39–44, 1986.
17. Short MA, Watson PJ, Ottenbacher K, Rogers C: Vestibular-proprioceptive functions in 4-year-olds: Normative and regression analyses. *Am J Occup Therap* 37:102–109, 1983.

18. Utley E, Pettit K, Robertson D: Southern California Postrotary Nystagmus Test: Adult normative data. *Occup Therap Mental Health* 3:29–34, 1983.
19. Fisher AG, Mixon J, Herman R: The validity of the clinical diagnosis of vestibular dysfunction. *Occup Therap J Res* 6:3–20, 1986.
20. Montgomery P: Assessment of vestibular function in children. In Ottenbacher K, Short MA (eds): *Vestibular Processing Dysfunction in Children*. New York, Haworth, 1985.
21. Scott AD, Trombly CA: Evaluation. In C. A. Trombly (ed): *Occupational Therapy for Physical Dysfunction*, ed 2. Baltimore, MD, Williams & Wilkins, 1983, pp 126–229.
22. Trombly CA, Quintana LA: Activity analysis: electromyographic and electrogoniometric verification. *Occup Therap J Res* 3:104–120, 1983.
23. Zemke R, Zemke WP: Electrogoniometry: A proposed research tool for the measurement of the asymmetrical tonic neck reflex. *Phys Occup Therap Pediatr* 2:51–62, 1982.
24. Helmstadter GC: *Research Concepts in Human Behavior*. New York, Meredith Corp, 1970.
25. Vasta R: *Studying Children*. San Francisco, WH Freeman, 1979.
26. Bohannon RW, LeVeau BF: Clinicians' use of research findings. A review of literature with implications for physical therapists. *Phys Therap* 66:45–50, 1986.
27. Payton OD: Clinical reasoning process in physical therapy. *Phys Therap* 65:924–928, 1985.
28. Rogers J, Masagantani G: Clinical reasoning of occupational therapists during the initial assessment of physically disabled patients. *Occup Therap J Res* 2:195–219, 1982.
29. Steger BH: Commentary: Clinical reasoning of occupational therapists during the initial assessment of physically disabled persons. *Occup Therap J Res* 3:50–53, 1983.
30. Campbell S: Editorial. *Phys Occup Therap Pediatr* 2 (2/3):1–2, 1982.
31. Christiansen CH: Editorial: Research: An economic imperative. *Occup Therap J Res* 3:195–198, 1983.
32. Gibson D: Guest editorial: The dearth of mental health research in occupational therapy. *Occup Therap J Res* 4:131–149, 1984.
33. Llorens LA: Guest editorial: A journal of research in occupational therapy: the need, the response. *Occup Therap J Res* 1:3–6, 1981.
34. Rogers JC: Guest editorial: Educating the inquisitive practitioner. *Occup Ther J Res* 2:3–12, 1982.
35. Ottenbacher KJ, Barris R, Van Deusen JV: Some issues related to research utilization in occupational therapy. *Am J Occup Therap* 40:111–116, 1986.
36. Smidt GL: Walking the trail of physical therapy research. *Phys Therap* 66:375–378, 1986.
37. Burkitt A: Development of research in physiotherapy. *Physiotherapy* 70:186–188, 1984.
38. Domholdt EA, Malone TR: Evaluating research literature: The educated clinician. *Phys Ther* 65:487–491, 1985.
39. Short-DeGraff MA, Ottenbacher KJ (Eds): *Collaborative Research in Developmental Therapy. A Model with Studies of Learning Disabled Children*. New York, Haworth Press, 1986.
40. Ottenbacher KJ: *Evaluating Clinical Change. Strategies for Occupational and Physical Therapists*. Baltimore, MD, Williams & Wilkins, 1985.
41. Ottenbacher K, Short MA: Publication trends in occupational therapy. *Occup Therap J Res* 2:80–88, 1982.
42. Wright HF: Observational child study. In Mussen PH (ed): *Handbook of Research Methods in Child Development*. New York, John Wiley, 1960.
43. Mussen PH, Conger JJ, Kagan J: *Child Development and Personality*, ed 5. New York, Harper & Row, 1979.
44. Pelletier JM, Short, MA, Nelson DL: Immediate effects of waterbed flotation on approach and avoidance behaviors of premature infants. *Phys Occup Therap Pediatr* 5 (2/3):81–92, 1985.
45. Hutt SJ, Hutt C: *Direct Observation and Measurement of Behavior*. Springfield, IL, Charles C Thomas, 1970.
46. Piaget J: *The Origins of Intelligence in Children*. New York, WW Norton, 1963.
47. Rothney JWM: *Methods of Studying the Individual Child. The Psychological Case Study*. Waltham, MA, Blaisdell, 1968.
48. Lucci JA: *Occupational Therapy Case Studies*, ed 2. Garden City, NJ, Medical Examination Publishing, 1980.

49. Cox RC, West WL: *Fundamentals of Research for Health Professionals.* Rockville, MD, American Occupational Therapy Association, 1982.
50. Line J: Case method as a scientific form of clinical thinking. *Am J Occup Therap* 23:308–313, 1969.
51. Fitzhugh ML: Sitting and sleeping habits of children. In *Growth and Development. An Anthology* Washington, DC, American Physical Therapy Association, 1975.
52. McCain G, Segal EM: *The Game of Science,* ed 3. Monterrey, Ca, Brooks/Cole, 1977.
53. Greene B: We need a hero. *Esquire* Jan: 29–30, 1986.
54. Hinde RA: *Animal Behaviour. A Synthesis of Ethology and Comparative Psychology,* ed 2. New York, McGraw Hill, 1970.
55. Blurton Jones N: Characteristics of ethological studies of human behavior. In Blurton Jones N (ed): *Ethological Studies of Child Behaviour.* New York, Cambridge University Press, 1972, ch 1, pp 3–36.
56. Michel GF, Moore CL: *Biological Perspectives in Developmental Psychology.* Monterey, CA, Brooks Cole, 1978.
57. Blurton Jones N: *Ethological Studies of Child Behaviour.* New York, Cambridge University Press, 1972.
58. Klaus MH, Kennell JH: *Parent-Infant Bonding,* ed 2. St. Louis, CV Mosby, 1982.
59. Sells SB: Ecology and the science of psychology. In Willens EP, Raush HL (eds): *Naturalistic Viewpoints in Psychological Research.* New York, Holt, Rinehart & Winston, 1964.
60. Walsh R: *Towards An Ecology of Brain.* New York, SP Medical & Scientific Books, 1981.
61. Klaus MH, Kennell JH: *Maternal-Infant Bonding.* St. Louis, CV Mosby, 1976.
62. Kielhofner G: Temporal adaptation: A conceptual framework for occupational therapy. *Am J Occup Therapy* 31:235–242, 1977.
63. Kielhofner G: A model of human occupation, part 2. Ontogenesis form the perspective of temporal adaptation. *Am J Occup Therap* 34:657–663, 1980.
64. Johnson DB, Deitz JC: Time use of mothers with preschool children: A pilot study. *Am J Occup Therap* 39:578–588, 1985.
65. Walsh RN, Greenough WT (eds): *Environments as Therapy for Brain Dysfunction.* New York, Plenum, 1976.
66. Walsh R: *Staying Alive. The Psychology of Human Survival.* Boulder CO, New Science Library, 1984.
67. Isaac S, Michaels WB: *Handbook in Research and Evaluation,* ed 2. San Diego, Edits, 1981.
68. Banus BS: Development and occupation. In Banus BS, Kent CA, Norton YS, Sukiennicki DR, Becker ML (eds): *The Developmental Therapist,* ed 2. Thorofare, NJ, Charles B Slack, ch 1, 1979.
69. Banus BS: *The Developmental Therapist.* Thorofare, NJ, Charels B Slack, 1971.
70. Ayres AJ: *Southern California Sensory Integration Tests.* Los Angeles, Western Psychological Services, 1972.
71. Ayres AJ: *Sensory Integration and Praxis Tests.* Los Angeles, Western Psychological Services, 1985.
72. Bernard HW: *Human Development in Western Culture.* Boston, Allyn & Bacon, 1970.
73. Bork CE, Francis, JB: Developing effective questionnaires. *Phys Ther* 65:907–912, 1985.
74. Goldman J, Stein C, Guerry S: *Psychological Methods of Child Assessment.* New York, Brunner/Mazel, 1983.
75. Murray HA: *Thematic Apperception Test Manual.* Cambridge, MA, Harvard University Press, 1943.
76. Ayres AJ, Reid W: The self-drawing as an expression of perceptual-motor dysfunction. *Cortex* 2:254–265, 1966.
77. Ottenbacher K, Haley D, Abbott C, Watson PJ: Human figure drawing ability and vestibular processing dysfunction in learning-disabled children. *J Clin Psychol* 40:1084–1088, 1984.
78. Scott LH: Measuring intelligence with the Goodenough-Harris Drawing Test. *Psychol Bull* 89:483–505, 1981.
79. Frankenburg WK, Dodds JB, Fandal AW, Kazuk E, Cohrs M: *Denver Developmental Screening Test Manual.* Denver, University of Colorado Medical Center, 1975.
80. Fitts WH: *Tennessee Self Concept Scale.* Nashville, TN, Counselor Recordings and Tests,

1965.

81. Palisano R, Short MA: Methods for assessing muscle tone and motor function in the neonate: A review. *Phys Occup Therap Pediatr* 4(4):43–54, 1984.

82. Kline P: Personality and individual assessment. In Dunkin EN (ed): *Psychology for Physiotherapists*. London, MacMillan, 1981.

83. Gardner H: *Frames of Mind. The Theory of Multiple Intelligences*. New York, Basic Books, 1983.

84. Folio MR, Fewell RR: *Peabody Developmental Motor Scales and Activity Cards*. Hingham, MA, Teaching Resources Corp, 1983.

85. Bayley N: *Bayley Scales of Infant Development*. New York, The Psychological Corp, 1969.

86. Achenbach TM: *Research in Developmental Psychology: Concepts, Strategies, Methods*. New York, Free Press, 1978.

87. Botwinick J: Intellectual abilities. In Birren JE, Schaie KW (eds): *Handbook of the Psychology of Aging*. New York: Van Nostrand Reinhold, 1977.

# 5

## Theories of Human Development
## Part 1. Cognition, Learning, and Ethology

A theory is difficult to define. Hill (1, p 23) notes, "In the broadest sense, a theory is a systematic interpretation of an area of knowledge." The functions of theories are to organize information, to provide descriptions and explanations, and to make predictions about bodies of information. Theories are not fact and are constantly subject to scrutiny and revision. One of the fundamental roles of research is to test hypotheses (predictions) that are derived from theories. Based on the results of such research, theories are supported or revised.

The field of human development is diverse, and is made up of many, sometimes similar and sometimes opposing, theoretical orientations. The purpose of this and the next chapter is to describe the more common theories in the field of human development and to illustrate how these theories also have application in the fields of physical and occupational therapy. Table 5.1 illustrates the major theoretical areas discussed in this chapter and in chapter 6.

# COGNITIVE THEORY

Cognitive theory examines the ability to reason, think, or solve problems, i.e., to gain knowledge about the world. This is the subject matter of the philosophical field of *epistemology*, the study of the nature of knowledge. The most famous cognitive theorist is the Swiss psychologist Jean Piaget (1896–1980), whose theory of development is drawn from information from the two fields of epistemology and biology.

# Piaget's Theory of Cognitive Development

Piaget's work reflects his biological origins. Although his theory of cognition is complex, it is often more clearly understood by drawing comparisons with familiar biological concepts (2, 3). For example, Piaget's overall view of the process of cognition can be compared to homeostasis. This is the body's process of maintaining a stable internal equilibrium by balancing energy input, internal stores, and energy requirements for existence. Piaget contended that the mind, like the body, also maintains a stable equilibrium; but, he refers not to physiological actions but to the abstract processes involved in thinking.

Table 5.1
Important Theories in the Study of Human Development

| Theory | Subject Matter |
|---|---|
| Cognitive | Thinking and reasoning, knowledge |
| Behavioral | Acquisition of behavior |
| Psychoanalytic | Feelings and personality |
| Ethological | Naturalistic study of behavior |
| Maturational | Sequential emergence of behavior |
| Humanistic | Uniqueness of human existence |

During homeostasis the body readily uses some simple substances (simple sugars) and modifies others (complex carbohydrates) for use. Similarly, in the process of gaining information about their worlds, people obtain knowledge that is either familiar or unfamiliar. Familiar information is readily adapted, but unfamiliar information must be modified for use. Piaget's theory states that in the same way that the body adapts, so also does thinking. This process of cognitive adaptation is termed *equilibration*, and Piaget's theory is referred to as the "cognitive theory of equilibration" (2, 3).

## Piaget's Concept of Equilibration

Equilibration consists of two complementary processes: *assimilation* and *accommodation*. Assimilation involves taking in information that fits with existing knowledge of the world. It requires incorporating existing cognitive structures into familiar or novel circumstances. Assimilation implies familiarity, whereas accommodation is a process of change. Accommodation is the development of new cognitive structures. It necessitates developing knowledge to accommodate to new or changed circumstances.

Thus, just as the body takes in substances and either readily uses them or changes them for use, so also does cognition involve taking in information or knowledge that is either readily assimilated or accommodated for use (2).

---

*Examples of Assimilation and Accommodation*

A 3-year-old girl who has a pet dog at home is out for a walk with her father. While walking through the park, they come across another dog. The little girl assimilates familiar information, points, and says, "Doggie!" Her father tells her, "Yes, there's a doggie."

As they continue walking, they come across a squirrel that has become tamed and is hunched by a tree watching the people. As the father and daughter approach the squirrel, it starts to hop toward them. Assimilating previous knowledge of furry animals, the girl points to the squirrel and says, "Bunny!" The father, however, says, "No, Genny, that isn't a bunny. It's a squirrel."

At this point, according to Piaget, Genny will enter a state of cognitive disequilibrium. Her existing knowledge of animals told her that furry, four-legged objects are either dogs or bunnies. Further, she may understand that dogs are big, and bunnies are small and hop. She saw a small animal that hops, assimilated that knowledge, and was wrong. This resulted in cognitive disequilibrium. Thus, to return to a state of balance, she must make accommodations to this new information about the world. She must learn that there are two kinds of small animals that hop, one called "bunnies" and one called "squirrels." Squirrels, she may also come to know, live in trees and tend to be gray.

When she accommodates to that new information, the process of equilibration will be complete until she encounters new and different animals. For now, she assimilates information about rabbits, squirrels, and dogs. That is, until she encounters other animals such as a cat, a raccoon, or a skunk. For each new animal, Genny will again need to accommodate to new knowledge of her world.

## Schema

The actual cognitive structure that undergoes adaptation during equilibration is termed a *schema* (plural, schemata). This is a complex concept that can also be simplified by comparing it to a biological concept. In the body, certain simple structures (such as the lens of the eye) or entire biological systems (such as the cardiovascular system) may undergo adaptations. Piaget proposed that just as physiological structures undergo adaptations, so also do cognitive structures. Also, just as physiological structures may be simple or complex, so also are schemata (2).

An example of a simple schema is a sucking reflex seen in infants, whereas an example of a complex schema is an adolescent's understanding of algebra. It is difficult to see what reflexes have in common with an understanding of algebra. To understand this parallel, we must first review Piaget's theory and stages of cognitive development.

### Sensorimotor and Internalized Intelligence

Piaget believed that one acquires knowledge by actively experiencing the world. Intelligence, according to Piaget, is a process of adapting to the world, and it changes during development. The earliest adaptations in life take the form of combined sensory *and* motor (i.e., sensorimotor) acts such as sucking, reaching, looking, and grasping. Thus, infants' early schemas are sensorimotor in nature. Later, however, through the processes of maturation and experience with the world, infants begin to combine schemata and then gradually begin to *internalize* their thoughts.

Internalized thinking involves forming mental representations for objects, events, or actions. It is the process that we typically refer to as "thinking." For example, sensory and motor schemata are combined and coordinated in the following sequence of actions. An older infant looks at a rattle, reaches for it, grasps it, brings it to the mouth, and begins to suck on it while releasing it from the fist. According to Piaget, each action in this sequence must initially be performed independently before all of these actions are coordinated together. At this point the emphasis is still on actions, but gradually with age and with interactions with the environment, the infant develops mental representations for all these movements.

As internal mental representations are formed, the infant will be able to actually think about events rather than having to always act them out. Sensorimotor acts that are eventually combined, organized, and represented internally in thought are termed *operations*. Operations, or internalized actions, form the basis for logical thought.

According to Piaget's cognitive theory, simple sensory and motor actions such as reflexes are the basic cognitive structures which make up *sensorimotor intelligence.* Sensorimotor intelligence gives way gradually to more sophisticated, formal, conceptual thought. As Piaget (4, p 119) noted, "Certainly sensori-motor intelligence lies at the source of thought, and continues to affect it throughout life." The continual reference to "action," as in "sensorimotor acts" or "internalized actions," is why Piaget's work is commonly recognized as a dynamic theory of cognitive development.

## Stages of Cognitive Development

There is a natural progression in moving from sensorimotor to internalized thought, and this progression makes up Piaget's well-known stages of cognitive development. Piaget proposed that there are four sequential stages of cognitive development that span birth to adolescence. Movement to a higher stage depends upon completion of each former level, as a child passes through an orderly, maturational sequence. The stages are not solely mat urationally determined, however, as Piaget recognized that children acquire knowledge of the world by active exploration. Thus, developmental progression results from a combination of maturation and interaction with the environment.

In regard to the nature-nurture issue that was introduced in chapter 2, Piaget is an interactionist. He does not adopt a dogmatic learning theory approach that claims that children learn by accumulating stimulus-response associations. In support of natural influences, Piaget contends that limitations are imposed on knowledge because of age; that is, an infant cannot be taught complex cognitive skills (such as mathematics) because the infant's cognitive system is not sufficiently mature to handle it. However, Piaget also recognized that knowledge is not passively obtained; it is a result of active exploration and interaction. Thus, equilibration is an interactionistic concept that, like homeostasis, involves an ongoing, constant interplay between the organism and the environment.

Each child's progression through the stages of cognitive development will depend on his or her own level of maturation and personal interactions with the world. Cognitive development is gradual and accumulative, not suddenly attained as one approaches a certain age. Thus, assigning ages to Piaget's developmental stages can be misleading, and caution should be exercised in doing so. In fact, different resources report varying ages. These approximate ages and the corresponding stages are listed in Table 5.2.

Table 5.2
**Piaget's Stages of Cognitive Development**

| Stage | Time Period | Approximate Age |
|---|---|---|
| Sensorimotor | Infancy | Birth to 18, or 24, months |
| Preoperational | Early childhood | 18, or 24, months to 7 years |
| Concrete operations | Late childhood | 7 to 12 years |
| Formal operations | Adolescence | 12 years on |

# Clinical Applications of Piaget's Theory in Occupational and Physical Therapy

Piaget's theory has been applied in occupational therapy, physical therapy, and other professions such as education, psychiatry, speech pathology, and psychology. Its widespread application is illustrated by the fact that it is so well adapted to so many disciplines. For example, the University of Southern California hosts an annual "International Interdisciplinary Conference on Piagetian Theory and the Helping Professions."

Cognitive theory finds its way into direct academic study as well as in application to clinical work. Piaget's theory is studied in allied health curricula as a basis for understanding normal human development. In addition, therapists have relied on this theory and its terminology as a basis for many principles of treatment.

Part of the appeal of Piaget's theory is its interdisciplinary focus. It is based on biological concepts, which are then integrated with information about sensory, motor, and cognitive development. In a similar fashion, both physical and occupational therapists utilize information regarding cognitive, sensory, motor, and other areas of development by pulling this information together and focusing treatment of the "whole" person.

There is a noticeable overlap between the language and scope of these two therapies and Piagetian theory; i.e., the mutual focus on adaptation, purposeful behavior, intentional movements, sensorimotor skills, cognition, function, and the development of all these abilities. Piaget has analyzed and provided details about the normal development of *praxis*, which is the coordination of movement (5). Therapists from many fields encounter patients who exhibit deficits in coordinated movements, or motor planning skills; this is referred to as *apraxia*. Thus, Piaget's work on normal development of praxis provides a reference for clinicians involved in the assessment and remediation of apraxic clients (e.g., 6–8). A comprehensive assessment developed and norm referenced by occupational therapists is the Sensory Integration and Praxis Tests (9), which is designed to examine a variety of sensorimotor and praxis skills.

## Mutual Focus on Adaptation and Purposeful Behavior

Piaget's theory is based on the concept of adaptation, and it stresses the significance of sensorimotor and purposeful behaviors as foundational for subsequent development. Similarly, the focus of occupational therapy is on occupation, or purposeful behaviors; both physical therapy and occupational therapy are oriented toward patient adaptation through sensorimotor activity. Wursten (10, p 214) notes, "One of the major psychological processes involved in retraining, rehabilitating, or helping an individual conquer an obstacle through occupational therapy is the process of adaptation. This process is one of the major cornerstones of Piaget's theory."

According to Piaget (4, p 5), "A perception, sensori-motor learning (habit, etc.), an act of insight, a judgment, etc., all amount, in one way or another to a structuring of the relations between the environment and the organism." Although this statement was made more than 20 years ago, it anticipated current ecological views regarding organism-environment relationships. Such views are compatible with occupational therapy and physical therapy, as they both help patients to restructure their relationships to their environments.

Physical and occupational therapists work with individuals who, for the first time, may be forced to relate to their worlds with only one hand, or with nonfunctioning limbs, or with visual impairments. Patients who once were independent may be unable to dress, feed, or clean themselves or to successfully adapt to their environments because of restricted motor or sensory functions. However, by altering their environments or the way they relate to their environments (e.g., with ramps, splints, adapted utensils, dressing aids, braces, work simplification) patients may be facilitated in adapting to their worlds and thereby restored to a previous level of independence.

## Mutual Focus on Interaction of Sensory and Motor Processes

Piaget's theory recognizes the interplay between cognition and sensorimotor interactions and the role that active involvement with the environment plays in development. This provides theoretical rationale and support for an integrated approach to treatment. A primary orientation of both physical therapy and occupational therapy treatment is on sensory and motor experiences designed to restore purposeful function. This can be found in clinical theories addressed primarily toward children with developmental delays as well as in rehabilitative theories oriented toward individuals with motor and cognitive deficits.

Sensorimotor approaches have been used in motor training and in treatment of individuals who have experienced cognitive, perceptual, sensory, and motor problems as a result of head trauma (11). Malkmus (12, p 1952) notes the importance of understanding normal cognition and cognitive disorders so that physical therapy can be aimed at "the cognitive, linguistic, emotional, and social consequences of head injury."

In regard to the use of sensorimotor approaches in pediatrics, Kopp (13, pp 217–218) explains, "Occupational therapists working with infants and children find Piagetian theory a cogent framework upon which to base a therapeutic program. No other theory incorporates infants' purposeful actions used in exploration and play within the context of developing intelligence."

Sensorimotor functions are the focus of therapy programs promoted by physical and occupational therapists (e.g., 14–17) and designed to facilitate normal motor and sensory functions in children whose development is delayed. Many of these programs assume that sensory and motor experiences will not only affect motor development but may carry over to other cognitive, social, emotional, or academic domains. For example, Ayres' sensory integration therapy is based upon the premise that the child becomes actively involved in sensory and motor activities that are designed to remediate cognitive or learning deficits. The theory behind this treatment approach recognizes that emotional, social, perceptual, language, and other skills are also affected by one's sensorimotor experiences (14).

### Piagetian Concepts in a Model of Treatment

An example of how Piagetian theory has been adapted to developmental therapy is found in *Children Adapt* (17). In that text, the authors—occupational therapists Elnora Gilfoyle, Ann Grady, and Josephine Moore—have directly apply Piagetian terminology and theory to a developmental therapy model of treatment. Gilfoyle and colleagues have constructed this developmental approach to pediatric rehabilitation that focuses on sensory, motor, and purposeful behaviors in children. These behaviors, the authors report, are attained (as is cognition according to Piaget) through the process of environmental interaction.

As Piaget reported on the significance of spatiotemporal relations to perception and intelligence, Gilfoyle and colleagues note the significance of spatiotemporal adaptation as a basis for the maturation of posture and movement. They report that adaptation consists of several components, two of which are assimilation and accommodation. In addition, these authors refer to a concept of the gradual development of motor behaviors that become automatic in performance (17, p 47). This process is similar in nature to Piaget's view of internalization. Thus, Piaget's principles are sufficiently useful that they can be adapted to forms of development other than cognition.

Although this section has emphasized sensorimotor aspects of Piaget's theory, it is not meant to diminish the additional social, emotional, and perceptual interactions with cognitive development. Piaget's theory is truly interactive. He recognized that cognitive abilities affect other developmental domains. For example, his theory includes explanations of play, language, and socialization, especially in regard to the infants', children's, and adolescents' relationships with the world around them. Therapists Gratz and Zemke (18) have pointed out the application of Piaget's theory for pediatric practitioners, in recognizing children's overall cognitive, social, and emotional reactions to illness or hospitalization.

Treating the whole person recognizes the significance of, and interaction among, all developmental domains, and Piaget's theory has acknowledged this interaction and provided a framework for therapists. In fact, from some therapists' perspectives, cognitive-perceptual and motor development cannot be considered in isolation and must be viewed from an integrated point of view (e.g., 16). These areas also relate to other behavioral, developmental, and clinical theories, as will be examined in Chapter 6.

The primary purpose of this section has been to provide a brief overview of cognitive theory. Its interactive implications and considerably more details about Piaget's stages of development are found in Chapter 11.

# Other Cognitive Theories

Piaget is not the only theorist in the field of cognition, but he is the most widely recognized and acclaimed (19). As discussed in detail in chapter 11, many other psychologists have contributed to a body of information about how humans think and acquire knowledge about the world. Some researchers have elaborated on or modified Piaget's theory, while other theorists, who are discussed next, offer ideas that may contrast with some of Piaget's concepts.

Piaget's theory is concerned with structures in the mind or the structure of thought. Thus, Piaget and others (e.g., Bruner and Kohlberg) have been called "structuralists" or "cognitive-structuralists," which refers to their proposal of specific structures in the mind that correspond to laws about the way the world is structured; i.e., that events and objects exist in time and space, that there are cause and effect relationships, and that these laws of nature regulate behavior (19).

According to cognitive-structuralists, thinking changes qualitatively as an individual progresses through the various stages of development. Thus, each of Piaget's stages represents very different types of thinking. Theories, such as Piaget's, that emphasize cumulative and sequential stages of development are referred to as *stage theories*. According to Piaget, a child must pass through one stage of development and accumulate the knowledge characteristic of that stage before progressing on to the next. Once achieved, that next stage represents for the child, new and qualitatively different forms of thinking that are a result of experiences and maturation.

In contrast, environmental-learning theorists focus on associations and how these are accumulated in knowledge or skill acquisition. Environmental-learning theorists see continuity in mental development, as increased knowledge is merely an addendum to previous associations and abilities (19, 20). While environmental-learning theory has been applied to cognitive development, it also exists as a general theory that explains how all behaviors are acquired.

# LEARNING OR BEHAVIOR THEORIES

Learning theorists (also referred to as "environmental-learning theorists" [19]) study observable, measurable behaviors and formulate principles about how those behaviors are acquired, i.e., learned. Some students confuse cognitive and learning theory, treating synonymously the terms "thinking" or "reasoning" and "learning". In Piaget's theory, thinking is an abstract process that differs markedly from the concept of learning as it is presented in this section. Learning theory and cognitive theory are very different, as will become apparent as you read on.

Learning theory has its roots in British empiricism (1700s and 1800s), which maintained that knowledge is acquired by experience. John Locke, an empiricist, viewed the mind as a blank slate or *tabula rasa* at birth, with knowledge being the accumulation of sensory experiences that are "etched onto" the mind. These sensory experiences become more and more meaningful when they are repeatedly associated with other sensations. According to the empiricists, associations are fundamental for learning, and combinations of associations form the basis for complex thinking and perceptual organization.

# What Is Learning?

We regularly use the terms *learning, stimulus,* and *response* without specific attention to their definitions. A stimulus (plural, *stimuli*) can be defined simply as some kind of sensation. Hill explains that in learning theory a sensation is "significant in that it elicits or controls a response." A response is an item of behavior (1, p 248). The terms "stimulus" and "response" are more easily and less controversially defined than is "learning."

As was pointed out in chapter 2, definitions of terms, especially "learning" and "maturation," can vary extensively and may have profound philosophical implications depending upon one's stance on the nature-nurture debate. In some cases, "learning" has been defined strictly as a change in behavior independent of innate, maturational influences. We know from chapter 2 that such a definition is too restrictive and has been replaced by more contemporary wording that acknowledges interactive influences in behavioral acquisition. With this in mind, a general definition is offered. Learning is defined by Hilgard and Marquis (21, p 2) as "a relatively permanent change in behavior as a result of practice." The term "relatively permanent" is added to the definition in order to discount temporary changes in behavior due to fatigue, drugs, or other short-term behavioral states.

Learning theorists emphasize the role that environment plays in development, contending that behavior is acquired by the formation and accumulation of associations. This does not mean that maturation or other forms of behavioral acquisition are not present, although at one time in the history of this field, learning was viewed as the primary, if not the only, way behaviors were acquired.

Learning theory, as it is presented here, does not defend nor advocate for one side of the nature-nurture debate. Instead, it exists to discover principles about how behaviors are acquired. As such, this theory is important to therapists because of their role in assisting patients to acquire new skills, to alter ineffective behavioral strategies, to change developmental task performance, and to discover new ways of interacting with the environment.

# Behaviorism

American psychologist John B. Watson (1878–1958) founded the American school of psychology known as *behaviorism*. Watson announced the behaviorist position in an article published in 1913. This was a time when much of psychology was studying the structure of thinking and abstract mental processes such as mental states and consciousness (22, 23). In contrast to analyzing how and what people think, Watson believed that psychologists should study behavior. Thus, while many psychologists, including Piaget, were interested in understanding the internal structures underlying thought, behaviorism was oriented toward the experimental investigation of observable, measurable behaviors.

In regard to the nature-nurture conflict, behaviorism has been associated with the extreme nurture position. For some psychologists and clinicians this has been a position too dogmatic and extreme for consideration, let alone for adoption in practice. For example, the extreme behaviorist position is characterized by Watson's famous and often-cited claim:

> Give me a dozen healthy infants, well-formed, and my own specified world to bring them up in and I'll guarantee to take any one at random and train him to become any type of specialist I might select—doctor, lawyer, artist, merchant-chief, and, yes, even beggar-man and thief, regardless of his talents, penchants, tendencies, abilities, vocations, and race of his ancestors (24, p 104).

Although Watson's claims have been used to illustrate an extremist nurture position, psychologist Skinner (25) points out that Watson was not as dogmatic as people have subsequently classified him:

> Watson himself had made important observations of instinctive behavior and was, indeed, one of the first ethologists in the modern spirit, but he was greatly impressed by new evidence of what an organism could learn to do, and he made some rather extreme claims about the potential of a newborn human infant. He himself called them exaggerations, but they have also been used to discredit him ever since (25, pp 5–6).

Influenced by British empiricism, Watson believed that learning is a result of experience and that behavior can be reduced to associations that occur between stimuli and responses. These associations are sometimes referred to as stimulus-response, or S-R, bonds or connections. "Learning theory" is sometimes used synonymously with the term "behavior theory," but the student should recognize that not all theories of learning are classified as "behaviorism."

For example, the terms "S-R bonds" and "S-R connections" derive from Edward L. Thorndike's (1874–1949) learning theory called *connectionism*, which antedated behaviorism (22, 23). In addition, Watson was influenced by the work of Russian physiologist Ivan Pavlov (1849–1936), whose work preceded behaviorism and whose work on conditioning is a major influence in learning theory.

# Classical Conditioning

His research on digestion won Russian physiologist Ivan Pavlov the Nobel Prize in 1904. Pavlov, however, is remembered in psychology for his work on the *conditioned reflex*, which he thus named (26). While examining physiological processes involved in digestion in dogs, Pavlov noted that in addition to the normal salivation that occurred with the presentation of food, the dogs sometimes salivated in *anticipation* of food. That is, they salivated when they saw the food or saw the individual who was bringing food to them.

Thus, as our everyday experiences might tell us, the presentation of a repeatedly used dish, the sound of a can opener, the noise associated with getting food from the refrigerator at a particular time of the day, all might have important associations for our family pets. Pavlov experimentally demonstrated this in what is now considered a classical study of learning. Thus, Pavlov's work is referred to as *Pavlovian* or *classical conditioning*. This is defined as the ability of a previously neutral stimulus to elicit a conditioned (learned) response.

Classical conditioning involves the formation of a relationship between unconditioned stimuli (UCS) and responses (UCR) and conditioned stimuli (CS) and responses (CR). In Pavlov's experiment, meat powder (UCS) was known to unconditionally elicit the reflex of salivation (UCR). Pavlov demonstrated that learning would occur when some neutral stimulus (the presence of the experimenter, for example) was repeatedly associated with the meat powder. Thus, a neutral stimulus would gradually take on the response-eliciting properties of the unconditioned stimulus (Fig. 5.1).

Figure 5.1 illustrates Pavlov's example of classical conditioning. In this demonstration, a tone, which is neutral stimulus, is paired with the delivery of meat powder (UCS) to the dog. Eventually the tone, by itself, elicits salivation, even in the absence of the meat powder. This happens because of classical conditioning. When the tone takes on the response-eliciting properties of the UCS, it is called the *conditioned stimulus* (CS). The response that it elicits is called the *conditioned response* (CR).

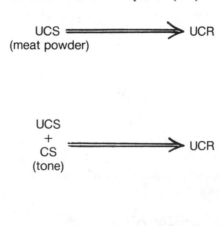

**Figure 5.1.**   In classical conditioning, a previous neutral stimulus (the conditioned stimulus) such as a tone, is associated with and gradually takes on the response-eliciting properties of an unconditioned stimulus (UCS).

Although the CR and UCR appear the same (e.g., in this case, the salivation responses), they are subtly different. For example, the CR and UCR have been found to differ in strength, duration, or *latency* (amount of time for response to be initiated) (21).

## Subtle Effects of Classical Conditioning

Classical conditioning is one way that learning occurs, and it is exhibited in very subtle ways in everyday life. Neutral stimuli in our lives take on many different values by their association with other powerful events such that certain sights, sounds, or smells can evoke in us very specific feelings or actions. For example, special songs, perfume or aftershave, crinkling candy paper, or the smell of burning leaves can produce powerful emotional responses because they have previously been associated with other potent unconditioned stimuli and responses. Thus, we all know how the smell and sound of popping popcorn can affect us when we are in a mall or a theater! Many grocery stores or shopping mall bakeries know that the smell of baking cookies has warm, pleasant, homey associations for most people. Thus, these odors are used not just to boost cookie sales but also to promote a warm ambience that signals the customers to feel good, to stay around, and to hopefully make more purchases than they would have without the baking cues.

Feinberg reportedly (27) demonstrated how subtle conditioning influences can be in everyday life. He used two groups of students in his experiment. All students were asked to give donations to the United Way charity. With one group a credit card was on a table when the students were queried, whereas the credit card was not evident for the second group. Nearly all the students in the first group offered donations compared with only one-third of the students in the second group. Feinberg (27, p 80) explains these results as evidence that the credit cards "may have become a conditioned stimulus that [provided] consumers with a cue to spend—even when the credit cards [weren't] their own!"

Thus, classical conditioning can be a powerful influence that accounts for many emotional responses in human behavior as well as for the acquisition of behaviors during development. Classical conditioning, however, is not the only form of learning. There are two basic types of conditioning. Each type is known by a variety of names (Table 5.3); for example, each

**Table 5.3**
**Different Names for Two Types of Conditioning**

| Pavlovian |
| Classical |
| Reflex |
| Respondent |
| Skinnerian |
| Instrumental |
| Operant |

type of conditioning is named after the scientist who demonstrated it or widely promoted its principles.

Pavlovian conditioning is of course named after the scientist, but since it was demonstrated early and is a classical form of learning, it is also referred to as *classical conditioning*. Other terms used to describe the same type of learning are *reflex* or *respondent conditioning*. The term "reflex conditioning" comes from Pavlov's term "conditioned reflex." In fact, the UCR (unconditioned response or unconditioned reflex) is typically a reflexive response, such as salivation, an eyeblink, increased heart rate, or crying. The term "respondent" comes from psychologist B. F. Skinner who wanted to differentiate between two types of conditioning and two types of behaviors: respondent and operant.

## Operant Conditioning

Like other behaviorists, Skinner recognized that interaction with the environment shapes behavior. However, Skinner noted that environmental interaction has different effects depending upon the temporal sequence of stimuli and responses. Skinner noted that there are two kinds of behaviors: *respondents* and *operants*. Respondents are responses that are either strengthened or weakened by stimuli that come *before* them. In classical conditioning, the meat powder and the tone come *prior to* the dog's salivation response. Salivation, in this case, is a respondent that is strengthened by the UCS that preceded it.

Skinner noted that in some cases of learning, behaviors emerge and are strengthened when no apparent stimuli elicit them. In such circumstances, the organism emits a behavior that in turn is strengthened by environmental consequences that *follow* it. Those consequences affect the organism such that the initial behavior will either be repeated or not. While this sounds very complicated, it simply means that if you do something and it ends up with good results, then you will tend to do it again.

An environmental consequence that causes a behavior to be repeated is what Skinner refers to as *reinforcement*, and the behaviors that are shaped by reinforcement are referred to as *operants*. Operants are defined as responses that are controlled by stimuli that follow them (28).

A simple example of an operant response is exhibited by a young boy who returns to a pay telephone to check the coin slot because a previous visit resulted in his finding a coin. Similarly, a cat may return to a particular neighbor's backyard because the initial visit resulted in its finding birds. The actions of finding a coin or locating birds altered the boy's and the cat's behaviors, respectively, such that both will repeat the same operant behaviors again (and again).

Skinner demonstrated operant conditioning using an experimental chamber referred to as a *Skinner box*. This is a controlled environment used most often for training laboratory animals such as pigeons and rats. The chamber consists of bare walls (sometimes fitted with Plexiglass to aid observation), a top, possibly some lights, and a lever that is attached to a feeding apparatus. Pushing the lever causes the feeding apparatus to dispense a small food pellet or a small amount of liquid.

In a demonstration of operant conditioning, a naive (untrained), hungry rat is placed into the Skinner box and allowed to explore it. Eventually, by accident, the rat will hit the lever and dispense a food pellet. This "accident" may happen several times such that the rat will associate proximity to a particular area of the Skinner box with food. Spending more time near the lever, the animal more frequently strikes the lever and receives food. Gradually, the rat makes an association between hitting the lever (the operant response) and receiving food (the stimulus). At this point, the rat has been operantly conditioned and will repeatedly hit the lever to receive food until either becoming sated or depleting the food supply.

Note that unlike classical conditioning, where the dog responded to some stimulus, in operant conditioning the rat in the Skinner box must first make a response and then a stimulus follows. Thus, respondents are shaped by stimuli that come *before* them and operants by stimuli that come *after* them.

Another way to differentiate operants and respondents is to note that "emitted responses" are operants, whereas "elicited responses" are respondents (26). In the Skinner box, the rat had to emit some behavior before obtaining a reinforcement. In respondent conditioning, conditioned behaviors are elicited by a specific stimulus or stimulus situation. This is explained by Schultz (23):

> Another difference between the two kinds of response behavior is that operant behavior operates on the organism's environment while respondent behavior does not. The harnessed dog in Pavlov's laboratory can do nothing but respond when the experimenter presents the stimulus. The dog cannot act "on his own" to deliver the stimulus (p 249).

Meat powder elicits salivation in Pavlov's experiment, whereas the rat in the Skinner box will not receive a food pellet unless he "acts on" the environment by striking the lever. The operant behavior of the rat in the Skinner box is actually instrumental in obtaining food. The fact that in operant, or Skinnerian conditioning, the organism is instrumental in obtaining reinforcement gives this type of learning an additional name, *instrumental conditioning* (22) (see Table 5.3).

# Applications of Conditioning and Learning Theory

Conditioning is not synonymous with learning. Learning is generally regarded as a change in behavior brought about by experience, whereas conditioning is more specific. Sometimes the term "learning" is used generally to cover a broad range of behaviors including conditioning and the general category of human problem solving (21). Learning theories are bodies of information designed to explain how learning occurs. The term "conditioning" is typically used to refer specifically to classical or operant conditioning.

## Connectionism

A learning theory developed by E. L. Thorndike was predominant in the early part of the 20th century and had considerable impact upon the work of Pavlov and Skinner as well as on contemporary clinical theories. Thorndike's work, known as "connectionism" because of its emphasis on the association between S-R bonds, or connections, contended that learning occurs according to certain laws (26). These laws were subject to considerable scrutiny in the fields of psychology and education, and in the 1930s, they were rejected and modified. While outdated and referred to as the "earlier laws" (Table 5.4), these laws continue to be applied in the fields of psychology and developmental therapy.

### Connectionism and Reinforcement

Thorndike's notion of a satisfying state of affairs in his law of effect (Table 5.4) and Skinner's concept of reinforcement are essentially the same. "Reinforcers" are events that increase the likelihood that a behavior will be repeated. Reinforcers and Thorndike's "satisfying state of affairs" both refer to environmental circumstances that enhance the probability that a behavior will be repeated and that learning will occur. (Thorndike's concept of an "annoying state of affairs" will be discussed later.)

While Skinner adopted Thorndike's earlier law of effect, he did not, however, adopt Thorndike's earlier law of exercise. Thorndike initially believed that practice, alone, would strengthen the connections necessary for learning. In contrast, Skinner's operant conditioning emphasizes reinforcement and not practice. Skinner claims that what alters behavior is the environmental consequence following it, not the action of behaving. Thus, according to Skinner, learning requires organism-environment interaction rather than action alone, as environmental consequences are necessary as feedback to sustain behavior.

Skinner does recognize that practice has some effect. He notes that although practice itself will not alter behavior, it may result in behavioral changes because of the increased opportunity of reinforcement during high rates of performance (26).

**Table 5.4**
**Thorndike's Early Laws of Connectionism**[a]

*Earlier Law of Effect*

When a connection is made and accompanied by, or followed by, a satisfying state of affairs, the strength of the connection is increased. When a connection is made and followed by an annoying state of affairs, its strength is decreased.

*Earlier Law of Exercise*

Law of Use: Connections will be strengthened with practice.
Law of Disuse: Connections will be weakened or forgetting will occur when practice ends.

[a] From Hilgard ER, Bower GH: *Theories of Learning*, ed 2. New York, Appleton-Century-Crofts, 1966, pp 19–20.

## Connectionism and Brain Organization

In the 1930s, Thorndike's laws were modified. Hilgard and Bower (26, p 25) note: "The law of exercise was practically renounced as a law of learning, only a trivial amount of strengthening of connections being left as a function of mere repetition. The law of effect remained only half true, the weakening effects of annoying consequences being renounced." Despite their renouncement in psychology, the laws still found their way into other fields. The law of exercise, especially, has an appeal, partly because of its simplicity and clarity and partly because of the generalizability of such terms as "exercise" and "practice" to the study of motor development and skill acquisition. This has led to its adoption in clinical theories regarding brain development and intervention.

Occupational therapist A. Jean Ayres has formulated a theory of clinical treatment called "sensory integration," which is aimed at children with deficits in learning skills. Sensory integration theory is based on the concept that some developmentally immature clients may gradually mature neurologically when they participate in meaningful sensorimotor experiments (14). In describing the theory underlying her principles of treatment, Ayres (29) compares muscle development to brain development in words that are reminiscent of Thorndike's law of exercise (Table 5.4).

Ayres (29, p 45) claims that just as exercise enhances muscle strength, so also can use of the brain enhance the connections between nerve cells, which are called *synapses*: "The more a muscle is used, the stronger it becomes. . . . If it is not used, it becomes weak. . . . Similarly the more a synapse is used, the stronger and more useful it becomes. As with muscles, the use of a synapse makes that synapse easier to use; and the disuse of a synapse makes it harder to use that synapse." Thus, Ayres claims that in brain organization (and by inference, learning), the connections between nerve cells are strengthened with use and weakened with disuse. This is the essence of Thorndike's earlier law of exercise.

Considering the fact that synapses are "connections" between nerve cells, it makes sense to apply connectionistic theory to explain how they develop. Despite the fact that Thorndike's earlier law of exercise was rejected and modified in learning theory in the 1930s, it has been widely applied (as in Ayres' work) to theories of brain organization. In regard to brain development, such explanations are termed "use/disuse theories" since they emphasize the need for exercise or practice in the development of synapses.

While these use/disuse theories were popular in psychology throughout the 1950s and 1960s, they have recently been replaced by other, contemporary formulations which view the brain from a systems, or holistic, perspective. Most contemporary views of brain organization now note that nerve cells are in constant activity and not just active during "exercise" or use (30). Additional details about use/disuse theories of brain organization and sensory integration theory are beyond the scope of this book, but the reader who is interested in such topics may refer to sources listed at the end of this chapter (e.g., 31, 32). At present, it should suffice to point out that while reference is often not made to Thorndike by name, his principles have been widely applied in many theories and in clinical practice.

## Learning in Development

Whether learning or behavioral theories are really developmental in nature is a topic for debate in the study of human development. When compared with Piaget's theory, learning and conditioning do not address a specific area of human development such as cognition, nor do they describe the progressive changes that occur during a certain time period in life. However, learning theories do explain how behaviors are acquired, and as such, explain how and why an organism's behavior may change over time.

A study conducted by Watson provides a clear example of how learning theory explains the acquisition of a specific behavior.[a] As a strict behaviorist, Watson contended that learning is a result of associations between stimuli and responses (24). Thus, he set up a classic case study to demonstrate how a behavior could actually develop in a child. His study of Albert was "the first laboratory attempt to condition an emotion in a child" (33, p 581).

---

### The Acquisition of Fear

Albert was a 9-month old boy who initially displayed no fear of white, furry animals. Psychologist Watson, however, demonstrated that Albert could learn this fear through classical conditioning. For this demonstration, an experiment was set up such that each time that Albert was presented with a white rat, a loud noise occurred. "After several associations of the startling sound with the presentation of the rat, Albert not only withdrew in fright from the rat but the negative reaction to the rat eventually persisted without reinforcement of the loud sound. . . " (33, p 581).

Through the process of classical conditioning, Albert associated the loud noise with the presence of the white rat so that eventually the rat alone caused him to cry.[b] In addition, Albert became fearful of other white animals, a rug, furry objects, and even a Santa Claus mask with a fuzzy white beard (33, 34).

While the experiment with Albert seems cruel, it must be evaluated according to the context in which it was conducted. At the time the study was

---

[a] Watson's study is also an excellent practical example of classical conditioning. As a review of the principles of classical conditioning, see if you can identify the UCS, CS, UCS, and CR in the study.

[b] In regard to identifying the UCS, CS, UCR, and CR: In this study, the neutral stimulus was the rat. It had no negative associations for Albert until conditioning took place. Once associated with the loud noise, however, the rat took on the fear-inducing properties of the noise and became a conditioned stimulus (CS). The UCS in this study is the loud noise because it unconditionally elicited a fear response in Albert each time it was presented. The UCR and the CR were fear reactions—crying, fidgeting, etc.

designed, emotional learning by association had not been demonstrated experimentally. While contemporary scientists would not consider conducting such a study today, their judgement rests on the hindsight gained from previous experimental work conducted by Watson and others who demonstrated the significance and power of associations. Watson explains why he perceived this experiment with Albert to be important in the field of human development:

> But this fear of the rabbit is not the only building stone we have laid in the child's life of fear. After this one experience, and with no further contact with animals, all furry animals such as the dog, the cat, the rat, the guinea pig, may one and all call out fear. He becomes afraid even of a fur coat, a rug, or a Santa Claus mask. He does not have to touch them; just seeing them will call out fear. This simple experiment gives us a startling insight into the ways in which our early home surroundings can build up fears. You may think that such experiments are cruel, but they are not cruel if they help us to understand the fear life of the millions of people around us and give us practical help in bringing up our children more nearly free from fear than we ourselves have been brought up. They will be worth all they cost if through them we can find a method which will help us remove fear (34, p 46).

## Removing Fears

One of Watson's students, Mary Cover Jones, reported that it was one of her greatest satisfactions that she could be associated with the removal of fear, rather than with its development (35, p 581). She demonstrated experimentally that a child's fears, as Watson predicted, could be alleviated by essentially reversing the procedure used with Albert. In this case, positive events were substituted for negative ones.

In her experiment (35), Jones used a 34-month-old child by the name of Peter as a subject. Similar to Albert, Peter had been found to exhibit strong fears of white rats, rabbits, fur coats, feathers, cotton, wool, etc. Thus, he provided an excellent illustration for reversing the experiment which had initially been conducted with Albert, i.e., rather than causing a new fear, eliminating an existing one.

The procedure for removing Peter's fears was started by bringing him into a playroom with three other children, all of whom had been tested and displayed no fear of a white rabbit. A rabbit was then introduced into the playroom and gradually placed closer and closer to Peter. In addition, Jones also placed the rabbit as close as possible to Peter while he was given food that he liked. The purpose of this was to enable Peter to become familiar with the rabbit in nonthreatening circumstances and to associate positive rather than negative events with the animal. The procedure worked. Eventually, Peter tolerated holding and playing with the rabbit (35).

The experiments with Peter and Albert were used as evidence that behaviors can be acquired or eliminated by forming or breaking associations. Thus, Albert developed a previously unacquired fear of white, furry objects, and Peter lost a similar fear. Watson and his colleagues clearly demonstrated the significance of associations and the influence of conditioning in the acquisition of behaviors early in development.

The fact that principles of learning theory can be used to change behavior has given it important clinical applications. Thorndike's work, for example, was generalized to the field of education (26), and Watson's concerns about the acquisition of fear responses during development have already been described. Learning and conditioning theory form the basis of a controversial but widely recognized and applied therapeutic intervention known as behavior modification therapy.

## Behavior Therapy

In a review of behavior modification therapy (also referred to as behavior therapy, behavior modification, behavioral techniques, applied behavioral techniques, etc.), O'Neill and Gardner (36) note:

> Behavior therapy has expanded prodigiously since the 1920's, when the Russian physiologist Ivan Pavlov discovered that reflexive behavior could be conditioned. The clinical application of behavioral techniques has grown so swiftly that terminology is often confused. No universally accepted definition of behavior therapy exists. For example, some writers use the term to refer only to those clinical techniques based on classical (Pavlovian) conditioning, reserving the term behavior modification for techniques predicated on operant conditioning or social learning theory. But a survey of the accumulating literature makes it clear that the term behavior therapy is used to label a wide variety of interventions with little reference to the specific principle on which the techniques are based. We use the term behavior therapy to refer to interventions based on principles of learning, including those derived from operant conditioning, classical conditioning, and social learning theory... (p 709).

This text will adopt O'Neill's and Gardner's position and use the term behavior therapy as they do. In addition, other definitions they offer will also be used: *Behavior modification* is defined as the clinical application of operant conditioning. The basic science of behavior modification is referred to as *applied behavior analysis.* O'Neill and Gardner (36, p 709) define this as "an attempt to understand the functional relationship between a response and the environment—that is, between behavior and its consequences."

There are many learning theory principles that are applied in behavior therapy. Some of these (e.g., reinforcement and punishment) will be briefly introduced in this section and then described with other concepts (e.g., discrimination, generalization, extinction) later, when it will be easier to see their application to specific instances of development. For more extensive examinations of learning theory principles and the application of behavioral techniques in the fields of human development, occupational therapy, and physical therapy, numerous resources in addition to other chapters of this text can be consulted (37–43).

Behavior modification principles have had enormous applications in a wide variety of settings and with diverse groups of patients. For example, in occupational or physical therapy settings, it has been effective in altering

*"I wish they'd make up their minds. One day I'm 'good boy,'*
*the next day I'm 'bad boy.'"*

**Figure 5.2.**   Drawing by Richter; © 1979 The New Yorker Magazine, Inc.

social skills in emotionally disturbed adolescents (44) and in adult patients with psychiatric problems (45); in enhancing dressing skills (46) and work skills during occupational therapy (47); in altering standing and walking abilities of a retarded child (48); and in enhancing the motor control or ambulation of individuals with motor deficits associated with cerebral palsy (49) or other neurological or orthopedic problems (50). In fact, Rapport and Bailey (51, p 87) refer to the "concurrent use of physical therapy and behavioral technology/assessment practices" as behavioral physical therapy.

Behavior therapy has been used with infants, children, adolescents, and adults and with individuals with psychiatric, emotional, muscular, orthopedic, neurological, or developmental problems. It is also applied in normal development such as in habit control or toilet training for children as well as in training parents of normal, disabled, or disturbed children (37, 52, 53). Many adults voluntarily participate in reinforcement-based behavior therapies designed to help them lose weight, quit smoking, or alter other undesirable habits or fears (43). Thus, behavioral principles from learning theory are important for clinical practice.

## Reinforcement

Reinforcement is an important process in Skinnerian conditioning and a fundamental tool in behavior therapy. There are misunderstandings, however, about the relationship between reinforcement and punishment and how they are applied in behavior therapy (Fig. 5.2).

First, it should be pointed out that there are no universal reinforcers. An important point for clinicians, and one to which we will return, is that reinforcement is defined by the effect it produces (17). The effect is to increase the probability of a response.

Reinforcement is some environmental consequence that causes a behavior to be repeated. Thus, each time the hungry rat in the Skinner box hits the lever, it receives a food pellet. If the food is reinforcing—which it should be to a hungry rat—then it will continue to hit the lever. Eventually the rat will work very hard, hitting the lever many times to receive food. Food is therefore a reinforcer for the hungry rat.

Note, however, that if the rat had just eaten, then food would not alter the rat's behavior and would not be considered a reinforcer. Similarly, a child who likes to be touched may increase the rate of certain behaviors by receiving a pat on the back. In this child's case, touching is reinforcing. However, there are some children who do not like to be touched. Cuddling, caressing, or hugging may not be effective in altering such children's behavior. What is reinforcing, therefore, must be determined by its effect on behavior. What is reinforcing to one person may not be to another.

Money, food, prestige, high grades, and attention can be positive reinforcers. In fact, much of our interactions with others rely on what is often called "social reinforcement." Psychologists distinguish between *primary* and *secondary reinforcers*. Primary reinforcers are those that increase the probability of behavior, whereas secondary reinforcers have the same effect only by being associated with primary reinforcers. Thus, a check for large sum of money is not intrinsically reinforcing except that it has been repeatedly associated with other reinforcers. Similarly, the words "Good job!" or "Thanks a lot!" do not have real value except for their previous association with other primary reinforcers.

Social reinforcers are tied to such important human values as approval, attention, and love; thus, they are very important in sustaining human behavior and interaction (Fig 5.3). For example, in conversation with friends, you often nod your head, or say, "Mm-hm," or "yes," in addition to paying close attention by smiling and looking at them while they talk. Your behaviors actually sustain the conversation by reinforcing your friends' conversation. Inasmuch as those behaviors increase the probability of your friends' talking, those behaviors are positively reinforcing.

If you want to test such reinforcers, try the following experiment. The next time you are with a friend who is talking excitedly, try out two different approaches. First, pay close attention, look at the friend, nod your head, and respond when appropriate during the conversation. Check your friend's level of conversation during those circumstances and compare it with a situation in which, when your friend talks, you look around but not at your friend, act distracted, and do not give any verbal reactions. If your friend is responding to your social reinforcers, then his or her rate of conversation should increase in the first case and diminish or terminate in the next. If your experiment worked, you actually performed behavior therapy based upon learning theory principles.

**Figure 5.3.**   A and B, Social reinforcers such as smiles, hugs, or approval serve to motivate behavior and to sustain social interaction.

## Positive and Negative Reinforcement and Punishment

Reinforcement can be either positive or negative, but its effect is to *increase* the probability of a response. Hilgard and Bower (26) provide the following definitions from their book, *Theories of Learning*:

> 1.  A positive reinforcer is a stimulus which, when added to a situation, strengthens the probability of an operant response. Food, water, sexual contact, classify as positive reinforcers.
> 2.  A negative reinforcer is a stimulus which, when removed from a situation, strengthens the probability of an operant response. A loud noise, a very bright light, extreme heat or cold, electric shock, classify as negative reinforcers (p 113).

Negative reinforcement is often confused with punishment, but these are really two different phenomena in learning theory. Strictly speaking, reinforcement increases the probability of operant behavior, whereas punishment tends to diminish the probability of responses that it follows. Punishment and negative reinforcement are similar in that they are both aversive, but there are subtle differences between them. "Electric shock, for example, [is used as] a negative reinforcer because the termination of the shock is reinforcing" (1, p 62).

While reinforcement increases the strength of behaviors, either through the occurrence of a positive reinforcer or through the termination of negative reinforcer, the goal of punishment is to reduce responses. Punishment, learning theorists point out, does not actually eliminate behaviors but may only suppress them. Thus, children who are spanked for reaching for the cookie jar will tend to reach for it *less often*, or they will learn not to reach for it when they are likely to be caught. A more effective procedure than punishing children is to positively reinforce other appropriate behaviors that can substitute for the negative one.

According to Skinner, punishment is not the method of choice for controlling behavior. The way to eliminate responses is with the absence of reinforcement, rather than with punishment (1). For the student who is interested in further details about the subtle differences between reinforcement and punishment, a variety of sources may be consulted (1, 21, 26, 54). To further clarify these terms and to illustrate the clinical applications of punishment, negative reinforcement, operant, and classical conditioning the following case example is presented. It is based upon a true case study reported in *Elementary Principles of Behavior* (55).

---

### A Clinical Example of Conditioning with Reinforcement and Punishment

Scotty was an institutionalized mentally retarded child who engaged in self-mutilation behaviors; that is, when his arms and hands were free so that he could touch his head, he would repeatedly hit and injure himself. Thus, he had to be restrained to prevent serious self-injury. Restraint, however, is not a desirable option, though in some cases it may be necessary. In Scotty's case, he had been institutionalized for 4½ years and had spent all of that time restricted to his crib and with his hands restrained to prevent self-abuse.

Clearly, the restraint prevented Scotty from developing adequate motor and occupational performance skills. Thus, an intervention was planned to eliminate Scotty's self-injurious behaviors, to free up his hands for manipulative activities, and to gain him greater freedom of movement on the ward.

It was decided that even a noxious intervention, if effective, would be justified if in the long run it prevented further self-abuse and freed Scotty's hands for other activities. Thus, it was decided that punishment would be used to reduce the incidence of Scotty's self-mutilation. However, before the intervention was started, the experimenters wanted to obtain some idea of the incidence of Scotty's negative behaviors so that they would have an objective measure against which they could judge his progress.

Initially, mittens were developed so that the experimenter could obtain a *baseline* (preintervention measure) of Scotty's hand-to-head behaviors. Baseline measures are essential in clinical studies if therapists want to objectively measure the effects of intervention.

After establishing a baseline, the therapists set up the initial intervention to deliver a small amount of electric shock to Scotty's leg each time he hit himself. Thus immediately after one of Scotty's hands hit his head, he received a small, brief amount of electric shock. The result of this was that the baseline measure of self-injuring behavior quickly reduced to zero. Punishment, however, as mentioned before, may temporarily suppress behavior but it is not a good option for controlling behavior. What was needed in this case was for the child to make some alternative response (operant) that could be reinforced. The eventual goal would be for Scotty to use his hands for skills other than hitting himself.

Once again, a noxious stimulus was used, but in this circumstance, it was used as negative reinforcement to increase the probability of a response.

Scotty was seated at a high chair, and a truck was placed in front of him. The experimenter-therapist took a baseline of the amount and number of times that Scotty's hands touched the truck. This was essentially zero, indicating that the toy was not reinforcing for Scotty. After baseline was established, the intervention consisting of negative reinforcement as well as classical conditioning was begun. This was done by taking Scotty's hand while at the same time sounding a buzzer and administering a shock. At the point that Scotty's hand was placed directly on the truck, the shock and buzzer were stopped. This procedure was continued for several associations, and Scotty quickly associated the buzzer and touching the truck with termination of the shock. In this case, the shock was a negative reinforcer that, when terminated, increased truck-holding behavior.

Then, the therapist stopped the shock and only sounded the buzzer. Since Scotty had associated the buzzer with the shock, the buzzer took on the reinforcing properties of the negative reinforcer (it had become classically conditioned), and the buzzer, itself, served to sustain truck-holding. Thus, shock was no longer used, but when the buzzer was repeatedly sounded, Scotty would immediately reach out to hold the truck or other toys.

Eventually, the buzzer was no longer required, other stuffed toys were substituted for the truck, and Scotty's self-injurious behaviors were under control. As the authors concluded:

> For over a year, Scotty has been maintained completely out of restraint and has not exhibited any self-injuring behavior whatsoever. He relies less on his stuffed animals, although during rest times he is generally seen holding one of the animals. Often, however, during recreation and games on the ward, he is completely without any stuffed animal. Since being out of restraint, Scotty has become completely toilet trained. He has learned how to feed himself with considerable dexterity and has developed more appropriate social behavior. Early in the treatment, he was seen to smile for the first time. Since that time, smiling has become one of his most prevalent behaviors (55, p 383).

The example of Scotty illustrates how a variety of conditioning procedures can be used to alter behavior. In other chapters, we will give other examples; here, however, since we are introducing and evaluating a variety of theories, it is important to briefly introduce some important points about learning theory and behavior therapy.

# Pros and Cons of Learning Theory and Applied Learning Theory

As was noted earlier, behavior therapy is not without controversy. Some clinicians take exception to such procedures that were used with Scotty. They find the use of punishment or negative reinforcement unjustified and feel that other, more humane, approaches are required. This will be discussed more fully in Chapter 6.

To counter the negative claims about using aversive stimulation (also called aversives, punishment, or negative reinforcement) in therapy, many behaviorists point out that in some cases, it is necessary to choose between long-term negative behavior or a short duration of aversive stimulation. Some behaviors, such as Scotty's self-injury, are too violent or destructive to let go, and such behaviors are often not susceptible to alternative therapies. One argument in favor of punishment or negative reinforcement is that the end result (e.g., ward freedom for Scotty) may justify the means, especially considering the brevity of aversive stimulation. As Whaley and Malott (55, p 383) note, "There can be no doubt but that Scotty's life has been enriched by the minimal use of electrical stimulation."

In some cases, behavior modification has come under severe criticism because of the clinical uses (and some therapists' justification for the use) of punishment in controlling behavior. In an interview regarding this issue, Skinner (56) pointed out, "I don't like the use of punishment. I am opposed to it. But there are people who are out of reach of positive reinforcement." It is important to note that behavior-modification principles are useful in many settings and with many patients, and in none of these cases is punishment used. In many cases of successful behavior management, only positive reinforcement has been used.

Physical therapists, occupational therapists, and speech pathologists often find that a knowledge of behavior therapy is important. Many therapists are unable to initiate their treatment programs until behavioral problems are corrected. For example, until Scotty's self-injurious behaviors were eliminated, it was not possible to free his hands for other daily living skills such as feeding, dressing, or other self-care tasks. Similarly, children who refuse to look at a therapist or a mirror often do not respond in speech pathology. Those who refuse to sit still, to stop repetitive head shaking or body rocking, cannot appreciably benefit from motor or skills training. Thus, the use of behavior therapy in combination with a practitioner's own specific therapies, or effective team intervention, can often eliminate maladaptive behaviors that prevent the introduction of effective developmental treatment.

Other criticisms leveled against learning theory in general claim that its emphasis on S-R bonds and the study of discrete behaviors is reductionistic. Learning theory is widely criticized because it promotes a mechanistic and deterministic view of humans; i.e., likening human behavior to that of a machine that operates with simple input (stimuli) and predictable, predetermined output (responses). Some claim that learning theory is too deterministic, giving too much power and control to environmental events (57). We will return to this topic in Chapter 6. For now, the reader may want to compare learning theory with the less mechanistic theories, such as humanism and cognitive theory, that are also described in Chapters 5 and 6.

Behaviorism has been criticized also for its dogmatic, polarized view of behavior, which has traditionally been sided with the nurture side in the nature vs. nurture controversy. Addressing these criticisms, Skinner (25, p 5) points out that they represent an "extraordinary misunderstanding of the achievements and significance of a scientific enterprise." He defends behaviorism as a "promising alternative" to traditional views of human behavior which so far have proven inadequate in solving world problems (25, p 8). Skinner further notes that behaviorism is not unidimensional, and he does accept the notion of innate or instinctive responses in behavioral repertoires. Such behaviors are the primary domain of study of ethological theory and are examined next.

Learning theory is probably the most controversial of the theories discussed in this chapter. In fact, some contend that it is not really a developmental theory at all. It is included in this chapter because it, as much as some other theories elaborated here, provides an approach for explaining how some behaviors may change over time. It provides explanations for the development of sex roles, emotions, language, and perception and thus will be elaborated again in chapters covering these specific areas of development.

# ETHOLOGY

Chapter 2 emphasized that a familiarity with the nature-nurture debate provides a basis for understanding important issues in human development. This is particularly true in regard to understanding the differences between learning and ethological theories.

Learning theorists have traditionally emphasized environmental experiences or learning as the key to the development of behavior, and as we pointed out in the section on "Learning and Behavior Theories," some behaviorists were extreme environmentalists. Watson, for example, initially accepted, but then later denied, the existence of instincts or inherited behaviors, capacities, or talents (22, 23). In contrast, ethologists have focused on innate behaviors, examining in detail behaviors traditionally classified as instinctive.

## Instincts and Learning

The term "instinct" is as difficult to define as is "learning." Just as there are extreme definitions of learning, so also are there extreme definitions of instinct. These emphasize the innate, inherited aspects of behavior in contrast with learned or acquired responses. For example, one definition of instinct is that it is a stereotyped, innate, unlearned behavior. This is comparable to saying that learning is behavior that occurs as a result of environment. All behavior occurs in an environment, and all organisms come into the world with certain predispositions to behave in certain ways.

Thus, as we have emphasized throughout this text, our definitions of terms must be selected cautiously, and our interpretations regarding behavior must recognize the significance of organism-environment interactions. With this in mind, instinct is defined as, a class of behaviors that tend to be similar and fairly stereotyped for specific species of animals (24).

Given the fact that learning theory has tended to emphasize environmental forces and to neglect the very behaviors that were of interest to ethologists, it is understandable that disputes occurred between these groups. This dispute also extended, as we pointed out in the last chapter, to differences of opinion regarding methodology. Ethology, we noted, is associated with a specific theoretical orientation in addition to a methodology that emphasizes the study of organisms in their natural habitats. Ethologists emphasize the use of methods from the natural sciences in the investigation of behavior (58, 59).

# The Scientific Study of Behavior

Konrad Lorenz, who along with Niko Tinbergen is considered a cofounder of ethology, states:

> Ethology, the comparative study of behavior, is easy to define: it is the discipline which applies to the behavior of animals and humans all those questions asked and those methodologies used as a matter of course in all the other branches of biology since Charles Darwin's time (58, p 1).

While ethology is considered the comparative study of behavior, so also is comparative psychology. In fact, many learning theorists were also comparative psychologists, since they examined the similarities and differences between different species' acquisition of behavior.

If both ethology and comparative psychology examine the same thing, how do they differ? As pointed out in the last chapter, ethology and psychology, especially comparative psychology, have been compared and contrasted over several decades. Both fields are interested in the similarities and differences between the behaviors of different animal species, including humans. Both fields share the aim of scientifically studying behavior. However, there are fundamental differences between the fields.

Contrasted with American-oriented psychology, ethology is European- and zoology-based. Lorenz' reference to Darwin in the previous quote is significant because ethological theory assumes an evolutionary continuity in behavior, based on the principles of natural selection and adaptation synthesized by Darwin. In contrast, some American learning theorists such as Skinner actually oppose theoretical orientations. Rather, they want merely to delineate laws and principles underlying the acquisition of behavior.

Both psychology and ethology contend that by learning what motivates other creatures, we may gain greater insight into human behavior, but the general topic matter of these two fields has differed. Many psychologists who studied animals tended to focus on learning and motivation and the application of information they gathered from animals to generate models for human behavior. On the other hand, ethologists are biologists and fundamentally interested in animal behavior for its own sake. They have traditionally examined behaviors—such as reproductive, courting, and parenting activities—that are common to a wide variety of species. Many of these activities fit within the category of instincts. They are, indeed, very similar within, and even across, different species. Many of these behaviors, such as conflict, aggression, social dominance, and territoriality, are social in nature and are referred to as *species-specific behavior patterns*, i.e., instincts (60, 61).

While historical differences exist between psychology and ethology, their fundamental aims are the same. These are to develop a science of behavior that examines organisms and their relationship to their environment. The focus of the fields, however, has been different. Learning theorists have tended to focus on environmental consequences of behavior (i.e., nurture), while ethologists have primarily explored organism's adaptations to environments (i.e., nature). Thus, the fields were at one time recognized as traditionally split based on nature vs. nurture.

Now, however, with the growing emphasis on interactionism, it is recognized that these fields do not focus on two separate phenomena but on different aspects of the same process. Thus, while there are still classical ethologists and psychologists, many contemporary scientists note that these fields may merge into a single scientific study of behavior (62). For example, Blurton Jones (63) notes that the interests and methods of ethology and developmental psychology have converged. In *Ethological Studies of Child Behavior*, Blurton Jones (63) includes a wide variety of studies that clearly illustrate ethological applications toward understanding human development. If and when these fields merge, they may adopt common approaches and language, referring not to the narrower ethological method we discussed in the last chapter, but to the broader ecological method to be used for investigating organism-environment interactions in the scientific study of behavior.

# Ethology and Developmental Theories

The purpose of this chapter is to point out some of the principles from many theories and to illustrate their significance to the study of human development and to occupational and physical therapy. Traditionally, three primary theories have been associated with the study of development, and ethology is not one of them. The three primary theories are *cognitive, psychoanalytic*, and *learning*, representing the major developmental areas of *thinking, feeling*, and the *acquisition of behavior*, respectively. Two of these theories, cognitive and learning, have already been introduced; psychoanalytic theory will be discussed in the next chapter.

Although ethological theory consists of some concepts that are infrequently used and outdated (64, 65), it still has widespread application to development (63). Ethology is a scientific study of behavior; thus its principles, as we will point out, naturally apply to the scientific study of behavior as it changes over time, i.e., to the study of development.

Neither ethological nor learning theories are developmental theories per se, but both do account for how behavior changes over time. Thus, it is important to introduce these theories in a text on human development. Both theories also have important applications to clinical approaches to human behavior.

# Fundamentals of Ethology

Ethology has its own history, jargon, and significant principles. The term "ethology" comes from the Greek *ethos*, which means habit or convention (59). This applies to ethologists' focus on "conventional" or instinctive behaviors. Blurton Jones (63) reports that although ethology is associated with outdated concepts, it still contains several important issues that are peculiar to the field (Table 5.5).

Robert Hinde (62) has written extensively about ethology and promoted its synthesis with comparative psychology. Hinde (62, 66) explains that the study of behavior involves identifying relations that occur between the behavior and any events that come before, during, or after the behavior. In addition, he notes that events can occur within and outside the organism. He refers to four major topic areas in which ethologists ask questions: immediate causation, development, function, and evolution (66). These four areas are elaborated next because they are fundamental to ethology and because of their potential application to a theory of clinical practice concerned with human adaptation.

## Immediate Causation

Immediate causation involves examining factors that are directly responsible for behavior. It is concerned with the question, "What causes an organism to behave in a certain way?" This area of ethology requires exploring stimulus and response variables; motivation, which affects responsiveness; and integrated activities. In addition, analysis of immediate causation can be conducted at various levels. It involves studying the organism, its internal and surrounding environments, and their relationships to each other.

Table 5.5
Basic Concepts of Ethology[a]

- Emphasizes simple features of behavior as raw data for science.
- Emphasizes description and natural history as starting point for study.
- Distrusts major behavioral categories that have not been clarified.
- Believes in an evolutionary framework for determining scientific questions about behavior.

[a] Modified from Blurton Jones N: *Ethological Studies of Child Behavior*. New York, Cambridge University Press, 1972.

### Releasers and Fixed Action Patterns

Hinde (42) refers to the environment as the "stimulus situation." Obviously different stimulus situations elicit different behaviors. Many stimuli, ethologists note, are the immediate causes of very specific responses. For example, a shadow or silhouette of a hawk will elicit avoidance or fear reactions in small birds. These stimuli that elicit or act to produce specific behaviors or behavior patterns are termed *releasers* or *releasing stimuli*, which is the terminology preferred to the conventional term "innate releasing mechanism."

Releasers elicit simple instinctive behaviors called *fixed action patterns* (or fixed motor patterns) (58, 67). As an example, certain colors and markings on parent birds act as "releasers" for the emergence of pecking responses by infant birds. The pecking behavior, in turn, is an immediate cause for the parent bird to regurgitate food that it brings back to the nest for the nestlings.

### Human Instincts

According to ethologists, releasers and fixed action patterns are found in all animal species, including humans. Lorenz (58) points out that young infants presented with a face or elements of a face, such as a balloon with eyes and a smiling mouth painted on it, will direct their gaze toward the face and smile. The face is considered a releaser for an (instinctive) social response (the smile). As further evidence that this behavior is common to humans and instinctive in origin, ethologists note that blind babies, who have no opportunity for imitation of such behaviors, often smile, laugh, and cry (68).

Eibl-Eibesfeldt (68), who was a student of Lorenz, has studied the components of greetings across many cultures and notes that they occur fairly uniformly across the human species (i.e., they are instinctive behaviors). The components of social greeting include smiling and an eyebrow flash (raising the eyebrows) (68, 69). Sometimes these occur simultaneously and at other times individually. To test this, next time you are passing an acquaintance who greets you, note his or her reaction. Most likely, your acquaintance will smile, flash an eyebrow, and offer some vocalization. Often, those who do not vocalize still smile and flash their eyebrows in greeting.

Thus, behaviors such as smiling and laughing are considered human fixed action patterns. Behaviors we classify as "reflexes" also fit our definition of instinctive behavior; i.e., they occur fairly standardly across a species. For example, nipples (releasers) consistently elicit *sucking reflexes* in human infants. The *rooting reflex* occurs when the side of the infant's cheek near the mouth is lightly stroked. The infant, in response, moves the head in the direction of the stimulus, orients toward, and will initiate sucking if the stimulus is appropriate. This rooting response is described by Lorenz (58) as a fixed motor pattern that is part of a chain of activities found in newborn mammals.

A sudden sound or loss of support will cause infants to startle and respond with a common motor response called a *Moro reflex*. If infants are supine (lying on their back), they will lift their arms out and up toward the midline of their body, as if to grab hold of something directly on top or in front of them.

### Motivation and Integration

Immediate causation involves examining the stimulus situation and the variety of response patterns the organism makes to those stimuli. One influence that affects organisms' responses is their level of motivation. As Hinde (66) points out, motivational changes are temporary and reversible, but their influence can be substantial.

Motivation means movement, and it is motivation that directs an organism toward a certain goal. Motivation is increased by internal physiological states; e.g., a hungry animal is motivated to seek food and a pregnant animal is motivated to build a nest. Thus, motivation affects an organism's readiness to respond to certain environmental situations. As we pointed out under "Learning Theory," a hungry rat will learn to bar press for food in a Skinner Box, whereas a sated rat may not. Similarly, ethologists point out that motivational variables affect instinctive behaviors.

All fatigued, hungry, or sexually aroused organisms will respond differently according to their motivational state, and if one organism experiences all three states at once, then it will be in *conflict*. Ethologists have described conflict behavior in detail and have offered theories regarding how conflicts are resolved in various species. Such information is important in determining what behavior predominates in certain circumstances and how species other than humans go about resolving interorganism and intraorganism conflict.

In addition, ethologists examine and explain complex behaviors that are actually integrations of activities. Hinde (66) describes how nest building in birds is actually a series of related behaviors, which include (a) gathering nest material, (b) carrying the material to the nest, and (c) building the material into the structure of the nest. These behaviors can be exhibited independently; however, given the correct motivational or environmental circumstances, they will be chained together into one continuous, integrated activity.

Thus, immediate causation, which examines the factors directly responsible for behavior, looks at stimulus variables (releasers); response variables (fixed action patterns and motivation); and combined, integrated activities. Immediate causation, however, is not the only factor affecting behavior. Other important categories of interest to ethologists are development, function, and evolution (66).

### Development

Hinde (66, p 89) tells us, "It must be remembered that development involves a continuing interaction between a changing organism and (usually) a changing environment." These are two areas of ethology that deal with behavior as it changes over time. One is *phylogeny*, which looks at the evolutionary origin and the development of behaviors; the other is *ontogeny*, which is the development of behavior in an individual organism. *Behavioral embryology*, considered a branch of ethology, is specifically concerned with one aspect of ontogeny, the prenatal development of behavioral response patterns (59).

In their study of ontogeny, ethologists traditionally focused on innate, instinctive behaviors rather than studying learned behaviors. Hinde reports that studies now tend to focus on *how* behavior changes over time, or *how* genes and learning influence development. (This is what psychologist Anastasi also promoted and which was discussed in Chapter 2.) One way of treating the nature-nurture issue, Hinde (66) notes, is to look at innateness and learning on a continuum rather than as separate or competing forces. Thus, ethological studies of development examine behaviors ranging along a continuum from those that are environmentally stable (i.e., not susceptible to extensive modification by the environment) to behaviors that are environmentally labile (i.e., very susceptible to environmental influence) (66, p 86).

Ethological contributions to the scientific study of development have been significant. One example of an area of ethological study that has been extensively applied to human development is the phenomenon known as *imprinting*. Within a certain time period after hatching, some immature birds respond to the parent birds (or to some organism that moves and makes noise) by following it when it moves. This following response is known as "imprinting." The significance of this behavior to development is that it has been widely applied to similar responses made by other immature organisms. Thus, imprinting responses initially described by ethologists in relation to birds has also been generalized to the attachment responses that human infants form to their caretakers and that are so essential to an infant's subsequent development. This, and other behaviors studied by ethologists, will be discussed in subsequent chapters of this book.

## Function and Evolution

In addition to immediate causation and development, function and evolution are two additional areas of study in ethology. Examination of function looks at the actual purpose or adaptive value of specific response patterns, while evolutionary study examines phylogeny, the course of the species' development over time.

Behavioral function is the area of study with which ethology is most commonly associated. Gardner (19), for example, describes ethologists as scientists who focus on the function of some behavior and attempt to trace its development through evolution. As Lorenz (58, p 102) points out, ethology is based on the fact that there are mechanisms of behavior (functions) that evolve during an organism's life exactly as do organs. To clearly understand the concept of function and its significance in adaptation, it is necessary to briefly review Darwin's theory of evolution, including the concepts of natural selection and survival of the fittest (70).

## Darwinian Theory of Natural Selection

Darwin's theory noted that organisms most adapted to their environments (i.e., the most fit) will be most apt to survive. Thus, over generations, the fittest organisms and their progeny will be naturally selected in greater proportion to those lacking adaptive traits. As an example of natural selection, the giraffes with the longest necks and legs will have access to food and will outsurvive smaller giraffes whose food supply is depleted by competitors. The fastest cheetahs will capture prey and outlive slower cats and, over generations, effect greater running speed in that species. Likewise, caterpillars, moths, or butterflies that are most camouflaged or that taste bad to their predators will outsurvive less camouflaged or more tasty members of their species. According to Darwin's theory, these organisms with specific traits that are adapted to environmental pressures will be naturally selected over time, maintaining the viability of their species over generations (70).

## Adaptive Behavioral Functions

Darwin's theory is significant to ethologists, as they apply it not only to anatomy and physiology but also to behavior. Just as some scientists examine the adaptation of organs, physiological systems, or structures and their continuity through evolution, ethologists examine the adaptation and continuity of behavioral functions.

Ethologists' study of behavioral function entails asking, "What is the behavior for?" or "What purpose (or function) does it serve the animal in promoting its adaptability?" (66). What ethologists are particularly interested in regard to function, is the similarity between behaviors exhibited by different species of animals; e.g., imprinting in infant birds and attachment in infant humans; nest building in rats and birds; flight in bats, birds, and insects; or courtship behaviors in birds, felines, and mammals.

## Homologous and Analogous Functions

There are two different mechanisms that may account for similar behavior patterns in different species. One mechanism is evolution from a common ancestor; the other is responses made by different species to similar environmental pressures. Thus, the similar prehension patterns of raccoons, monkeys, apes, and humans may be due to descendance from a common ancestor or may have occurred as a result of similar adaptive responses to environmental pressures. Those behaviors or organs that are similar due to descendance from a common ancestor are referred to as *homologous*; whereas *analogous* functions are those that are similar but evolved independently, from what is referred to as *convergent evolution* (58, p 102).

Studying and comparing analogous and homologous functions in evolution has enabled ethologists and other scientists to better understand *how* and *why* certain behaviors have developed as well as their possible adaptive value. For example, the function of the Moro response, which was described under "Human Instincts," is easy to ascertain when one examines other species that share a common ancestor with humans. Many primates carry their young supported on their abdomens. The young, in turn, cling to their supporters. The adaptive function of behavioral components of the Moro reflex is that when there is a threat or loss of support, the infant reflexively reaches out and clings to the object above it. In nature, that object is most likely going to be a caretaker or safe object in its environment. Although the Moro reflex in humans no longer serves adaptive value for support, it is reasonable to assume that it and the reflexive fist and clinging responses made by human infants are remnants of adaptive behaviors that were at one time very important for survival.

## Summary of Ethological Principles and Their Significance to Therapists

Ethologists are interested in the *evolution* of behavior, its *immediate causes* and adaptive *functions*, and how behavior changes over the *development* of the organism. These four areas about which ethologists ask questions also have relevance for clinicians seeking to learn the causes and consequences of human behavior.

Consider the area of human functional hand skill and prehension. Human prehension is a result of a long history of adaptive change during evolution of the primate species. In humans, manipulative hand use is associated with tool making, communication and gestures, self-care, touching, and numerous other special skills and abilities. As Napier (71, p 22) has aptly noted, "There is nothing comparable to the human hand outside nature. We can land men on the moon but, for all our mechanical and electronic wizardry, we cannot reproduce an artificial fore-finger that can feel as well as beckon."

Thus, the anatomy and physiology of the hand and its special evolution in nature are important to study if one wants to fully understand human manipulative skills. In addition to evolution, therapists also need to know about the ontogeny or development of the hand. We know, for example, that hand skills change during human ontogeny. An infant's hand use is crude and gradually emerges through maturation and experience into skilled, precision use. Thus, the preschool child can manipulate utensils and some articles of clothing. The school-age child finally acquires the dexterity to zip, button, tie shoes, and perform other precise, demanding tasks with the hands. And the skilled musician or surgeon develops extraordinary hand use with years of practice and training.

A specific stimulation to the sole of an infant's foot or palm of the hand will elicit predictable motor responses. With maturation, these responses diminish; however, they remain evident in individuals with delays in development or with damage to the nervous system. The Babinski reflex, for example, is a flexion response of the great toe when the sole of the foot is stimulated on the lateral side. This response occurs as a part of an immature flexion, withdrawal response in infants, and it can also be seen in individuals with specific forms of neural injury (72). Knowing the immediate causes of the Babinski sign and the conditions for its elicitation will provide therapists with sensory, motor, and developmental information as well as information about the integrity of the nervous system.

Finally, as therapists, we are most concerned about the adaptive functions of behavioral response patterns. All humans adopt specific habits and behavior patterns because of the "functional significance" to their particular life-style. Most of the patients we see have found that certain ways of acting or performing specific skills serve "adaptive functions" for them, and they have integrated these activities into their daily lives. Therapists often must analyze these specific responses in order to determine what is important to each individual patient and his or her daily routine.

In such analyses, therapists determine why patients move or perform skills in certain ways, and determine whether these responses are the most effective and adaptive given the patient's trauma or disease. By such analyses, we learn the immediate causes of behaviors as well as the adaptive functions those behavior patterns serve for individuals. Our therapy is designed to promote the most functional adaptive responses, i.e., those that preserve the integrity of sore or injured muscles and joints, those that conserve energy, and those that are most integrated with the patients' previous and existing life-style.

The significance of understanding organism-environment relationships is apparent when occupational and physical therapists recognize that one of their fundamental roles is adapting environments geared toward producing specific functional responses in patients.

Thus, the four areas of concern to ethologists, immediate causation, function, development, and evolution can also be used to inform and improve clinical approaches to human intervention. Hinde (66) illustrates the relevance of each of these four concepts to the study of prehension:

> Suppose you were asked, 'Why does your thumb move in a different way from the other fingers?' You might give an answer in terms of the anatomy of the hand—the differences in skeletal structure and muscle attachments between the thumb and the other fingers: that would be an answer concerned with the immediate causation of thumb movement. You might give an answer in terms of the hand's embryology, describing how, as the finger rudiments developed, one came to have a different structure from the others. Or you might give a functional answer—an opposable thumb makes it easier for us to pick things up, climb trees, and so on. Or finally you might say that we are descended from monkey-like creatures, and monkeys have opposable thumbs, so of course we do too. This would be an answer in terms of evolutionary origin. All of these answers would be correct: no one would be complete.
>
> In the same way, ethologists are interested in questions of all four types about behaviour. Indeed they believe that, although logically distinct and independent, questions concerning immediation causation, development, function and evolution are inter-fertile (66, p 21).

# Ethology as a Scientific Field Important to Occupational and Physical Therapy

Both fields of physical therapy and occupational therapy have lately been focusing on developing theoretical bases that are founded on sound clinical research (e.g., 73–75). Since both fields are so diverse and focus on so many aspects of human behavior, they have, of necessity, tended to "borrow" theories and data from many other fields (75). Now, however, the tendency in many clinical fields is to move away from adopting the scientific data from other disciplines and to move toward developing their own secure research base. For example, the field of occupational therapy lately has been promoting research oriented around the unifying concept of occupation or occupationology (76).

Even though ethology is relatively new itself, it has a sound basis not common to many other new areas of study. As a branch of biology, ethology takes its methods and fundamental information from subject matter already comprising the older, broader discipline. Thus, ethology, as Blurton Jones (63, p 6) points out, can be subdivided into "studies of causation (physiology), development (embryology), survival value (ecology), and comparative studies (taxonomy and comparative anatomy). These are of course linked around an interest in the evolution of both the form and the behaviour of an animal ... In practice, ethologists move freely across these fields, and across anatomy and behavior."

Ethology has managed to develop a unified field of study that has incorporated research and theories about issues that are of major interest to both occupational and physical therapy. These issues include such areas as physiology, development, ecology, function, adaptation, and purposefulness (67). The fact that another field has integrated various topic areas may be instructive to clinical fields, which are similarly concerned with developing a research base around the causes, development, ecology, and taxonomy of behavior. Both physical and occupational therapy, like ethology, have a strong foundation in the biological sciences, and may, like ethology, find some successful integration of the field by adopting some of biology's classification schemes and methodological approaches.

## Methodology

There are obvious fundamental differences between ethology and clinical fields. For example, the central unifying theme of ethology is evolution, and occupational therapy's unifying principle is occupation. In addition, while occupational and physical therapy have been primarily clinical in orientation, ethology has been methodological. Ethology also has focused on animal behavior or comparative studies, whereas occupational and physical therapy have been centered primarily on human behavior.

Despite these differences, clinical fields dealing with human behavior may benefit from ethological methods. Some of the skills required to become effective ethologists and therapists overlap. Both ethologists and clinicians require acute observational skills in the study of behavior. The late Dian Fossey was an occupational therapist who became a noted ethologist and whose observational studies of gorilla behavior informed and fascinated the scientific and public communities (77).

As was explained in the last chapter, ethology refers to theory as well as method. As zoologists, ethologists come from a discipline that emphasizes observing, describing, recording, and classifying as initial activities involved in any study (63). Immelmann (59, p 2) illustrates this point: "The starting point and basis of the scientific study of behavior is a precise and detailed compilation of behavior patterns that are typical for a species that is as complete as possible." This is referred to as an ethogram.

Such compilations of behavior, or ethograms, are particularly useful in the study of human development as well as in clinical study of human pathology, and they may be particularly applicable in the study of human movement or occupation. In regard to a scientific study of child behavior and development, Tinbergen notes:

> It is clear that this simple, careful observation of normal children is going to be a very demanding task . . . but it will give a wider scope and more purpose to human studies . . . and no less important by gradually building up an ethogram of our species work will provide a yardstick by which behavioral pathology can be measured(78, p ix).

The need also exists in occupational and physical therapy for general classifications of behaviors to be used in theory construction as well as a basis of normative data to be used in evaluation. The field of ethology investigates and describes many of the same behavioral functions of interest to therapists, thus we may benefit from ethological procedures and existing methodology, as well as from their descriptive data. Table 5.6 lists some specific behaviors that are particularly significant to therapy and upon which extensive ethological research has already focused. For example, ethologists classify and describe in detail simple types of coordinated movements and study their significance and development in specific environ-

**Table 5.6**
**Some Topics under Study in Ethology and of Potential Interest to Therapists**

Motor learning
Voluntary movement
Purposeful behavior
Balance between activity, sleep, and rest
Tool use and its relation to cognition
Nonverbal behavior
Development changes in hand function and prehension
Grip use and adaptation to shapes
Gesture development
Posture
Play

ments. This is also an important area of study in both physical and occupational therapy.

Vandenberg and Kielhofnner (79) have illustrated how ethological concepts and principles relate to occupational and developmental behaviors. Their article describes the ontogeny and evolution of play and illustrates the adaptive significance (function) of play behavior and its relationship within a context of human occupation. Ethological concepts and applications are evident throughout this study of human occupational behavior.

## Common Themes and Models

King (80) and Kleinman and Bulkley (81) provide evidence for a notable similarity of central topic material between ethology and health professions. They note that the major function of all health disciplines is to foster adaptation (81, p 17). Ethology, too, is a science of adaptation. King has proposed for the field of occupational therapy the unifying theoretical model of "a science of adaptive responses" (80). She notes that she was influenced by the work of Lorenz and the implications of his work for the field of occupational therapy. Both fields of physical and occupational therapy may be able to derive a wider science of adaptive responses by drawing some basic data, principles, methods, and models from ethology.

### Model of Sensorimotor Feedback

Notable in all three fields of occupational therapy, physical therapy, and ethology is the common use of the sensorimotor-based feedback loop. This involves a model of input and output, based on the principle that organisms receive sensory information (input) on sense receptors, transfer that information to some central receptor (the brain), and react with some overt output behavior, a motor response called an *effector*.

This basic concept of input/output has been liberally generalized to human behavior in both simple and complex models. As a simple model, it is used to describe the reflex arc, a simple reflexive response found in humans and other animals. The reflex arc is said to occur at the "spinal level" because it is a response that occurs automatically to sensory stimulation which is processed by the spinal cord rather than centrally by the brain. It is, indeed, a simple model of "sensory input/motor response."

A more complex model incorporates the concept of feedback with this initial input/output design (Fig. 5.4). Feedback is important, as it provides sensory information relevant to the initial response. Just as a frog receives the feedback that it missed a targeted fly, so also does a tennis player "feel" the action of a serve, note by sound and vision whether the ball was served with accuracy and force, and receive feedback from others watching the performance.

Feedback is an important concept in therapy as well as in ethology, as it is an essential component of organism-environment interaction. Feedback comes from internal cues, such as information from the limbs, muscles, and

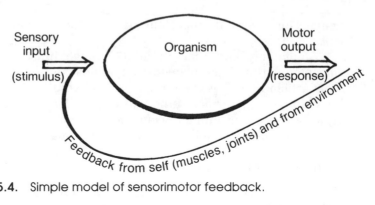

**Figure 5.4.**    Simple model of sensorimotor feedback.

joints, as well as from external cues. External feedback may include therapists' comments or other auditory, visual, or sensory information received from the environment. Organisms making functional, adaptive responses successfully use feedback. This is true of animals adapting to their environments as well as patients making progress in therapy. As Loomis (82, p 172) notes, "Crucial to motor learning is the need to have a knowledge of results." That knowledge is feedback, and it is necessary to help a patient or learner develop the capacity to detect errors as well as to strengthen correct responses.

The sensorimotor feedback model provides a framework for sensorimotor-based treatment programs in occupational and physical therapy (e.g., 14–17, 29, 82) and is also significant for ethological theory. Describing the history of ethology, Eibl-Eibesfeldt (68) notes the significant influence provided by experiments conducted by J. V. Uexkull in the 1920s. Studying the interrelations between organisms and their environments, Uexkull developed a schema of a functional cycle, (Fig. 5.5) that, although it was at the time applied to insect and animal behavior, can be equally applied to human behavior (68).

More complex systems models of human occupation consist of the same basic concept. These latter models have incorporated the concept of feedback as a link in the chain from stimuli (input) to responses (output) and illustrate human behavior in interaction with the environment. The models of occupational therapists Kielhofner and Burke (83) and Barris (84) are similar enough that they are combined into one model summarized in Figure 5.6. (See also Kielhofner's and Burke's illustration of the open system of occupation behavior [from 85] in Fig. 3.9). The model in Figure 5.6 illustrates how the environment interacts with and provides feedback for human occupational behavior.

# Human Ethology and Sociobiology

The application of ethological principles to a science of human behavior is called *human ethology*. One of the goals of this field is to identify natural human behaviors (64), that is, human species specific or instinctive re-

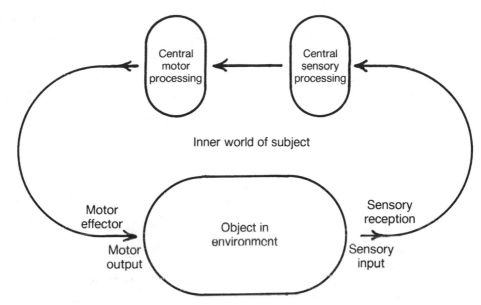

**Figure 5.5.**   Uexkull's 1921 functional cycle. (Modified from Eibl-Eibesfeldt I: *Ethology. The Biology of Behavior.* New York, Holt, Rinehart & Winston, 1970, p 6.)

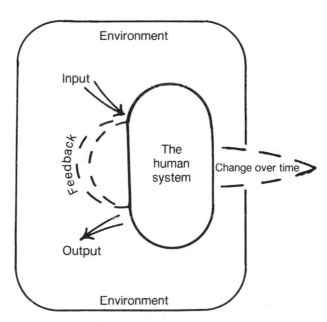

**Figure 5.6.**   Model of human occupation in interaction with the environment. (Modified from Barris R: Environmental interactions: an extension of the model of occupation. *Am J Occup Therap* 36: 641, 1974, and Kielhofner G, Burke JP: A model of human occupation, part 1. Conceptual framework and content. *Am J Occup Therap* 34: 575–576, 1980).

sponses. Such ethologists as Lorenz, Tinbergen, and Eibl-Ebesfeldt as well as popular authors such as Ardrey (86) and Morris (87, 88) have described and classified such behaviors, delineating numerous human behaviors such as aggression, affection, status-seeking, territoriality, and nationalism, that seem to have evolutionary, biological origins.

Human ethology, like ethological theory itself, is controversial. Michel and Moore (64, p 104) claim, "There is no generally accepted theory or set of constructs that can be said to define ethology today. Therefore, any attempt to carve out a new discipline of human ethology by applying the remnants of classical ethological theory to human behavior is doomed from the beginning." In contrast, Medawar and Medawar (89, p 84) claim, "If there is human physiology, human anatomy, and human genetics, why should there not also be human ethology?"

Since instincts are traditionally studied by ethologists and associated with one side of the nature-nurture debate, there are many who object on principle to the application of such concepts to human behaviors. Some scientists and philosophers note that humankind is a rational, intelligent species that engages in conscious decision making and "higher level"[c] cognitive skills that underlie human language and culture.

Still others contest the notion of human ethology on religious principles, denying Darwin's concept of evolutionary continuity between humans and animals and claiming that humans, not animals, possess a spiritual element or level of higher consciousness. Both of these groups, for different reasons, would claim that humans are not driven by animal instincts and that comparison of humans to other animals species is invalid.

A field of study that is equally as controversial as human ethology is *sociobiology*, which is the application of biological principles to social behavior (Fig. 5.7) (94, 95). Sociobiology originated with work by E. O. Wilson (95) and is included by some as a subarea of ethology. Immelman (59) claims that sociobiology, can also be termed "socioecology", and is a subdivision of *ecoethology*, the study of the relationships between the behavior of some species and other living and nonliving aspects of the environment.

One of the premises of sociobiology is that principles applied to individuals and their behavior can also be applied to social groups and social behavior. Thus, sociobiologists generalize the notion of instincts to account for some behaviors of societies as well as for individuals. This becomes particularly controversial when it is applied to human behavior (60, 94). In

---

[c] The term *higher level* is commonly used in clinical fields to refer to the phylogentically newer (cortical) regions of the brain. "Higher level" activities are cortically directed and most often associated with human beings because of their great amount of cortical tissue. In some cases (e.g., Ayres [7, 14]), "higher level" is used to imply the superiority of human skills and abilities over those of other species. The use of such terms, however, is not without controversy and is not common in ethology, where it is recognized that species are "different than," not "better than," "higher than," or "lower than" other species. For a discussion of this issue and its relationship to evolution and animal behavior see Hodos (90) and Wallace (91, pp 74–75). In addition, for a discussion of nonhuman animal cognition and intelligence, see Griffin (92, 93).

*"What's even more astonishing is it coincides exactly with the World Series."*

**Figure 5.7.** Human ethologists and sociobiologists assume that a continuity exists between the functional behavior patterns of other species and humans. Drawing by Stevenson; © 1984 The New Yorker Magazine, Inc.

addition, many who object to ethology also object to sociobiological notions of the genetic determinism of behavior.

Implicit in sociobiology is the assumption that human social behavior can be shown to have evolved by Darwinian principles of natural selection (88). Thus, the group behavior that has naturally been selected out over time is that which is functional for the survival of the group, the family, or more largely, the species. As an example, altruism, which is defined as a selfless regard for others, is cited as such a behavior that, although it often jeopardizes the individual, serves to maintain the viability of the species. Sociobiologists cite examples of altruistic acts where a parent, or even a stranger, will put his or her life at risk to save another human being.

Both fields of human ethology and sociobiology provide insights regarding behaviors conventionally studied in the field of human development, e.g., the development of aggression, parental care-taking behaviors, the development of sex roles and gender identity, moral identity, and other behaviors that help to maintain the continuity of the species (64). Thus, they are important to examine in a study of human development.

The final contributions of these fields to the study of human development and human behavior, however, cannot yet be determined. There can be no simple conclusion regarding the significance and application of these fields. Both are very new, both are controversial, and, in some cases, widely misunderstood. Just as with the nature-nuture controversy, there is no one correct dogma to accept. Instead, it is fruitful to be educated about new theories and the controversies they incite and to adopt conglomerations of principles and concepts from a variety of theoretical perspectives.

Human development is so divergent that clearly one line of thinking will never adequately cover it. However, it is probable that theories that take ecological perspectives will be more seriously applied or adapted, as scientific approaches to human development and occupation become more ecological in concern. Michel and Moore (64) note how sociobiology's contemporary ecological emphasis makes it potentially more widely acceptable than other narrow theories. One of the more important aspects of sociobiology, Michel and Moore (64, p 109) claim, is that it "calls on people to seriously consider adjustment of their societies to a more 'conservationist' relationship with the ecosystems of the planet." But as for their contributions to an integrated theory of human development, both sociobiology and human ethology must continue to be explored.

## Ethology and Interactionism

As noted in Chapter 4, the existing disciplines of human development, occupational therapy, physical therapy, and psychology are focusing on contexts and interactions. Despite its historical association with innateness, ethological study has focused on organism-environment interactions. In ethology, Hinde (66) notes, the behaviour of each species is seen in relation to the environmental context to which it has been adapted. Lorenz (58, p 7) points out that as early as the 1930s an eminent scientist recommended abondoning "the conceptual separation of the innate and the acquired . . . All behavior, in his opinion, consisted of reactions to stimuli and these reflected the interaction between the organism and its environment".

Fox illustrates how a field such as ethology can have widespread applications, not only for the study of human development but also for fields with clinical orientations.

Because of its holistic and integrative orientation, ethology can also serve as a core discipline linking studies in areas such as veterinary medicine, agricultural science, and biology and psychology. In veterinary medicine courses in ethology can serve to integrate courses in physiology, anatomy, and biochemistry, so that the dynamics of the whole animal in relation to other animals and to its environment may be conceptualized . . . Similarly in human medicine ethology could offer the student a more holistic view of disease where social and environmental factors must be considered in addition to the behavior, role, and personality of the patient himself (96, pp 5, 6).

Perhaps, over time, a science of human occupation and human adaptation will find similar generalizations to a similarly wide variety of fields related to human behavior.

# CHAPTER SUMMARY

This chapter dealt with three major theoretical areas that have contributed to an understanding of behavior as it changes over time. Cognitive theory, associated primarily with Jean Piaget, focuses on the development of reasoning abilities. Learning theory is associated with a number of individuals, such as the behaviorist John Watson, physiologist Ivan Pavlov, connectionist E. L. Thorndike, and contemporary behaviorist B. F. Skinner. Collectively, learning theories focus on environmental consequences associated with behavior and have resulted in a controversial clinical approach known as "behavior therapy".

While learning theorists primarily have been American psychologists who used experimental methods to explore behavior, European ethologists such as Nikko Tinbergen and Konrad Lorenz used observational or ethological methods in the study of the behavior of different species in their natural habitats. Ethology has been concerned with adaptive functions and the evolution of behavior. The chief concern for ethologists has not been the stimulus-response associations studied by learning theorists but instead species-specific behaviors such as courtship, territoriality, and parental care.

Each of these theories—cognition, learning, and ethology—has contributed to our understanding of human behavior and to how that behavior changes over the course of an individual's life. But these theories are not sufficient. Human beings are complex; they not only have thoughts, acquired behaviors, and integrated response patterns, but they also have feelings, attitudes, personalities, and other complex and special behaviors associated with being human. Thus, it is necessary to explore additional theories that round out our profile of human behavior and development. These are explored in the following chapter.

## References

1. Hill W: *Learning. A Survey of Psychological Interpretations.* Scranton, Chandler, 1971.
2. Baldwin AL: *Theories of Child Development.* New York, John Wiley, 1967.
3. Phillips JL: *The Origins of Intellect. Piaget's Theory,* ed 2. San Francisco, WH Freeman, 1975.
4. Piaget J: *Psychology of Intelligence.* Paterson, NJ, Littlefield, Adams & Co, 1963.

5. Piaget J: *The Child and Reality. Problems of Genetic Psychology.* New York, Grossman, 1973.
6. Roy EA: Current perspectives on disruption to limb praxis. *Phys Therap* 63:1998–2003, 1983.
7. Ayres AJ: *Developmental Dyspraxia and Adult Onset Dyspraxia.* Torrance, CA, Sensory Integration International, 1985.
8. David KS: Motor sequencing strategies in school-aged children. *Phys Therap* 65:883–889, 1985.
9. Ayres AJ: *Sensory Integration and Praxis Tests (SIPT).* Los Angeles, Western Psychological Services, 1984.
10. Wursten H: On the relevancy of Piaget's theory to Occupational Therapy. *Am J Occup Therap* 28:213–217, 1974.
11. Umphred D: Conceptual model of an approach to the sensorimotor treatment of the head-injured client. *Phys Therap* 63:1983–1987, 1983.
12. Malkus D: Integrating cognitive strategies into the physical therapy setting. *Phys Therap* 63:1952–1959, 1983.
13. Kopp CB: An application of Piagetian theory: Sensory-motor development. *Am J Occup Therap* 28:217–219, 1974.
14. Ayres AJ. *Sensory Integration and Learning Disorders.* Los Angeles, Western Psychological Services, 1973.
15. Montgomery P, Richter E: *Sensorimotor Integration for Developmentally Disabled Children: A Handbook.* Los Angeles, Western Psychological Services, 1977.
16. Morrison D, Pothier P, Horr K: *Sensory Motor Dysfunction and Therapy in Infancy and Early Childhood.* Springfield, IL, Charles C Thomas, 1978.
17. Gilfoyle EM, Grady AP, Moore JC: *Children Adapt.* Thorofare, NJ, Charles B Slack, 1981.
18. Gratz RR, Zemke R: Piaget, Preschoolers, and Pediatric Practice. *Phys Occup Therap Pediatr* 1(1):3–9, 1981.
19. Gardner H: *Developmental Psychology. An Introduction.* Boston, Little Brown & Co, 1978.
20. Gardner H: *Frames of Mind. The Theory of Multiple Intelligences.* New York, Basic Books, 1985.
21. Kimble GA: *Hilgard and Marquis' Conditioning and Learning,* ed 2. New York, Appleton-Century-Crofts, 1961.
22. Watson RI: *The Great Psychologists. From Aristotle to Freud.* Philadelphia, JP Lippincott, 1963.
23. Schultz D: *A History of Modern Psychology,* ed 2. New York, Academic Press, 1975.
24. Watson JB: *Behaviorism.* Chicago, University of Chicago Press, 1930.
25. Skinner BF: *About Behaviorism.* New York, Knopf, 1974.
26. Hilgard ER, Bower GH: *Theories of Learning.* New York, Appleton-Century-Crofts, 1966.
27. Cordes C. Don't leave home without it. *Psychol Today* Dec:80,1984.
28. Bijou SW, Baer DM: *Child Development: A Systematic and Empirical Theory.* New York, Appleton-Century-Crofts, 1961.
29. Ayres AJ: *Sensory Integration and the Child.* Los Angeles, Western Psychological Services, 1980.
30. John ER: *Mechanisms of Memory.* New York, Academic Press, 1967.
31. Ottenbacher K, Short MA: Sensory integrative dysfunction in children: A review of theory and treatment. In Wolraich M, Routh DK (eds): *Advances in Developmental and Behavioral Pediatrics.* Greenwich, CT, Jai Press, 1985, vol 6.
32. Eccles JC: Possible synaptic mechanisms subserving learning. In Karczmar AG, Eccles JC (eds): *Brain and Human Behavior.* New York, Springer-Verlag, 1972.
33. Jones MC: Albert, Peter, and John B. Watson. *Am Psychol* Aug:581–583, 1974.
34. Watson JB: *Psychological Care of Infant and Child.* New York, Norton, 1928.
35. Jones MC: A laboratory study of fear: The case of Peter. *Pedagog Semin* 31:308–315, 1924.
36. O'Neill GW, Gardner R: Behavior therapy: An overview. *Hosp Commun Psychiat* 34:709–715, 1983.
37. Harris SL: *Families of the Developmentally Disabled: A Guide to Behavioral Interventions.* New York, Pergamon Press, 1983.

38. Gaines BJ: Goal-oriented treatment plans and behavioral analysis. *Am J Occup Therap* 32:512–516, 1978.
39. Lovitt TC: *Managing Inappropriate Behaviors in the Classroom*. Reston, VA, Council for Exceptional Children, 1978.
40. Norman CW: Behavior modification: A perspective. *Am J Occup Therap* 30:491–497, 1976.
41. Sieg KW: Applying the behavioral model to the occupational therapy model. *Am J Occup Therap* 28:421–428, 1974
42. Stein F: A current view of the behavioral frame of reference and its application to occupational therapy. *Occup Therap Mental Health* 2:35–62, 1982.
43. Rosen GM: The development and use of nonprescription behavior therapies. *Am Psychol* Feb:139–141, 1976.
44. Jodrell RD, Sanson-Fisher R: An experiment involving disturbed adolescent girls. *Am J Occup Therap* 29:620–624, 1975.
45. Bellack AS, Hersen M, Turner SM: Generalization effects of social skills training in chronic schizophrenics: An experimental analysis. *Behav Res Therap* 14:391–398, 1978.
46. Diorio MS, Konarski EA: Evaluation of a method for teaching dressing skills to profoundly mentally retarded persons. *Am J Mental Deficiency* 89:307–309, 1984.
47. Ogburn KD, Fast D, Tiffany D: The effects of reinforcing working behavior. *Am J Occup Therap* 26:32–35, 1972.
48. Hester SB: Effects of behavioral modification on the standing and walking deficiencies of a profoundly retarded child. *Phys Therap* 61:907–911, 1981.
49. Hill LD: Contribution of behavior modification to cerebral palsy habilitation. *Phys Therap* 65:341–345, 1985.
50. Gouvier WD, Richards JS, Blanton PD, Janert K, Rosen LA, Drabman RS: Behavior modification in physical therapy. *Arch Phys Med Rehabil* 66:113–116, 1985.
51. Rapport MD, Bailey JS: Behavioral physical therapy and spina bifida: a case study. *J Pediatr Psychol* 10:87–95, 1985.
52. O'Dell S: Training parents in behavior modification: a review. *Psychol Bull* 81:418–433, 1974.
53. Tyler NB, Kogan KL: Reduction of stress between mothers and their handicapped children. *Am J Occup Therap* 31:151–155, 1977.
54. Holland JG, Skinner BF: *The Analysis of Behavior*. New York, McGraw-Hill, 1961.
55. Whaley DL, Malott RW: *Elementary Principles of Behavior*.New York, Appleton-Century-Crofts, 1971.
56. Dietz J: Psychologist BF Skinner comes to defense of autistic center leader. *The Boston Globe* Oct 17, 1985, p 33.
57. Bandura A: Behavior theory and the models of man. *Am Psychol* Dec:859–869, 1974.
58. Lorenz KZ: *The Foundations of Ethology*. New York, Spring Verlag, 1981.
59. Immelman K: *Introduction to Ethology*. New York, Plenum Press, 1980.
60. Wolkomir R: Five decades of discovery. *Natl Wildlife* April-May:1986.
61. Gleitman H: *Psychology*, ed 2. New York, WW Norton, 1986.
62. Hinde RA: *Animal Behavior. A Synthesis of Ethology and Comparative Psychology*, ed 2. New York, McGraw-Hill, 1970.
63. Blurton Jones N: *Ethological Studies of Child Behaviour*. New York, Cambridge University Press, 1972.
64. Michel GF, Moore CL: *Biological Perspectives in Developmental Psychology*. Monterey, CA, Brooks Cole, 1978.
65. Hutt SJ, Hutt C: *Direct Observation and Measurement of Behavior*. Springfield, IL, Charles C Thomas, 1970.
66. Hinde RA: *Ethology*. New York, Oxford University Press, 1982.
67. Tinbergen N: An attempt at synthesis. In Birney RC, Teevan RC (eds): *Instinct. An Enduring Problem in Psychology*. Princeton, NJ, Van Nostrand, 1961.
68. Eibl-Eibesfeldt I: *Ethology. The Biology of Behavior*. New York, Holt Rinehart & Winston, 1970.
69. Eibl-Eibesfeldt I: *Love and Hate. The Natural History of Behavior Patterns*. New York, Holt Rinehart & Winston, 1972.

70. Darwin C: *The Origin of the Species.* New York, Collier Books, 1962.
71. Napier J: *Hands.* New York, Pantheon, 1980.
72. Carr JH, Shepherd RB: *Physiotherapy in Disorders of the Brain.* London, Aspen, 1980.
73. Currier DP: *Elements of Research in Physical Therapy*, ed 2. Baltimore, Williams & Wilkins, 1984.
74. Ottenbacher K: *Evaluating Clinical Change. Strategies for Occupational and Physical Therapists.* Baltimore, Williams & Wilkins, 1985.
75. Ottenbacher K, Short MA: Publication trends in occupational therapy. *Occup Therap J Res* 2:80–88, 1982.
76. Christiansen CH: Editorial: Toward resolution of crisis: research requisites in occupational therapy. *Occup Therap J Res* 1:115–124, 1981.
77. Fossey D: *Gorillas in the Mist.* Boston, Houghton Mifflin, 1983.
78. Tinbergen N, Foreward. In Blurton Jones N (ed): *Ethological Studies of Child Behavior.* New York, Cambridge University Press, 1972.
79. Vandenberg B, Kielhofner G: Play in evolution, culture, and individual adaptation: implications for therapy. *Am J Occup Therap* 36:20–28, 1982.
80. King LJ: Toward a science of adaptive responses. *Am J Occup Therap* 32:429–437, 1978.
81. Kleinman BL, Bulkley BL: Some implications of a science of adaptive responses. *Am J Occup Therap* 36:15–19, 1982.
82. Loomis J: Model to integrate sensorimotor treatment approaches for neurological dysfunction. *Physiother Can* 37:170–176, 1985.
83. Kielhofner G, Burke JP: A model of human occupation, part 1. Conceptual framework and content. *Am J Occup Therap* 34:572–578, 1980.
84. Barris R: Environmental interactions: an extension of the model of occupation. *Am J Occup Therap* 36:637–644, 1982.
85. Kielhofner G, Burke JP: Components and determinants of human occupation. In Kielhofner G (ed): *A Model of Human Occupation. Theory and Application.* Baltimore, Williams & Wilkins, 1985.
86. Ardrey R: *The Territorial Imperative.* New York, Dell Publishing, 1966.
87. Morris D: *The Human Zoo.* New York, Dell Publishing, 1969.
88. Morris D: *The Naked Ape. A Zoologists Study of the Human Animal.* New York, McGraw Hill, 1967.
89. Medawar PB, Medawar JS: *Aristotle to Zoos. A Philosophical Dictionary of Biology.* Cambridge, MA, Harvard University Press, 1983.
90. Hodos W: Evolutionary interpretation of neural and behavioral studies of living vertebrates. In Schmitt FO (ed): *The Neurosciences. Second Study Program.* New York, Rockefeller University Press, 1970.
91. Wallace RA: *The Ecology and Evolution of Animal Behavior*, ed 2. Santa Monica, CA, Goodyear Publishing, 1979.
92. Griffin DR: *Animal Thinking.* Cambridge MA, Harvard University Press, 1984.
93. Griffin DR: *The Question of Animal Awareness. Evolutionary Continuity of Mental Experience.* New York, Rockefeller University Press, 1981.
94. Ruse M. *Sociobiology: Sense or Nonsense?* Boston, D Reidel Publishing Co, 1979.
95. Wilson EO: *Sociobiology. The Abridged Edition.* Cambridge MA, Harvard University Press, 1980.
96. Fox MW: *Concepts in Ethology. Animal and Human Behavior.* Minneapolis, University of Minnesota Press, 1974.

# 6

# Theories of Development
## Part 2. Maturation, Psychoanalytic Theory, and New Approaches in Development and Health

The study of human development must consider all aspects of behavior. In the last chapter three theories were introduced. Each of these offers information relevant to the study of development. However, these three theories alone are inadequate in dealing with all aspects of human behavior. The purpose of this chapter is to follow up chapter 5 and to introduce additional theoretical information required to round out a complete picture of human development.

# THE IMPORTANCE OF THEORY TO CLINICAL PRACTICE

Theories tend to be abstract and not particularly highly regarded by students of clinical practice (1). Despite this, the professions of both occupational and physical therapy have lately been promoting the development of theoretical inquiry and research within undergraduate and graduate curricula (2–8). For example, in 1984, the American Occupational Therapy Association identified research priorities for the 1990s that included the development of an integrated theory of occupation (2). Similar priorities are found in the field of physical therapy (3–5). As Dean (4, p. 1061) notes, "A primary goal within the profession is defining a body of knowledge and establishing physical therapy as a unique clinical science."

The value of theories is that they systematically organize wide bodies of information into coherent, cohesive wholes. In clinical fields, this type of integration is valuable. It ties together and gives a sense of identity to the field. It gives therapists a sense of confidence by providing a unified approach to practice, while systematic research provides direction toward the development and testing of new and existing theories of intervention (7).

One of the ways to facilitate the construction of new theories is to understand the synthesis of other fields. A large portion of any field of clinical practice involves understanding human behavior and how it changes over the life span. Thus, knowledge of theories about human development will be particularly relevant to therapists interested in a science of clinical practice. The study of development is the study of behavior *as it changes over time.* In comparison, therapeutic intervention is systematically causing or supporting *change over time.* Knowledge of the principles and theories of human development is directly relevant to a science of intervention.

The purposes of this chapter and the previous one are to provide an introduction to theories that are important in understanding human development and to explain how these theories are also relevant to therapists. In this chapter, three areas are explored: maturation, psychoanalytic theory, and a collection of new approaches including humanism and ecology. All of these, as will be discussed here, have particular applications to clinical practice and to the scientific development of the occupational therapy and physical therapy professions.

# MATURATIONAL THEORIES

Maturation, as we pointed out in chapter 2, is the behavioral and physiological "unfolding" or expression of one's genetic potential. There are many theories that emphasize this influence on development, and these are clustered under the term "maturational theories." Although they do not comprise a large organized body of knowledge like cognitive or learning theories, they can be compiled into a category of information that emphasizes the significance of maturational influences in human development.

At one point in time, maturation was viewed as a force directly opposed to or exclusive of learning. Now, however, as maturational theorist McGraw (9, p. 130–131) states, it is recognized that "maturation and learning are not different processes—merely different facets of the fundamental process of growth."

---

*Learning vs. Maturation*

"The fundamental problem about learning vs. maturation is not the question about which is more important (whatever that means). It is the problem of going beyond typologies of either. It is a problem of amassing evidence, large masses of missing evidence, about what the child does or does not do with experience, of admitting correlative evidence about animal behavior, brain structure, physiological development, of proposing and slowly perfecting a general theory or system that articulates their interaction (10, p. 69)."

When maturation and learning were dichotomized, those scientists and philosophers who opposed learning theory and behaviorism often naturally endorsed maturational concepts. Ethologists, for example, whose work concentrated on instinctive, innate behaviors, therefore emphasized genetic or maturational influences to the exclusion of the effects of learning. In addition, because of its widespread effect on early growth and development, maturation was emphasized by those researchers who explored prenatal, early infancy, and early childhood development.

Thus, most maturational theories have to do with early development, and collectively they have had an impact on the scientific study of children as well as clinical applications in pediatrics. One reason why such theories are so significant to allied health fields such as occupational and physical therapy is that much of maturational study has focused on brain and motor development, areas of major significance to clinical practice. Physical and occupational therapists, in turn, have directly applied these maturational concepts in their models of treatment for patients with brain and motor (neuromotor) or with other developmental impairments.

Some of these theories, such as those of occupational therapists Lela Llorens (11) and A. Jean Ayres (12, 13) and occupational and physical therapist Margaret Rood (14, 15), will be discussed in this section. It must be emphasized that the specific clinical models or theories proposed by these therapists are treatment rather than maturation oriented. However, they are discussed in this chapter because of the maturational-developmental concepts incorporated within them.

# G. Stanley Hall and Recapitulation

G. Stanley Hall was one of the leading contributors to the maturational viewpoint in child development. Considered the father of the psychology of adolescence, Hall developed a theory of child development based on maturational principles (16). Hall applied evolutionary principles to his theories of development and was particularly influenced by Charles Darwin's work, which during Hall's time was popularized by German evolutionist Ernst Haeckel (1834–1919) (17, 18).

Haeckel's work is significant because, although it was found to be erroneous, it has been widely applied in human development and in clinical practice. This is the concept of evolutionary recapitulation, which asserts that each individual's development (*ontogeny*) passes through the same stages as the species did during evolution (*phylogeny*). The biogenetic law of recapitulation states "that phylogeny repeats ontogeny [and] that all creatures recapitulate their evolutionary history" (18, p. 110).

While Haeckel's view appears true in gross observation, especially in regard to embryo development, it is an inaccurate and misleading overstate-ment of the details of development (19–21). Haeckel's ideas were in vogue, however, in the late 1800s, and when Hall endorsed them, he brought together the fields of embryology and psychology.

Hall (10, p. 63) asserted that "the development of the mind of the child recapitulated the evolutionary development of mankind." Similarly, Hall posed that children's play and social development go through stages corre-sponding to the actual course of human evolution. He believed that chil-dren's love for water and play at the beach, their fears, and adolescence itself, are all "replays" of earlier ancestral behaviors (22, 23). Thus, according to Hall, in recapitulating the development of their ancestors in both brain and social development, humans are more significantly affected by matura-tional than by environmental forces.

Hall's theories and the recapitulation concept itself were subsequently discounted. For example, Piaget (24) explained the inaccuracies of Hall's theories of play. Piaget reported that it is now widely recognized that games are influenced by one's social and natural environment and are not subject only to embryological laws, as Hall had claimed. Referring to Hall's theory, Piaget (24, p. 157) claimed that "all the indications are that play is rather a matter of participation in the environment than of hereditary resurrection."

Thus, it was recognized that genetic influences are not the sole effects on social and cultural behavior, and Hall's theories were subsequently rejected by the academic community. The concept of recapitulation, how-ever, has had a long history and a vast influence that goes beyond Hall's work. This influence can be found in other theories of human development as well as many approaches in clinical treatment.

# The Vast Influence of Recapitulation

Anthopologist Gould (22, p. 13) notes that the comparison between ontogeny and phylogeny "may be the most durable analogy in the history of biology." Although the founder of the recapitulation theory of embryology was Etienne Geoffroy Saint Hilaire (1722–1844) (25), Haeckel is more commonly associated with the biogenetic law and with the phrase "ontogeny recapit-ulates phylogeny" (17, 18, 22, 23).

Thus, with the influence of Darwinian theory and Haeckel's law, it was widely believed that "individuals in their own embryonic and juvenile growth, repeat the adult stages of their ancestors—that each individual, in its own development, climbs up its family tree " (23, p. 216). Even though this biogenetic law of recapitulation was, as Gould (22, p. 1) describes it, "abandoned by science" years ago, it has occupied a significant position in evolutionary, developmental, social, educational, and clinical theories.

## Recapitulation and Clinical Theories

In clinical theories, the concept of recapitulation has been modified to refer to patients' recovery of functional abilities rather than to the process of development. Thus, the traditional biogenetic law that "ontogeny recapitulates phylogeny" is modified in treatment approaches to what Mosey has termed "the recapitulation of ontogenesis" (26). This view assumes that human development is a normal progression through certain sequential, maturational stages and that rehabilitation should follow that same process.

### Llorens Developmental Theory for Health and Rehabilitation

The recapitulation approach has been consolidated by occupational therapist, Lela Llorens, into a theory for health and rehabilitation (11). This theory explains how disease or trauma interrupts development and how therapy is aimed at restoring it.

Llorens notes that some individuals who experience injuries or degenerative conditions are prevented from following a normal developmental course because of delays or because they actually regress in functional skills. Therapy is therefore aimed at assisting these individuals to make up for delays or to actually recapitulate previous development by restoring lost functions. Llorens' describes this process in the following excerpt from her theory:

> When the organism experiences overwhelming physical or psychological trauma related to disease, injury, environmental insufficiencies, or intrapersonal vulnerability, the normal growth and development process may be interrupted. Such growth interruption can cause a gap in the developmental cycle resulting in a disparity between expected adaptive, coping behavior and the necessary skills and abilities to achieve it.

> At the time of growth interruption, through skilled application of activities and relationships, occupational therapy can provide growth and development experience-links to assist in closing the gap between expectation and ability by increasing skills, abilities, and social relationships in . . . development (11, p. 3).

Following is a description of how a young man's developmental level may be altered and how therapy is directed at recapitulating normal development.

---

*Recapitulation and Developmental Level*

Joe was a popular high-school senior. He was very athletic, participated in team sports, and belonged to many civic organizations. He had worked all through high school and with some help from his parents had made enough money to buy a used car and to go to college away from home. He was proud of his accomplishments and was looking forward to a life of independence at college.

The night of high school graduation, Joe was involved in a car accident in which he severely injured his spinal cord. As a result of the injury his lower extremities were paralyzed. In addition, because of the trauma of the accident and the time spent recovering, Joe's arms and upper body were initially very weak. Although he would eventually recover adequate use of his arms and hands, he was initially dependent on health care staff for meeting all of his personal needs. At first, Joe was unable to feed, dress, toilet, or even move himself. In addition, the trauma, the losses he had to face, his altered interactions with his family and friends, and his dependency and disappointments caused Joe to be temporarily depressed.

Joe's life radically changed in one day. Joe had been functioning as an independent 18-year-old, and one day later he became physically, socially, and possibly emotionally dependent on other people for most of his needs.

The long-range consequences of Joe's injury will depend upon his medical care, rehabilitative therapies, his health, his family and other support groups, and mostly upon Joe's own level of motivation and drive. His dependency and depression are hopefully temporary, and his goal will be to return eventually to a normal independent life-style in which he manages his own needs, either by meeting them himself or by hiring others to assist him.

At present, Joe's developmental level has changed from one of independence to dependence on others for assistance with self-care and mobility. In fact, Joe may have to physically relearn (or repeat) the same motor skills that he had previously developed when he earlier learned to walk, to dress himself, toilet himself, write, etc. Thus, *in a sense*, Joe has regressed in development, and therapy will be aimed at going back through development so that he can return to his previous skill levels and independence.

---

Recapitulation approaches to Joe's rehabilitation assume that his physical and psychosocial progress over time will follow a sequence similar to the stages that he had already passed through during his childhood and adolescent development. For example, in emotional development, it is normal for a child to become gradually less dependent upon his family and more dependent upon himself and on the support of his peer group. Joe had already passed through those stages of emotional and social development; however, because of his injury, Joe has had to reestablish dependence on his family or on supportive nursing staff. The same is true of Joe's activities of daily living (ADL) skills and other motor functions.

Thus, the goal of therapy is for Joe to become more skillful in taking care of his own physical needs, which in turn will enable him to be more emotionally and functionally independent again. According to recapitulation theory, Joe will undergo those same processes of development again (i.e., recapitulate his previous development) in reestablishing his independence. This refers to all aspects of development, including physical, ADL, as well as psychological, social, and emotional processes (11, 26).

According to Llorens (11), rehabilitation therapy provides "experience links" that assist Joe to develop specific occupational and movement skills as well as psychosocial adjustment to enable him to return to his previously achieved physical, social, and emotional levels of development and independence.

### Rood's Ontogenetic Motor Patterns

Another example where recapitulation theory was generalized to sensorimotor treatment is found in the principles endorsed by physical and occupational therapist Margaret Rood, which are summarized by Trombly (14) and Huss (15).

Rood has proposed specific ontogenetic motor patterns (Fig. 6.1) for habilitating or rehabilitating patients who have problems with motor control. Treatment is based on the assumption that an individual's normal development (ontogeny) follows certain sequences and that therapy should be directed at repeating (recapitulating) those sequences. Ontogenetic treatment principles are geared toward mimicking normal maturation. They rely on active stimulation and involvement of the patient in order to progress through sequences that recapitulate normal development.

## Principle of Proximodistal Development

Ontogenetic approaches such as Rood's are based upon principles of normal maturation, one of which is called the *principle of proximodistal development*. This was derived from observations that whole-body movements involving the trunk and gross motor skills developmentally precede precise,

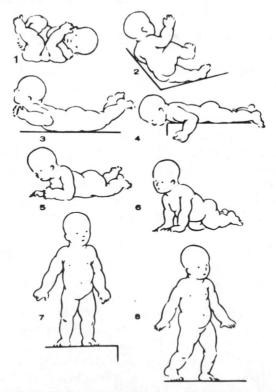

**Figure 6.1.** Ontogenetic motor patterns according to Rood. (1) Supine; withdrawal flexion; (2) Supine to prone; unilateral flexion. (3) Prone; dorsal flexion and extension, elbow flexion. (4) Cocontraction of the neck. (5) Cocontraction; head, neck, shoulder girdle and glenohumeral joint. (6) Creeping. (7) Standing. (8) Walking. (From: Ayres AJ: Occupational therapy directed toward neuromuscular integration. (In Willard HS, Spackman CS (eds): *Occupational Therapy* ed 3. Philadelphia, JB Lippincott, 1963, p. 399.)

fine motor skills. Thus, an infant lifts the head and establishes good head control, rolling, and sitting skills before developing good eye-hand coordination and refined dexterity for manipulation.

Whole-body movements use muscles that are located near the center of the body (e.g., those of the trunk and shoulders), which are referred to as *proximal muscles. Distal muscles* are those located away from or distant from the body, e.g., the muscles of the hands and feet. The proximodistal principle states that development progresses from proximal to distal directions, and that proximal precedes distal motor control. A further assumption based on this principle is that distal motor control *depends* on previous proximal motor development (27). Thus, treatment of individuals with motor weaknesses or delays is aimed at following (recapitulating) that same sequence of motor development exhibited in human ontogeny.

Such treatment approaches are illustrated by Rood's previously described ontogenetic model (Fig. 6.1) and by a similar one described in the 1950s by Ayres (28). In an article entitled "Ontogenetic Principles in the Development of Arm and Hand Functions," Ayres describes a treatment sequence that recapitulates ontogeny by following the proximodistal principle. Treatment, according to this approach, is initially aimed at control of the head and eyes and then is aimed at the body parts according to the following sequence: the trunk, the shoulder girdle, elbow, gross hand grasp, wrist, hand release, forearm rotation, and then manipulative skills by the fingers.

The proximodistal principle, which forms the basis of both Rood's and Ayre's treatment approaches, was generated in the 1920s and 1930s. Loria (27, p. 167) notes that despite the fact that this principle was based on early phylogenetic, embryologic, and ontogenetic observations, and has served as a basis for many therapeutic procedures, the data supporting it "were controversial and often scant."

Thus, just since the 1980s Loria (27) and others (29) have been testing this principle of development and its applicability to normal maturation as well as to clinical practice. For example, based on a study of distal and proximal muscle activity in a small group of subjects, Wilson and Trombly (29, p. 16) note that there is some question "whether the proximal-distal sequence of development provides an adequate basis of therapeutic principles of treatment." A similar question is raised by Loria's study of the relationship between proximal and distal functions during normal motor development. She notes that contemporary studies of motor control indicate that perhaps two different systems may be involved in the development and control of proximal and distal abilities.

Thus, whether treatment should focus on gross motor *prior to* fine motor development, as recapitulation theory might assume, or whether treatment is most effective when gross and fine motor abilities are addressed simultaneously or in different sequences for different patients are subjects for continued occupational and physical therapy research. A temporary resolution of this issue is one that most therapists tend to adopt in practice.

That is the tendency to adapt treatment approaches according to the responses of patients. Proximal and distal functions are both addressed, and in cases where proximal control is inadequate and therapy is directed at distal functions, support is provided by the use of such aids as slings, bolsters, braces, and wheelchairs.

At present it appears that the traditional reliance on maturation and recapitulation approaches to treatment, especially in regard to the proximodistal principle, may be subject to continued scrutiny in contemporary clinical research. This serves as an excellent example of how research enables us to question and test important clinical and developmental theories and how it may ultimately show us the direction for the most effective approaches in practice.

## Jacksonian Views of Brain Levels and Treatment

The traditional ontogenetic views of treatment such as Rood's (14, 15) and Ayres' (28) were influenced by the work of neurosurgeon Temple Fay. As Ayres notes, Fay promoted the therapeutic application of sequential maturational steps in phylogenetic development (30). Fay's basic premise, Huss (15, p. 125) points out, was "that ontogeny recapitulates phylogeny. Therefore an individual's neurological development parallels the evolution from fish to amphibian, to reptile, to anthropoid."

Based upon such recapitulation approaches, Fay proposed that therapeutic intervention start with simple patterns of movement and that these "lower" levels of movement be developed before "higher" movement patterns are addressed (15). Although the terms "higher" and "lower" are not universally accepted (31), they are widely used, especially in clinical practice (13, 15). These terms stem from a hierarchical view of brain organization that was posed by Hughlings Jackson (1851–1911) in the late 1800s. Jackson described a model of brain function that was based on evolutionary principles and which applied Haeckel's biogenetic law (18, 32). Updated by Paul MacLean (33) in the 1960s and 1970s, this controversial theory assumes that levels of the brain correspond to certain stages of evolution. In addition, the assumption is made that normal brain development actually recapitulates the phylogeny of the nervous system.

According to Jacksonian views of brain organization and development, phylogenetically older subcortical brain regions, considered "lower" levels of the brain, regulate behavior early in development. These regions are associated with involuntary behaviors and with fundamental drives and motor skills, such as posture, equilibrium, and simple sensorimotor and gross motor functions that humans share with other species. Jackson claimed that with maturation, phylogenetically newer brain regions (the cortical areas) gradually take over and regulate voluntary functions and those skills conventionally regarded as human, e.g., complex oral and written language and problem-solving abilities, and skilled manipulative abilities involving dexterity.

These newly evolved or "higher" brain regions are also presumed to integrate (i.e., pull together) and "oversee" the overall functions of the brain. According to Jackson, disease or trauma in humans interrupts normal ontogeny and causes a reversal of the normal developmental progression of the nervous system (32, 34).

Jackson's hierarchical view of the brain and his recapitulation view of recovery from brain damage found their way into clinical approaches to the treatment of patients with brain or neuromotor damage. They also formed the basis for Temple Fay's view of motor control. Fay's work is considered to be the forerunner of sensorimotor approaches to treatment (15).

Thus, Jackson's influence has indeed been profound. It serves as a basis for sensorimotor approaches to treatment (15) and rehabilitation (35). Jackson is referred to in the works of Bobath (36) and his ideas, as translated by Paul MacLean, form the basis for Ayres' sensory integration theory (12, 13). Fioretino's (37) approach to testing reflexes is also based upon the notion of specific levels of the central nervous system regulating motor behaviors in a hierarchical manner (Table 6.1). Physical therapist Joan Loomis (38, p. 171) points out that Jackson's hypothesis of a hierarchically organized central nervous system "has served as the basis for all approaches to therapeutic exercise." In addition, as we will discuss later in this chapter, Jackson's ideas about recapitulation form an essential part of Sigmund Freud's psychoanalytic theory (32).

## Integration of Jackson's Views with Contemporary Models of the Brain

Physical therapist Joan Loomis (38) points out that current research indicates that systems theory, which conceptualizes the various levels of the central nervous system as an interacting network, more aptly describes brain functions than does a Jackson-based levels approach. Contemporary views of brain function (38–41) portray it as working as a whole (i.e., holistically, with integrated systems) rather than according to specific phylogenetic levels as portrayed in Table 6.1. Further detailed analyses of Haeckel's and

**Table 6.1**
**Hierarchical View of Normal Sequential Development**[a]

| Level of CNS Maturation | Corresponding Level of Reflex Development | Resulting Level of Motor Development |
|---|---|---|
| Spinal and/or brainstem | Apedal | Lying prone |
|  | Primitive reflexes | Lying supine |
| Midbrain | Quadrupedal | Crawling |
|  | Righting reactions | Sitting |
| Cortical | Bipdeal | Standing |
|  | Equilibrium reactions | Walking |

[a] From Fiorentino MR: *Reflex Testing Methods for Evaluating C.N.S. Development*, ed 2. Springfield IL, Charles C Thomas, 1973, p 5.

Jackson's theories of development and brain organization and their clinical applications are beyond the scope and purpose of this chapter. Interested readers may consult other sources for varied approaches to and discussions of this topic (e.g., 17, 18, 21, 22, 31–35, 39).

Hierarchical views of brain development and their corresponding maturational approach to behavioral development provide therapists with a convenient frame of reference on which to base treatment. This is illustrated in the following quote from therapist Jennifer Couper:

> The rationale for using dance therapy as a treatment for the psychologically and neuromuscularly impaired is based on the theory that ontogeny recapitulates phylogeny. This theory involves a repetition of the experiences that govern an individual's development. The repetition involves having the person undergo certain environmental and social interactions to enhance normal brain function. The goal is to organize and integrate components of psychomotor maturation that are usually learned in the normal growth process.

> Ayres theory of sensory integration is similar to the rationale for using dance therapy, in that Ayres' theory is also based on ontogeny's recapitulating phylogeny. (41, p. 23)

Sensory integration theory (12, 13) assumes that the brain is immature at birth and is also immature in some individuals with learning problems. Treatment for these individuals is aimed at recapitulating normal neuromotor development by providing therapeutic sensory and motor experiences. The goal of sensory integration therapy is to provide stimulation that will address certain brain levels (primarily subcortical), enabling them to mature, and thereby assisting the brain to work as an integrated whole (12).

Ayres notes that contemporary holistic views of brain organization are now widely accepted, but it is easier for therapists to think about brain function from a hierarchical, levels approach. She claims that it is difficult to plan treatment "when conceptualizing the brain as performing as a whole although it does indeed do so (12, pp. 38–39)." Thus sensory integration theory is based on a combined approach which acknowledges Jacksonian levels view of neuromotor control as well as a holistic or systems view of brain function. Both views may be helpful to guide students of normal development as well as to direct therapists in approaching treatment and in selecting therapeutic modalities and activities.

## Evaluation of Maturation and Recapitulation in Clinical Practice

Maturational views about development and their applications to treatment are less popular now than they were even 20 years ago. Recapitulation, it seems, has come under scrutiny from nearly every direction. Thus, just as maturation and learning have been integrated into interactional views of development, so also is brain organization commonly viewed holistically from a systems view, rather than hierarchically, as recapitulation would assume (38, 39).

Although the biogenetic law was proven inaccurate early in this century, it has still been a guiding force in developmental and clinical studies.

Such maturational principles have profoundly influenced occupational and physical therapy sensorimotor as well as psychosocial treatment approaches. They provide a framework for various interventions, illustrated by Mosey's "recapitulation of ontogenesis" (26) as well as Rood's (14, 15) and Ayres' (28) ontogenetic approaches, and Llorens' (11) developmental approach. Recapitulation concepts have been applied in controversial approaches to brain development (13, 33), in dance therapy (41), and in Freudian theory (21, 32).

Now, however, many assumptions based on recapitulation theory are being questioned. For example, in the 1980s, the traditional proximodistal view of motor development is being examined in clinical research (27, 29), as is Jackson's 19th century levels view of the brain and its resulting clinical approach to treating neuromotor-injured patients (38, 39). One potential resolution of this issue is to think of development from a maturational view but to also keep in mind systems views of brain and behavioral organization.

There are additional cautions in regard to overapplying hierarchical approaches in treatment. Such views may lead to preset or prejudicial views of the interrelationships between the elements in the hierarchy. As long as hierarchies are established, they lead to such dichotomies as "higher vs. lower." These, in turn, lead to comparisons such as "better than" or "simpler than." Such comparisons can be destructive and misleading when used to refer to human behavior or to compare humans with one another or with other species (22, 23, 31).

Gould (22) notes that such comparisons have occurred as a result of misusing or misapplying recapitulation approaches to human groups. In some cases, scientists have used recapitulation theory to erroneously account for the superiority of some people over other "underdeveloped" or to what are referred to as "primitive" individuals or races. In similar fashion, clinicians who adopt recapitulation views need to guard against the assumption that patients are "undeveloped," regressed, or immature because of interrupted occupational behaviors and skills.

Although some patients may lose skills customarily associated with maturity (e.g., language, walking, bowel and bladder control, writing, dressing), these individuals are not necessarily cognitively, emotionally, or socially immature. Versluys (42) points out that some patients with physical disabilities may indeed emotionally regress or return to earlier developmental levels in response to trauma. But this is certainly not the case with all disabled people, nor is it necessarily a permanent effect for those individuals who do regress.

Thus, clinicians must be careful not to assume that patients who lose maturational skills become "children" who are expected to recapitulate their maturity. Many former patients who may never recover the ambulatory or occupational skills that are normally associated with maturity and independence still live mature and productive lives.

In a critique of Mosey's theory of recapitulation (26), Bing (43, p. 435) notes many advantages of such an approach but also observes that "care must be exercised . . . so that sequence and hierarchy are not so rigidly viewed as to negate the still extremely significant factor: unique self perceptions of the individual." The significance of human individuality and the importance of looking at the whole person is stressed in this chapter's final section regarding humanistic and holistic theories of development.

## Hall, Gesell, and Maturation Theory

As noted earlier in this chapter, the recapitulation view in human development was championed by Hall. Hall also emphasized the significance of genetic forces in development. He claimed that, up until adolescence, environmental influences were essentially ineffective, as maturation predominated (16). Although Hall's work was subsequently discounted, the belief in the predominance of maturation early in development persists. White describes what happened to Hall's work:

> Hall was not followed—his synthesis was personal . . . and his views did not have any overt line of descent. We remember him causally today as the advocate of Recapitulationism; we associate to his name the slogan "ontogeny recapitulates phylogeny." After all, his ideas were not 'testable' (the word seems puny as applied to the vast scope of his ideas); he proposed no program of research, no outline, or organization, no method. [But] part of his outlook on developmental psychology may have survived in the work of Arnold Gesell, his student. (10, p. 64)

Arnold Gesell, as we previously noted in chapter 2, had a significant impact in the field of child development as well as in the clinical fields of occupational and physical therapy. Gesell's contributions to these fields is theoretical as well as practical. For example, Gesell's studies of the normal development of children resulted in age-related standards for motor, language, and social skills which are still widely used or referred to in developmental assessments today.

Gesell, like Hall, viewed maturation from a conventional point of view. He described maturation conventionally as "those phases and products of growth which are wholly or chiefly due to innate and endogenous factors." He defines growth as "the total complex of ontogenetic development (44, p. 25)." Although Gesell's theories recognized the contribution of environment and external factors in development, he emphasized maturation as the primary influence.

## The Primacy of Maturational Influences in Early Development

This belief in the primacy of maturational influences, which was held by Hall and by Gesell, is controversial but still widely supported, especially in regard to early development. For example, some life-span theorists note that early life may in fact be more subject to maturational than to environmental influences because of the enormous physiological and primarily internal changes that occur during early development. Evidence suggests that, compared with other times during development (with the possible exception of old age), early life tends to demonstrate greater similarities among people. This is due to their common experience of internally controlled (maturational) events such as bone and muscle growth and development, the appearance and growth of teeth, neuromuscular maturation, etc. Over the life span, however, people have greater exposure to such a variety of events that can alter physiology, brain, behavior, and attitudes. Thus environmental changes will account for broader individual differences later, as opposed to earlier, in development (see "Canalization" chapter 11).

To illustrate this point, consider five normally developed 2-month-old infants who were born to different families. One family lives in the United States, one in the Soviet Union, one in Greece, one in Norway, and one in Japan. The social, emotional, and motor behaviors of all these infants will be quite similar. They will all communicate their needs by crying or fussing, and they will all charm their families by smiling at, and trying to reach toward, their parents. They will all tend to roll over and crawl at the same age, and each infant's first tooth will emerge at around the same time as the other infants'.

Over time, however, these infants will become progressively more dissimilar as cultural influences are exerted. By 3 years of age, these children will all speak different languages and have a slightly different perspective on the world based on their differing cultures and environments. By age 56, for example, assuming the infants stayed and grew up in their own countries, they will be fully acculturated and very different from one another as a result of exposure to different political, educational, and religious systems. Over time in life, there seem to be more opportunities and more variables that can account for broad individual differences and diversity in people.

## Maturation and Prenatal Development: Problems in Assessment and Treatment

The significance of maturation in early development is very apparent in the field of assessment and treatment of infants and children. A particularly troublesome problem for therapists is how to assess, and what behaviors to expect from, an infant that is born early, i.e., is preterm (see chapter 7, "Pre-term or Low-Birth-Weight Infants).

In the past 10 years, the population of preterm and small infants has expanded considerably because of tremendous advances in medical technology. With the growth of this new population of infants, occupational and physical therapists are increasingly being called upon to assess and to design treatment programs for younger and smaller infants (45). In addition, prematurity is associated with risk status for later developmental problems, thus preterm infants may be followed for months or years by their pediatricians and may often be evaluated by therapists in outpatient clinics or in early education/intervention programs. The process of tracking these infants over time to be on the lookout for potential deficits is called *developmental follow-up* (46).

Consider the following cases to see the difficulties that therapists face when called upon to evaluate preterm infants.

Josh and Travis were both born on June 1. Josh, however, was born 2 months premature. Because of his prematurity, Josh had to stay in the hospital for 1 month. During this time he was placed in an isolette, which helped him keep a constant body temperature as well as to stabilize his breathing. In addition, he received medical attention that helped meet his nutritional and other health and medical needs. While he was in the isolette, he was visited by his family, who occasionally were able to take him out and hold and feed him and who also were able to stroke him in the isolette. Later, when he left the isolette, the family continued to visit him for meals and to take care of him during brief visits. Since he slept a lot and since medical procedures seemed to take a lot of his energy, the family did not want to disturb him by handling him too often. Once his health and weight stabilized after a month, he was discharged home.

In comparison, Travis left home 2 days after birth. Since he was born full term, he did not have the additional developmental and medical needs that Josh exhibited. Thus, Travis was able to stay in his mother's room immediately after birth. There he was nursed by his mother and visited by his father and sister. Two days later, he was at home, receiving care from his parents, his sister, and his grandparents, who all took turns comforting, feeding, changing and playing with him.

On December 1, both Josh and Travis become 6 months of age, but the two infants appear slightly different. Josh seems to be slightly slower than Travis. His development seems to lag, as he seems to display the same motor skills as Travis did a few months earlier.

What are the potential reasons for the differences between Josh and Travis? One possibility is that because of all of his early experiences, Travis' development is accelerated in comparison to Josh's. Or that Josh's necessary medical care and separation from his family deprived him of the early stimulation that Travis received and actually delayed his development. Another interpretation is that Josh is really 2 months behind Travis in normal motor development since he was born prematurely. Thus, although both infants are considered 6 months of age, this is accurate only for Travis, the full-term infant. Josh, having been born 2 months early, should really be considered 4 months of age to account for the 2 lost months.

## Problems of Assessment

Therapists who are called upon to assess infants like Josh face numerous difficulties depending upon when they administer their assessments. The first problem they face is determining how to assess a preterm neonate.

Developmental assessments typically include sequences of behaviors and the approximate chronological age associated with the accomplishment of each behavior (47). Most assessments start with normal development at age 0, i.e., at birth, which for most infants is full term, 9 months gestational age.

Therapists must ask themselves these questions: "Are such assessments appropriate for an infant that is premature?" "Does 2 months of maturation in the uterus make a significant difference in motor, social, and adaptive skills so that 9 months GA is significantly different from and should be evaluated differently than 7 months GA?" "Are there specific behaviors to expect from a preterm infant that are not seen in a fullterm infant, and vice versa?" "If there are differences between these infants, then do we need evaluations specific to preterm infants?" "Furthermore, if there are fundamental differences in the behavior of preterm and fullterm infants, how long, if ever, do those differences last?"

This last question points to the second issue confronted by therapists. If preterm infants are qualitatively different from full-term infants, how long do those differences last? Will preterm infants always be 2 months behind their full-term counterparts, or will they catch up, with differences gradually washing out over time (Fig. 6.2)?

Unfortunately, there are not clear-cut answers to these questions, and those answers that do exist tend to be controversial. This controversy, once again, finds its origins in the nature vs. nurture debate. For example, scientists who support maturational concepts take one side in the controversy, arguing for the predominance of internal over environmental influences. Gesell and Amatruda (48) consider maturation to be the most important determinant in an infant's development. They claim that environmental events early in life will neither accelerate nor delay development, but

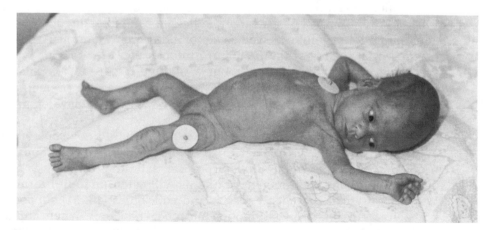

**Figure 6.2.** Controversies exist regarding how to assess and treat preterm infants and whether they are fundamentally different from full-term neonates. Should preterm infants be allowed to mature "on their own," or should sensorimotor interventions be started to address existing or potential deficits? (See Fig. 6.3.) (Courtesy of Neonatal Intensive Care Unit, University of Connecticut Health Center's John Dempsey Hospital, Farmington, CT.)

that, in 2 to 3 years, a preterm infant who initially exhibits delays, should eventually catch up to his or her full-term peers (46).

Others do not agree with the view that maturation will somehow normalize development. Instead, they contend that learning and other environmental stimulation are necessary to compensate for developmental differences exhibited between preterm and full-term infants (Fig. 6.2). Still others contend that preterm and full-term infants are very different and that simple consideration of gestational age and maturation are inadequate in understanding the differences between them. Thus, these infants must be evaluated and treated differently (46, 49–51).

If indeed preterm infants are qualitatively different in their maturation than their full-term counterparts, then assessment must address this. One way to address this is to take into account differences in maturation between preterm and full-term infants. Thus, different evaluation instruments, such as the Assessment for Premature Infant Behaviors (50), need to be used for preterm neonates than are used for full-term neonates.

Another way to address potential differences between these infants is to adjust for prematurity during developmental follow-up. Thus, if a developmental assessment is being used for a preterm infant, the evaluator should credit the infant with the amount of maturation time missed. This is advocated by Gesell and Amatruda (48), who recommend that for the first 2 to 3 years of a premature infant's life, developmental assessments should be adjusted to account for the premature infant's earlier gestational age. Adjusting for age is based upon the assumption that a preterm infant actually misses out on maturation. Thus, if an infant, who is 2 months premature, is tested at 6 months of age, the infant should be treated as a 4-month-old infant instead of a full-term 6-month-old (46).

Whether or not therapists should adjust for prematurity is under exploration. Research studies by physical and occupational therapists as well as those by psychologists have indicated that adjusting for age seems appropriate for cognitive as well as motor assessments administered during the first 2 to 3 years of the premature infant's life. Clearly, this will vary according to the type of assessment (cognitive, gross motor, reflex, fine motor) and whether it is designed specifically to assess premature and neonatal behaviors or whether it is administered later in development (46, 51).

Up to what chronological age of the infant should developmental age adjustments be made is unclear to clinicians and researchers. Some environmental as well as maturational theorists claim that differences eventually wash out, and that premature infants will catch up to full-term peers. Work by physical therapist Palisano and colleagues (46, 51) supports this but points out that this assumption may be valid only when applied to healthy preterm infants. Since we are increasingly seeing younger and smaller infants, therapists may have to make special accommodations in developmental follow-up of these very special infants. In addition, there may be fundamental differences in the rate of maturation of various developing systems (e.g., cognitive, fine motor, gross motor) as well as between preterm males and females (46).

## Problems of Treatment

Even more subject to debate than assessment is the issue of how to most adequately support the development of preterm infants. If one believes that at this time in prenatal life, maturation has a normalizing and stabilizing influence, then treatment of infants born early should be geared at re-creating an environment as similar as possible to "in utero" existence (Fig. 6.3). According to this view, there should be as little intervention as possible, on the assumption that maturation will take care of development until (or even after) the infant achieves a full 9-month term of gestation.

This maturational view has been extended not only to preterm infants but also to other children with developmental delays and motor handicaps. As Cunningham and Mittler (52, p. 310) explain, some maturational approaches assume "that the development of a handicapped child follows essentially the same pattern found in normal children." Thus, the handicapped child goes through the same sequences and stages although at a slower rate. The authors note, "An extreme practical consequence of such a view is that no special measures are necessary for these children; it is only necessary to provide the 'normal' range of experiences and learning opportunities appropriate for comparable levels of development (52, p. 310)."

This point of view is illustrated in the following quote by Gesell, in which he claims that even in children who are exposed to physical handi-

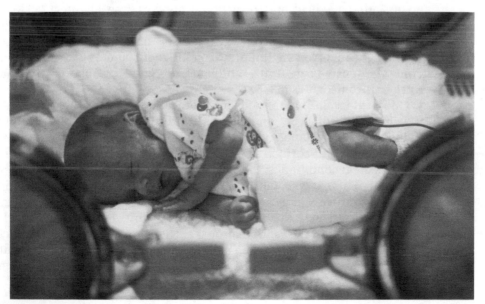

**Figure 6.3.** Maturation theory assumes that treatment for preterm infants should simulate, as much as possible, their natural, uterine environment. (Courtesy of Neonatal Intensive Care Unit, University of Connecticut Health Center's John Dempsey Hospital, Farmington, CT.)

caps, malnutrition, or prematurity, maturation serves a normalizing purpose to restore the fundamental genetic patterns of development:

> When certain areas of the nervous system are actually damaged by disease or injury, maturation cannot make amends, but the maturation of the nervous system seems to proceed toward the optimum in the areas unimpaired, even though lacking the stimulus of exercise of the functions controlled by the impaired areas. It is for this reason that certain clinical types of profound motor disability attain none the less considerable approximation to normality in certain patterns of behavior. . . (44, p. 29).

> Although it is artificial to press unduly a distinction between intrinsic and extrinsic factors, it must after all, be granted that growth is a function of the organism rather than of the environment as such. The environment furnishes the foil and the milieu for the manifestations of development, but these manifestations come from inner compulsion and are primarily organised by inherent inner mechanics and by an intrinsic physiology of development (44, p. 30).

Many theorists and health professionals, however, disagree with this line of thought and believe that the earliest possible intervention is critical, to prevent disabilities or deformities that may be present. In addition, they believe that they must intervene in special care nurseries with preterm infants because of the complications associated with many of the medical interventions necessary to keep these ill or very young infants alive. Thus, many treatment programs, in general, are started as early as possible. Interventions for preterm infants are aimed at altering their environments by providing various forms of sensory and motor stimulation programs in addition to providing proper positioning, splints for paralyzed or weak limbs, and essential medical and nutritional support (see chapter 7. "Roles for Therapists in Special Care Nurseries"). These interventions are provided by occupational therapists, physical therapists, and others who are becoming increasingly involved in important assessment and intervention roles in neonatal intensive care units, which are designed to support very small, ill, or premature neonates (53–57).

Most intervention programs for preterm infants are aimed at combining environmental intervention and maturation principles. Thus attempts are made to strike a balance between maintaining as natural an environment as possible but still correcting or preventing physical abnormalities as soon as the medical condition of the infant stabilizes and the health of the infant enables it. The extraneous sensorimotor programs for very young infants are most controversial as are the numerous medical interventions and the disruptive ecology of the neonatal medical unit. It is unknown whether such interventions are beneficial or whether they are destructive, actually interfering with what should be a relative rhythmic, protected intrauterine existence, influenced primarily by maturational than by environmental forces (58–61).

### Separating Maturation and Treatment

A separate but related issue has to do with evaluating the significance of clinical intervention. There is, at times, difficulty in sorting out the effects of maturation and treatment. This is especially true when treating young children since maturation plays such a significant role in early development, especially in the emergence of neuromotor abilities, which are often addressed by physical and occupational therapists. The following case report illustrates the difficulty of teasing apart maturation and intervention.

*Maturation or Intervention?*

Sally is a 3-year-old child attending an early intervention preschool program because of developmental delays. During Sally's birth delivery, her umbilical cord became wrapped around her neck, temporarily interrupting the blood supply carrying oxygen to her brain. Sally's physician was concerned with this event because anoxia (lack of oxygen) at birth may cause brain damage and result in cognitive and motor abnormalities. Sally's physician has been following the course of her development and has recommended periodic evaluations. Sally underwent a thorough team evaluation at the age of 36 months, and the results indicated that her motor development was significantly delayed to warrant intervention by specialists in motor-oriented treatment.

As part of her early intervention program, Sally receives occupational therapy, physical therapy, and speech pathology three times per week. Since Sally has been attending the early intervention treatment program, she has made gains in her motor skills. The question is: Were those gains a result of therapy and the stimulation she received in the treatment program or are the gains a normal consequence of maturation?

Questions such as these are difficult to answer. How, in fact, can we determine if Sally's current gains are a result of treatment or of normal development? Certainly we do not want to deprive her of treatment experiences "just in case" she may make up her delays on her own. However, there are some who note that we should question the amount and intensity of our early treatment programs because of the vast influence of maturational principles.

Ongoing clinical research is attempting to provide such answers by comparing treatment gains with normal gains as a result of maturation. Such studies are typical of the research being conducted in developmental therapy, and is a topic addressed in advanced clinical treatment, pediatrics, and research classes.

## Maturation and the Concepts of Readiness and Critical Periods

The concepts of readiness and critical periods also focus on the significance of maturation in early development. The concept of readiness has been widely applied in psychology, human development, and education. *Readiness* refers to some optimum time for learning or for training to occur.

Magnussen and Allen (62, p. 26) define readiness to respond as "the condition of the organism at a particular age period with respect to optimal responsiveness to a certain element in the environment." In ethological theory, that element is considered a releaser.

The concept of readiness underlies the general acceptance that a 1-week-old infant cannot be toilet trained, cannot eat solid food, and cannot sit up in a chair and work an algebra problem. Nash (63, p. 129) explains how maturation and the concept of readiness interrelate, "Presumably the acquisition of any behavior pattern depends on maturational readiness (that is, the organism must have achieved at least the minimum state of development that permits that pattern to be performed and retained)."

Most theorists would agree that there is a probably an age limit that can be placed on the earliest time that certain skills can be acquired by infants and children. As McGraw (9, p. 127) states, "It seems evident that a certain amount of neural maturation must take place before any function can be modified by specific stimulation." Objections to the concept of readiness occur when it is extensively applied to a large variety of skills or to most learning situations. Instead, development is viewed as a gradual building process; children are seen as resilient and able to pick up a diverse array of skills over an extended period of time during life, not just during some "special" time when all conditions are prime and readiness occurs. The controversy over readiness will be examined in more detail in following section and in chapter 11.

## Critical Periods

A concept closely related to "readiness" is that of critical or sensitive periods. The critical period hypothesis has a long history. It originated in the field of embryology where it had been noted that during embryogenesis, periods of rapid growth and development of certain organs, structures, and systems occurred. Experiments conducted in the late 1800s and early 1900s repeatedly demonstrated that during these critical periods of growth and development, the embryo (prenatal organism) was particularly susceptible to external, potentially injurious influences (63).

Consistent with their taking laws about physical functions and applying them to behavior, ethologists took the concept of critical periods in structural embryogenesis and applied it to specific behaviors they had observed. Most commonly, the critical period concept was applied to imprinting behaviors and then later generalized to emotional and social attachment processes in other species, including humans (see "Attachment" in chapter 12). As Rutter (64, p. 21) notes, "Imprinting was originally regarded [by ethologist Lorenz in the 1930s] as a unique phenomenon which could develop only during a very short critical period in early infancy and which was irreversible once it had developed."

## Sensitive Periods

Critical periods came to be defined as *fixed time intervals* during which *irreversible* learning or instinctive behavior patterns occur. This definition has been widely debated (e.g., 65–67) and is most controversial when applied to human behavior. Subsequent studies of imprinting and other behaviors to which the critical period hypothesis has been applied indicate that it was too narrow a concept and that it required revision (see chapters 9 and 12). Fixed intervals were found to be more flexible than originally believed, and the notion of irreversible learning was too strict. The term "sensitive period" rather than the stricter term "critical period" has been widely adopted in response to evidence negating *fixed* periods of time and *irreversibility* of learning (68).

Imprinting was caught up in the nature-nurture debate, and the critical period hypothesis also became embroiled in that issue. Some researchers suggested that imprinting or other behaviors that seem to emerge during relatively narrow intervals of time are actually conditioned and no different from what learning theorists had already described (66). Others suggested that a variety of types of learning exist and that one type which explains critical periods is very rapid and relatively fixed. This latter notion retained a critical period concept but integrated maturational and learning theories as well.

While some scientists have contended that the notion of critical or sensitive periods is highly exaggerated, Hess (25) has suggested that actually three types of sensitive periods exist (Table 6.2). These vary in the degree of reversibility, the emphasis on innate response patterns, as well as in the length of time during which certain behaviors are supposed to occur. In contrast, other developmental theorists have noted that existing evidence does not support the notion of sensitive periods at all (66).

### Sensitive Periods and Developmental Tasks

The general concept of "sensitive periods" does not differ from the way we regard developmental stages. These are commonly regarded as times during which certain developmental tasks are commonly addressed and potentially resolved. As was pointed out in chapter 3, certain ages or stages of life are commonly associated with specific psychological, social, and physical tasks. For example, adolescence is commonly associated with rebellion against

**Table 6.2**
**Three Classifications of the Sensitive Period in Behavioral Development**[a]

Critical Period
  Very brief period of time in development during which the organism must undergo certain experiences.
Susceptible Period (also Sensitive Period, Sensible Phase)
  Period of time during which the organism is extremely susceptible to certain stimuli and during which it makes innate responses to these stimuli.
Optimal Period
  Period of time during which the organism has the greatest sensitivity to certain aspects of the environment and thus can respond most readily to certain kinds of learning situations.

[a] Adapted from Hess, EH: *Imprinting: Early Experience and the Development Psychology of Attachment*. New York, Van Nostrand Rheinhold, 1973.

authority and establishing a clear sense of self; old age is associated with loss of family and friends as well as coping with the recognition that time in life is running out; and early development is associated with the acquisition of language as well as gross and fine motor skills necessary for independently performing ambulation and daily living skills. Thus, developmental tasks can be seen as being addressed during time periods when one is particularly sensitive to specific issues or skills, i.e., during sensitive periods.

Developmental tasks, however, apply to general gross classifications of behavior, and they are typically associated with events across the life span. Sensitive periods, as with most maturational concepts, are most commonly applied to *early* growth and development, and often they are used in reference to very specific responses made to specific stimuli, or releasers, in the environment. This is studied extensively in the field referred to as "early experience" and will be covered in more detail in chapters 9 and 12.

## Sensitive Periods and Treatment

The concepts of readiness and sensitive periods also have important implications for clinical practice. For example, clinicians often assume that associated with various emotional or physical traumas, there exist certain time periods when patients are more susceptible to treatment than they are at other times. The brain, various organs, tissues, and neuromotor systems may go through some natural recovery process akin to the process of maturation. Thus, some treatment approaches take into consideration and try to maximize each patient's "natural" process of recovery. This, in part, provides the basis for applying to patient rehabilitation such concepts as maturation and recapitulation.

## Maturation and Early Development

Maturation has been viewed as a predominant force in early life. For example, Hess illustrates why sensitive periods are more commonly applied to early development:

> A young organism differs from the not so young not only in amount of experience but also in neurological and physiological structure and functioning. This means that its susceptibility to different types of experiences can differ radically from the susceptibility of an older organism to the same experiences. It can be more sensitive to certain types of events and less sensitive to other classes of events because of biological, not just experiential, differences. (29, p. 41)

In a current review of developmental theory, Thomas notes that the notion of sensitive periods has been modified and that it does not necessarily have to be associated with early life:

> Certainly, there may be optimal periods in life for different kinds of learning. But this is a far cry from asserting that the optimum period is always in early childhood, or that learning at a less than optimal time cannot be successful. (69, p. 585)

With growing numbers of life-span approaches to the study of human development, it is possible that maturational concepts such as readiness and sensitive periods may continue to be revised as well as extended beyond just the early stages of life.

# Summary Regarding Maturational Theories

Maturational theories have provided important information that clinicians and students of human development find useful in studying and in assessing the emergence of normal developmental processes, especially in infants and children. Principles such as readiness, maturation of premature infant behaviors, proximodistal development, and recapitulation have provided useful insights into normal developmental processes and in clinical approaches to treatment. We have learned, however, that each of these maturational concepts must be conservatively interpreted. They cannot be viewed as either-or principles of development but instead as maturational forms of guidance that combine with experience or environment to effect behavior.

With the growth of clinical research exploring the best methods for assessment and treatment, our understanding of the extent and nature of maturational influences should expand.

# PSYCHOANALYTIC THEORY

Those same biological, evolutionary concepts that are important in maturational theory were also influential to the work of Sigmund Freud (1856–1939), the father of psychoanalytic theory. Like the ethologists, learning theorists, Piaget, and many maturational theorists, Freud also took a biological view of human behavior. Freud, however, was trained as a physician in Vienna, and during his clinical practice, he explored areas that were not typical of American psychology nor of European ethology. Freud was interested in the internal, unconscious forces and conflicts that were responsible for people's actions and feelings (17).

Freud's work completes the traditional threefold division in human developmental theory. This includes thinking (associated with Piaget), learning (associated with Pavlov, Watson, and Skinner), and personality (associated with Freud). These are also referred to as cognitive, behavioral, and *affective* (emotional) theories, respectively. Because of the different subject matter in these areas, obvious fundamental differences exist between them. There are other major differences between Freud's and the other developmental theories, and these have to do with his methods of data collection and the nature of the subjects he investigated.

The ethologists, behaviorists, and maturational and cognitive theorists explored the normal processes of development. Their subsequent theories offered either norms, descriptions, or explanations for the progression of *normal* development. Learning theorists typically examined behavior in the laboratory, and ethologists examined animal and human behavior in natural settings, just as maturational theorists explored normal infant and child development. Although Piaget used clinical methods of study, his subjects were his own children, who exhibited normal development. Piaget's theory describes normal cognitive processes.

In contrast, Freud, a physician, gathered his data from the medical clinic. Rather than investigating theoretical suppositions about normal subjects, Freud's purpose was to help patients live more normal lives (70). Initially, Freud used hypnosis to treat patients, but he became dissatisfied with the results and subsequently began using a method that came to be called *psychoanalysis*. This involves a technique called the "talking cure" (17), which Baldwin describes in this manner:

> The patient is instructed to report every thought and every idea that comes to mind. He is told that some of the ideas will be unpleasant and repugnant, others will seem too trivial to mention, and still others may be anxiety arousing, but that his job is to express them all, withholding nothing, and not try to make the stream of his ideas logical or coherent. These instructions are not easy to carry out; in fact, the realization that there are barriers of certain kinds to free expression of thought is an important part of the therapy. (70, pp. 305–306)

This talking procedure is also known as *free association*. It is based upon the assumption that in normal and neurotic people, unconscious thoughts may emerge during dreams or during the process of letting the mind "wander" as it recalls whatever comes to mind, i.e., free association. A trained therapist, using psychoanalytic methods, can begin to understand a patient's problems by listening to dreams or free association and can eventually help the patient to explore, recognize, and ultimately alter thoughts and behavior that cause problems.

Thus, Freud listened to thousands of hours of patients' thoughts, feelings, and reports of their dreams. These data and similar findings from other physicians undergoing psychoanalysis themselves, formed the data for psychoanalytic theory (70).

The term "psychoanalysis" actually has three different meanings. First, it refers to Freud's technique of psychotherapy. Second, it describes his method of gathering data about behavior and his method for studying behavior. Third, it describes the theoretical system for understanding human personality development and functions that emerged in the 1920s (71).

Psychoanalytic theory is in part developmental, but it also incorporates many now familiar concepts used to describe personality functions. It is more commonly recognized as a personality theory, with Freud's personality stages recognized as developmental in nature.

Some of the basic concepts important to psychoanalytic theory will be elaborated here, but it is difficult to cover Freudian theory in brief because of its vast influence and because of the existence of so many volumes of critical analysis. Freudian theory, like learning theory and behaviorism, cannot be treated neutrally. There are fervent Freudian followers who dogmatically endorse and defend his work; there are those who strongly oppose it; and there are *neo-Freudians*, who endorse some traditional psychoanalytic theory but have modified it to fit more contemporary knowledge of human behavior and pathology. In this small section of this chapter, an attempt will be made to survey some of the more salient aspects of Freudian theory, primarily as it relates to the field of human development.

# Three Systems of Personality

Freud believed that there were fundamental biological drives that motivated people to think and behave in certain ways. These drives are referred to as "instincts," but they are somewhat different from the specific instinctive behaviors studied by ethologists. Like ethologists, Freud believed that humans are biological organisms, driven by the same forces as animals, but Freud's theories are specifically oriented toward human behavior. Human instincts are basic, primarily sexual and aggressive unconscious forces that affect the human personality.

*Personality* refers to fairly stable traits that tend to characterize an individual and make him or her distinct from other people. Personality is fairly consistent and stable over time differentiating it from *emotions*, which are transient feelings (see chapter 12). Freud's theory assumes that there are different fundamental personality elements that operate together in a *dynamic* interactive system. Thus Freudian theory is said to study *personality dynamics*.

One of the most important and unique, as well as highly criticized Freudian concepts, is the notion of the *unconscious*. Freud claimed that many mental activities such as ideas, feelings, drives, urges, and perceptions occur largely without our awareness. These unconscious mental activities, in turn, are responsible for directing human thoughts and behavior. These unconscious drives are expressed through the personality element called the *id*.

## The Id

Freud claimed that the human personality was divided into three distinct systems, each regulated by different forces. He called the simplest personality element the *id*. This is the part of the human personality that is driven by what Freud termed *primary process* and the *pleasure principle*. These two latter terms can best be described by referring to the behavior of an infant.

Infants, as any casual observer is aware, are quick to express their feelings and impatient to have them met. This is characteristic of primary process behavior, which is irrational, impulsive, and oriented toward immediate gratification of needs. The pleasure principle is simply the drive that avoids pain and aims toward immediate satisfaction of simple needs. The id, which is regulated by primary process and is oriented toward gratification of pleasure, is responsible for primitive sexual, aggressive urges that operate at primarily an unconscious level.

All human behavior, however, is not infantile in nature. As humans mature, they are governed by other forces that regulate and subordinate primary process and pleasure seeking. The reality principle contrasts with the pleasure principle and is the force that governs the second personality system, which Freud called the *ego*.

### The Ego and Superego

The ego operates according to *secondary process* thinking which is rational, logical, and practical (72). The purpose of the ego is to satisfy the demands of the id, but it does so by operating rationally. Thus, the ego acts as an intermediary between the primitive, immediate demands of the id and the need to function "maturely" according to the demands of the real world. Such "mature" behavior typical of secondary process thinking is characterized by delaying drives, by inhibiting id forces, and by the gradual satisfaction of personal needs and interests by rational, logical means. The ego actually serves many functions. It controls access to threatening, powerful unconscious thoughts; provides logical thought patterns; guides mature behavior toward acceptable and timely goals; and regulates and controls the impulses of the id (70).

The ego and id are, in turn, regulated by another personality system, the *superego*. This part of personality is what we often refer to as "conscience." It is the social, moral, ethical, and cultural aspect of personality that provides us with a sense of right and wrong. The superego is each individual's internalized notion of cultural norms, thus it is personal, individual, and ranges widely among people.

From a developmental perspective, the personality system associated with infant behavior is the id. With development, however, reality principle takes hold, and the ego begins to exert control over the gratification-oriented primary process. Then, with acculturation, the superego also comes to exert control over the id so that the individual's basic drives are regulated by two forces, those that are realistic and those that are sociocultural.

## Personality Development

There are two areas of Freudian theory that specifically relate to human development. One was just described. It involves the gradual evolution of reality principle, as the individual develops and the ego emerges from a primary process being. Since reality principle is logical and rational and the superego represents sociocultural norms, this trend in development concerns emotional, social, cultural, moral, and cognitive aspects of development. According to Freud, this aspect of ego development is primarily maturational in nature (70).

The second developmental trend has to do with psychosexual development. In Freudian theory, special significance is assigned to the basic drive of sexuality. *Libido* is the term Freud used to refer to this sexual energy. Freud's emphasis on sexuality has, in fact, been highly opposed by many other theorists, but it is the basis for personality development according to psychoanalytic theory.

According to Freud, during development, sexual energy becomes centered around (or focused on) specific parts of the body. These focal parts of the body are areas that are associated with the greatest degree of pleasure at that particular time in life. Freud claimed that there are five stages of personality development named for the changing location of sexual energy during life. Thus, these are referred to as *psychosexual stages of development* (Table 6.3).

## Stages of Psychosexual Development

During the first 2 years of life, infants pass through the oral stage of psychosexual development. At this time, gratification is associated with the mouth and involves such activities as sucking, eating, and oral-motor exploration of objects. Later, gratification shifts away from the mouth and becomes centered around the anal region of the body.

At this point, the child enters the second or anal stage of psychosexual development. During this time, the child takes pleasure in and is preoccupied with the elimination or retention of feces. Toilet training is often a major focus by the family, and the child often becomes involved in complex interpersonal dynamics that may involve issues of control, independence, shame, and inhibitions. All of these issues symbolically involve anal-related concerns. For example, children who are being toilet trained are often praised for learning to control themselves, are made to feel shameful for soiling their clothes, learn that inhibiting impulses often brings praise, and develop a sense of pride associated with being "grown up" and independent.

When the child is about age 4, gratification shifts away from the anal region and becomes centered on the genitals. The third stage of psychosexual development is referred to as the *phallic stage* (also sometimes called the *early genital period*). During this time, children often stimulate themselves, masturbate, and exhibit curiosity about the genital regions of their own and of others' bodies.

Freud placed considerable emphasis on the significance of the penis, claiming that the phallus was valued by both boys and girls. Thus, according

Table 6.3
Freud's Psychosexual Stages of Development

| Stage | Age |
| --- | --- |
| Oral | 1st year |
| Anal | 2nd and 3rd years |
| Phallic or early genital | 4th–6th years |
| Latency | 7th year to puberty |
| Genital | adolescence and adulthood |

to Freud, during this stage of development, with their increased awareness of the differences between the sexes, boys and girls become aware that boys possess a valuable object and girls do not. Since the penis is valued, girls develop penis envy, and boys develop anxiety over its potential loss. This concern over the potential loss of the penis is what Freud referred to as *castration anxiety*.

### The Oedipal Complex

Psychoanalytic theory claims that during this third stage of psychosexual development, a boy's interest in his genitals becomes overtly sexual. He fantasizes about sexual relations with his mother but is also threatened by the power and the already central position occupied by his father. Castration anxiety, which is already apparent, is exacerbated by fears that the father might punish the boy for his incestuous feelings toward his mother. Thus, because of his sexual feeling for his mother and competition with his father, the male child develops considerable mixed and confusing feelings toward his father. These are a combination of hostility and jealousy, along with fear and powerlessness.

Apparently influenced by literature, Freud saw a reflection of his own life in the drama by Sophocles and named this series of conflicts the *Oedipus complex*. This includes feelings of love for the mother, competition and hostility toward the father, and the fear of paternal retaliation (73). These Oedipal conflicts are the source of considerable confusion and concern, which the young boy tries to resolve.

The Oedipal complex is resolved by the male child's actual turning toward his father. Thus, rather than keeping up a confusing and threatening competition, the boy begins to identify with and emulate his father. This process, called *identification with the aggressor*, eliminates the boy's fear of retaliation and subsequent castration anxiety and reduces the real and imagined fears of the Oedipal period.

### Identification

The process of identifying with his father provides the male child with a fundamental source of sexual identity, a sense of self, and of superego development, all of which will be described in more detail in chapter 12, "Social-Emotional and Personality Development."

It should be noted that in this description of the phallic period of development, emphasis has been placed on male, but not female, development. This is one of the weaknesses of Freud's theory. Although he did theorize about female development, these theories are weak and incomplete compared with the work regarding males (74, 75).

### Latency and Genital Period of Development

Following the turbulence of the phallic period of development is a period of quiet called the *latency period*. At this time (elementary school age up until adolescence) sexual feelings are inhibited or at least are not central.

They become latent as the child focuses on other aspects of development. Attention is shifted away from sexuality toward the acquisition of skills and abilities. The superego, by now, has emerged, and the child possesses a value system for regulating his or her own behavior.

Then, with adolescence, the child enters the fifth and final stage of psychosexual development. This is the *genital period*, where interest is again centered on the genitals but now is directed in a mature and permanent manner. Because of maturation of the ego, genital interest is no longer self-centered but instead well-integrated with other social experiences, enabling the individual to form meaningful sexual relationships with other people. This is the basis for mature heterosexual behavior.

# Personality Dynamics and Pathology

Psychoanalytic theory was derived from clinical practice, thus only part of the theory describes normal development. Freud was concerned with sick patients who came to him because of troubles with their thoughts and feelings. Freud's psychodynamic theory explains how and why personality forces can cause such troubling behavior.

### Defense Mechanisms

According to Freud, the id, ego, and superego are in constant interaction. The primary process that activates the id is always in force, so the ego and superego are constantly acting to regulate it. The ego keeps the id in check and prevents its ongoing forces from entering consciousness. Still, humans may at some level be aware of these primary forces and feel threatened by uneasy feelings that they may act out and be subsequently be punished for hasty, immature, or inappropriate actions. This fear of punishment results in tension or *anxiety* which is central to Freudian theory. Anxiety, in most people, is effectively regulated by *defense mechanisms*; however, in others anxiety becomes an overriding influence resulting in *neuroses* or *neurotic behavior*.

As Baldwin (70, p. 335) explains, the common feature of defense mechanisms "is that they distort the individual's consciousness in a way that prevents or alleviates the pain and anxiety that would be caused by a more realistic awareness of his environment or his own ideas and feelings." *Repression* is an example of a defense mechanism that accompanies to some degree, all the other defenses. Repression is the blocking of consciousness so that unpleasant events are not recalled. It is an active process in which normal individuals effectively keep from consciousness events and behaviors that are too discomforting to think about. According to psychoanalytic theory, repressed thoughts emerge during dreaming or during free association. Thus, a study of the content of one's dreams can divulge normally inaccessible unconscious thoughts and feelings.

Freud's theory of personality dynamics seeks to explain the differences between nonneurotics and neurotics, i.e., how nonneurotics effectively use mechanisms to defend themselves against anxiety and how, on the other hand, neurotics end up poorly defended.

### Fixation

Another clinical Freudian concept that is especially important in the study of development is *fixation*. According to Freud, an individual could be prevented from normal development by failing to proceed beyond a particular psychosexual stage. This blockage of development, or fixation, is due to exaggerated emotional associations that occur during a particular aspect of development. Thus, for example, children that are severely punished at the early genital stage of development may fixate at that level and stay there without further progression. The result is that they can only engage in self-centered types of sexual interactions and are therefore unable to form a lasting, mature, heterosexual relationship. This, as well as other aspects of personality development, will be further explained and illustrated in subsequent chapters of this book.

# Critical Discussion of Freudian Theory

Freudian theory has had an enormous impact on our society. Sigmund Freud is cited as one of the most controversial figures of the 20th century (75). Those who praise Freud place him in a category with Galileo, Einstein, and Darwin, while others take exception to and severely criticize his work.

While many attempts have been made to systematically and objectively assess Freudian concepts, they cannot help but reflect the theoretical biases of their authors. Some authors claim that Freud's influence is negligible, while others point to a profound and pervasive impact. Most would agree that Freud's theories are now outdated, but many psychoanalytic concepts have made their way into literature and general conversation. People often speak of "fixations," "unconscious ideas," and "Freudian slips." In addition, psychoanalytic theory has affected parent-child relationships, altering parents' methods of toilet training and ways of educating their children regarding sexuality (74).

One of the difficulties with evaluating Freud's work is that it is so extensive and diverse. With his developmental theory and unique ways of explaining behavior, Freud's work had an impact on scientific disciplines and philosophies of development. Freudian theory also profoundly influenced psychiatry and the clinical treatment of neurotic and other patients.

### Applications of Freudian Theory to Physical and Occupational Therapy

Freudian theory has provided a frame of reference for studying and for treating patients. While this frame of reference is controversial, it has been adopted by many clinicians who work in the mental health and physical health fields.

Occupational therapy includes a primary treatment component in the field of mental health, and Freudian concepts have been important to some treatment approaches and models adopted in that field (76, 77). Kielhofer and Burke point out that the field of occupational therapy had been criticized as being vague: "[In an attempt to define a theoretical framework], occupational therapy aligned itself with the psychoanalytic theory in an attempt to establish a scientific basis for practice. . . . Within the psychodynamic framework, primary importance was placed on the unconscious phenomena, with exploration of here and now feelings and behavior as a means to arrive at an awareness and understanding of intrapsychic conflict" (78, p. 683). Activities in occupational therapy were (and still are) used as a means of expression that provide opportunities for therapists to explore patients' unconscious thoughts and feelings as well as a context for therapist-patient communications (76, 79).

Although occupational therapy is traditionally oriented more toward psychiatric treatment than is physical therapy, clinicians in both fields require an essential understanding of personality dynamics. Freud's theories round out a picture of human behavior by incorporating affective and personality elements with cognitive and learning theories. Knowledge of affective development is an area with which all clinicians must be comfortable in order to gain self-understanding as well as effective interaction with patients and colleagues.

The close interaction between psychological health and physical health is widely recognized (4, 80–82), and many mental health programs include services that address exercise and conditioning, movement-based or sensorimotor therapies, (e.g., 82, 83) as well as traditional programs oriented around productive occupation and transitional self-care, home care, and socialization skills (e.g., 76–79, 84–86). Thus, a significant role exists for physical therapists as well as occupational therapists in the fields of psychiatry and mental health. In fact, physical therapists are increasingly participating in treatment and prevention roles in the mental health arena (87, 88). Thus, familiarity with Freud's work is important for those clinicians, either in adoption in their own clinical frames of reference or in understanding the psychodynamic approaches of various colleagues.

## Pros and Cons of Freud's Concepts

Freudian concepts that have been cited as particularly influential include the notion of unconscious mental activity, the interpretation of dreams and significance of dreams as an outlet for anxiety, the process of free association in psychotherapy, the significance of repression, oral- and anal-based concepts, the significance of early life experiences, and some basic notions from the Oedipal complex (74, 75).

Regarding oral- and anal-based behavior, contemporary clinical jargon and some developmental observations support the use of such terms to describe specific personality types. For example, clinicians often refer to people who perform an endless succession of oral activities (e.g., talking, drinking, eating, chewing gum, smoking cigarettes, or chewing fingernails and pencil erasers) as being orally fixated. According to Freudian approaches, such individuals may have been prevented from normal development or normal sucking responses or may have been overindulged during early development. Similarly, individuals found to be "uptight," to withhold kindness or generosity, and to be miserly and strict are often regarded as anal-retentive by psychoanalytic popularizers; that is, miserly people are viewed as too strictly toilet trained, which results in their overly inhibited, potentially neurotic temperament.

While such terminology may be useful to describe some patients, Freudian concepts about anal and oral fixations are probably too severe. For example, the use of Freudian theory to explain stingy, miserly behavior as a result of supposedly strict toilet training does not hold up. It appears that no consistent long-term effects of toilet training procedures can be found (74, 89).

While data do not support the existence of strict psychosexual stages as Freud proposed, there are, according to some authors, fantasies and attitudes that loosely correspond with these phases. Further, Fisher and Greenburg (74) point out that some developmental studies indicate that in early childhood, boys do, indeed, display affection for their mothers and show elements of competition with their fathers. This provides some validation for the psychoanalytic concept of the Oedipal complex. However, the Oedipal complex cannot be taken too literally, as the basis for boys' normal identification with his father is promoted by a nurturant, not a fearful, competitive, or hostile parent (74).

One very strong criticism of Freudian theory is its inadequacy in dealing with female development. With the growing interest in women's studies in the past 20 years, Freudian theory has been hotly debated. Some individuals claim that his work provided an impetus for the sexual liberation of women, while others have attacked Freud for sexist theories that demote women, claiming they are developmentally inferior to men (74, 75). Recent studies integrating Freudian theory and feminist theory and literature illustrate that issues raised by psychoanalytic theory are not merely of historical interest, are not simply resolved, and continue to be reexamined with unique, contemporary views of behavior and pathology (e.g., 73–75).

The most obvious theoretical objections to Freud's work have expectedly come from American psychology. Behaviorism developed in response to a trend away from investigating "mind" and other vague, immeasurable phenomena. Watson and other learning theorists wanted to develop a natural science of behavior and attempted to do so by empirical investigations that reduced elements of behavior into objective, measurable, and observable elements. Understandably, they objected to the introduction of such elusive, untestable, nonoperationalized, and unobservable concepts as "the unconscious," "fixation," and "defense mechanisms."

Freudian theory, therefore, is strongly criticized for many reasons: First, it is based on the study of neurotic, not normal, individuals. Despite the emphasis of his theory on early development, Freud studied adults, not children. In addition, his theory is *not* derived from empirically generated assumptions. In fact, Freud's theory was quite personal and occasionally modified as he saw fit. Schultz (17, p. 312) explains that "Freud's theories were formulated, revised, and extended in terms of the evidence as he alone interpreted it. Thus, his own critical abilities were the predominant, indeed the only, guide in his theory building. He seemed to ignore criticism from others, particularly from those not sympathetic to psychoanalysis. . . ."

## Contemporary Psychoanalysis

While traditional American psychologists objected to such a subjectively developed theory, other contemporary scientists point out that many psychoanalytic assumptions are available for empirical testing (71, 74). Some researchers have promoted the synthesis of psychological theories, such as psychoanalysis and behavioral approaches, for the clinical treatment of individuals or couples with diverse psychological disorders. In the preface to their book, *The Interface Between the Psychodynamic and Behavioral Therapies*, Marmor and Woods (90, p. xii) explain that disturbing "symptoms can result from relatively uncomplicated conditioned learning or from complex psychodynamic factors or both." They advocate an interdisciplinary approach to mental health treatment.

Psychoanalysis is still practiced as a form of psychotherapy, and many psychoanalytic therapists are much less progressive than Marmor and Woods. In many cases, very traditional Freudian approaches to treatment are maintained. Free association, dream analysis, and hypnosis are still used as are other specific procedures or principles of treatment originally promoted by Freud. Such procedures are adamantly defended by Freudian followers and criticized by opponents of psychoanalytic theory (e.g., 91, 92).

Most contemporary supporters of psychoanalytic theory, however, would concur with the criticism that Freud placed too much emphasis on the role of sexuality in human personality. Thus, groups of practitioners and theorists set out to modify and "tone down" some of Freud's work. These *neo-Freudians* or *neopsychoanalytic theorists*, such as Carl Jung, Alfred Adler, Anna Freud and many others, have altered the traditional psychoanalytic methods of clinical treatment or have provided substitute theories of development. The most notable of these individuals is Erik Erikson, who not only altered Freud's work but provided a unique and important life-span view of human development.

Erikson, who is discussed later in chapter 12, posed that human development involved a resolution of social, rather than sexual, issues. Thus, he proposed that humans go through stages of *psychosocial* rather than psychosexual development. These stages, which will be elaborated later, are significant because they span birth to death. Contrary to Freud's and Piaget's stages of development, which stop at adolescence, Erikson's stage theory recognized the significance of developmental tasks throughout life.

# Summary Regarding Biological Views of Behavior

As was pointed out in previous sections of this chapter, one of the pervasive ideas in developmental theories is Haeckel's biogenetic law, or as it is often called, "the theory of recapitulation." This law of evolution assumed that the ontogeny of individuals repeated the actual entire developmental history of their species (phylogeny). It assumed that immature organisms of one species actually passed through the adult stages of development of other "inferior" or more primitive species.

## Freud and Recapitulation

As we have already noted, maturational, educational, and other developmental theories endorsed these views, which provided a basic framework for the clinical treatment of individuals with developmental delays and with neuromotor problems. Many of Freud's ideas also took root from these views. Freud literally interpreted Haeckel's biogenetic law, and he popularly disseminated Hughlings Jackson's notions of brain levels in the field of psychiatry (32). In fact, one of the criticisms of Freudian theory is that it relies on these inaccurate, outdated, 19th century biological concepts.

Hughlings Jackson, as we have described, applied evolutionary views of development to functions of the brain and of mind. He portrayed the brain as an organ that evolved in specific, hierarchically arranged levels, corresponding to particular stages of evolution. In clinical applications, Jackson assumed that with brain damage, evolution actually reversed itself. Thus, more recent brain areas lost control and older parts of the brain took over (32).

Jackson also claimed that during sleep and dreaming, the older brain regions naturally became temporarily "unleashed." This and Haeckel's work provided a basis for Freud's view of subconscious thoughts emerging in dreams. Freud is quoted as believing that through the study of dreams and neuroses, one could ultimately perceive, "a phylogenetic childhood—a picture of the development of the human race, of which the individual's development is in fact an abbreviated recapitulation influenced by the chance circumstances of life" (32, p. 272).

Psychoanalytic theory shows a pervasive influence of Jackson's and Haeckels' views. Freud's notion of fixation assumes that ontogenetic development actually stops and is prevented from further progression. The concept of *regression* in psychoanalytic theory contends that in response to severe stress, individuals may actually return to prior ancestral or infantile stages of development. Thus, development actually reverses itself.

One of Freud's followers, Carl Jung, posed the existence of a human collective unconscious, which is "the inherited, unconscious memory of humanity's ancestral past. It is a dim memory of all the communal events, fears, beliefs, and superstitions from throughout human history" (72, p. 332). Thus, in Jung's neo-Freudian concept, the human unconscious is perceived as actually containing all of phylogenetic memory.

Freud's clinical practice and theories focused on sexual deviations, which he assumed were a result of retardation or regression. The notions of regression and fixation were particularly applied to sexual problems, e.g., regression being a return to more primitive, ancestral forms of sexual behavior. Thus, Freud's application of the biogenetic law is particularly evident in his views of human sexuality. He assumed that if human development was a recapitulation of the entire history of the race, then sexual development also recapitulated the sexual history of the race (32). Thus, a child's sexual development actually repeated the main phases of the evolution of sex, with infantile sexuality containing the potential for all future sexual development (85). According to Freud, the pregenital stage actually represented the child's display of more primitive, ancestral adult levels of sexual development (32).

## Recapitulation Revisited

Clinical approaches in occupational and physical therapy also relied on recapitulation concepts about the brain and behavior. Recapitulation theory applied to clinical practice assumed that individuals who experience trauma or disease become emotionally or physically regressed. Treatment is therefore aimed at promoting their development or, as Mosey (26) claims, recapitulating ontogenesis.

Such views have been passed down in many fields, most notably in physical therapy and occupational therapy through physical medicine and rehabilitation (35, 38) and through sensorimotor approaches to treatment as advanced by neurosurgeon Temple Fay (15, 30). It appears that occupational therapy may have also received this influence through the field of psychiatry, much as did Freud.

Before immigrating to the United States, the Swiss psychiatrist Adolf Meyer studied with Hughlings Jackson and endorsed his popular, hierarchical views about brain organization and regression. Jackson's views came to the United States through Meyer (32), who is considered one of the pioneers and founders of the field of occupational therapy (94).

The significance of Haeckel's and Jackson's views can be found in their reliance on biological, evolutionary concepts associated with Darwin. This is important because the major theories of human development and behavior are based on biology. The theories on thinking, personality, and learning that have been discussed so far and that have dominated the field of human development in the past have been associated with Piaget; Freud; and Watson, Pavlov, and Skinner, respectively. All of these men, including the ethologists and many of the maturational theorists discussed so far, either viewed humans as fundamentally biological organisms or organized their theories around biological concepts. The most influential ideas that affected all of these individuals came from Charles Darwin and late 19th century views of evolution and biology.

With the exception of Skinner, some ethologists, and maturational theorists, most of these major contributors to traditional developmental theory are no longer alive. Their works originated primarily in the first half of the 20th century and reflect ideas that were predominant during that time. Now, however, since developmental theories have thoroughly assimilated these views, there is a gradual shift toward different perspectives that combine and integrate fields. These newer, predominantly 20th-century views of development, have tended toward life-span analyses of organism-environment systems. These new views are considered interactive, ecological, humanistic or holistic in approach and will be discussed in the remaining pages of this chapter.

# NEW, INTEGRATED THEORIES AND APPROACHES

During the entire 20th century, there have been groups of scientists and philosophers who have opposed the views of human behavior that have dominated American psychology. The basis of their criticism is that these dominant theories are mechanistic, reductionistic, and deterministic. Mechanistic views are those that compare human behavior to a machine, i.e., when given certain input, machines respond mechanically with a predictable (and therefore determined) output. This is characteristic of stimulus-response learning or behavioral theories as well as ethologists' views of innate releasing mechanisms and instincts.

The instinctive views of behavior, characteristic of Freud, ethology, and maturational theories, relied on the concept of "automatic responses" that organisms make to specific stimulus events. In extreme, these views portray instinctive behaviors as emerging according to a genetic, predetermined timetable (e.g., imprinting and attachment).

## Reaction to Traditional Theories of Development

Opponents to such views of behavior claim they are too deterministic, that humans are really much more complicated, variable, and unpredictable than instinct theories portray. Such critics promote *humanism*, an awareness that human behavior cannot be reduced to chains of stimulus-response associations, reflexes, instinctive responses that occur in reaction to innate releasing mechanisms, nor to predetermined behaviors that emerge according to some predictable genetic timetable.

*Reductionism* is another term used to refer to the process of examining the simplest elements or relationships during the scientific study of behavior. Thus, psychology's emphasis on experimentation promoted reductionism by simplifying complex behaviors into elemental relations between independent and dependent variables. In addition, the reference to behaviors in terms of stimulus-response chains, reflexes, or instincts also "reduces" complex behaviors down into simple, predictable events. For example, one opponent, who strongly criticizes American psychology's emphasis on behaviorism states:

The attempt to reduce the complex activities of man to the hypothetical "atoms of behaviour" found in lower mammals produced next to nothing that is relevant—just as the chemical analysis of bricks and mortar will tell you next to nothing about the architecture of a building. Yet throughout the dark ages of psychology most of the work done in the laboratories consisted of analyzing bricks and mortar in the hope that by patient effort somehow one day it would tell you what a cathedral looked like. (95, p. 9)

Thus, the same characteristics of early developmental psychology that appealed to many behaviorists and empiricists (i.e., it was systematic, predictable, and subject to experimental investigation) made it unacceptable to others. This was especially true of biological models of human behavior, on which most traditional developmental theories relied.

## Criticism of Biological Views of Behavior

As has been noted, traditional theories of human development relied on popularized Darwinian views of evolution. While evolution is not a linear event (22, 31, 96), it was, however, popularly viewed that way. Once recapitulation was applied to normal human development, the common view of life became that of a single continuum, with maturity as the progressive goal, immaturity as lack of progress, and pathology as regression. Such views had an enormous and profound influence on the developmental and clinical theories we have discussed so far.

Yet, despite the impact of these evolutionary views, they have been severely criticized by some theorists. The criticism is based on the belief that biological or evolutionary views of humankind actually dehumanize or oversimplify complex and unpredictable behavior. For example, Sulloway (32, 497) claims that "Freud's theories reflect the faulty logic of outmoded nineteenth-century biological assumptions, particularly those of a . . . biogenetic nature."

Critics of traditional reflex, stimulus-response, or linear views of human behavior point out the simplistic assumption that environmental events occur, and that the human organism, in adapting, responds appropriately. Action and reaction are viewed from a simple, linear perspective (Fig. 6.4). The critics of these views explain that not only do organisms respond *to* their environments, but the organisms also act *on* their environments. In addition, the environments react back and at the same time the organism is modifying or correcting its behavior. To complete a complex interaction, it must be recognized that within each organism internal systems are in constant flux, and they in turn affect how the organism responds and reacts to the environment. The environment, which includes other living orga-

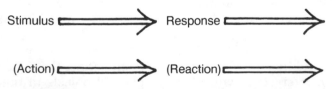

**Figure 6.4.** A Simple Linear View of Behavior

nisms and nonliving objects, also comprises multiple, interacting, and changing systems.

Thus, instead of a simple, linear view of behavior or development, a new and much more complex approach is posed. This increasingly popular approach attempts to describe and explain behavior in terms of dynamic, interacting organism-environment systems.

# Integrated Views of Behavior

These interactive views of behavior, which are discussed in this section, were a natural step beyond the separate compartments of cognitive, behavioral, and psychoanalytic theories. Many of these views have been around for decades, but they did not seem to really gain acceptance until the 1970s. Now, however, syntheses of theories or interactive views are increasingly making their way into developmental and clinical models of human behavior.

Many of these interactive views are not theories per se, but they provide an approach to behavior that is important in the study of human development. These approaches are interactive, ecological, systems, humanistic, or holistic and will be discussed in more detail next. They are all closely interrelated and based upon the "obvious reality that the bio-social-psychological nature of man is such that no one theory or discipline is likely, in the foreseeable future, to explain, much less predict, all of the complexities of human behavior. (90, p. xi)"

## Interactive and Holistic Views

As already noted in earlier chapters of this book, the nature-nurture debate provided a foundation for much theory and research in the field of human development. Many scientists such as Schnierla (97), Anastasi (98), Lorenz (99), and others recognized that nature and nurture could not be dichotomized but must be viewed interactively. As Plomin (100, p. 91) describes it, "Interactionism merely connotes the truism that behavior cannot occur unless there is both an organism (nature) and an environment (nurture)."

One of these interactive views was proposed in the 1960s by Schnierla (97), whose definitions of learning and maturation have already been discussed in chapter 2. Schnierla's interactive definitions make up part of what he referred to as a "holistic" theory of behavior. As Michel and Moore (101, p. 45) note, "It is a *holistic* theory in the sense that it stresses that the object of interest in behavioral investigation is *the entire living and active organism as it exists in its species-typical environment.*"

Organism-environment interactions were not new to developmental theories. They form the basis of von Uexkull's work foundational to ethology (see chapter 5), were certainly a component of maturational theories, and are fundamental to Piaget's theory of sensorimotor and cognitive development. Additionally, Freudian theory emphasizes the significance of individuals in interaction with other people and objects in their environments.

The relevance of organism-environment interaction is clarified by Michel and Moore:

> When we speak of an organism, we are speaking of a creature that has a past; it has already been through a series of interactions with its environment, and it is interacting with its environment while we talk about it or look at it. Schnierla argues that one may separate an organism from its environment for analytic purposes but that, in fact, for the organism they are inseparably fused. There can be no organism without an environmental context; such an eventuality is, quite simply, impossible. (101, p. 46)

## Ecological Views

The actual study of organism-environment interaction is ecology, and Urie Bronfenbrenner (102, 103) is the widely recognized promoter of the ecological view applied to human development. Bronfenbrenner defines the ecology of human development as "the scientific study of the progressive, mutual accomodation, throughout the life span, between a growing human organism and the changing immediate environments in which it lives" (102, p. 514). This process of interaction, he notes, is affected by relations that occur within and between immediate settings and in larger social contexts.

In Bronfenbrenner's descriptions of the ecology of human development (102), he explains how the developing person relates to environmental contexts, emphasizes the necessity for reciprocal interchange between organisms and their environments, describes levels of systems (Table 6.4) important to the developing person, and then explains how all of these complex elements interact with one another as well as with systems of the environment. Clearly, such an ecological approach is a much more comprehensive and complex view of behavior than the simple linear or mechanistic relationships characteristic of learning theory (Fig. 6.4).

When one considers the relative complexity of human social and nonsocial environments, it is possible to appreciate the difficulty of graphically portraying, let alone studying, all the human biological, environmen-

Table 6.4
**Bronfenbrenner's Varying Levels of Ecological Systems that Affect the Developing Person**[a]

| |
| --- |
| Microsystem |
|     The complex of relations between the developing person and his or her immediate environment (e.g., the family, school, peer group) |
| Mesosystem |
|     The interrelations among the various microsystems of the developing person |
| Exosystem |
|     An extension of the mesosystem including environmental systems which affect the developing person (e.g., the neighborhood, media, government agencies) |
| Macrosystem |
|     The institutional patterns of the developing person's culture as reflected in economic, social, educational, legal, and political systems |

Adapted from Bronfenbrenner U: Toward an experimental ecology of human development. *Am Psychol* July: 513–531, 1977.

tal, and cultural systems in interaction with one another. Some such models applied to human behavior have already been illustrated in Chapter 3 under the discussion of interactional models of development. Most of these models have emphasized human subsystems in interaction with one another and with other systems in the environment. This emphasis on systems, which has become increasingly popular in many disciplines, is also now often applied to human behavior and development.

## Systems Theory

A comprehensive systems view applied to many different disciplines originated with the biologist von Bertalanffy (104). His ideas, which were originally posed in the 1950s, were not greeted with much enthusiasm then (95). However, since his book on general systems theory (104) was published in the late 1960s, his views have become increasingly popular and now are referenced frequently in developmental and clinical work.

Bertalanffy opposed the mechanistic views of behavior typical of behaviorism and instinct theories and objected to the biological view of organisms *reacting to* environments. In response to what he viewed as the traditional portrayal of humans as robots, he posed an alternate view of a human organism as a system undergoing active adaptation and comprising a dynamic collection of parts and processes in mutual interaction (104).

Such a systems view of human behavior was consistent with those holistic or ecological approaches posed by Schnierla (97), Bronfenbrenner (102, 103), and others. For example, Bronfenbrenner (Table 6.4) described varying levels of systems that could be examined according their individual effects on the developing person and according to their combined effects. Michel and Moore (101, p. 55) point out how holistic theory, such as Schnierla's, recognizes many different levels of systems and how this "concept of levels is important to an understanding of the development of the individual and to an understanding of the role of behavior in that individual's development." They describe the meaning of "different levels of systems":

> Any organism can be characterized as consisting of many levels of organization, or systems, ranging from those characteristic of single cells to those involving the relationship of the whole organism to its environment. The simpler systems at each level become important components of the higher-level systems." (101, p. 55)

## Levels of Analysis

Many different systems exist within, and outside, organisms. Understanding these different systems contributes to a greater awareness of why and how humans behave the way they do. As an example, to enhance their clinical and interpersonal abilities, physical and occupational therapists must study biology, anatomy, physiology, psychology, and sociology. These fields analyze human abilities from these different levels: cellular, structural, functional, personal-emotional, and sociocultural.

This textbook provides an excellent example of levels of analysis. The purpose of this book is to introduce and describe how the many different systems of human behavior undergo changes over the life span. While the overall plan of the book is to explore the behavior of complete human beings, the book is divided into different levels of analysis. Thus, there are separate chapters on motor development, genetics, social development, perception, and cognitive development. While it is recognized that none of these areas of development occurs independently of the others, each is examined separately for ease of analysis. In the long run, however, the reader (and the author) recognize that all these human systems work together in a whole human organism.

Lerner and Busch-Rossnagel (105) point out the many different levels of systems that must be recognized in understanding human behavior from an appropriate context. Human development must be understood as it is linked to family and society. In addition, "the inner and outer syntheses that compose the human condition have to be integrated as well" (105, p. 9). This includes biological, cultural, historical, and evolutionary changes.

An example of the varying levels of systems in human interaction is described next:

### Infant's Ecology and Social Systems

Belsky and Tolan (106) have applied Bronfenbrenner's different levels of systems (Table 6.4) toward understanding the ecology and behavior of an infant. For example, Figure 6.5 illustrates the varying levels of an infant's existing and potential social contexts. The smallest level of analysis in the infant's social sphere involves the actual infant, whereas the next level includes the infant's immediate social agents, the mother and father. Given these three individuals, analysis can focus on three different two-person relationships, called *dyads:* mother-infant, father-infant, and father (husband)—mother (wife). The mother and father may interact with one another in different capacities in parental or in spousal roles; in addition to considering the immediate dyads, analysis must consider that each of these three individuals brings his or her own developmental history (i.e., genetic, evolutionary, cultural).

All these family variables are involved in the infant's immediate environment, the *microsystem*: the parents' attitudes about, and prior experience with, child-rearing; their expections of infant behavior; the duration and the quality of their (spousal) relationship; whether they are both alive and live together; their ages; their health; and all their reciprocal interactions. Additional variables include prenatal and maternal care, maternal and family nutrition, the conditions of the mother's pregnancy, the conditions of the birth process, and the health and temperament of the infant. These and many more factors will affect how the infant responds to the parents and how the parents, in turn, respond to the infant.

The next level in the infant's ecology is the *mesosystem*. In this case, the mesosystem is the infant's hospital or home environment (depending on where

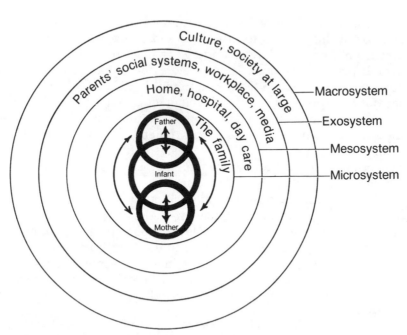

**Figure 6.5.** Levels of systems in the ecology of the developing infant. (Adapted from Belsky J, Tolan WJ: Infants as producers of their own development. An ecological analysis. In Lerner RM, Busch-Rossnagel NA (eds): *Individuals as Producers of Their Development. A Life-Span Perspective.* New York, Academic Press, 1981.)

birth occurs) and the infant's day-care location. As we have noted before, the duration of hospital stay, especially for a preterm or sick infant who may require extensive care, will have effects, possibly long-range, on the infant's development and on the family-infant social system. In addition, the infant's experiences in day care—whether it is custodial or very stimulating, whether the infant receives considerable one-to-one attention or whether attention must be shared with numerous other children, whether the other children are far or near in age to the infant, where care is provided at home or away from home, and the type of nutrition received—affect the infant and how the infant responds in that setting and, later, toward the parents.

Other home factors include the location of the home (i.e., urban, rural, suburban), which can provide or limit opportunities for interacting with other children or families; type of and availability of toys; the degree of safety, stimulation, and variety offered by the home environment; and the number of other individuals living in the home, their ages, and the nature of their relationships. Extended family members living in the home may offer comfort, provide day-care and support for the parenting and spousal relationship, or cause greater stress and burden of care for the family. Similarly, sibling relationships and their spacing and ages may result in positive and supportive infant care or may put additional strain on the family system.

Beyond the mesosystem is Bronfenbrenner's (102) *exosystem*, which includes the various social networks that are not immediate but that still affect the infant and family. The exosystem includes the social networks of the family such as friends, parents, and other extended family, as well as employment choices, activities, and plans. Family financial status, Belsky and Tolan (105) note, may affect whether or not a parent takes a day off when the infant is ill; stability of employment or type of employment may affect whether the family must relocate often; and job stress and demands, satisfaction with and actual wage earned, will all affect the parent's relationships with one another and with the infant.

An example of how the mesosystem (e.g., day-care) and the exosystem (workplace) interact and ultimately affect the family is illustrated by the following questions: Must the mother quit work to take care of her infant, and does she do so willingly? Must a mother return to work for financial reasons and unwillingly leave her child in day-care? What is the policy of the parents' employers regarding maternal or paternal leave? Do the parents' employers offer on-site day-care? Do the parents' employers allow for leave or flexible makeup time if parents need to make doctor's or school appointments? How far is day-care located from the home or place of employment? Do the parents receive family insurance from their employers? What is the financial status of the family, how do the parents feel about their level of income, and how do parents' work schedules affect their relationship with one another and with the infant? For example, the infant who interacts with both parents at the same time will experience a different relationship than the infant who sees mother alone, when the father is at work, and then sees father alone while the mother is at work.

An additional element that Belsky and Tolan (106) include in their example of an infant's exosystem is the media, which includes television and popular literature. Television may directly or indirectly affect child development, either through the types of programs watched or by affecting parent's attitudes and reactions to their child. Television and popular magazines carry information about child health and development, and numerous child-care books pass on information that may directly affect parents' attitudes, policies, and actual behavior during child rearing. In addition, attitudes or images passed on through the media may affect the parent's attitudes about the care and designs of their home or the types of food, clothes, furniture, or toys they select for themselves or children.

The *macrosystem* is the largest level of analysis, involving the culture and social system surrounding the family and community. Attitudes toward child rearing and sex roles, such as reacting differently to a male infant compared with a female child, are passed on through social values. The resources of the community and its social, religious, and political policies toward such issues as child labor, child abuse, abortion, pregnancy, and sex will affect the parents; the possible timing, spacing, and number of children; and the children's education and aspirations. Certainly, children raised in middle-class families in the United States have different dreams and relationships to their families than do children currently raised in war-torn or severely economically and nutritionally deprived nations.

To illustrate the value of such a view of interacting systems, consider how this model could be used in analysis of the ecology of an ill or a disabled infant. Each of the different levels of systems would be affected. For example, consider an infant born with a physical deformity such as paralysis of the right arm. Consider all the various systems and how they might be affected by this occurrence. Some of the factors to be considered might include the following:

The sex of the infant and parent's expectations for the infant in terms of productivity;
Parents' attitudes toward disability;
Parents' attitudes toward handedness;
Cause and nature of the infant's paralysis;
Attitude of other family and friends toward the infant and parents;
Relationship of siblings and their physical and health status;
Whether the infant must be repeatedly hospitalized;
Extent of the parents' health insurance;
Amount of available time for locating and participating in the child's therapy;
Attitudes toward therapy;
Assistance from family and friends;
Relationship of the parents to members of the health care team;
Community resources for the child and for the family in support of the child.

Such descriptions and considerations of infant ecology are not specific to infants. Ecological views of different interacting systems can be used to refer to infant, child, adolescent, or adult behavior and can be appropriately applied across the life span (Fig. 6.6). Clearly, an ecological or a systems model enables developmentalists and clinicians to obtain a rich, expansive, and realistic profile of methods of intervention and potential for client or family change.

In the following quote, Laslo (107) illustrates the utility of the systems view for different levels of investigation:

> The systems approach does not restrict the scientist to one set of relationships as his object of investigation; he can switch levels, corresponding to his shifts in research interest. A systems science can look at a cell or an atom as a system, or it can look at the organ, the organism, the family, the community, the nation, the economy, and the ecology as systems, and it can view even the biosphere as such. A system in one perspective is a subsystem in another. (p. 14)

# Applications of Ecological, Systems Views to Team or Community Treatment

Auerswald (108) has combined views into an "ecological systems approach," which he claims provides an excellent reference point in clinical practice as well as in understanding human development. He explains that this view easily incorporates a developmental model of an individual's life cycle as it accommodates, "various larger human systems as they move through time in the ecological field of the environment". (108, p. 205)

**Figure 6.6.** While an older sister may affect the development of her younger brother, the infant also affects the development of his older sibling. Life-span ecological approaches consider the dynamics of the family system and the reciprocal effects of all family members on one another.

An ecological systems perspective is valuable in clinical practice for many reasons. First, it is realistic because it considers all the potential variables in an individual's life rather than narrowly focusing on one theoretical or developmental point of view. This is particularly valuable considering that many disorders are being recognized as attributable to multiple, interacting, individual-social-cultural factors. Second, an ecological systems approach requires input from many professionals, thus it necessitates (and provides) a common frame of reference for different disciplines so that they can effectively work and talk together.

## Ecological Systems in Family and Community Health

The emphasis on family and community ecology is increasingly important in the health care field today. In the 1980s, requirements for hospitals necessitate that patients go home "sicker and quicker." The phenomenal costs of in-hospital health care have caused our society to focus once again on the value of, and need for, home care. Thus, new models of health care delivery are being created and oriented around family, home, and community support systems. This is occurring in programs addressed to individuals across the life span. For example, there are home-based programs for infant and child intervention as well as home-based health care and therapy programs for the elderly.

Parallel with the growth of home care is the growing recognition that in the United States, numerous socially related problems are occurring in alarming numbers, e.g., malnutrition; drug abuse, alcoholism; homelessness; teen, or unwanted, pregnancy; and increasing rates of suicide, stress- and diet-related diseases, and radiation- and pollution-related illnesses. These problems concern and affect individual as well as social and community health and require a community health approach well beyond the auspices of acute medical or hospital-based care.

As Auerbach notes, the ecological systems approach easily incorporates these new trends in health care together with life-span views of human development:

> It provides an emphasis which stresses the organization of events in time and traces the movement of the developing infant-child-adolescent-adult-aged individual's degree of participation versus his isolation in relation to his family and to the flow of surrounding community life—such an emphasis makes it possible to determine with much more clarity in what life arenas the individual, the family or a group of individuals needs assistance, and thus to more effectively combat the anomie and dehumanization characteristic of our age. The result is that the targets of therapeutic activity are much clearer and therapeutic work is more clearly focused on forces and situations that are truly etiological in a given problem situation. Techniques of producing therapeutic change can be brought to arenas much larger than the therapy room or even the home. (108, p. 206)

## Ecological Systems Approach as a Common Frame of Reference

As therapists become increasingly involved in problems that extend beyond the local clinic, hospital room, or classroom, they are looking for a treatment model that incorporates the individual-environment relationship. As Auerswald (108, p. 204) notes, the ecological systems approach is one such model: "[K]nowledge that has been accumulating from the study of specific ecological systems, such as the family and small groups, the development of which lagged until recently because the systems did not fit neatly into the bailiwick of any one traditional discipline, can also be included without strain [in an ecological systems view of behavior."]

The appeal of this view is that it provides a common frame of reference for the entire health care team. This common reference point can lead to effective communication across or between disciplines, in increasing popular *transdisciplinary* or *interdisciplinary* team approaches (see the section in chapter 9, "Interdisciplinary and Transdisciplinary Approaches to Early Intervention"). Thus, rather than treating various "parts" of individuals, the team can work cooperatively in identifying and treating various biological, social, and cultural systems and resulting aspects of illness and disease.

## Systems in Physical and Occupational Therapy

Just as systems views have been applied to human development and behavior, they have also found extensive application in allied health fields. As already discussed in chapter 3, both fields of occupational therapy and physical therapy are becoming more systems oriented. This includes an ecological focus as well as an orientation to different levels of analysis in human biological, psychological, or social systems.

For example, Reilly (77) and Kielhofner (108) proposed the application of Bertalanffy's general systems theory to a science of human occupation. Such a systems view applied to human behavior is ecological in emphasis because it stresses the relationship between organisms and their environments. Kielhofner and Burke's (110) application of the systems model of human occupation illustrates the various levels of interactions that go on within the organism and between the organism and various aspects of the organism's environment. This view, they note, is necessary given the complex nature of human behavior.

Occupational therapists Howe and Briggs (111) point out how the work of Bertalanffy (104), Bronfenbrenner (102), Auerswald (108), and others can be pulled together in a context for theory construction in occupational therapy and for understanding the individual in relation to a wider social environment through an ecological systems perspective.

Physical therapy research has also addressed systems views but at a different level of analysis. The work of Keshner (112) and of Loomis (38) illustrate the application of a systems view in enhancing our understanding of neural motor control. Keshner, and others (39), have pointed out that the traditional Jacksonian, hierarchical views of brain function have been replaced by systems models of neural control. Keshner explains the implications of such systems views on treatment approaches in physical therapy, specifically on neurodevelopmental theory (NDT), a treatment orientation based on traditional hierarchical views of the central nervous system (CNS):

> Contrary to the vertically designed model of the CNS as seen by the NDT theorist, research results in motor control suggest that the CNS does not operate in a strictly descending manner. The concept of descending neural inhibition of phylogenetically older behaviors by those higher on the evolutionary scale implies a localization and storage of concrete behaviors and an inefficiency and inflexibility that is not characteristic of the human system. (112, p. 1036)

Loomis (38) further explains how the selection of physical therapy modalities such as vibration, tapping, icing, and other sensory stimulation can be understood within a systems model of motor control. She, as have many others, advises that therapists be receptive to a wide variety of theories and incorporate many different approaches to treatment because of the diversity of human behaviors and clinical problems. A systems view, which incorporates ecological considerations as well as varying levels of analysis, naturally accommodates such diversity.

# Combined Theoretical Views in Clinical Practice

One of the ways that clinical theorists have adjusted to the need for different treatment approaches to accommodate the diversity of human behavior has been by combining approaches into one collective view. Such combinations are attempts at a holistic approach toward human behavior inasmuch as these views pull together varied human functions into one collective view of a whole organism in interaction with the environment.

## Mosey's Biopsychosocial Model

An example of such an approach was proposed in 1974 by occupational therapist Anne Mosey (81). In her biopsychosocial model of human behavior, she integrates many different approaches of human function into a combined clinical practice view of the whole person. Mosey's proposal was progressive since it anticipated similar trends that were to occur in other fields.

For example, in the journal *Physical Therapy* in 1985, physiologist Elizabeth Dean (4) explains how occupational therapy and nursing, in addition to some areas of medicine, have adopted philosophies of practice that include a psychosocial, instead of a strictly biological, perspective. Dean proposes that the field of physical therapy also adopts a psychobiological adaptation model for practice. She notes that this approach to practice enables health care professionals to view patient management "based on the individual as a whole rather than on a conglomeration of organ systems" (4, p. 1061). Dean's proposal is based on Engel's (113) similar proposal and challenge for a new model of patient care for the medical profession.

## Biopsychosocial Model of Medical Treatment

In 1977, psychiatrist George Engel (113) published an article in the widely disseminated, interdisciplinary journal, *Science*. In this article, he challenged the medical community to adopt a new approach to practice, the biopsychosocial model. This approach, he noted, was necessary because of medicine's increasingly reductionistic view of human disease based on the *biomedical model*.

The biomedical model of disease assumes that medical conditions are a result of some biological cause. Treatment is therefore aimed at locating the offending biological organism (e.g., bacteria) and eliminating it through treatment. Such an approach reduces disease into a mechanistic, oversimplified condition, comparable to the S-R models of learning theory discussed earlier. For example, in the biomedical model, bacteria or toxins (stimuli) are seen as causing disease (response). Certain vaccines and drugs, (stimuli) are recognized as causing cures (response).

Capra (114) explains the reductionistic tendency of medicine's biomedical model:

> Throughout the history of Western science, the development of biology has gone hand in hand with that of medicine. Naturally then, the mechanistic view of life, once firmly established in biology, has also dominated the attitudes of physicians toward health and illness ... The biomedical model ... constitutes the conceptual foundation of modern scientific medicine. The human body is regarded as a machine that can be analyzed in terms of its parts; disease is seen as the malfunctioning of biological mechanisms which are studied from the point of view of cellular and molecular biology; the doctor's role is to intervene, either physically or chemically, to correct the malfunctioning of a specific mechanism. (p. 123)

Engel (113) makes a similar point:

> The dominant model of disease today is biomedical, with molecular biology its basic scientific discipline. It assumes disease to be fully accounted for by deviations from the norm of measurable biological (somatic) variables. It leaves no room within its framework for the social, psychological, and behavioral dimensions of illness. (p. 130)

Thus, to counter this overreliance on biology, Engel proposed a new medical model, the biopsychosocial approach, which accounts for all aspects of behavior. It unites mental and physical illness by recognizing the body and the mind as part of one whole organism.

In his description of this model, Engel (113), (115) further suggests the application of Bertalanffy's general systems theory (104) to medicine because it "provides a conceptual approach suitable not only for the proposed biopsychosocial concept of disease but also for studying disease and medical care as interrelated processes" (113, p. 134). Much like Bronfenbrenner's view of ecological systems in human development, Engel has pointed out different environmental systems useful for diagnostics and treatment. These systems range from subcellular components in patients' microenvironments to social factors in their macroenvironments (115).

# Humanistic and Holistic Approaches to Human Behavior and Health Care

It appears that we have come full circle. At the beginning of this section, new integrated theories and approaches to human behavior and development were introduced. For example, these included Schnierla's holistic theory (97) and Bronfenbrenner's ecological model with its different levels of systems (102). Just as these integrated theories in psychology and human development began to emerge, trends toward holistic, systems, or ecological views were taking shape in other fields. Similarly, as opposition occurred in reaction to psychology's overreliance on biological models of behavior, there was an opposition to the biomedical model of treatment in medicine, which resulted in the integrated biopsychosocial approach to health care promoted by Mosey (81), Engel (113), and by others in the physical and mental health professions.

Similar trends also occurred in two other areas, and although they are closely related to the fields already discussed, they developed separately and have had influences of their own. One of these areas is commonly called *humanism* or *humanistic psychology*, and the other is *holistic health care* or *holistic medicine.*

## Humanistic Psychology

Many of the practitioners in clinical psychology in the 1940s and 1950s were also disenchanted with Freudian and behavioral models and methods of study. In reaction to the strict biological views of human behavior, they proposed an alternative approach. Such individuals as Abraham Maslow and Carl Rogers set forth *humanistic approaches*, in contrast with the traditional mechanistic, approaches to human interaction. As Capra explains:

> To counteract the mechanistic tendency of behaviorism and the medical orientation of psychoanalysis, Maslow proposed as a "third force" a humanistic approach to psychology. Rather than studying the behavior of rats, pigeons, or monkeys, humanistic psychologists focused on human experience and asserted that feelings, desires, and hopes were as important in a comprehensive theory of human behavior as external influences. Maslow emphasized that human beings should be studied as integral organisms and concentrated specifically on healthy individuals and positive aspects of human behavior—happiness, satisfaction, fun, peace of mind, joy, ecstasy. (114, p. 365)

While humanism is sometimes recognized as a theory of human behavior, it is actually more of a prescriptive approach for clinicians than an organizing, academic theory (116). As Pellegrino (117) notes, humanism is less a doctrine than it is a focus on the significance of human values in the relationships that individuals form with one another. It emerged from a clinical context, and it therefore focuses on the person, rather than person-environment interactions. As such, it is more useful in guiding a clinical approach than in explaining behavior from an ecological context.

One effect of the humanistic approach in psychotherapy has resulted in what Capra (114, p. 365) explains as "a significant shift in terminology. Instead of dealing with 'patients' therapists were now dealing with 'clients,' and the interaction between therapists and client, rather than being dominated and manipulated by the therapists, was seen as a human encounter between equals." This distinction is noted in occupational therapy as well: As Mosey (118, p. 6) reports, the term "patient" has come to refer to someone who is ill and who is often a passive recipient of prescribed services. In contrast, the word "client" usually designates an individual who is involved with the therapist in a collaborative relationship. This relationship is typically in a community setting; e.g., the home, outpatient clinic, day treatment, etc.

Thus, while the humanistic approach is not necessarily ecological in orientation, it effectively accommodates the ecological view required for contemporary therapists engaged in clinical work. The humanistic approach promotes a collaborative, open, mutual problem solving between clients and therapists and thereby respects the uniqueness and rights of each of these individuals involved in a special, though temporary, relationship. As such, humanism is an effective approach that has also been espoused in medicine, where it is promoted for several reasons: as a means of helping the physician and other health care professionals to obtain a deeper understanding of their roles in health care; as a means for understanding the client as a unique and special person; as a means for helping clients use untapped, internal resources to promote their own health; and as a means for developing treatment approaches that are more sensitive to individual clients' needs (119, pp. 9–10).

## Holism and Health

Closely related to humanism, and springing from the same tendency to move away from reductionistic views of human behavior, is the field of holisitic medicine, or as it is also called, "holistic health care." As most authors reviewing this topic have noted, the terms "holism," "holistic health" (sometimes "wholistic health"), and "holistic medicine" are commonly popular. This has resulted in many misperceptions, controversies, and possible misapplications of the view of holism in the health care field today. As a result, several recent analyses have been developed to define holistic theory and health care. These analyses (e.g., 120–123) have concentrated on describing the origins of holistic health as it is known in the United States in the 1980s and also in pointing out the common principles of, and resources for, recognized holistic health care approaches.

Physician James Gordon (124, p. 3) notes that in the 1980s holistic medicine has "come to denote both an approach to the whole person in his or her total environment and a variety of healing and health-promoting practices. This approach, which encompasses and is at times indistinguishable from humanistic, behavioral, and integral medicine, includes an appreciation of patients as mental and emotional, social and spiritual as well as physical beings."

The holistic approach in the health care field seems to have developed along the same lines as the biopsychosocial approaches of Mosey (81) and

Engel (113, 115) as well as the holistic, ecological systems approaches of other psychologists and clinicians. Gordon (123) explains that practitioners and theorists became dissatisfied with the overuse of mechanistic approaches to human behavior. The biomedical model of disease, appropriate for some bacterial infections and diseases, was inappropriately applied to many other forms of illness. He describes the result:

> Birth and death, those most inevitable of human processes, were taken from the familiar home context in which they had alway taken place to the hospital ...{and}... later conditions that had been viewed in religious, moral, economic, or political terms acquired medical metaphors and demanded medical intervention. Juvenile delinquency, social protest, the struggles of Blacks and women to obtain their rights, and the activities of our children were variously diagnosed and, where occasion permitted, treated as illnesses. Young people who could not sit still in school rooms, prisoners who protested jail conditions, and occasionally, civil rights workers who wanted to vote were all given pills to treat the mental illnesses that presumably caused them to behave in this fashion. (124, p. 6)

With this overemphasis on the biomedical model in medicine in addition to numerous other sociocultural and scientific changes (Table 6.5), a backlash resulted in what is in the 1980s a popular "holistic health care movement" in the United States.

### Characteristics of Holistic Medicine

Holistic approaches tend to be ecological and humanistic in nature. In addition, they recognize the various interacting levels of systems in human

**Table 6.5**
**Scientific, Medical, and Sociocultural Changes Resulting in a Shift Toward Holistic Health Care**[a]

Increase in chronic, stress-related disorders
Concern about unnecessary medical procedures
Awareness of close body/mind relationship
Awareness of the significance of one's ecology (family, home, work environment)
Information from modern physics (relativity) that shows us that we alter the events we study, thereby shaping patients' illness and health
Information from modern physics regarding the continuity of time and space and interconnection of life events
The close relationships between modern physics, religion, mysticism, psychology, and health
Political and ecological emphasis on coexistence and adaptation instead of expansion
Political concerns about the health care system and increasingly knowledgeable and active consumerism
Health professionals' dissatisfaction with bureaucracies and fragmented health care
Awareness of health care alternatives from other cultures
Concern about health care costs and increasing aged population

[a] Adapted from Gordon JS: The paradigm of holistic medicine. In Hastings AC, Fadiman J, Gordon JS (eds): *Health for the Whole Person. The Complete Guide to Holistic Medicine.* Boulder, CO, Westview Press, 1980, p 3.

behavior. Thus, some of the fundamental assumptions of holistic health care are also common to the human development and clinical practice fields.

It is difficult to sort out the fundamental differences between humanistic and holistic approaches to health care (e.g., 125). If a distinction must be made between these two areas, it would be that humanistic approaches tend to be narrower in focus. They emphasize respect for humaneness, but they do not necessarily incorporate all the additional components that are often included within the domain of holistic health care. Thus, holistic health care includes a humanistic component and many other characteristics as well.

While these other characteristics of holistic medicine vary from source to source and are wide-ranging, some tend to be cited more often than others. These common characteristics of holistic medicine include a multidisciplinary approach that (a) recognizes the importance of the physical, mental, and spiritual components interacting together in the whole person; (b) focuses on health and wellness and recognizes that these are positive, ongoing states rather than just the absence of disease; (c) recognizes that each person is actively responsible for his or her own health; (d) focuses on prevention through education, nutrition, exercise, and stress reduction; and (e) considers organism-environment relationships by viewing each person and his or her relationship within a family, community, and social culture (120–125).

### Origins of Holism

While holistic health is widely popular in the 1980s, holism is not a new concept. The concept of holism is attributed to South African philosopher Jan Christian Smuts, who used the term in his 1920s book, *Holism and Evolution* (122–123).

While the term "holism" may be contemporary, the concept that it has grown to represent has been common in psychology for a century and in philosophy and religion for even longer. For example, holism is essential to Gestalt psychology, which emphasizes the concept that "the whole is greater than the sum of its parts." This ability to perceive "whole" images or form is a fundamental human experience (Fig. 6.7) which has been studied widely for decades by Gestalt psychologists. As Cmich (122, p. 30) notes, "holism is a new name for a very old concept." She explains how four primary principles of holism (Table 6.6) can be traced to ancient Chinese philosophy and religion. These principles are also fundamental to other religions and cultures (e.g., Native Americans) whose practices and ecological beliefs do not differ from many principles of contemporary holistic health care practice.

### Holistic Applications in Occupational and Physical Therapy

The concept of holism is not new to occupational therapy nor to physical therapy. There is a widespread recognition in both fields of the necessity of

**Figure 6.7.** Gestalt psychology has to do with the perception of whole forms and with finding meaning and organization in forms. For example, this figure is actually a conglomeration of parts, i.e., a circle surrounding two dots and a curved line. However, most people perceive the parts united as a whole form and label this a smiling face.

Table 6.6
Principles of Holism[a]

- Entities and systems in the universe exist as unified wholes.
- The parts of a whole are dynamically interdependent and interrelated.
- A whole cannot be understood by the examination of its isolated parts.
- The whole is greater than the sum of its individual parts.

[a] Adapted from Cmich DE: Theoretical perspectives of holistic health. *J School Health* 54: 30–32, 1984.

treating "the whole person." Thus, as one of the fundamental principles of holistic medicine, this approach to the whole person is consistent with contemporary practice in both physical and occupational therapy. However, there may be some resistance as well as misunderstanding about the effect of holistic medicine in both of these allied health fields.

Holistic medicine had as its primary impetus a reaction against the established biomedical model. Thus, some view it as "antimedical." Physical therapist Gee (126) points out that in some circles, holistic health care is seen as "the alternative" in contrast to "traditional" (biomedical) approaches to medicine. In the preface of her book, *A Holistic Approach to the Treatment of Learning Disorders* (127), occupational therapist Barbara Knickerbocker notes, "the term holistic as I have used it is in no way related to the recently emerging holistic approach to health care, advocating a naturalistic, antimedical point of view." She notes that in her book the term "holistic" means comprehensive and cohesive.

Holistic health care is an emerging field. It incorporates a widespread view of human behavior and health. Thus, it has attracted a wide variety of practitioners and followers—some who are legitimate and some not. As

Moore and Moore (120) note:

> Since holistic health is the current "trend" in medicine, beware! Currently, there is a great deal of "prostitution" of the word *holistic*, undoubtedly because it appeals to a large number of consumers. There are fields and practitioners that are less than holistic. Holistic healers may be physicians, other health care professionals, or lay persons. Their methods must both treat (find the source of) and prevent disease, although many holistic areas cannot make use of the words *diagnose* and *treat*. Some practitioners in holistic fields prescribe drugs, some fields are composed only of physicians, and other modalities rely strictly upon natural measures for treatment and prevention. This does *not* detract from the particular field, because it is the unique realm of the individual that is important. (p. viii)

Holistic medicine includes many concepts from "traditional" medicine and integrates them with a positive, dynamic view of health and human behavior. For example, The American Physical Therapy Association (APTA) has adopted from Dorland's illustrated Medical Dictionary the definition of holistic health as a "system of preventive medicine that takes into account the whole individual, his own responsibility for his well-being and the total influences—social, psychological, environmental—that affect health" (128, p. 1).

As physician Gordon explains, "[holistic medicine promotes the integration of] the ecological sensitivity of ancient healing traditions and the precision of modern science, techniques whose effectiveness has already been extensively documented and techniques we are just beginning to explore." Thus, holistic health can be seen not as "antimedical" but instead as a positive, dynamic view of preventive health.

Since holistic health care is a relatively new field, it has not been extensively studied nor integrated into occupational therapy or physical therapy theories or models of treatment. However, many of its principles are common to those already discussed here, e.g., ecological systems, humanism, and psychobiosocial approaches. In regard to holistic health, specifically, it has been addressed more directly by the physical therapy profession, whereas the concept of holism has been more widely examined in theories of occupational therapy.

### Holistic Health and Physical Therapy

The APTA has compiled an educational resource guide (128, p. 1) regarding holistic health, and in it notes that the profession endorses the concept of health promotion and has charged their legislative body to "study, design, develop and implement plans that promote physical therapy in the areas of holistic health and health promotion by incorporating them into programs of the Association's strategic planning process." Other clinical researchers have illustrated the significant role for physical therapists in health promotion and prevention (88, 126, 129) as well as advocating for a physical therapy endorsement and approach to the whole person (4, 80, 126, 128).

Gee (126) notes that five primary concepts in holistic health (self-responsibility, prevention, wellness, looking at the whole person, and viewing illness as a potential learning experience) are concepts with which physical therapists are intuitively aware. Thus, he points out that by applying holistic concepts to treatment, therapists can enhance and benefit themselves, their patients, and their profession (126, p. 21).

## Holistic Theory and Occupational Therapy

Occupational therapists, while not specifically addressing holistic health care, have traditionally tended to emphasize humanistic and holistic approaches to treatment. Llorens (131) points out that as early as 1967, a theory-building group in occupational therapy proposed an ecological and holistic theory that integrated cognitive-perceptual-motor theories with psychosocial theories and then interrelated these with emotional, environmental, and neurological systems views of human behavior. Mosey has consistently promoted a holistic clinical approach, with her 1974 biopsychosocial model (81) and with an update to this model in 1980 (132). In this update, she proposed a preliminary structure for upgrading her previous model because it "did not provide sufficient structure and content to give a holistic view of occupational therapy" (132, p. 11).

In their analysis of the historical development of occupational therapy, Kielhofner and Burke (78) describe how the field was traditionally holistic in orientation and has become progressively less so over time. They note that in the 18th and 19th century, humanistic philosophies resulted in a moral treatment approach for the mentally ill. Their description of this approach bears repeating because it sounds like a liaison between the present-day concept of human occupation and holistic health care:

> Engagement in normal daily activities within a cheerful and supportive environment was the model of treatment. Moral treatment was a grand scheme for activities of daily living, which placed a patient in a *total program* with the goal of arranging healthy living. It employed the moral remedies of education, daily habits, work, and play as therapeutic processes. . .(78, p. 678)

Kielhofner and Burke (78) further point out how in the early 1900s the use of occupation in the treatment of the mentally ill "is impressive . . . from the perspective of its holistic theoretical orientation." They note that, then, the newly developed field of occupational therapy was ahead of its time because "[i]t embraced the recognition that an individual's health was bound up in the intricacies of daily experience in a complex physical and social world. It was also a bold concept: the occupation paradigm proposed that Man had not only the right to freedom from disease but also to respectability and self-satisfaction in his existence" (78, p. 679).

However, as Kielhofner and Burke explain, the field of occupational therapy began to move away from this holistic-humanistic unified approach. It, like many other fields, became fragmented and specialized so that now there is no one unifying theory, despite recent attempts by Mosey (26, 81,

132), Llorens (11), and others (77, 78, 131). As a partial solution, Slaymaker (133) points out that occupational therapists must integrate the holistic approach with clinical specialization. She claims that it is essential that therapists be both generalists and specialists at the same time and that they remain always cognizant of a total view of the patient.

### Summary Regarding Holistic Health Care

Thus, both the occupational and physical therapy professions have addressed some, if not all, of the principles of holistic health care. The actual integration and viability of holistic principles in occupational therapy and in physical therapy would be difficult to predict at this point in the discipline's development. Like any new area of science or health care, holistic health approaches will take some time to fully develop. It will also take some time for the basic, durable principles of holistic health to emerge and for us to assess its potential long-range effects on the fields of physical and occupational therapy, on the medical/health care community, as well as upon society at large.

Clinicians and consumers are advised to be cautious about holistic health practices. As Ferguson, a supporter of the new movement in health care, notes:

> Any wide-open, fuzzy field like "holistic health" offers abundant opportunity for fraud and overpromise. Ground rules include making sure that the unorthodox procedures are used only to complement proven conventional treatments rather than subjecting consumers to needless risk. Consumers are warned against practitioners who make unwarranted promises or charge outrageous fees.

> There have been some calls for licensure, but the debates usually come to this: Holistic health is a perspective, not a specialty or discipline. You can't license a concept. And you can't even know for sure what works. (p. 262)

Gordon points out the potential wide-ranging social implications for holistic medicine:

> The future of our medicine and our health as a people will in part be determined by the ways this approach {to holistic medicine} comes to shape the larger health care system, the training of the professionals who will work in that system, and the education of the citizens who must ultimately learn to take care of themselves. (124, p. 23)

# CHAPTER SUMMARY

This, and the last, chapter have described the traditional theories of human development. These include cognitive, behavioral, ethological, maturational, and psychoanalytic theories. In addition, some new approaches in human development and in clinical, health care fields have been examined. One of the salient characteristics of these new ecological and holistic approaches is the focus on different levels of analysis in human behavior.

The major premise of the levels approach is that behavior can be examined from a variety of perspectives. Thus, it can be reduced to biological analyses, or it can be broadened and examined in terms of complex sociocultural interactions (see chapter 12). Clearly, multiple levels of analysis must be included to provide an accurate profile of human behavior.

While different levels of analysis are appropriate for examining human behavior, they are also appropriate in terms of theory adoption. One particular theory cannot describe, explain, and predict all human behavior. Instead, multiple theories are necessary just as are multiple levels of behavioral analysis.

Thus, while some of the traditional theories (e.g., behaviorism, psychoanalytic theory) may be considered reductionistic, they are still appropriate for a certain level of analysis of behavior. Just as some clinicians and psychologists will still focus on cellular and physiological aspects of behavior, so it is necessary to maintain biological models of human function. Similarly, some clinicians and researchers may focus on social aspects of behavior to the exclusion of biological bases. This is appropriate as long as it is recognized that each level of analysis represents one, not the only, view of behavior.

The contemporary ecological, systems, or holistic theories discussed here do not replace their traditional forerunners. They are just broader in scope and offer a wider, additional level of analysis for continuing explorations of human development and interactions.

*"Our philosophy here at Dandelion Motors is to treat the total car."*

**Figure 6.8.**   Holistic approaches are increasingly popular in today's culture. Drawing by Koren; ©1979 The New Yorker Magazine, Inc.

# REFERENCES

1. Fox JV: Occupational therapy theory development: Knowledge and values held by recent graduates. *Occup Therap J Res* 1:79–93, 1981.
2. Research priorities for the 1990's. *Occup Therap Newspaper* 38:1, 12, 1984.
3. Smidt GL: Walking the trail of physical therapy research. *Phys Therap* 66:375–378, 1986.
4. Dean E: Psychobiological adaptation model for physical therapy practice. *Phys Therap* 65:1061–1068, 1985.
5. Currier DP: *Elements of Research in Physical Therapy*, ed 2. Baltimore, Williams & Wilkins, 1984.
6. Christiansen CH: Toward resolution of crisis: research requisites in occupational therapy (editorial). *Occup Therap J Res* 1:115–124, 1981.
7. Reed KL: Understanding theory: the first step in learning about research. *Am J Occup Therap* 38:677–682, 1984.
8. Ottenbacher K: *Evaluating Clinical Change. Strategies for Occupational and Physical Therapists.* Baltimore, Williams & Wilkins, 1985.
9. McGraw MB: *The Neuromuscular Maturation of the Human Infant.* New York, Hafner, 1963.
10. White SH: The learning-maturation controversy. Hall to Hull. In Charles DC, Looft R (eds): *Readings in Psychological Development through Life.* New York, Winston Holt & Rinehart, 1973.
11. Llorens LA: *Application of a Developmental Theory for Health and Rehabilitation.* Rockville, MD, American Occupational Therapy Association, 1976.
12. Ayres AJ: *Sensory Integration and the Child.* Los Angeles, Western Psychological Services, 1980.
13. Ayres AJ. *Sensory Integration and Learning Disorders.* Los Angeles, Western Psychological Services, 1973.
14. Trombly C: Neurophysiological and developmental treatment approaches. The Rood approach. In Trombly C (ed): *Occupational Therapy for Physical Dysfunction*, ed 2. Baltimore, Williams & Wilkins, 1983.
15. Huss AJ: Sensorimotor approaches. In Hopkins HL, Smith HD (eds): *Willard and Spackman's Occupational Therapy*, ed 5. Philadelphia, JB Lippincott, 1978.
16. Hall, GS. *Adolescence. Its Psychology and its Relations to Physiology, Anthropology, Sociology, Sex, Crime, Religion, and Education.* New York, Appleton, 1904.
17. Schultz D: *A History of Modern Psychology*, ed 2. New York, Academic Press, 1975.
18. Taylor GR: *The Natural History of the Mind.* New York, Dutton, 1979.
19. Gregory RL: *Mind in Science.* Cambridge, UK, Cambridge University Press, 1981.
20. Romer AS: *The Vertebrate Story*, ed 4. Chicago, University of Chicago, 1959.
21. Luria SE, Gould SJ, Singer S: *A View of Life.* Menlo Park, CA, Benjamin Cummins, 1981.
22. Gould SJ: *Ontogeny and Phylogeny.* Cambridge, MA, Harvard University Press, 1977.
23. Gould SJ: Racism and recapitulation. In Gould SJ (ed): *Ever Since Darwin.* New York, WW Norton, 1977.
24. Piaget J: *Play, Dreams and Imitation in Childhood.* New York, WW Norton, 1962.
25. Hess E: *Imprinting. Early Experience and the Developmental Psychobiology of Attachment.* New York, Van Nostrand Reinhold, 1973.
26. Mosey AC: Recapitulation of ontogenesis: A theory for practice of occupational therapy. *Am J Occup Therap* 22:426–432, 1968.
27. Loria C: Relationship of proximal and distal function in motor development. *Phys Therap* 60:167–172, 1980.
28. Ayres AJ: Ontogenetic principles in the development of arm and hand functions. In The *Development of Sensory Integrative Theory and Practice.* Dubuque, Iowa, Kendall/Hunt, 1974.
29. Wilson BN, Trombly CA: Proximal and distal function in children with and without sensory integrative dysfunction: an EMG study. *Can J Occup Therap* 51:11–17, 1984.
30. Ayres AJ: Occupational therapy directed toward neuromuscular integration. In Willard HS, Spackman CS (eds): *Occupational Therapy*, ed 3. Philadelphia, Lippincott, 1963.
31. Hodos W: Evolutionary interpretation of neural and behavioral studies of living verte-

brates. In Schmitt FO (ed): *The Neurosciences. Second Study Program.* New York, Rockefeller University Press, 1970.

32. Sulloway FJ: *Freud, Biologist of the Mind. Beyond the Psychoanalytic Legend.* New York, Basic Books, 1979.

33. MacLean PD: The triune brain, emotion, and scientific bias. In Schmitt FO (ed): *The Neurosciences. Second Study Program.* New York, Rockefeller University Press, 1970.

34. Rozin P: The psychobiological approach to human memory. In MR Rosenzweig, EL Bennett (eds): *Neural Mechanisms of Learning and Memory.* Cambridge, MA, MIT Press, 1976.

35. Kottke FJ: From reflex to skill: The training of coordination. *Arch Phys Med Rehabil* 61:551–561, 1980.

36. Bobath K: The facilitation of normal postural reactions and movements in the treatment of cerebral palsy. *Physiotherapy* 50:246–262, 1964.

37. Fiorentino MR: *Reflex Testing Methods for Evaluating C.N.S. Development,* ed 2. Springfield IL, Charles Thomas, 1973.

38. Loomis J: Model to integrate sensorimotor treatment approaches for neurological dysfunctions. *Physiother Can* 37:170–176, 1985.

39. Ottenbacher K, Short MA: Sensory integrative dysfunction in children: a review of theory and treatment. In Wolraich M, Routh DK (eds): *Advances in Developmental and Behavioral Pediatrics.* Greenwich, CT, JAI Press, 1985, vol 6.

40. Luria AR: Human brain and psychological processes. In Pribram KH (ed): *Mood, States and Mind.* Baltimore, Penguin, 1969.

41. Couper JL: Dance Therapy. Effects on motor performance of children with learning disabilities. *Phys Ther* 61:23–26, 1981.

42. Versluys HP: Psychosocial adjustment to physical disability. In Trombly CA (ed): *Occupational Therapy for Physical Dysfunction,* ed 2. Baltimore, Williams & Wilkins, 1983.

43. Bing RK: Discussion of Mosey's Recapitulation of Ontogenesis. *Am J Occup Ther* 22:433–435, 1968.

44. Gesell A: Maturation and Infant Behavior Patterns. In Stendler CB (ed): *Readings in Child Behavior and Development,* ed 2. New York, Harcourt Brace & World, 1964.

45. Connolly BH: Neonatal assessment: An overview. *Phys Ther* 65:1505–1513, 1985.

46. Palisano RJ, Short MA, Nelson DL: Chronological vs. adjusted age in assessing motor development of healthy twelve-month-old premature and fullterm infants. *Phys Occup Ther Pediatr* 5(1):1–16, 1985.

47. Henderson A: Occupational Therapy. In Levine MD, Carey WB, Crocker AC, Gross RT (eds): *Developmental-Behavioral Pediatrics.* Philadelphia, WB Saunders, 1983.

48. Gesell A, Amatruda CS: *Developmental Diagnosis.* New York, Haeber, 1947.

49. Als H, Lester BM, Tronick EC, Brazelton TB: Toward a research instrument for the assessment of preterm infants' behavior (APIB). In Fitzgerald HE, Lester BM, Yogman MW (eds): *Theory and Research in Behavioral Pediatrics.* New York, Plenum Press, 1982, vol. 1.

50. Als H, Lester BM, Tronick EZ, Brazelton TB: Manual for the assessment of preterm infants' behavior (APIB). In Fitzgerald HE, Lester BM, Yogman MW (eds): *Theory and Research in Behavioral Pediatrics.* New York, Plenum, 1982.

51. Palisano RJ: Use of chronological and adjusted ages to compare motor development of healthy preterm and fullterm infants. *Dev Med Child Neurol* 28:180–187, 1986.

52. Cunningham CC, Mittler PJ: Maturation, development and mental handicap. In Connolly KJ, Prechtl HR (eds): *Maturation and Development: Biological and Psychological Perspectives.* Philadelphia, Lippincott, 1981.

53. Girolami GL: Improving preterm motor control: The forgotten area of infant stimulation. *Phys Occup Ther Pediatr* 3:69–89, 1983.

54. Pelletier JM, Palmeri A: High-risk infants. In Clark PN, Allen AS (ed): *Occupational Therapy for Children.* St. Louis, CV Mosby, 1985.

55. Stern FM: Physical and occupational therapy on a newborn intensive care unit. *Rehabil Nurs* Jan–Feb:26–27, 1986.

56. Redditti JS: Occupational and physical therapy treatment components for infant intervention programs. *Phys Occup Ther Pediatr* 3:33–44, 1983.

57. Anderson J, Auster-Liebhaber J: Developmental therapy in the neonatal intensive care unit. *Phys Occup Ther Pediatr* 4:89–105, 1984.
58. Speidel BD: Adverse effects of routine procedures on preterm infants. *Lancet* April 22:864–866, 1978.
59. Ferry PC: On growing new neurons: Are early intervention programs effective? *Pediatrics* 67:38–41, 1981.
60. Newman LF: Social and sensory environment of low birth weight infants in a special care nursery. An anthropological investigation. *J Nerv Ment Dis* 169:448–455, 1981.
61. Campbell SK: Effect of developmental intervention in the special care nursery. In Wolraich ML, Routh D (eds): *Advances in Developmental and Behavioral Pediatrics*. Greenwich, CT, JAI Press, 1983, vol. 4.
62. Magnusson D, Allen VL: An interactional perspective for human development. In Magnusson D, Allen VL (eds): *Human Development. An Interactional Perspective*. New York, Academic Press, 1983.
63. Nash J: *Developmental Psychology. A Psychobiological Approach*. Englewood Cliffs, NJ, Prentice-Hall, 1970.
64. Rutter M: *The Qualities of Mothering, Maternal Deprivation Reassessed*. New York, Jason Aronson, 1972.
65. Bateson PPG: The characteristics and context of imprinting. *Biol Rev* 41:177–220, 1966.
66. Hoffman HS, Ratner AM: A reinforcement model of imprinting: Implications for socialization in monkeys and men. *Psychol Bull* 80:527–544, 1973.
67. Lehrman DS: A critique of Konrad Lorenz's theory of instinctive behavior. *Q Rev Biol* 28:337–363, 1953.
68. Gardner H: *Developmental Psychology. An Introduction*. Boston, Little, Brown, 1978.
69. Thomas A: Current trends in developmental theory. *Am J Orthopsychiatr* 607–609, 1981.
70. Baldwin AL: *Theories of Child Development*. New York, John Wiley 1967.
71. Sarnoff I: *Testing Freudian Concepts. An Experimental Social Approach*. New York, Springer, 1971.
72. Houston JP, Bee H, Rimm DC: *Essentials of Psychology*. New York, Academic Press, 1985.
73. Garner SN, Kahance C, Sprengnethe M (eds): Introduction. In *The Mother Tongue. Essays in Feminist Psychoanalytic Interpretation*. Ithaca, NY, Cornell University Press, 1985.
74. Fisher S, Greenberg RP: *The Scientific Credibility of Freud's Theories and Therapy*. New York, Basic Books, 1977.
75. Freeman L, Strean HS: *Freud and Women*. New York, Frederick Ungar, 1981.
76. Fidler GS, Fidler JW: *Occupational Therapy. A Communication Process in Psychiatry*. New York, MacMillan, 1963.
77. Sharrot GW: An analysis of occupational therapy theoretical approaches for mental health: Are the profession's major treatment approaches truly occupational therapy? *Occup Ther Mental Health* 5:1–16, 1985/86.
78. Kielhofner G, Burke JP: Occupational Therapy after 60 years: An account of changing identity and knowledge. *Am J Occup Ther* 31:675–689, 1977.
79. Kielhofner G, Barris R: Mental health occupational therapy: Trends in literature and practice. *Occup Ther Mental Health* 4:35–50, 1984.
80. Davis H, Kenyon P: Psychology: its relevance to the practice of physiotherapy. *Physiotherapy* 67:67–69, 1981.
81. Mosey AC: An alternative: The biopsychosocial model. *Am J Occup Ther* 28:1137–1140, 1974.
82. Ben-Shlomo LS, Short MA: The effects of physical exercise on self-attitudes. *Occup Ther Mental Health* 3:11–28, 1983.
83. King LJ: A sensory-integrative approach to schizophrenia. *Am J Occup Ther* 28:529–536, 1974.
84. Talbot JF: An inpatient adolescent living skills program. *Occup Ther Mental Health* 3:35–45, 1983.
85. Brady JP: Social skills training for psychiatric patients, I: Concepts, methods and clinical results. *Occup Ther Mental Health* 4:51–68, 1984.
86. Schwartzberg SL: Motivation for activities of daily living: A study of selected psychiatric patients' self-reports. *Occup Ther Mental Health* 3:1–26, 1982.

87. Mead P: The advisory role of the physiotherapist in mental health. *Physiotherapy* 71:261–262, 1985.
88. Davison K: Physiotherapy group work on an acute psychiatric ward. *Physiotherapy* 69:309–310, 1983.
89. Orlansky J: Infant care and personality. *Psychol Bull* 46:1–48, 1949.
90. Marmor J, Woods SM (eds): *The Interface Between the Psychodynamic and Behavioral Therapies.* New York, Plenum, 1980.
91. Malcolm J: *Psychoanalysis: The Impossible Profession.* New York, Alfred Knopf, 1981.
92. Malcolm J: *In the Freud Archives.* New York, Knopf, 1984.
93. Wolman BB: *Contemporary Theories and Systems in Psychology,* ed 2. New York, Plenum Press, 1981.
94. Meyer A: The philosophy of occupational therapy. *Am J Occup Ther* 31:639–642, 1977.
95. Koestler A: *The Ghost in the Machine.* New York, MacMillan, 1967.
96. Lockard RB: Reflections on the fall of comparative psychology. *Am Psychol* 26:168–179, 1971.
97. Schnierla TC: Behavioral development and comparative psychology. *Q Rev Biol* 41:283–302, 1966.
98. Anastasi A: Heredity, environment, and the question "How?" *Psychol Rev* 65:197–208, 1958.
99. Lorenz KZ: *The Foundations of Ethology. The Principal Ideas and Discoveries in Animal Behavior.* New York, Simon & Schuster, 1981.
100. Plomin R: *Development, Genetics and Psychology.* Hilsdale NJ: Lawrence Erlbaum, 1986.
101. Michel GF, Moore CL: *Biological Perspectives in Developmental Psychology.* Monterrey, CA, Brooks Cole, 1978.
102. Bronfenbrenner U: Toward an experimental ecology of human development. *Am Psychol* July:513–531, 1977.
103. Bronfenbrenner U: *The Ecology of Human Development. Experiments by Nature and Design.* Cambridge, MA, Harvard University Press, 1979.
104. von Bertalanffy L: *General Systems Theory, Foundations, Development, Applications.* New York, George Brasiller, 1968.
105. Lerner RM, Busch-Rossnagel NA: Individuals as producers of their development: conceptual and empirical bases. In Lerner RM, Busch-Rossnagel NA (eds): *Individuals as Producers of Their Development. A Life-Span Perspective.* New York, Academic Press, 1981.
106. Belsky J, Tolan WJ: Infants as producers of their own development. An ecological analysis. In Lerner RM, Busch-Rossnagel NA (eds): *Individuals as Producers of Their Development. A Life-Span Perspective.* New York, Academic Press, 1981.
107. Laslo E: *The Systems View of the World. The Natural Philosophy of the New Developments in the Sciences.* New York, Braziller, 1972.
108. Auerswald EH: Interdisciplinary versus ecological approach. *Fam Proc* 7:202–215, 1968.
109. Kielhofner G: General Systems Theory: Implications for theory and action in occupational therapy. *Am J Occup Ther* 32:637–645, 1978.
110. Kielhofner G, Burke JP: A model of human occupation, Part 1: Conceptual framework and content. *Am J Occup Ther* 34:572–581, 1980.
111. Howe MC, Briggs AK: Ecological systems model for occupational therapy. *Am J Occup Ther* 36:322–327, 1982.
112. Keshner EA: Reevaluating the theoretical model underlying the neurodevelopmental theory. *Phys Ther* 61:1035–1040, 1981.
113. Engel GL: The need for a new medical model: A challenge for biomedicine. *Science* 196:129–136, 1977.
114. Capra F: *The Turning Point. Science, Society, and the Rising Culture.* New York, Bantam, 1983.
115. Engel GL: The clinical application of the biopsychosocial model. *Am J Psychiatr* 137:535–544, 1980.
116. Yussen SR, Santrock JW: *Child Development. An Introduction,* ed 2. Dubuque, IA, 1982.
117. Pellegrino ED: *Humanism and the Physician.* Knoxville, TN, University of Tennessee Press, 1979.

118. Mosey AC: *Occupational Therapy. Configuration of a Profession.* New York, Raven Press, 1981.
119. Belknap MM, Blau RA, Grossman RN: *Case Studies and Methods in Humanistic Medical Care. Some Preliminary Findings.* San Francisco, Institute for the Study of Humanistic Medicine, 1975.
120. Moore MC, Moore LJ: *The Complete Handbook of Holistic Health.* Englewood Cliffs, NJ, Prentice-Hall, 1983.
121. Hastings AC, Fadiman J, Gordon JS (eds): *Health for the Whole Person. The Complete Guide to Holistic Medicine.* Boulder, CO, Westview Press, 1980.
122. Cmich DE: Theoretical perspectives of holistic health. *J School Health* 54:30–32, 1984.
123. Bamberg DL: Holistic health, human ecology, and you. *Occup Health Nurs* 30:21–24, 1982.
124. Gordon JS: The paradigm of holistic medicine. In Hastings AC, Fadiman J, Gordon JS (eds): *Health for the Whole Person. The Complete Guide to Holistic Medicine.* Boulder, Co, Westview Press, 1980.
125. Garrott SS, Garrett R: Humaneness and health. *Topics Clin Nurs* Jan:7–12, 1982.
126. Gee R: The physical therapist as a holistic health practitioner. *Clin Manag* 4:18–21, 1984.
127. Knickerbocker B: *A Holistic Approach to the Treatment of Learning Disorders.* Thorofare, NJ, Charles B Slack, 1980.
128. Quinn P: *Holistic Health Educational Resource Guide.* Alexandria, VA, American Physical Therapy Association, 1984.
129. Baker KE: Health promotion: Promoting health to prevent disease. *Clin Manag* 4:10–12, 1984.
130. Huhn RR, Volski RV: Primary prevention programs for business and industry. Role of physical therapists. *Phys Ther* 65:1840–1844, 1985.
131. Llorens LA: Theoretical conceptualizations of occupational therapy: 1960–1982. *Occup Ther Mental Health* 4:1–14, 1984.
132. Mosey AC: A model for occupational therapy. *Occup Ther Mental Health.* 1:11–31, 1980.
133. Slaymaker JH: A holistic approach to specialization. *Am J Occup Ther* 40:117–121, 1986.
134. Ferguson M: *The Aquarian Conspiracy. Personal and Social Transformation in the 1980's.* Los Angeles, JP Tarcher, 1980.

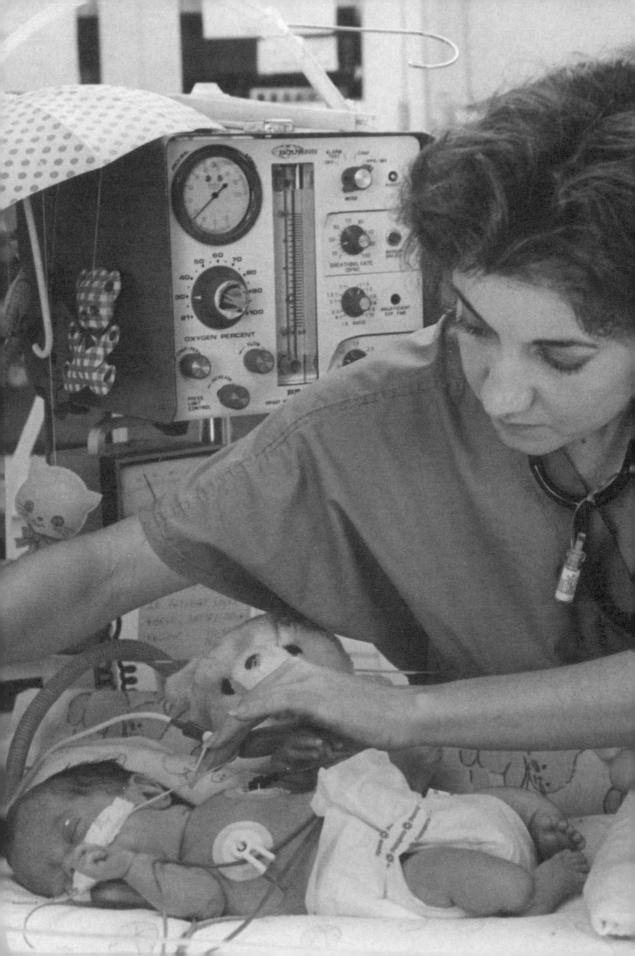

# 3

# PRENATAL DEVELOPMENT AND BIRTH

# Section 3

Section 3 discusses prenatal development and the birth process. The ecological approach is highlighted as it relates to the continuum of reproductive casualty for the normal or at-risk infant. Prenatal development is examined, and of particular interest to therapists will be the discussion of prenatal motor behaviors that have only recently been explored. The prenatal diagnostic procedures that enable study of fetal behavior are described. New approaches to childbirth, and the various genetics and environmental conditions that may contribute to risk conditions are also discussed. The long-term consequences of birth or prenatal complications are examined, as they may affect the attachment relationship and subsequent interactions between the infant and family.

# 7

# Genetics, Prenatal Development, and the Birth Process

In the last chapter, the importance of using different levels of analysis in the study of human development was discussed. Genetic considerations are an essential level of analysis; these involve an understanding of how very specific mechanisms—genes and chromosomes—are responsible for the actual formation, organization, and regulation of human structure, function, and behavior. This chapter will examine the genetic regulation of behavior within the context of prenatal development. It will also discuss potential developmental problems that may emerge as a result of genetic irregularities and some of the major developmental events that occur within the human organism from the time of fertilization to the time when the infant leaves the womb and begins life in a different environment on its own. In addition, this chapter will explore various risk factors that may affect pregnancy, the normal birth process, and potential problems that may occur during delivery, as well as some of the consequences of prenatal or birth complications.

# FERTILIZATION

Prenatal development starts at *fertilization*, the process that unites the father's sperm and the mother's ovum. During sexual intercourse, sperm exit from the male's penis upon ejaculation into the female's vagina. The ejaculate contains an enormous number of sperm that, if conditions favor it, will propel themselves through the vagina and cervix, through the uterus, and into the Fallopian tubes (also termed *oviducts*) that extend outward on each side of the uterus. If the female has recently ovulated, a ripe egg released from the ovary is trapped by the end of the oviduct and carried downward where it may encounter a sperm and start the 9-month process of organizing a new human being.

# STAGES OF PRENATAL DEVELOPMENT

Once the sperm and egg unite, they form a fertilized egg, which is called a *zygote*. The zygote begins a voyage down the oviduct toward the uterus, where it will lodge in the uterine wall. During the process of moving toward the uterus, the zygote remains extremely active. This activity comprises what is known as the first stage of prenatal development.

## Stage of the Zygote

The first stage of prenatal development has many names; it is alternately known as the "stage of the ovum", the stage of the zygote", or the "germinal stage". Since the zygote is a result of a special interaction between the germ plasm of both parents' gametes, the term "the stage of the zygote" is preferred. Following fertilization, the zygote does a number of things. First, it forms a protection against repeated penetration from additional sperm. In addition, the zygote begins to repeatedly divide through the process of mitosis, or cell division. Thus, the zygote initially becomes two cells, then four, then 16, etc. The cells tend to congregate in different layers, forming an inner cell mass and an external cell layer called a *trophoblast*. Once this cell mass develops a cavity inside so that it is like a hollow ball, it is called a *blastocyst* (1).

By day 4 after fertilization this blastocyst has arrived in the uterus, and by about day 6 or 7, the outer wall has formed small projections called *villi*. These tiny villi function as extensions to "fix" the blastocyst into the uterine wall. This process from fertilization to implantation takes approximately 2 weeks; once implantation has occurred, the second stage of prenatal development begins.

## Stage of the Embryo

The villi of the zygote actually take root in the walls of the uterus, firmly attaching the embryo and serving as the foundation of the *placenta*. The placenta is a network of membranes that connect the prenatal organism and its mother, providing nourishment and oxygen and removing wastes. It is important to note that the placenta does not provide a direct neural nor a direct blood supply connection between the mother and the embryo. The bloodstreams of the mother and the embryo are adjacent, separated by semipermeable membranes, which typically let pass only those materials that are small enough or absorbent enough to penetrate the membranes or that are essential for the ongoing maintenance and survival of the embryo. Thus, the placenta delivers nutrients such as sugars, fats, and some proteins from the mother to the embryo and serves as a locus for gas exchange, e.g., providing oxygen and removing carbon dioxide from the embryo. Other substances such as vitamins, some disease germs, and drugs also pass through the placenta, and their effects will be examined later in this chapter (2).

Three major functions occur in embryo development: growth, differentiation, and morphogenesis (1). Hetherington and Parke (3) note that during the time from fertilization to the end of the embryonic period, the prenatal organism increases in size 2,000,000%! In addition to its incredible growth rate, the embryo also demonstrates extraordinary cell differentiation. This is the process whereby cells become specialized for specific functions, e.g., as liver, brain, or muscle cells. In addition to cells becoming specialized, similar types of cells come together in the formation of tissues and organ systems. The unfolding of these various structures is known as morphogenesis (1).

Thus, during the stage of the embryo, cell proliferation and differentiation take place on a massive scale, forming thousands of cells that will perform different characteristically human functions (1). The blastocyst continues to differentiate, forming layers of cells that give rise to different structures. For example, two layers, the amnion and chorion, will form the outer protective sac in which the embryo is suspended. This structure, called the "amniotic cavity", is filled with fluid and surrounded by the chorion and the amnion as well as the mother's uterine wall. Thus, except for the connection through the placenta, the prenatal organism is suspended in amniotic fluid, which acts as a buffer for shock as well as maintaining a regular temperature (2).

Other primitive embryonic layers give rise to all the structures characteristic of a mature human being (1). There are three basic layers that arise from the inner cell mass that was part of the blastocyst: the outer layer called the *ectoderm*, which gives rise to skin, nerves, hair, and nails; a middle layer called the *mesoderm*, which becomes muscles and bone; and the *endoderm*, which will differentiate into internal organs such as the gastrointestinal tract, the liver, glands, and lungs.

The embryonic stage represents a time of structural development which results in an organism that has the foundational characteristics of its species (1). Thus, during this time period from the 2nd to the 8th or 9th week of gestation, the embryo beings to take on an appearance and traits that are human-like, e.g., a beating heart; external genitals; a face with mouth, eyes and ears; and arms and legs with tiny hands and feet. Nerve cells have developed sufficient to allow rudimentary responses to be made. Gross movements are made in response to touch stimulation, and major portions of the trunk and limb musculature are formed (4). Although at 8–9 weeks, the embryo is only 1 inch long and its systems are very incomplete, it is clearly on its way to becoming a structural and functional human.

# Stage of the Fetus

The appearance of bone cells signals the start of the fetal period, which is the final prenatal stage and lasts from the 3rd month to birth. As Kopp and Parmelee (5) have described, by 12 weeks of prenatal life, the fetus is structurally complete, thus further development is not structural but functional. This involves cell growth and maturation of body systems, movement, and behavior. By the 3rd month, the fetus has additional rudimentary behaviors. This is Gardner's description:

Various reflexes have developed. For instance, if the fetus is dropped a few centimeters, its muscles automatically contract. Stimulating the palm of the hand causes the fingers to close quickly though incompletely; electrical stimulation makes the limbs move. And the heartbeat is so regular that patterns characteristic of adults can be discerned. Instead of moving in unison, legs now move separately. And, during motion, the arm on one side moves more extensively backward than the arm on the opposite side. These independent movements testify to the differentiation of bodily functions (4, p 95).

Continued growth and maturation occur, so that by the 4th and 5th months, bones and external genitals are discernible. The sex of the fetus can be determined at this time, and the nervous system becomes more specific. Self-initiated and localized movements occur. For example, the head can move independent of the body so that flexion, extension, and rotation of the neck occurs. A fetus of 20 weeks is about 10 inches long and weighs about 10 ounces. It swallows amniotic fluid, hiccups, can open and close its eyes, make facial movements that accompany head and mouth movements, and can flex and extend its limbs and fingers, and suck its thumb. It is sensitive to self-produced movements, receiving considerable input from stimulation of the skin and from movement of the joints (4, 6).

Although the 5-month-old fetus closely resembles a newborn infant, it cannot survive independently. It still requires considerable development of the lungs, endocrine glands, and circulatory and nervous systems. The 28-week-old fetus may survive outside the womb, although probably requiring supportive care to assist with immature digestive and respiratory systems as well as with temperature regulation. Continued growth in the womb during this time results in maturation of the lungs in preparation of respiring oxygen, development of an insulating layer of fat for temperature control, development of muscle tone and sensory systems, as well as integrated, coordinated behaviors evident in regular sleep-wake cycles, sucking, and responses to environmental stimulation (4, 6).

Because of the significance of fetal responses to later development, clinicians and developmental researchers are very interested in obtaining descriptive information about age-related changes in fetal behavior. This information is particularly relevant to physical and occupational therapists, who are concerned with normal sensory and motor development throughout life. Thus, the following section describes in greater detail some of the recent information obtained about prenatal capabilities.

# FETAL AND EMBRYONIC BEHAVIOR

In the past, different procedures were used to gain information about prenatal behavior. Information from studies of nonviable embryos and fetuses and from living preterm infants was typically used to draw conclusions about normal prenatal development. There are, however, several problems with such procedures. First, numerous reasons exist for preterm delivery so that many of these organisms may be quite different from age mates still in utero. Second, there is considerable variability in the medical complications associated with prematurity. That fact, and the tremendous changes in medical technology that have so radically altered conditions for preterm infants, make it difficult to obtain a comparable group of subjects about which to draw generalizations (7). Therefore, normative studies of preterm development and behavior have been rare.

Denenberg (8, p 85) reported in 1972 that the behavioral effects of prenatal stimulation is the least investigated of any of the areas involved in the biobehavioral basis of development, and he suggests caution in making generalizations from studies that do exist. Only recently in the 1970s and 1980s, however, have alternative, noninvasive methods been developed that enable physicians and scientists to actually observe ongoing fetal behavior. Because of the previous difficulties in studying prenatal behavior, research in embryology tended to focus on immediate physiological, rather than behavioral, changes. Now, however, new technologies enable researchers and physicians to obtain new and exciting profiles of fetal movement and behavior. These new technologies, such as ultrasound, will be discussed in more detail later in this chapter. They are important to mention here, however, since they have given us an actual "window on the womb" (9) so that whole patterns of fetal movements can now be visualized and described.

## Sensorimotor Responses

Consistent evidence from studies with pregnant women indicate that fetuses are active and responsive to various forms of stimulation. In addition, research with animals has found that some forms of fetal stimulation have such long-range consequences that they last over generations. While some of this research and the issues it brings up will be explored in more detail in Chapters 8 and 9, some descriptive studies of fetal behavior will be reported here.

Normative studies of fetal behavior are only recently forthcoming because of new technology that enables physicians to actually obtain profiles of prenatal movement. Some of the data from these studies are inconsistent, either due to differences in levels of sophistication of equipment or due to methods physicians and researchers may use to gauge fetal age. For example, fetal age can be considered equivalent to *gestational age* or *conceptual age*; i.e., the number of days from the time of fertilization. However, many pregnant women are unable to accurately pinpoint the actual time of conception. Thus, physicians may use a different measure. This is called *menstrual age*, which is measured in terms of the mother's last menstrual cycle. Since this normally occurs 2 weeks before ovulation, menstrual age is typically 2 weeks earlier (or longer) than conceptual age (10).

Considerable information about fetal motor responses has come from the work of Humphrey (11) and Milani Comparetti (12). From a long series of studies of human embryos and fetuses, Humphrey (11) has observed entire developmental sequences of motor responses. Both she and Milani Comparetti, who most recently has used ultrasound analyses, claim that many of the prenatal responses they have seen are quite similar to later postnatal behaviors. Humphrey believes that it may be common for similar behavior patterns to be elicited over and over as different regions of the brain mature and begin to influence other neural areas.

Some of the behavior patterns Humphrey (11) has described are of particular interest to occupational and physical therapists. For example, she has noted how the maturation of grasp response in the hand parallels voluntary grasp movements in postnatal development. Like grasp, the other behavior patterns she has observed seem to be initially relatively gross and undifferentiated, but become gradually more regulated and discrete with time. For example, the earliest reflex she has observed begins during the embryonic period at 7.5 weeks menstrual age,[a] or 5.5 weeks conceptual age. At this time the embryo bends its head and upper trunk away from a stimulus of light stroking of the *perioral area* (the area around the mouth). This stimulation is usually administered to one side of the mouth, and the embryo reacts with an *avoidance response*; i.e., an action that effectively pulls it away from the source of stimulation. Since the action is made on the side of the body away from the point of stimulation (contralateral) and since it eventually results in a flexed position, it is called a *contralateral flexion reflex.*

One week later, at 6.5 weeks, this movement consists of even more muscle activity. There is neck, trunk, and pelvic flexion with extension of the arms as well as some pelvic rotation. In addition, movement of the lower extremities causes separation of the soles of the feet. By 7.5 weeks, mouth opening also occurs as part of the total pattern reflex. Then, by 8.5 weeks, the contralateral flexion reflex is replaced by a much less complete (and possibly more controlled) response, which is simple head and trunk extension.

These observations differ somewhat in age when compared with Milani Comparetti's (12) work, which concludes that before 10 weeks, fetal movements are not yet organized into patterns. However, both researchers find parallels between fetal responses and later postnatal behavior. For example, Humphrey (11) notes that it has been commonly observed in motor development that rotation toward (approach) some stimulus is consistently seen later than rotation away (avoidance). She has confirmed this in observations that avoidance behaviors as part of the contralateral flexion response are seen as early as 5.5 weeks, whereas the approach response of ventral head flexion (bringing the head toward the chest) is not first seen until 9 weeks.

Humphrey (11) describes other behavioral patterns—such as flexion responses in the hands and feet, and fetal hand to mouth reflexes—that seem to parallel similar motor behaviors seen postnatally. She describes the similarities between the *Babkin reflex*, which is seen around the time of

---

[a] For consistency, Humphrey's calculations are converted to conceptual age.

Table 7.1
Classifications of Fetal Movements

Rolling movements[a]
    Sustained, associated with whole body
Simple movement
    Short, thought to originate from an arm or leg
High-frequency movement
    Short, easily felt through entire abdomen, sometimes rhythmic
Respiratory movement
    Breathing-like, sometimes felt by mother
Fetal locomotion[b]
    Movement within the uterus leading eventually to the correct presentation
    for birth
Fetal propulsion
    Ability of fetus to "push off" with feet, necessary for birth collaboration

[a] Data from Timor-Trisch I, Zador I, Hertz RH, Rosen MG: Classification of human fetal movement. *American Journal of Obstetrics and Gynecology* 126: 70–77, 1976
[b] Data from Milani Comparetti A: Pattern analysis of normal and abnormal development: The fetus, the newborn, the child. In Slaton DS (Ed) *Development of Movement in Infancy*. Chapel Hill, University of North Carolina at Chapel Hill, 1981.

birth, and responses elicited from palmar stimulation of a fetus. The Babkin reflex is elicited by applying pressure to both palms of a neonate's hands. Following this stimulation, the infant opens the mouth, flexes the head forward, and closes the eyes. In addition, rotation of the head toward the midline is also seen. Humphrey reports that a similar reaction was obtained by palmar stimulation of a 12-week-old fetus (11).

Milani Comparetti (12) and Timor-Trisch and associates (13) have devised general classifications of fetal movements (Table 7.1). Milani Comparetti notes that these movements are actually preparatory for the fetus' participation in the birth process as well as for normal activity and development following birth. He refers to activity patterns that are necessary for normal fetal development as "fetal competencies"; he refers to those patterns of movement that will support postnatal development as "emerging competencies". Milani Comparetti (12) concurs with Humphrey (11) that numerous fetal behaviors are related to subsequent behaviors that emerge after birth.

# U-Shaped Growth of Behavior

Humphrey's (11) and Milani Comparetti's (12) work have important implications for clinicians and developmentalists interested in understanding how motor control matures. Humphrey's observations and conclusions about motor patterns support the existence of considerable repetition in behavior during development, a phenomenon reported by others in regard to other developmental domains, e.g., perception and cognition (14, 15). Strauss (16) describes this cyclic repetition of behavior as *U shaped behavioral growth*. This is manifested as the appearance of some behavior, which

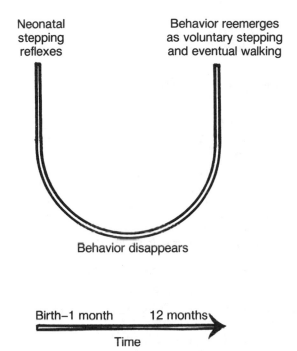

**Figure 7.1.** According to some theorists, behavioral growth is cyclic; it appears, disappears, and reappears.

during development disappears, and then later reappears (Fig. 7.1). Many examples of behavioral cyclicity have been described in *Regressions in Mental Development*, edited by Thomas Bever (17). In a review of this book, physical therapist Suzanne Campbell notes:

> In recent years, the idea that child development consists of a gradual accumulation of skills and behavior leading to the ultimate achievement of adult form and function has been repeatedly denied by the identification of numerous behavioral patterns that undergo apparent regression[b] at various points in time. For example, locomotor movements of the legs are present in the neonate, disappear during the early months of postnatal life and eventually reappear as voluntary stepping behavior. Similar regressions in development during infancy have been identified in reaching, visual following, and vocalization, and further examples are available in the development of older children (18, p 116).

Thus, development, at least in some regards, is not continuous but is characterized by progression, regression, and progression again. Such information also indicates that at earlier times in development, actions and behaviors are ways of "practicing" or laying a foundation for subsequent behavior. As Milani Comparetti (12, p 1) notes, "No neonatal motor pattern can be considered to originate at birth. The fetus already has the full repertoire of movements found in the newborn baby". The full implications

---

[b] "Regression" here refers to a normal disappearance of behavior, not to some clinical condition.

of these findings for the birth process and for clinical intervention will be discussed later in this chapter.

# DEVELOPMENTAL RISKS

Although the processes of cell division and differentiation that occur during embryonic development are genetically regulated, they are not exclusively under genetic control. Luria and colleagues explain:

> The mere existence of a genetic program is not sufficient to produce an adult organism. A seed containing all the necessary genetic information will never give rise to an adult plant if it does not experience conditions that cause it to germinate—to open up and produce a growing seedling. Generally speaking, development involves an interaction between the genetic information of an organism and information from the environment ... In a sense, the miracle of organismic development is that genetic programs actually do encounter the specific circumstances needed to bring them to full fruition (1, p 314).

Genetic *and* environmental conditions must be appropriate for the prenatal organism to survive at any stage. Determining what conditions are appropriate and what are harmful to the embryo and fetus has been a challenge for scientists for centuries.

Many conditions may deleteriously affect development. These have been referred to as "risk factors" because their presence signifies potential risk for developmental delays, for subsequent neurological and behavioral problems, or for other forms of abnormal development for the prenatal organism. Such factors may be genetic or environmental or may involve an intricate interaction between them. Other factors may be structural, such as malformation of the placenta, or may relate directly to the health, age, and nutrition of the mother (5). Some of these risk factors are listed in Table 7.2.

Kopp and Parmelee (5) suggest that we look at the effects of risk factors on three different time periods: prenatal, perinatal, and postnatal. Prenatal includes fetal and embryonic development, whereas perinatal refers to the time period around birth, i.e., at birth as well as the first 3–4 months of extrauterine life. Postnatal refers to the rest of the time after birth.

Table 7.2
**Developmental Risk Factors Associated with the Prenatal and Perinatal Periods**

| |
|---|
| Maternal infections |
| Maternal drugs, toxic agents |
| Spinal cord defects |
| X-linked genetic disorders |
| Metabolic disorders |
| Autosomal disorders |
| Anoxia in the neonatal period |
| Preterm birth |
| Low birth weight |

Willemsen explains the limits and the value of attending to at-risk conditions:

> The terms "at risk" and "perinatal stress" have been coined to refer to all babies whose histories contain any factors known to be related to psychological or educational handicaps. The term "at risk" implies that there is some greater than average chance that the infant so designated may later have cerebral palsy, mental retardation, psychosis, learning disabilities, or any of a host of other handicaps. But the term does not indicate that the infants so labeled will inevitably develop a handicap. It simply implies that a child's history—the parents' genetic make-up, the mothers' pregnancy, and the birth process—contains one or more factors that have appeared in the histories of children who have these handicaps more often than in the histories of normal children.
>
> The primary advantage of labeling a baby "at risk" is that it alerts the baby's doctor and parents to watch for signs of any problem. Also, there are now programs for babies "at risk" that attempt to prevent the problems from developing (19, p 26).

# Critical Periods of Development

The first 12 weeks of prenatal development are particularly sensitive to risk and are crucial for the subsequent development of the human organism. This first third of the gestational period, called the *first trimester*, is so important because of the foundation of organ and structural systems that occur during this time. Since embryogenesis is crucial for normal structural development, the first trimester is referred to as a "critical period" in human ontogeny.

Noxious substances crossing the placental barrier or physical trauma occurring at this time can interfere with morphogenesis such that development of organ systems is impeded or interrupted. The result may be some birth defect such as a structural or orthopedic deformity or impairment of the function of an entire system. Thus, as a result of interference with very rapid cell differentiation and growth, the following may result: blindness, deafness, poor development of the spinal cord and resulting paralysis or limb weakness, limb deformities, and even spontaneous abortion. Clearly, embryogenesis is critical for subsequent sensorimotor and other forms of development. Figure 7.2 illustrates the most common sites of birth defects and their times of occurrence. It illustrates the sensitivity of embryonic, compared with the fetal and zygotic, stages of development.

While environmental risk factors tend to have more critical effects during embryogenesis, they still can affect ongoing fetal and subsequent perinatal and postnatal development. Some of these environmental, as well as other genetic and perinatal, risk factors will be examined following a brief review of brain development. The development of the brain is particularly important to understand because of its susceptibility to damage during certain sensitive periods and because of the widespread implications of neural damage, before, during, and after birth.

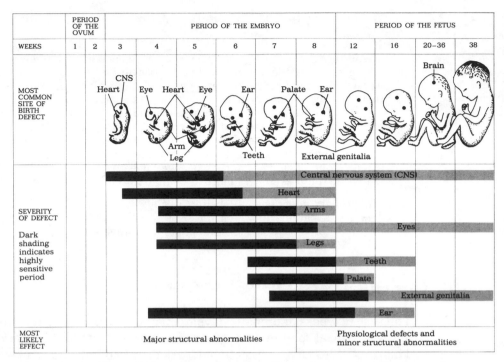

**Figure 7.2.** The most serious damage from drugs, chemicals, and some diseases is likely to occur in the first 8 weeks after conception. However, damage to many vital parts of the body, including brain, eyes, and genitals, can occur during the last months of pregnancy as well. (From Berger KS: *The Developing Person Through the Life Span.* NY, Worth Publishers, Inc., 1983, p 90. Reproduced with permission from the publisher.)

# Development of the Brain

While the brain has been studied for centuries and is important in the regulation of all human behavior, its development still remains a mystery. We do have some basic descriptive information about brain ontogeny, but as Lecours (15, p 267) explains, "The process of brain maturation, which in man does not reach its term before at least 15 and conceivably many more years of life have elapsed, is reflected in biological changes of several orders (chemical, electrical, anatomical, etc.). Contemporary knowledge of these changes is indeed scarce and fragmentary; understanding of their significance is still more so."

## Maturation of the Brain

The nervous system originates from cells in the ectoderm of the embryo. These cells form a ridge that begins to differentiate. It folds into a long, hollow tube, one end of which swells up into three areas that will subsequently become the forebrain, midbrain, and hindbrain. By the end of the

4th week, the embryo has a recognizable brain with two lobes, but it will continue to develop well into adolescence and possibly adulthood (20, 21).

Numerous factors must be examined in brain maturation; e.g., the development and continued growth of nerve cells; the development of synapses between cells; the formation of different types of synapses; and the development of *glial cells*, which fill in between neurons and serve important functions for providing nourishment to the brain cells. In addition, glial cells are responsible for manufacturing myelin, a fatty substance that coats the long branches (called *axons*) of some long neurons. Myelin is particularly important because it serves as an insulating material that facilitates nerve conduction and efficiency. Cowan (20, p 113) explains that brain development actually goes through eight major stages. The first four stages encompass the beginnings of the long neural tube in the embryo and the proliferation of cells and their migration to appropriate locations, where they aggregate with other similar cells to form identifiable parts of the brain.

Eventually, according to Cowan (20), the immature brain cells begin to specialize and to make connections with other neurons. Then, and contrary to expectation, the final two stages of brain development involve the elimination of many already established synapses and the selective death of many cells. Thus, while the normal adult brain contains about 100 billion neurons and about 100 trillion synapses (22), the fetal and infant brain contains many more that are selectively eliminated in early development. As Cowan explains:

> It has become evident in recent years that the development of many structures and tissues is sculptured by highly programmed phases of cell death. This is true also of the developing brain. In many regions of the brain the number of neurons originally generated greatly exceed the number of neurons that survive beyond the developmental period (p 132).

Of all the organs at birth, the brain is the closest in weight to that of an adult brain (about 25% of its adult size and weight). This gives newborns a disportionate appearance that, when compared with that of an adult, emphasizes the head in comparison to the body. The brain continues to grow in size and weight until adolescence. In the first 2 years of infancy, this growth is very rapid (Fig. 7.2), then it becomes more gradual but continues slowly until adolescence when adult brain size and weight are reached (10, 23).

The increase in size and weight of the brain is a result of the continued development of glial cells and the growth in size of existing neurons. What is striking about brain tissue is that many different parts of the brain form at different stages of development. Thus, there is no one predictable period of brain sensitivity to damage since different parts become more sensitive than others at different times of maturation (21, 24, 25). This has implications for understanding motor and behavioral control and responsiveness of neural tissue to brain injury. While some have referred to specific ages for critical periods in brain development (26), Rodier (24) points out that there are probably many different critical periods for differing brain regions.

Obviously, one of the most important times of brain development is the prenatal period, at which time the basic structure of the central nervous system is being laid down and the cells are being built. The fetal brain is constructed from food provided to it from its mother. As Teyler (27) describes, "the mother's body alone cannot provide all of the necessary proteins at this time." Thus, some of the fats and proteins necessary for fetal brain development must be obtained from the mother's diet, which points out the necessity for adequate maternal nutrition. However, prenatal neural development cannot be considered "the most important" time since brain cell formation or myelination continue into childhood and adolescence. Dobbing and Sands (25) have noted that about ⅚ of the brain growth spurt in humans, (i.e., a period of very rapid brain changes) occurs postnatally; this is a period when neural tissues are especially vulnerable to nutritional restrictions as well as environmental and physiological risk factors such as irradiation or hormonal problems. Thus, an adequate balanced diet is essential for neural development during infancy, childhood, and adolescence (21, 27).

## Special Characteristics of the Brain

One of the phenomenal characteristics of the brain is its *plasticity*, i.e., its adaptability. This plasticity is apparent in the remarkable capacity of neural tissues to reorganize, so that the brain often self-corrects in cases of developmental errors or injury (20). Another way in which the brain is adaptive is in response to environmental stimulation. In appears that functional stimulation in some cases can direct brain organization so that experiences may in fact alter neural tissue (20, 28). On the other hand, many brain regions appear to be directly regulated by a genetic program with specific sensitive periods of development characteristic of each particular species (29). The intricate and special relationship between species-specific neural development and environmental stimulation has been studied closely by scientists, and their work comprises a large body of research in the fields of developmental psychology and developmental neurobiology. This book will return to this topic in greater detail in chapter 9, which discusses environmental stimulation.

Other special developmental characteristics of the brain will also be examined more closely in other chapters of this book. For example, prenatal hormonal effects on the brain are responsible for the ultimate expression of sexual behavior and sex differences that appear in adolescence. This will be examined in chapter 12, which discusses gender differences.

## Central and Peripheral Nervous Systems

The brain is the most important and least understood organ in the body. It essentially controls everything we do, thus it is important to study from both a developmental and a clinical point of view. One of the important clinical implications of development is the relative immaturity of the nervous system early in life. When speaking of the nervous system, there are two different areas that are referred to. One is the *central nervous system*, comprising the brain and the spinal cord. The second, although not really separate, is the *peripheral nervous system*.

The peripheral nerves are so named because they run from the brain to the exterior (or periphery) of the body. They are responsible for carrying sensory information from the distal areas (e.g., from the fingers or toes) toward the brain and then relaying motor signals outward to the muscles, where some action results. While the central nervous system does the "thinking", the peripheral system does the "acting".

## Implications of the Immaturity of the Nervous System

Because of the large size of the adult brain and the limits on the size of infant that can be delivered, humans must continue their growth and neural development well beyond gestation. Since the peripheral and central nervous systems are responsible for behavior, certain actions will be immature as a reflection of the immaturity of nerves at birth. Thus, some areas of the body that are essential for survival tend to be mature at birth, while others remain immature for years. Some of the behavioral immaturity or incompetence of infants is a result of the lack of myelination of the nervous system. For example, the peripheral nerves around the facial region are quite sophisticated at birth, allowing the newborn to engage in complex multistep, oral activities such as locating a nipple, sucking, and swallowing. Other, precise, coordinated, voluntary movements may take years before they are managed, and their achievement will reflect practice as well as gradual myelination of peripheral nerves (11).

Some motor skills, such as arm and head movements, develop quite rapidly, while others, such as bowel and bladder control, walking, and hopping are slower to emerge. Some of these skills will emerge rapidly as muscles develop and as nerves continue to form synapses with muscles. Other characteristically human activities, such as cognitive skills and reasoning, may take decades to emerge and may reflect a very gradual maturation of the outer shell of the brain (called the *cortex*) in interaction with long-range environmental effects. The study of the gradual emergence of these various skills is the subject of developmental psychology and the clinical field of developmental motor control.

The fact that brain development continues through the prenatal period and beyond has important implications for the maturation of normal behavior as well as for the special case of the infant delivered before the full term of pregnancy. Much of the activity in the nervous system during the last months of gestation prepares the fetus for survival on its own. Thus, the timing of the maturation of feeding, breathing, sensory, and motor abilities in central and peripheral nerves is important to survival of the fetus (see

"Preterm or Low-Birth-Weight Infants"). In the following section, some of the developmental risks to brain and behavioral development will be examined.

# Prenatal Risk Factors

Separating maternal from fetal conditions is difficult since the fetus and mother are so interconnected. They share much the same internal environment; e.g., nutrition, infections, bacteria, drugs, hormones, and nicotine that the mother may have been exposed to and that filter through the placenta. Age and health are factors that are specifically characteristic of the mother but they still affect the fetus.

## Maternal Age and Health

Very young mothers and older mothers experience different kinds of risks when pregnant. Teen mothers have a special set of problems due to their physical, emotional, social, and cognitive immaturity. Adolescence is when many organ systems are undergoing rapid growth and development in addition to the common emotional and social stresses of this time period. Therefore, the addition of pregnancy at this time in life may be threatening to the health and physical and emotional development of the mother as well as the fetus.

### Adolescent Motherhood

Immature mothers are more likely to have complicated deliveries, to experience perinatal or infant mortality, or to deliver infants who have birth complications; i.e., who are preterm, or low birth weight, or exhibit birth defects. For example, babies born to adolescents under age 16 are two to three times more likely to die in the first year, and premature and low birth weight infants are twice as likely to occur among adolescent mothers as among young adults (30).

Tracing direct causes for some of these problems is very complicated because teen mothers are less likely than are older mothers to seek prenatal care and to provide adequate prenatal nutrition. Thus, it is not possible to determine whether some of these birth complications are a result of the actual age of the mother or due to other prenatal factors. There is, however, some evidence that indicates that very young motherhood can be deleterious for the pregnant girl and her child. Since some teenagers are still growing, they may be competing with their fetuses for essential vitamins and minerals. Chan (32) reports that teenagers typically experience increased bone growth, thus they need extra calcium at this time. If, in addition, an adolescent is breast feeding, then she may be putting her infant in competition with her own needs for calcium and other minerals.

Subsequent problems in child rearing also may result with teen parents because they are more likely than older parents to experience dissatisfaction with their marriages, to have higher divorce rates, and to have more children, which, in turn, may place greater financial burden and emotional stress on the family (32). In addition, children born to teen-age mothers are more likely to experience health risks and to suffer from abuse. Elster and colleagues (33) report that teen-age mothers tend to behave differently with their children. Because of their immaturity and lack of knowledge of child development, teens often have unrealistic attitudes about child rearing, tend to interact less often with their children, tend to be less sensitive to their children's needs, and tend to punish their children more than other mothers (33).

While such findings sound pessimistic, Elster and colleagues (33) point out they should be considered tentative. Since many studies of adolescent mothers are compromised by methodological problems as well as confounded by the numerous variables that must be studied, developmentalists are cautioned against overgeneralizing these findings. Elster and colleagues note that not all adolescent parents are at risk for parenting disorders. Furthermore, not all adolescent mothers have pregnancy difficulties or give birth to infants with special risks for developmental problems. Lester and colleagues (34) have confirmed this in their studies of teenage childbearing, in which they suggest that age alone is not a determinant of newborn behavior. Instead, they suggest that sociocultural and medical factors may work together in affecting pregnancy and neonatal outcome. Continued study of pregnant adolescents who have emotional and social support as well as adequate financial and educational resources for obtaining prenatal care may even out what now appears to be a pessimistic prognosis for teen pregnancy.

### Women Over 35

Older mothers (typically those over 35) tend to have many age advantages over teen mothers, but they exhibit some genetic and medical risks associated with pregnancy and delivery. The statistics on maternal risks for mothers over 35 are less formidable than they were a decade ago because of the large numbers of healthy women in their 30s and 40s who have recently been opting to have babies. Now, rather than emphasizing the risks of childbirth, many advantages are also noted. For example, women over 35 tend to be more content and resolved about their pregnancies, many of which are planned and deeply desired. As Fay and Smith (36) note, many of these older parents are mature, financially secure and able to develop realistic expectations for child care based on life experiences. Such couples may be closer than other younger families and may be willing to commit greater amounts of time and preparation for child rearing at this time in their lives because of previous experiences and successes in the work force.

The primary disadvantage for a mother giving birth when she is over 35 is the risk of chromosomal problems. Older women, for reasons not fully understood, are more likely to give birth to infants with genetic defects, most commonly infants with Down syndrome (see "Birth Defects").

## Maternal Nutrition

As explained under "Development of the Brain", adequate nutrition during pregnancy is critical for the health and development of the fetus. In the 1950s and 1960s, it was customary for obstetricians to recommend that women restrict their weight gain during pregnancy. Nowadays, however, with the recognition of the significance of nutritional demands for both the fetus and the mother, it is not uncommon for pregnant women to gain 3–4 lbs/month, thus gaining 25–30 pounds during pregnancy.

Malnutrition has been studied extensively in the last two decades, and it is difficult to judge long-range effects because of the additional psychological and social detachment that accompanies severe malnutrition. The effects of malnutrition on prenatal development will depend upon various factors: its severity and extensiveness over generations (36), whether it occurs during critical periods of neural growth (37), the mother's health, other factors that often accompany maternal malnutrition (teenage pregnancy, drugs, infections), whether nutritional supplements are given after birth or whether the malnourished mother continues to nurse her infant, and the degree and type of malnutrition and environmental stimulation following birth (38, 39). There is evidence that while chronic malnutrition may seriously compromise the developmental potential of some children (40), negative effects in less severe cases can be combatted by well-balanced meals and interactive, environmental stimulation during infancy and childhood (41–43).

## Maternal Stress

Another area that is also very difficult to assess because it interacts with so many other factors is maternal stress. During times of emotional stress for the mother, numerous chemical changes occur. For example, her adrenal glands release adrenalin and other hormones into the bloodstream, and these are eventually transmitted to the fetus (44). Thus, as Thompson first demonstrated in 1957 in a classic study with pregnant rats, maternal *psychological*, and not just physical factors, can affect her offspring.

Maternal stress during pregnancy has been linked to various measures of offspring emotionality. This work, however, has been conducted primarily with animal subjects and is difficult to generalize to humans. Studies of human neonates corroborate these findings, as infants born to stressed mothers tend to cry more (45) and to perform more poorly on neonatal assessments (46) than do controls. Other deleterious effects that have been examined in animal studies but are not clear in regard to work with humans are the potential effects of prenatal stress hormones on the brain and sexual organs of the fetus. As will be explained later in chapter 12, extreme amounts of adrenal hormones during certain periods of neural maturation can affect subsequent sex organ development and behavior in adolescence.

While maternal emotionality can have subsequent negative effects on the newborn, a certain amount of early perinatal stimulation has been found to be positive, at least for neonatal animals. For example, in a number of studies, Levine (46, 47) has demonstrated that stimulation of infant rats may actually enhance maturation of adaptive responses. Similarly, Ader and Conklin (48) found that stimulation of pregnant rats resulted in positive behaviors in the pups; in a series of studies, Denenberg and colleagues (49, 50) noted that the experiences of mother rats during pregnancy can have positive effects on birth weight and activity levels of their grandchildren.

It is not known what causes these effects, e.g., stress-mediating systems such as the adrenal glands may be changed; hormones produced prenatally by the mother may alter fetal brains; or maternal stress may affect milk supplies or other nutritional needs of the fetus (49–51). Regardless of the causes, these findings point to what Denenberg (50) called a "non-genetic transmission of information" as well as to potential positive effects on their offspring of some forms of stimulation administered to pregnant animals. These data must be interpreted cautiously, however, because Fride and colleagues (52) have reported the opposite findings. They found that prenatal stress actually increases the vulnerability of female offspring to stressful situations in adulthood and have therefore suggested that researchers pay attention to the following: the types and times of stressors that are used, the kinds of outcome measures that are examined, and the type of organism that is studied.

While it is not suggested that direct evidence from animal studies be applied to humans, such work regarding prenatal stress points to possible areas for continued exploration. Clearly, it is difficult to separate a human mother's level of emotionality and stressful experiences during pregnancy from her subsequent attitude about that pregnancy and her behavior toward her infant. Mothers experiencing stressful divorces or severe psychological losses during pregnancy will have different interactions with their infants than will mothers who experience a manageable level of stress to which they have accommodated within the context of their household, career, or everyday living responsibilities. Assessing either short-term or long-term consequences of prenatal stress is difficult, although it is presumed that any excess stimulation is not beneficial for the mother or for the fetus. Continued studies may clarify if there can be "positive" stressors as well as delineate the types of factors that may be deleterious to the pregnant mother or her offspring.

## Infections and Diseases

Various bacteria and viruses can cross the placenta and cause a variety of different effects on fetal development. For example, if the mother contracts German measles (rubella) during the first trimester, the fetus is at risk for severe sensory, neural, or cardiac damage, depending upon what systems are developing at the time the infection is acquired. The types of problems associated with rubella are numerous and have been clustered together under the name *rubella syndrome*. A *syndrome* is a group of related problems, which in the case of rubella include, among others, deafness, heart defects, mental retardation, cataracts, and deficits in immunological systems. There is clearly a sensitive period for the contraction of rubella, as some data indicate that 50% of the fetuses that survive the mother's infection during the 1st month of development exhibit some form of damage. This contrasts with only 20% of the babies who exhibit defects if they contract the infection during the 2nd month (10).

Numerous other infections may have similar effects on organogenesis or may alter structural or behavioral development in various ways. For example, syphillis bacteria affect the fetus primarily after the 16th–18th week of gestation. After this time, the bacteria may cause lesions (literally holes) that destroy the eyes, skin, or mucous membranes. Other infections cause damage during the time of birth. Herpes simplex virus, for example, frequently affects the cervix of the infected mother. Although this infection is often dormant, it may be activated in about 15% of women during pregnancy; their infants may contract the infection during passage through the cervix at birth. In these cases, delivery other than vaginal, e.g., cesarean section, may be conducted. Those infants that do contract herpes simplex may die or may exhibit sensory, intellectual, or motor deficits later in development (10).

Although there are varying effects of different infections, the rule seems to be that the earlier in embryological development the infection is passed on, the more destructive it may be. Clearly, not all mothers who are infected with viruses or bacteria pass them on to their fetuses. Viruses, which are smaller than bacteria, have a greater chance of crossing the placenta, although some bacteria (e.g., syphilis) are also able to do so. Many women are immune to different infections, and others can eliminate needless risks during pregnancy by receiving inoculations against diseases for which they do not have immunity. In the case of syphilis, diagnosis is often recognized before pregnancy because of mandatory blood tests required for marriage certificates. In other infections, good prenatal care and maternal health often eliminate the occurrence of needless problems.

## Maternal Drugs

Prenatal effects of various maternal drugs have been examined over the years, but conclusions about their effects are hampered by the numerous variables that must be considered. Many women who take illicit, prescription, or over-the-counter drugs may tend to consume a variety of different substances during their pregnancies. In addition, they may have other problems such as malnutrition, stress, and infections that make it difficult to point to one specific variable as "the" potential offending source. In addition, as will be explained in more detail later, many drug effects on the fetus may be long-range, subtle behavioral or intellectual deficits that are not manifested until later on in childhood (53).

Other methodological problems associated with prenatal drug investi- gations concern the longitudinal studies that are necessary for exploring long-range effects. Such studies are costly and difficult to control. Once these studies are conducted during postnatal periods, it is most difficult to distinguish environmental interactions that occur between possibly drug-addicted mothers and their children. Another methodological problem is that most of these studies are "correlational", meaning that if negative effects are found, they can only be *related to* prenatal maternal drug consumption. And finally, illicit drugs of choice tend to change rapidly. This means that studies will not be "pure" and that long-range studies of new drugs such as the cocaine-derived "crack", popular in the mid-1980s, must be forthcoming.

### Thalidomide

For some drugs, clear-cut effects have been found. For example, as a result of thalidomide consumption in the 1960s, it is now widely recognized that drugs that may be good for the mother may have untoward side effects for the fetus. In the 1960s, thalidomide, a widely used tranquilizer that could be purchased over the counter in Europe was subsequently found to cause severe physical defects if taken during prenatal sensitive periods of limb formation. Children born to such mothers had a variety of deficits, such as missing ears, and malformed arms or legs. There were severe effects or no effects at all, depending upon when the drug was taken and which organ system was developing at the time it was taken.

### Nicotine and Alcohol

Studies of the effects of maternal smoking and alcohol consumption on fetal development are inconsistent. This is due in part to the fact that many women who smoke may also eat less and may drink substances such as coffee and alcohol more than mothers who do not smoke. Nicotine is known to constrict maternal and fetal blood vessels, thereby reducing the amount of oxygen and nutrients transported to and by the fetus. In addition, nicotine speeds up maternal and fetal heart rate, and increases the amount of carbon monoxide in the maternal and fetal circulatory systems. The overall effects of maternal smoking may be malnutrition and oxygen deprivation, placental abnormalities, low birth weight, prematurity, death before or around the time of birth, or behavioral or intellectual deficits in childhood (10, 35, 40).

Like nicotine, alcohol also enters the maternal and fetal bloodstreams and therefore must be processed by the immature fetal liver. This area, however, is also difficult to study. Mothers who drink large quantities of alcohol may also have nutritional deficits, different attitudes toward their pregnancies, and different prenatal care when compared with mothers who are not heavy alcohol consumers. The degree of deficits associated with alcohol consumption clearly depends upon the quantity and duration of the mother's drinking. A recently identified complex of problems called *fetal alcohol syndrome* (FAS) is seen in infants born to mothers who are chronic alcholics. This syndrome is characterized by visual, joint, and cardiac abnormalities; retarded growth and intelligence; impaired motor development; emotional withdrawal; and abnormal cranial or facial features (54). Streissguth and colleagues (54) note that although FAS is receiving considerable attention lately, many other adverse effects of maternal alcohol consumption can be found, e.g., higher rates of neonatal mortality, low birth weight, or other neurological problems. Alcohol consumption has been found to have direct effects on fetal behavior, causing reduced fetal breathing movements. In addition, a sensitive period may exist for heavy alcohol consumption during the last trimester, when brain growth is very rapid. Neonates born to mothers who reduced their alcohol consumption during this time tended to be larger than the infants from women who continued to drink (40). The effects of moderate alcohol consumption are unclear.

Many other variables that interact with maternal alcohol consumption during pregnancy are currently under investigation. These are referred to as "prepregnancy factors" and include three different variables: potential genetic effects of the father's heavy drinking, previous heavy drinking on the part of the mother, and a genetic susceptibility of the offspring in regard to later alcoholism. The interaction between genetics and subsequent alcoholism is currently under study and poses a potential explanation for variations in different individuals' tolerance for the substance (10).

### Illicit Drugs

A more serious consideration is maternal consumption of illicit, or street, drugs ranging from hallucinogens (LSD), to amphetamines ("uppers"), heroin, and cocaine. In two separate reviews of maternal heroin addiction, Davidson and Short (53) and Householder and colleagues (55) indicate serious long-range consequences for the offspring of heroin-addicted mothers. One alarming concern during the perinatal period is the fact that fetuses, themselves, may become addicted and must then undergo withdrawal at birth. In addition, even though some of these mothers are no longer taking heroin, they may be receiving a synthetic narcotic, methadone, which also is sold illegally on the streets and may also cause negative effects for the pregnant user and her fetus.

Examining prenatal and long-range developmental effects of maternal drug abuse is particularly difficult because of the numbers of confounding variables. Some of these, Davidson and Short point out, include difficulties in controlling for purity and dosages of drugs used; variability in maternal health and prenatal care; multiple drug abuse; difficulty maintaining follow-up with many of these mothers for purposes of longitudinal studies; interaction between drug effects and socioeconomic status, nutrition and similar factors; and the numerous psychological problems that are also associated with drug abuse (53).

In addition to drug addiction and withdrawal found among infants born to heroin- or methadone-addicted women are many other perinatal problems, including high infant mortality, prematurity, low birth weight, accelerated liver and lung development, brain damage, and difficulties with labor (53, 55). Subsequent developmental problems during infancy and childhood are also common and include motor, cognitive, speech, and learning deficits; developmental delays; reflex deficits; and interrupted interactions or attachment behaviors between parents and infants. (The implication of interrupted attachment will be discussed later in this chapter.) Drugs other than heroin and methadone have also been studied. For example, there is some question whether LSD causes genetic damage in the offspring of users. One study linked chromosomal abnormalities to the purity of the substance, which, when obtained on the street, is always questionable (56). It appears that, with the exception of genetic defects with LSD, the effects of drugs such as marijuana, heroin, and amphetamines may be limited to altering fetal, neonatal, or infant *behavior* or *function* as opposed to causing serious structural damage or birth defects (57). The potential deleterious effects of other drugs used during the birth process (e.g., anesthetics) will be examined later under "Birth Process".

### Other Substances

The effect of maternal caffeine intake on fetal development is less understood than alcohol or other drugs. Caffeine has been found to pass into the fetal blood system. Obviously, there is greater risk for fetuses exposed to large amounts compared with those fetuses whose mothers have a moderate caffeine intake; these risks included stillbirth and prematurity (35). Numerous other substances have come under recent scrutiny because of their potential prenatal destructive effects or because they may be carcinogens. For example, concerns have been expressed about stillbirths, miscarriages, and childhood cancer found among families that have been exposed to hazardous wastes or to toxins such as Agent Orange, a chemical defoliate used during the Vietnam War to clear dense foliage so that the enemy would be more easily spotted from the air. Some veterans exposed to this substance have become ill from different cancers, while some of their children have exhibited a variety of birth defects or later developmental problems. Other concerns have been expressed about prenatal problems as a result of exposure to toxins in the workplace, pesticides, food additives, lead poisoning, aerosols, video display terminals in computers, and the potential skin absorption of chemicals and hormones in cosmetics and hair dyes (35). Since many of these substances have only been developed recently, research on their potential harmful effects is only tentative but questionable enough that pregnant women should be cautious in their exposure to them.

### Radiation

An additional grave concern under increased scrutiny is the effect of radiation on embryo and fetal development. It has been known for decades that irradiation can cause chromosomal damage or affect embryo or fetal structural development. The central nervous system is particularly at risk, and irradiation of the fetus can result in microcephaly (small skulls) or other brain malformations, mental retardation, and other potential long-range deficits. Data about these effects have been obtained from women who were pregnant and near the drop sites of the atom bombs during the

Second World War. From studies of these and other women exposed to radiation, it has been found that the extent of the damage to the fetus depends upon the stage of pregnancy in which the exposure occurs as well as the quantity and the duration of exposure (58).

## Conclusion About Environmental Effects

Clearly, the negative consequences of many of these prenatal factors have far-reaching implications for physical and occupational therapists as well as for society. It should be pointed out that while this section has appeared pessimistic in tone, such is not the case in terms of statistics about development. Based upon observation of the numbers of healthy people around us, the prenatal period is not fraught with severe, hopeless hazards that will interfere with normal development. On the contrary, prenatal interruptions are the exception and not the rule. However, we do know that some prenatal factors can be deleterious, and developmentalists and clinicians, should be aware of the potential negative effects of these substances. As Spezzano remarks in a review of prenatal stress:

> All the evidence to date encourages parents to avoid "pregnancy paranoia" in which they begin to fear that every wrong move they make will doom their child to an abnormal future.

> It appears more productive, therefore, to turn our attention to the larger number of children who appear to develop normally despite early problems, whatever the cause. The prenatal period is merely one stage in the total life span of an individual, and infants show great plasticity. Each new developmental stage brings reorganization of past experiences and the chance for neutralizing or aggravating the effects of previous traumas (59, pp 56, 57).

## Roles for Therapists

Infants who do exhibit many of the problems discussed above are likely at some point to be referred for developmental services. They may be seen in a neonatal intensive care unit, in a preschool screening program, in a school program, or in an outpatient clinic. The roles for occupational and physical therapists with these infants and children will be to assess the extent of their deficits and to recommend and initiate various types of treatment in cases where it is warranted.

Beyond the direct therapeutic role, there is an additional preventive role for those in social services, educational, clinical, and health care fields. Many of the deaths or birth defects caused by prenatal hazards can be eliminated by educating pregnant women or prospective parents about the need for adequate health, nutrition, self-care and prenatal care. However, as Spezzano (59) pointed out in the previous quote, the prenatal period is only one part of the entire life-span. Thus, the type of intervention that will be most effective is one that addresses people's needs throughout life, that supports families and promotes parenting skills, and that educates individuals about proper child care, not just during the prenatal period but throughout childhood and adolescence. (This topic will be discussed in greater detail later in this chapter after additional risk factors are noted.)

Genetic defects and problems encountered during the birth process are other areas where developmental problems may occur and which concern therapists. These two areas are discussed next.

# Genetics and Developmental Risks

While some developmental risks occur as a result of clear-cut environmental problems, many are a result of genetic-environmental interactions. However, before these are introduced, it is necessary to review some basic concepts and terminology from the field of genetics. Some of these concepts were introduced in Chapter 2, thus the following section will expand on the information already introduced there.

## General Principles of Genetics

The germ cells (i.e., the ova or sperm) are a product of *meiosis*, also called *reduction division*. Thus, the ova or sperm contain only half the amount of genetic material necessary for a human organism. When these two germ cells unite during fertilization, their 23 chromosomes combine to form a zygote with the normal complement of 23 pairs. Segments of DNA along the chromosomal strands are termed *genes*, and it is these that contribute to specific characteristics of individuals. Genes are typically located in certain positions (called *loci; locus*, singular) on specific chromosomes. Since one of each pair of chromosomes is inherited from each parent, the new infant will inherit genes, and therefore certain traits, from each parent.

We know from casual observation as well as from scientific study that it is the nature of characteristics that some will overshadow others. For example, some children tend to look more like one parent or relative than another. This tendency of one characteristic to express itself over another is termed *dominance*. The opposite of dominance is *recessivity*. Only during certain circumstances will these subordinate, or *recessive*, traits be fully expressed. (This will be discussed later in more detail in this section.) Scientific studies in the field of genetics have shown us that actual gene expression is a blend of dominant and recessive characteristics. An individual's external appearance, termed the *phenotype*, is therefore a result of a combination of influences from the environment as well as his or her actual genetic composition, called the *genotype*.

People have different kinds of chromosomes in their cells. The *autosomes*, of which there are 22 pairs, are the regular body chromosomes. The remaining set, called the *sex chromosomes*, is partly responsible for determining an individual's gender. Human males possess 22 pairs of autosomes and two sex chromosomes, an X and a Y. This XY combination of sex chromosomes is what makes an individual a genetic male. In contrast, females possess 22 pairs of autosomes and two X sex chromosomes. Thus, the XY genotype normally results in a male phenotype, and the XX genotype results in a female phenotype.

## Birth Defects

Many genetic problems result in what are collectively linked together under the term "birth defects." Howell defines this term in this manner:

> A problem or disability is referred to as a birth defect if it is present at birth and is a disorder of body chemistry of other body function (including mental function), a malformation, or a defect of structure. At one extreme the defect can be inconsequential, with no disabling aspects. At the other it can be severely disabling, affecting multiple body systems and functions. Birth defects often have profound effects not only on the infant or child but on the lives of others involved (60, p 235).

### *Reasons for Birth Defects*

The reasons for many birth defects are unclear. Those with known etiologies occur as a result of specific environmental (20%) or genetic (20%) causes or as a result of their combined effects (60%). Some of the more clear-cut environmental problems (such as viral or bacterial diseases, maternal malnutrition, maternal drug addiction) have already been discussed. In this section, inherited or combined environmental-genetic birth defects will be examined.

Genetic defects are associated with either the autosomes or the sex chromosomes. Inherited birth defects are those that are actually passed on to the offspring by the ova or sperm of one or both parents (60). Other, genetic or environmental-genetic interactions that result in birth defects may be due to a variety of causes. Some genetic abnormalities are a result of *mutation*, which is a mix-up of the protein that comprises the genetic material. A mutation may occur without cause; i.e., randomly, during cell division, or it may occur as a result of environmental events such as exposure to radiation or certain drugs. Radiation actually garbles the genetic information so that cells typically cannot survive. If, however, the cells responsible for forming eggs or sperms are irradiated and survive, the garbled genetic program may be passed on to offspring (61). Similar effects result from *translocations*. A translocation occurs when two chromosomes split and recombine with each other rather than back in their original pair. Such a combination will not affect the organism but can have severe consequences for the offspring to which the translocated chromosomes are passed (62).

Thus, some of the potential reasons for genetic defects include spontaneously occurring translocations that are passed on to progeny; random chromosomal mutation that may occur during meiosis, resulting in inadequate or too much chromosomal material passed on by the ova or sperm; random mutation or error that occurs when the zygote is first formed and begins cell division; mutations as a result of irradiation or drugs; problems with the ova or sperm of older individuals, resulting in some genetic anomaly; and many other problems—some of which are understood and some not.

Most severe chromosomal anomalies result in organisms that do not survive. For example, Milunksy (63) reports that about 90% of the recognized conceptions that have some chromosomal abnormality are spontaneously aborted; Zellweger and Simpson (62) note that the earlier a spontaneous abortion occurs, the more likely some kind of chromosomal abnormality is found. Yet despite the widespread loss before birth of these abnormal genetic combinations, there are still many different types of birth defects that can be found in surviving infants. For example, in 1978, Howell noted that of the 3 million babies born in the United States each year, about 250,000, or 8%, are born with defects. She further estimates that at any one time, about 15 million residents of the United States have one or more birth defects (61).

Similar problems are found across the wide variety of birth defects. For example, mental retardation is the most common birth defect, as it accompanies many different syndromes. Other common deficits are visual or auditory impairments and metabolic disturbances that if untreated may lead to cognitive deficits or even death. Other environmental or genetic problems may result in limb deformities, malformation of the kidneys or heart, or other structural abnormalities that lead to subsequent physical or functional deficits. As was discussed in regard to environmental defects, many of the structural problems arise as a result of some interference during organogenesis (Fig. 7.2).

The following section illustrates how one kind of birth defect may be transmitted from one generation to another. It is included for two major reasons: to introduce some of the principles of inheritance and to describe how birth defects may arise as a result of a genetic condition. The condition described below occurs as a result of an autosomal inheritance. Other types of defects associated with sex chromosomes will be described later.

### An Example of a Genetically Determined Birth Defect

*Osteogenesis imperfecta*, which means imperfect development of the skeleton, is a general skeletal deficiency characterized by bones that are soft and fragile and therefore subject to frequent fractures and deformities (64). This condition is typically autosomal dominant and is attributable most likely to spontaneous mutation. How this happens is not completely understood, but it is believed that the gene that carries the dominant trait for osteogenesis imperfecta appears spontaneously in a sperm, ova, or in the zygote.

The genetics of such a condition are described in the following example: As Figure 7.3 illustrates, neither the father nor the mother have osteogenesis imperfecta. Both their genotypes are recessive (OO) for the condition; however, at some point during cell division a random mutation occurs, resulting in a gene that carries the dominant trait that is responsible for the degenerative skeletal condition. Thus, one half the germ cells of one partner will contain the dominant gene, and the other half will remain recessive. The potential offspring of this couple with have a 50% chance of inheriting the dominant, mutant gene and developing osteogenesis imperfecta.

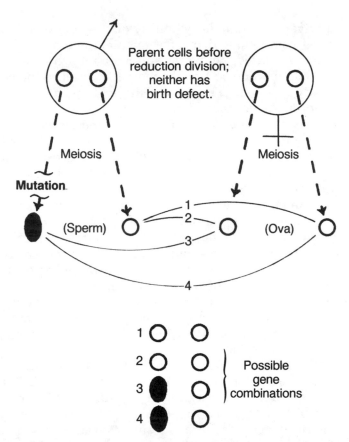

**Figure 7.3.** Possible genetics of a birth defect. As a result of a random mutation during meiosis, a father, who does not have a birth defect, ends up with a 50% chance of transmitting it to his offspring. Both parents are homozygous recessive (∞) for the genetic disorder, but the mutation produces a dominant gene (●) carrying the birth defect. There is a 50% chance that the offspring will be homozygous recessive and not inherit the disorder and a 50% chance that the offspring will be heterozygous (●○) and will inherit the birth defect.

Now consider what would happen if a spontaneous mutation occurred during meiosis of both the sperm and the ova. The potential zygotes that could result from the union of these germ cells are illustrated in Figure 7.4, which indicates that there is a 75% chance that the offspring will express osteogenesis imperfecta. Note that there are two genotypes containing genes that are identical to one another: (●●) or (∞). These are called homozygous (from *homo* meaning "same" and *-zygous*, referring to zygote). To differentiate these two, the genotype with the dominant trait is called *homozygous dominant*, whereas the recessive trait is called *homozygous recessive*. In contrast, ●○ genotype is referred to as *heterozygous* (from *hetero* = different). The phenotypes of these individuals will vary. Those with the heterozygous (●○) or with the homozygous dominant (●●) combinations will have osteogenesis imperfecta. Those with the homozygous recessive (∞) combination will not end up with the

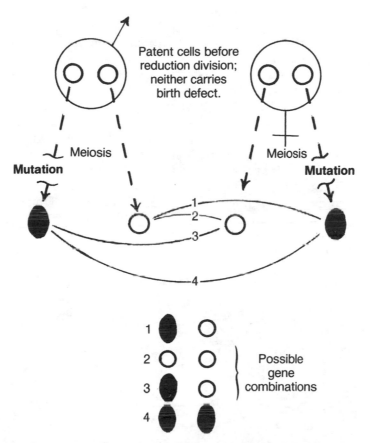

**Figure 7.4.** Possible genetics of a birth defect. Random mutations have occurred during meiosis involving both sperm and ova. Now both parents produce half their germ cells with the mutated dominant gene. Both the homozygous dominant (●●) and heterozygous (●○) conditions, which have a 75% chance of occurring, will result in inheritance of the birth defect. There is a 25% chance that the potential offspring (homozygous recessive (○○)) will not inherit the disorder.

degenerative condition. *It is the nature of dominant traits that they express themselves in heterozygous combinations. Recessive characteristics express themselves in homozygous combinations.*

## Autosomal Birth Defects

Osteogenesis imperfecta is an example of a birth defect caused by a gene located on one of the autosomes. This is not the only kind of autosomal problem, however. Other problems occur when there is an abnormal amount of chromosomal material, as in *trisomy*, the most frequently encountered autosomal condition. Trisomies result in an additional chromosome beyond the normal pair.

A familiar example of trisomy occurs with the 21st chromosome; this is referred to as trisomy 21, or more commonly, Down's syndrome. It most often occurs when an extra, or parts of an extra, chromosome appear spontaneously in the sperm, egg, or zygote during cell division. Down's syndrome is notable because it is the most common form of mental retardation. If affects the total development of the individual, resulting in altered facial features (which is why it is sometimes referred to as "mongolism"), mental retardation, delayed motor development, and sometimes other functional or structural problems such as heart problems (62, 65) (Fig. 7.5).

Down's syndrome is also notable because the condition is linked to the mother's age. A high proportion of infants with Down's syndrome are born to mothers who are 35 years of age or older. Thus, it appears's that some age-related events may alter the older mother's ova or her corresponding older spouse's sperm such that trisomy 21 is more apt to occur (63). Although Down's syndrome occurs as a result of a genetic problem, it is not necessarily

**Figure 7.5.** A child with trisomy 21, also known as Down's syndrome. (Courtesy of March of Dimes, Birth Defects Foundation.)

"inherited" from a parent. Thus, Down's syndrome, like radiation, provides an example of a potentially environmentally related (or caused) genetic condition since it is believed that some environmental effect alters the germ material to include an extra 21st chromosome (60).

### Genetic Disorders Associated with the Sex Chromosomes

In some cases, a birth defect may be linked to a sex chromosome instead of an autosome. Just as with autosomes, there may be specific gene-linked disorders or there may be syndromes associated with an abnormal quantity of chromosomal material. The gene-linked disorders are typically referred to as "X-linked", because they are associated most often with the X sex chromosome. Milunsky (66) notes that there are relatively few sex chromosome disorders and they are typically not associated with fatality or severe mental retardation. In contrast, there are about 200 X-linked disorders, some of which may result in serious effects.

Because males and females have different sex chromosomes, X-linked disorders can have differential effects for the sexes. The Y chromosome contains very few genes. Thus, for a male, the majority of genes associated with the sex chromosomes will be located on the X. A male receives only one X chromosome and few genes on the Y, thus many recessive conditions that normally would not be expressed will be evident in males (Fig. 7.6). Since females possess two X chromosomes, a single recessive condition will not be expressed with them. For the same recessive characteristic that

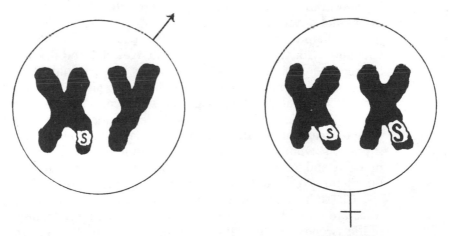

**Figure 7.6.** Males are much more susceptible to inheriting X-linked disorders because they lack gene locations on their Y chromosomes. Disorders which are carried by a recessive gene (labeled "s") are *normally* only expressed in the unlikely homozygous case (ss). Males, however, because of their XY gene combination, can inherit a genetic condition with only one recessive gene on their single X chromosome. In contrast, females have two X chromosomes. Thus, if they inherit one recessive gene on an X chromosome, they are often protected by a dominant gene (S) on their other sex chromosome. The heterozygous (Ss) condition, which can only occur in females, will not result in a genetic disorder.

males may exhibit with a single gene, females will require the less common homozygous recessive condition. Males, therefore, are more likely to exhibit sex-linked diseases or conditions than are females (67).

Other important implications regarding sex-linked problems are relevant for both parents. For example, if a father possesses an X-linked problem, he will transfer that X chromosome to all his female progeny but not to any of his sons because they inherit only Y chromosomes. Females, on the other hand, are typically *carriers* of genetic disorders. This means that, for a recessive trait associated with the X chromosome, the female may often carry the gene but not express it. In heterozygous combinations, females may actually possess the gene for some defect on one X chromosome but not exhibit the defect because it is blocked by another dominant gene on the second X chromosome. If the mother is a carrier of a birth defect on one of her X chromosomes, there will be a 50% chance that her sons will inherit the birth defect and a 50% chance that her daughters will also be carriers (67).

## X-Linked Problems

Two familiar X-linked genetic problems illustrate the variety and destructiveness of these conditions. Both are found significantly more often, if not exclusively, in males. One of these is the blood coagulation disorder known as *hemophilia*. The other is *Duchenne's dystrophy* (also called *pseudohypertrophic muscular dystrophy*), the most common serious X-linked disease in humans. Boys with this birth defect are often seen and treated by occupational and physical therapists.

Duchenne's muscular dystrophy occurs in 1 of 3000 male births. It becomes evident in boys between the ages of 1 and 6 years and is characterized by muscle wasting and weakness. It is gradually progressive, affecting first the proximal muscles of the pelvic girdle and later the shoulder girdle. The result is altered gait and eventual wheelchair mobility. Gradually, with continued weakness, cardiorespiratory problems emerge and may lead to death at 10–15 years after onset (68). Physical and occupational therapists who work with such individuals often focus on wheelchair mobility and on promoting strength and independent functional skills while taking into consideration the progressive muscle weakness.

Classic hemophilia is also an X-linked disorder, but its clinical symptoms are quite different from those of muscular dystrophy. Classic hemophilia is one of many different types of coagulation disorders and affects about 1 of 120,000 males in the United States and Europe (66). In some it goes undiagnosed until an individual experiences an injury or surgery where bleeding becomes uncontrolled; in other cases, it is previously recognized because of family history. Hemophilia can be accompanied by mild to severe deficits. Those with mild bleeding problems may lead a normal life but may seek advice from a health professional for maintaining muscle mass to prevent internal injuries. Those hemophiliacs with severe deficits are uncommon; such individuals may bleed spontaneously or as a result of emotional or physical trauma. Clearly, their life-styles must be monitored and restricted to activities compatible with a minimal amount of emotional and physical stress.

Duchenne's muscular dystrophy and classic hemophilia are only two examples of approximately 200 X-linked defects. Many of these disorders can be treated, and many can be prevented if individuals are aware of their family histories and previous genetic disorders. Later in this chapter, continued discussion will examine the topic of genetic counseling and prenatal diagnostics to prevent or identify genetic risk conditions.

## Sex Chromosome Disorders

The number of sex chromosomal problems are much fewer than X-linked disorders. Problems with sex chromosomes typically result when there are too few or too many X or Y chromosomes. Two such disorders are Turner's syndrome and Klinefelter's syndrome.

Turner's syndrome, also called "monosomy X," is a low-incidence sex chromosome anomaly that results when only one X chromosome is present. This genotype is represented as XO, X', or X-. In this case, the individual is a female, who is often short and stunted in stature, has undeveloped ovaries, may exhibit edema (swelling) in the hands and feet in infancy, and sometimes will display cardiovascular and kidney deficits. Contrary to earlier reports, intelligence is often within normal limits in individuals with Turner's syndrome although their verbal measures tend to exceed performance IQ. In addition, there may be visual-spatial-perception problems. While the actual characteristics of the syndrome vary from case to case, such females fail to develop in puberty. Thus, they tend to be sterile, and unless they are administered hormones, lack breast and hip development, pubic hair, and menstrual cycles (62, 69).

While Turner's syndrome results from a lack of a sex chromosome, Klinefelter's syndrome is a result of too many chromosomes; i.e., two or more X and one or more Y chromosomes. The genotype is typically XXY, and more rarely XXXY, XXXXY, XXYY, or XXXYY (64). The general rule is, the more extra chromosomes that are evident, the more severe the intellectual deficits associated with this disorder; but like Turner's syndrome, there is considerable variability from case to case. Sometimes normal or superior intelligence is found. Affected individuals are nonreproductive males, with small testes and other problems such as epilepsy, tremors, breast cancer, and endocrine or pulmonary problems. In most cases, identification of Klinefelter's syndrome goes undetected until adolescence, at which point these individuals often achieve above-average height, may have previously exhibited language problems, and may also tend toward apathy or flat emotional responses (62).

Other sex chromosomal problems include XYY males. This anomaly has received considerable attention because of studies that indicated a high incidence of this genetic combination among institutionalized, specifically prison, populations. While at one time it was thought that the added Y chromosome was associated with aggressive or antisocial behavior, this seems to be an overgeneralization that has not held up to close scrutiny from other investigations (70).

### Summary

Birth defects can occur for various reasons, such as autosomal or sex chromosome disorders as well as the previously discussed maternal and environmental risk factors. As was discussed in regard to environmentally caused deficits, the extent and severity of some birth defects can be controlled to some degree. Since the mid-1960s, numerous sophisticated prenatal diagnostic and genetic tests have been developed, and these offer a partial solution for limiting or for early diagnosis of many birth defects. Some of these tests will be discussed in the following section.

# Prenatal Diagnostics

In the 1970s and 1980s, numerous methods were developed and refined to assist physicians evaluating prenatal risks. These range from methods used to actually visualize the fetus to methods used to obtain fetal blood or skin samples. Two relatively familiar procedures are *amniocentesis* and *ultrasound*. Other procedures such as *fetoscopy* and *fetal blood sampling* as well as *chorionic villi sampling* are recently developed and therefore less studied.

## Amniocentesis and Ultrasound

Amniocentesis and ultrasonic procedures are used together in fetal diagnostics. Amniocentesis is the process of actually puncturing the maternal abdominal wall, entering the amniotic sac, and drawing out fluid that contains fetal cells (Fig. 7.7). The cells then can be examined for chromosome composition. Thus, amniocentesis can be used to confirm or verify the presence or absence of suspected genetic (autosomal or sex chromosome) disorders. This procedure has become routine for women at risk for certain genetic problems because of their age or because of their own or their husband's family history of previous genetic abnormalities. The procedure is done in an outpatient clinic with local anesthesia, but it cannot be done until the 14th–16th week of gestation, at which time there should be sufficient fluid for taking a sample (71).

Some risks occur with amniocentesis, although there are indications that as technology improves, the risks decline. Risks involve bleeding and the possibility of mixing fetal and maternal blood, which can lead to birth complications (see "Blood Incompatabilities). Other problems arise when using amniocentesis with a twin pregnancy; diagnostics may also be compromised. For example, sometimes it is unclear if fetal or maternal cells have been sampled. Thus, if an XX genotype is obtained, it is questionable whether this was obtained from the maternal or from fetal cells. Another risk of amniocentesis is the actual puncture of the fetus; however, this problem is diminished by using ultrasound to locate the fetus before the needle is inserted and while the procedure is being conducted (62, 71).

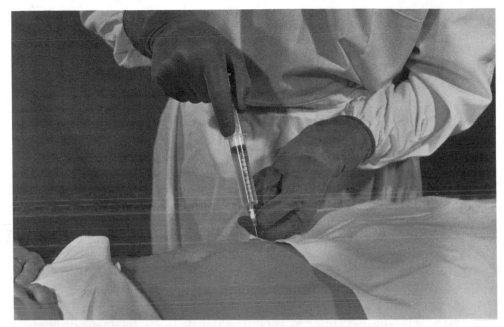

**Figure 7.7.**  Amniocentesis as a prenatal diagnostic procedure. (Courtesy of the March of Dimes, Birth Defects Foundation.)

**Figure 7.8.**  Ultrasound as a prenatal diagnostic procedure. (Courtesy of the March of Dimes, Birth Defects Foundation.)

Diagnostic ultrasound is the process of transmitting low-intensity sound waves through the mother's abdomen and down through maternal and fetal tissues. The sound waves are reflected back, picked up, and processed into visual signals that can be photographed and visually recorded (Fig. 7.8).

**Table 7.3**
**Uses of Ultrasound**[a]

| |
|---|
| Enhance safety and effectiveness of amniocentesis |
| Identify multiple pregnancies |
| Determine fetal age |
| Detect abnormalities in fetal growth rate |
| Detect structural abnormalities in fetus |
| Visualize the site of attachment of placenta |
| Visualize the position of umbilical cord |
| Visualize the position of the fetus |
| Detect changes in volume of amniotic fluid |
| Visualize pockets of amniotic fluid for safe amniocentesis |
| Identify tumors |
| Observe and obtain norms of fetal behavior and development |

[a] Data from Mahoney MJ, Hobbins JC: Fetoscopy and fetal blood sampling; Campbell S: Diagnosis of fetal abnormalities by ultrasound. In Milinsky A (ed): *Genetic Disorders and the Fetus: Diagnosis, Prevention, and Treatment.* New York, Plenum Press, 1979.

Ultrasound has no known risks, and it can be used for a number of important reasons (Table 7.3), one of which is the assessment of fetal age (72). The accurate definition of fetal age is essential for safe and effective amniocentesis. Thus, with ultrasound a variety of measures can be used to judge the size of the fetus, either in length from crown to rump or in head or abdominal circumference. These measures can then be used to gauge the actual age of the fetus. In addition, since ultrasound enables one to actually visualize external and internal structures, it can be used to diagnose structural birth defects (72).

Ultrasound is also used diagnostically in regard to fetal behavior. As research data are compiled, norms are obtained about fetal motor development. With these norms, physicians are able to look for developmental milestones in prenatal behavior. For some mothers who are concerned about fetal inactivity, ultrasound may be used as reassurance that the fetus is exhibiting normal behavior patterns. In other cases, it may be used to signal developmental deviations and potential complications during delivery. Ultrasound may then also be used during the actual delivery if potential problems are suspected or actually arise (12, 73, 74).

A limitation of ultrasound is that a three-dimensional object is actually translated into a two-dimensional image, so some visualization is lost. Visualizing small structures such as fingers, toes, and genitalia is also limited. Thus, if there is suspicion of small external defects that may signify other serious anomalies, then other, more invasive diagnostic procedures can be used. One of these other procedures might be amniocentesis. However, some defects must be explored with the use of a diagnostic procedure other than biochemical analysis of the amniotic fluid. In that case, fetoscopy may be recommended (72).

## Fetoscopy and Fetal Blood Sampling

Fetoscopy is a relatively new procedure involving the insertion of a telescopic device through the uterine wall. The scope, which is guided by ultrasound, can be used to directly visualize the placenta and the fetus as well as to guide the actual sampling of fetal blood from the placenta or skin. Skin samples are typically removed from the fetal scalp, where circulatory and muscle damage will not result. Once fetal skin and blood are obtained, they can be subject to laboratory examination for potential metabolic or genetic abnormalities. In addition, the fetoscope can sometimes be used to actually visualize some small body parts, such as genitalia or digits. Thus, it may be used for sex determination or for diagnosis of *polydactyly* (the presence of extra digits on the hands or feet). Extra digits are often considered markers for birth defect syndromes. Thus, if these are sighted, continued diagnostics may be used to confirm or deny the presence of other structural and metabolic deficits (75).

Clearly, greater risks occur with this procedure than with the relatively harmless ultrasound. Ultrasound is certainly the method of choice for visualizing internal and external anatomy, however, when it is limited, fetoscopy offers additional information. The complete risks of fetoscopy have not been fully explored inasmuch as it is a new procedure. Mahoney and Hobbins (75) indicate that fetoscopy and fetal blood sampling have other untested potentials. For example, it may be possible at some time to sample fetal muscle tissues for the diagnosis of neuromuscular problems. At present, this has not been done because of concern about causing nerve or muscle tissue damage when the sample is taken. Mahoney and Hobbins also suggest that the technology that underlies fetoscopy and fetal blood sampling may lead to potential therapeutic uses such as fetal administration of drugs, cells, or surgery used to eliminate or correct potential birth defects.

## Chorionic Villi Sampling

Another use of ultrasound is to guide the relatively new diagnostic procedure called "chorionic villi sampling" (CVS). This is conducted by using visualization from ultrasound to guide a catheter through the vagina and cervix and into the uterus. A sample of chorionic tissue is then obtained and analyzed much as amniotic fluid is with amniocentesis. The advantage of this procedure is that unlike amniocentesis, which cannot be performed until the pregnancy is sufficiently advanced to safely draw a sample of fluid, CVS can be performed as early as the 8th–10th week of pregnancy. Thus, diagnostics can be performed earlier, and if termination of the pregnancy is warranted, there is less chance of risk for the mother. Because of its usefulness early in pregnancy and because it does not carry the risks associated with puncture of the abdomen, CVS may come to replace amniocentesis in the near future.

*Summary*

While all the fetal diagnostics discussed here and their subsequent interventions may serve to reduce birth defects, the actual "intrusion" into fetal
development by using these varied methods is regarded by some individuals
as "unnatural." Fetal diagnostics may cause some real ethical and moral
dilemmas that, at a time when technology is advancing so rapidly, society
and the health professions must address.

# Dilemmas Associated with Prenatal Diagnostics

Prenatal diagnostic procedures are used to verify the presence or the
absence of genetic, structural, or metabolic birth defects. For example,
amniocentesis can be used to confirm the presence of trisomies such as
trisomy 21, Down's syndrome. Prenatal blood assays can determine the
presence of a variety of chromosomal disorders including X-linked Duchenne's muscular dystrophy or autosomal problems. Fetoscopy and ultrasound can indicate when physical and structural deformities are present,
and many of the diagnostic procedures can be used to determine the sex of
the fetus.

A real dilemma is faced by the family for whom tests indicate the
presence of such defects. In countries where abortions are legal and in cases
where families' religion or personal ethics do not eliminate abortion as an
option, mothers and fathers are faced with the difficult decision of whether
to terminate a pregnancy or to give birth to a child whom they know may
face severe physical and mental handicaps or even life-threatening or
painful disorders.

While some diagnostics do lead to termination of pregnancy, many
more, physicians point out, are used for preservation of the pregnancy (75,
76). For example, they may be used to reassure families that have chosen
to have children despite a certain probability that a birth defect may occur.
Given the family history and anticipated disorders, prenatal diagnostics can
be used to confirm the absence of clinical signs. This, in turn, alleviates
anxiety and concern on the part of the expectant parents. In addition, other
genetic deficits that are picked up can be used as an early warning to the
mother and to her relatives regarding their potential risks as carriers of
such problems as X-linked disorders (75). Milunsky (76) points out that no
prenatal diagnostic procedure should be conducted without thorough communication with the family members involved. This communication process is known as genetic counseling.

## Genetic Counseling

Before genetic diagnostics are conducted, clear communication should
occur between the health professional and the family members involved
(Figs. 7.9 and 7.10). The purpose of the communication process is multifold:
to discuss the potential occurrence and risks associated with a genetic

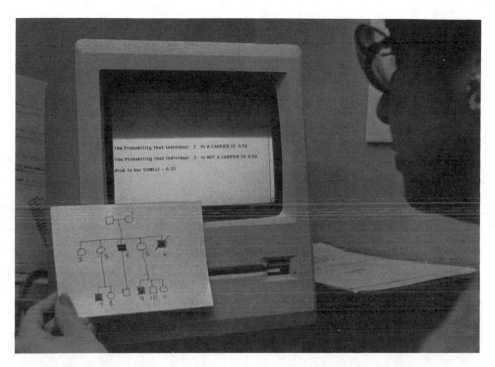

**Figure 7.9.** Genetic counseling involves determining the potential occurrence and risks associated with a genetic disorder within a family. (Courtesy of the March of Dimes, Birth Defects Foundation.)

**Figure 7.10.** Genetic counseling is a clear communication between health professional and family. Such counseling both informs and supports the family. (Courtesy of the March of Dimes, Birth Defects Foundation.)

disorder within a family, to provide the family with an understanding of the nature of the disorder that the child may inherit, to inform them of the possible options available if they are at risk for genetic problems, and to assist and support them in making a final decision about whether to have children, to adopt, or to terminate a pregnancy (76).

Clearly, genetic counseling should occur prior to the parents' decision to have children, especially for couples with the following problems: women of advanced age; women who have previously given birth to a child with a serious birth defect; parents with a history of genetic disorders; and parents who know they may be carriers of genetic diseases. Milunsky (76) notes that genetic counseling not only apprises the family members of their potential genetic risks but also informs them of the risks and limitations of diagnostic procedures that may be used.

The moral and ethical dilemmas faced by individuals with genetic or pregnancy-related problems are great, and counseling should not only be educational but also helpful and supportive (76). In a recent study of the reactions of women who chose to have abortions because of fetal malformations, Jorgensen and colleagues (77) noted that all the prospective mothers experienced psychological trauma, which in some cases lasted an extended period of time. The authors suggest that physicians be sensitive to these problems and be prepared to provide active psychological support for women experiencing such pregnancy-related difficulties.

### Genetic Engineering

A recent area of biomedical research, genetic engineering, may at some time reduce or eliminate many birth defects. Genetic engineering is the actual manipulation of genes, resulting in new combinations. While this technology is in its infancy, actual gene maps are being constructed. These provide loci for some of the genes or combinations of genes responsible for specific diseases/characteristics, e.g., muscular dystrophy. While only a small proportion of the large number of human genes have been located, genetic technology is proceeding rapidly and making strides in this field. A future result of this work is that genetic engineering will enable scientists to actually alter the genotype, thereby substituting or altering a gene associated with a birth defect with a different, nondefect-related one.

Genetic engineering has the potential to offer numerous benefits for medicine, yet the field, which is not fully understood by many health professionals or the public, is controversial. For example, some concerns exist regarding the ethics of manipulating or altering human potentialities. Other concerns exist about the consequences of mismanaging unknown or known gene combinations. The history of the human race has witnessed horrors in the name of *eugenics*, the science of "improving" human species by changing genes or manipulating gene pools. Thus, the field of genetic engineering, like prenatal diagnostics, offers benefits as well as sizable ethical considerations which science and health fields must address.

## External Fertilization

External fertilization, i.e., combining a sperm and an egg outside of the body, is another area that has resulted from current technology and that poses similar ethical dilemmas. In vitro literally means "in glass," and in vitro fertilization refers to what is commonly called "test tube baby" procedure. This procedure has been used for women with certain types of fertility problems such as when the ovaries and uterus are healthy but the oviducts are blocked, preventing conception. Thus, the mother's ova may be removed and, in a container outside of the body (hence the name "test tube"), joined with a sperm contributed by the father. Once fertilization occurs, the zygote is placed into the mother's uterus or into a surrogate mother, where implantation should occur (78, 79).

In vitro fertilization is relatively recent. In July 1978, Louise Joy Brown of Bristol, UK became the first test tube baby to be delivered. Since then in vitro fertilization has been quickly accompanied by a variety of other approaches to fertility problems. These include surrogate motherhood and embryo transfers as well as the older technique, artificial insemination by donor sperm.

Artificial insemination is used as a remedy for male infertility, whereas the other procedures are used for female infertility or gestational problems. A surrogate mother is a woman who "rents" her uterus for the gestation of another couple's child. This woman may provide her own ova and accept artificial insemination from the prospective father or she may accept a zygote from in vitro fertilization. Embryo transfers are very recent procedures where the embryo is conceived in a fertile mother and then transferred to a mother with a healthy uterus but with fertility problems (78–80).

All these procedures provide solutions for couples with fertility problems or for single individuals who want to have children, but these new approaches to conception also pose other dilemmas. As Marx (79) and Grobstein (78) both note, since these procedures are so recent, the risk factors for the prenatal organism have yet to be determined. In addition, Andrews (80) points out some of the unimaginable social and legal complications and psychological effects that may arise. She points out that for some children, five separate individuals could potentially be involved as "parents": a sperm donor, an egg donor, a surrogate mother who bears the child, and the infertile couple who want to raise the child. The legal dilemmas of these multiple relationships pose problems, such as concerns about screening donors for medical, genetic and psychiatric problems; determining liability and responsibilities in cases where the child may be disabled; the problems of keeping medical and genetic histories on children conceived by "anonymous" donors; and concerns about determining "ownership" (legally and humanely) when surrogate mothers change their minds about giving up a child that they "fostered" but that is actually a product of another couple's germ cells.

Genetic and fertilization technology also pose other ethical issues that must be examined in regard to eugenics. In external fertilization procedures, numerous sperm and ova are united, and out of several embryos, one is selected to be implanted. Concerns have been expressed about the death of the embryos that are not selected as well as the fact that some individual actually chooses which zygote will be the viable one (78, 79). With gene-mapping technology available, that kind of selective decision making could have long-range social implications; e.g., How will scientists—or prospective parents—determine what are positive and what are negative traits to be eliminated or retained? Furthermore, as Marx (79) notes, there are additional considerations regarding the medical creation of lives at a time when the earth is overpopulated by many other children who lack families and adequate care. Clearly, genetic engineering, gene mapping, and alternative fertilization procedures offer exciting possibilities if they are used appropriately, and they also offer challenges to the health/medical/scientific communities regarding long-range ethics and controls.

# BIRTH PROCESS

What actually starts the birth process is still not fully understood, although increased use of prenatal diagnostics has given us more information about the physical events and clinical complications that may arise at this very important time. The changing perspectives about childbirth and some of the problems that may arise for the mother or for the neonate will be examined in the following sections.

## Stages of Labor

There are three stages of labor. The first stage starts with uterine contractions and lasts until the fetus passes through the cervix, which is the opening between the uterus and the vagina. Considerable variability exists in the amount of time this stage takes, ranging from 3 or 4 to 18 hours or more. In general, first-time mothers tend to spend longer amounts of time in this stage. The second stage of labor involves the movement of the fetus through the vagina in what is typically considered the actual birth process, although there is still one more stage, the expulsion of the placenta.

What actually starts the birth process is not fully understood, however most fetuses tend to shift position in the weeks before birth. Thus, they are generally situated with their heads resting against the cervix, so that they will be born head first (19). As will be described in more detail later, Milani Comparetti (12) notes that the period of labor is actually a mutually interactive process between the mother and fetus. He notes that some hormonal message is relayed from the fetus to the mother, thus preparing her for the uterine contractions that are made in response to fetal kicking and movements. After that, reciprocal responses between the healthy fetus, which moves and pushes, and the mother, whose uterus periodically contracts, form the basis for labor and the birth process. While this typically proceeds normally for most mothers and infants, for some there are complications that require special medical intervention. Some of these special problems as well as contemporary approaches to the birth process are discussed next.

## Changing Trends in Delivery

In the United States and in some other countries, the natural process of giving birth has become progressively more technical and medical. In the past few decades, however, this tendency has begun to reverse. In the early 1900s in the United States, all births occurred at home and were assisted by a midwife or physician. However, with the emphasis on medical technology and anesthesia and the inability to transport laboratories and special equipment to homes, in-hospital care became the vogue. In fact, as Gillespie (81, p 101) reports, "The field of obstetrics grew out of the medical profession's efforts to devise ways to deal with those unusual situations that were life-threatening. As time passed, these interventions became accepted as normal for even uncomplicated deliveries. The public was educated to believe that mother and baby were safer in the hospital, thus turning a normal physiologic event into a pathologic process."

Now, although alternative childbirth practices are increasingly available, many women still believe that hospital birth is the best option. Gillespie (81) reports that the hospital is still the location for most births in the United States. Shepperdson (82, p. 406), who conducted a survey of women's attitudes about the birth process, notes "The majority of women clearly feel they are embarking on something which is rather dangerous and in which complications are likely to occur. This can, of course, be true but should not be so for the majority of women."

Numerous alternatives to hospital birth have been promoted in the United States since the 1950s. The "natural childbirth movement" took hold in the 1960s, with the promotion of women's control over their own bodies and their own labor and birth process. Some families objected to the increasingly technical and medical approach to childbirth, e.g., to unnecessary cesarean births, the use of anesthetics, possible complications with fetal monitoring equipment, and other procedures such as the mother's supine position during birth, which may be used at times more for the convenience of the facility or health care team than for the mother (83, 84).

Thus, a variety of reasons exist for this growing movement toward more natural childbirth experiences (Table 7.4). These have led to home births and to in-hospital or out-of-hospital birthing centers, which are home-like locations for the mother and close family or supportive companions. The purpose of these centers is to provide an alternative to nonrisk pregnant mothers, so that they experience childbirth as a natural, family process and a positive emotional experience (81).

Table 7.4
Advantages and Disadvantages of Various Approaches to Childbirth[a]

| Advantages | Disadvantages |
|---|---|
| *Home Birth* | |
| Mother in a familiar, supportive place—does not have to be moved | May require transfer to hospital |
| Avoids unfamiliar and/or unsupportive attendants | Medically trained individuals to assist in birth difficult to find |
| Family and friends can be present at any point for any amount of time | Emergency equipment not readily available |
| Least chance of medical intervention | Insurance eligibility varies |
| Best environment for bonding | |
| *Out-of-Hospital Birthing Center* | |
| Individualized care | May require transfer to hospital, posing risks for mother and baby |
| Labor and delivery in one room | |
| Minimal intervention by staff | |
| Parents and baby not separated after birth | Only low-risk patients who remain low risk can deliver in this setting |
| Early discharge | Unfamiliar environment |
| | Insurance coverage uncertain |
| *In-Hospital Birthing Room/Center* | |
| Presence of husband, friend, or relative throughout | Possibility of subjection to unnecessary technologic intervention or unwanted hospital routines |
| Emergency care immediately available | |
| Opportunity to have a professional stay throughout labor and delivery | Care by regular hospital staff who may not share philosophy of birthing room attendants after birth |
| Labor and delivery in same room | |
| Emphasis on normal rather than emergency aspects of care | |
| Warm relaxed atmosphere | Risk of in-hospital infections |
| Comprehensive health insurance coverage | Possible limitations of visiting hours |
| *Traditional (Hospital) Birth* | |
| Widely available, traditional, accepted | Greatest likelihood of routine use of medical intervention |
| Qualified physicians | |
| Equipped to handle high-risk patients and complicated labor and delivery | Least control over environment |
| | No choice on position for delivery |
| Resuscitation and infant intensive care for newborns | Unfamiliar labor attendants |
| Hospital staff give mother physical help, rest, and support | Likely to remain in bed |
| Limited visiting hours may keep new mother from becoming overtired | Customary separation of mother and baby after birth |
| Comprehensive insurance coverage | |

[a] From Gillespie, S.A. Childbirth in the 1980's. What are the options? *Issues Men Health Nurs* 3(2) 114–115, 1981. Reprinted with permission from Hemisphere Publishing Corporation.

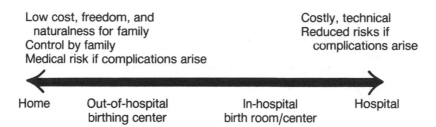

**Figure 7.11.** A continuum of childbirth options.

Advantages and disadvantages exist for all these options. Home and out-of-hospital birthing centers are apt to be more home-like for the pregnant woman and her family, but these offer the least amount of medical support if complications arise. In-hospital birthing rooms or centers stand in between home and hospital births (Fig. 7.11). Hospital birth offers the least amount of naturalness for the birth process, but it also provides the greatest amount of preventive medical assistance and intervention. However, sometimes medical procedures themselves cause risks such as infection and emotional strain, which actually complicate the birth process and subsequent interaction between the neonate and the mother (83, 84).

While birthing centers exist in many areas of the country, they are options only to certain women who meet center criteria. Such criteria typically include ongoing prenatal care, specific maternal age, physician involvement and commitment to the family's birth choice, and lack of prenatal risks. Gillespie (81) reports that an increase in home births has occurred due in part to choices made by many white, middle-class, college-educated families who have decided to take greater participatory roles in birthing experiences. Home births are also increased because of women who feel they have no other choice; i.e., they cannot afford hospital care and do not meet the criteria established by many birthing centers.

Tremendous changes have occurred in the attitudes of individuals in the private sector as well as in the health care community regarding approaches to childbirth. For families to make appropriate decisions about these new options available to them, they must consider all potential prenatal risks, the mother's health and support, and other factors (Table 7.4). With the growing emphasis on holistic health approaches and people's active roles in their own health care (see Chapter 6), with changes in the field of midwifery (see 85), and people's concern about increasing technology in medicine, childbirth procedures and options should continue to change during the remaining decades of the 20th century (see 86).

# Birth Complications

While the birth process is a natural event for most women and fetuses, there are cases where complications arise. These complications may occur as a result of previous prenatal risks, which have already been discussed (e.g., maternal drug abuse, birth defects, chromosomal syndromes), or as a result of perinatal events. Events that are sometimes associated with a complicated birth include anesthesia administered during the birth process, instruments such as forceps used during birth, *breech delivery* (i.e., feet- or bottom-first instead of head-first), prematurity, the knotting of the umbilical cord during birth, and other events. The consequences of such events are varied in severity and in duration over the life span. These consequences and some of their causes will be explored in this section.

## Prolonged Labor

Labor, for a variety of reasons, may become unusually long, resulting in fatique on the part of the mother and her inability to expel the fetus. This may be a cause for intervention because the placenta can gradually lose its ability to supply oxygen, resulting in *asphyxia* (reduced oxygen) or *anoxia* (lack of oxygen) to the fetus. Thus, the delivery team may select any number of alternatives to facilitate the birth process. *Forceps*, for example, may be used to exert pressure on the infant and actually extract it from the vagina.

Forceps are tong-like instruments that are clamped to the infant's head in order to ease it out of the birth canal. Forceps can be classified as high, medium, or low, depending upon their length. High forceps are inserted deep into the birth canal, whereas low forceps are shorter and used when the infant's head is close to emergence. High forceps, in particular, may cause head or neck damage to the fetus. The damage may be superficial or may be severe enough to cause internal hemorrhaging. As a result, high forceps are now rarely used, and cesarean section is used as an alternative (10).

There seems to be controversy regarding the use of medium and low forceps, as well. One short-term study (87) indicated that when compared with babies from normal deliveries, infants with whom low forceps were used tended to display less organized behaviors and were seen by their mothers as more bothersome. Whether long-range consequences of medium and low forceps exist has yet to be determined, although many physicians have opted for cesarean sections as an alternative (10).

## Cesarean Section

Cesarean section is the surgical removal of the infant through an incision made in the mother's abdomen and uterus. This procedure is used instead of vaginal birth for the following reasons: if the fetus is too large for the mother's pelvis, if the pelvis is contracted, if the fetus is abnormally situated for birth (e.g., breech), if the fetus seems to be in distress, or sometimes if the mother has previously given birth with this procedure. While cesarean sections were very common in the past (e.g., up to 25–30% births in some locales) (82), their popularity may be leveling off. Concerns have been expressed about the risks of the procedure for both the mother and the fetus. For example, there is a greater probability of the death of the mother when cesarean, compared with vaginal, delivery occurs. In addition, there is increased chance of infection of other maternal organs, increased probability of depression because of the mother's inability to have vaginal delivery, greater recovery time for her, and a potential for reduced interaction between the infant and family immediately following birth (10).

Some facilities, sensitive to the possibility of interrupting family-infant interaction and bonding use local anesthetics during this surgery. Thus, the mother is alert during the surgery and able to interact with the newborn and attending family members immediately after birth (83). However, there is some evidence that because of the anesthesia and lack of physical exchange between mother and fetus, which is typical during vaginal delivery, cesarean sections may result in an infant whose behavioral responses are temporarily suppressed. The extent of neonatal inactivity following cesarean delivery will depend upon the amount and type of anesthesia used as well as other perinatal procedures and complications (10, 83).

## Blood Incompatibilities

One problem that may occur during delivery has to do with an incompatibility between the elements of the mother's and the fetal blood. As most people are aware, humans possess different blood types, i.e., A, B, or O. An additional characteristic of blood is the presence of Rh factor, which can be either positive (Rh+) or negative (Rh−). Normally blood factors will pose no problems unless the father has the dominant Rh+ and the mother, the recessive Rh−. In this case, the fetus may inherit the Rh+ and ends up with a different Rh factor from its mother.

In some cases this incompatibility causes no problems; however, if the maternal and fetal bloodstreams intermingle, then the Rh+ blood from the fetus will enter the mother's bloodstream. The mother's body will treat this Rh+ blood as a foreign substance, called an *antigen*, and will begin to set up a defense against it. This defense takes shape in the form of antibodies that build up in the mother's bloodstream and that will then "attack" the red corpuscles in the fetal blood. The fetus may then become anemic and develop numerous complications associated with anemia.

Rh factor incompatibility can pose risks, but many medical interventions have been set up to combat the problem. First, such incompatability rarely occurs during the first pregnancy because of the low probability of blood mixture. However, after the first birth, the mother has been exposed to fetal blood so that she will develop antibodies that may operate in subsequent pregnancies. The risks of this happening can be reduced or eliminated by prenatal testing, knowledge of previous deliveries, blood transfusions of the neonate or the fetus, or by treatment of the mother with gamma globulin, which blocks antibody formation. Thus, existing risks of blood incompatibility have been markedly reduced by current health care and obstetric practices.

# Fetal Monitoring

The process of checking on the status of the fetus before and during delivery is known as "fetal monitoring". This is conducted by ultrasound or by using electrodes placed on the mother's abdomen or actually put through the vagina and cervix and attached to the fetus. The electrodes will pick up information relevant to fetal heartbeat and circulation as well as brain activity. Sometimes scalp needles are also used to take samples of blood (10, 88).

The uses of the fetal monitor are controversial. Some feel that it is too invasive, that it increases the chance of spreading a vaginal infection to the fetus, that it may interrupt labor by restricting the mother's mobility during a difficult and painful time, that the information it obtains may be unreliable, that it is costly, and that it is sometimes used for the wrong reasons (i.e., to avoid malpractice or to monitor delivery when a facility is short staffed.) However, at times the information obtained from the monitors is invaluable as a direct gauge of fetal status, which may be essential when prenatal risks are present or when complications arise during the delivery (10, 84, 88).

# Drugs Used During Delivery

Different classifications of drugs are used during the birth process. *Preanesthetic medicines* are those used prior to the administration of anesthesia during delivery. Such drugs may be used to stimulate uterine muscles (oxytocin) or as sedatives, analgesics, or tranquilizers. In addition to these, a variety of anesthetics may also be used during delivery.

Concerns have been expressed about the uses of such drugs perinatally. In addition to their suppression of maternal activity, many of these substances provide particular risk to the fetus. The actual effects of perinatal anesthetics have been studied by several investigators, and their effects are unclear. While some studies report no short- or long-term effects of perinatal anesthetics, others find potential physiological and developmental risks for the infant. In a comprehensive review of this area, Brackbill (89) concludes that drugs given to the mother during labor and delivery do have effects on infant behavior, and that these effects are more severe as the dose is increased.

The developmental domains of the infant that seem to be most affected by perinatal anesthetics are cognitive and gross motor, and delays in these areas have been found as long as 1 year after birth (89). In addition, physiological hazards are posed for the neonate. As Brackbill explains, once the newborn is detached from the placenta, he or she must metabolize all toxins. The neonate is poorly equipped to deal with such substances because many of the immature organ systems such as the liver, kidney, and brain are still undeveloped and may be damaged by "adult" doses of different substances. Anesthetics may also suppress maternal and neonatal interactions during and following birth, thereby interrupting or interfering with many of the potentially positive interactive processes that occur during normal childbirth (87, 89). Some of the implications of these interruptions will be more fully covered under "Parent-Infant Interaction".

# THE AT-RISK NEONATE

Labor and birth involve a close, mutual interaction between the mother and the fetus. Thus, it is difficult to determine where, in examining the birth process, attention should shift in the direction of one member of the partnership to the exclusion of the other. The previous sections of this chapter have focused on prenatal and pregnancy-related risks. A concept very closely related to those is the risk status of the newborn. However, as Robertson (90) has pointed out, there is a difference in approach to these areas:

> A high-risk pregnancy need not necessarily give rise to a high-risk infant, since careful management of a patient with risk factors can reduce the hazard to the fetus so the baby may be normal at birth. In addition, there are many pregnancies in which there are not detectable risk factors during the prenatal period, although labor supervenes prematurely. An infant is then born who is at high-risk for subsequent physical and developmental deficits. Thus, the definition of high-risk for the fetus and mother differ from the definition of high-risk for the neonate (p 3).

Robertson (90) points out that certain conditions constitute high risk for the neonate. These are the presence of some structural defects that may affect the development of organ systems, low birth weight, or some trauma or lack of oxygen causing an acute impairment of neonatal functions at, or around the time of, birth.

## Anoxia and Asphyxia

The terms "anoxia," "asphyxia," and "hypoxia" are used synonymously to refer to "reduction in the oxygen level below the biophysiologic requirement of the organism" (91). Literally, anoxia refers to lack of oxygen, whereas asphyxia or hypoxia are a reduction of oxygen. Reduced oxygen can occur because of several reasons during the perinatal period. At this time, the umbilical cord may become tied or squeezed, or the placenta may detach so that the fetus is cut off from oxygen delivery. In other cases, the neonatal respiratory system fails to function independently when the baby is delivered.

Short- and long-range effects of oxygen deprivation on neonatal behavior and development have been studied by many investigators, and their conclusions vary. The severity of the consequences of oxygen deprivation are difficult to assess because so many other variables may be involved; e.g., the presence and variety of anesthetics used perinatally, duration of oxygen deprivation, other prenatal risks, genetic problems, and the health and maturity of the infant. The critical effect of such deprivation is on the fetal central nervous system, and prolonged lack of oxygen can lead to death or to severe motor impairment or mental retardation. Less deprivation may lead to mild motor deficits and incoordination; mild intellectual impairments evidenced in learning disabilities; epilepsy; or behavioral problems (40, 91, 92).

The duration of these negative consequences is unclear. While motor and mental handicaps may be evident perinatally, they seem to gradually diminish over time. Some longitudinal studies, for example, have reported that by the time childhood or adolescence is reached, a reduction of deficits is seen, possibly reflecting the plasticity of the central nervous system (40, 91, 92). Clearly, the diminution of these problems will depend upon their severity. Mild deficits may be eliminated over time via compensatory activities or neural plasticity, whereas frank brain damage resulting in *cerebral palsy*, which although it may not get progressively worse, will also not disappear.

## Cerebral Palsy

One nonprogressive condition associated with anoxia and with which many therapists are familiar is that of cerebral palsy. It is a condition associated with motor deficits due in some cases to perinatal oxygen deprivation. Individuals with cerebral palsy may show a variety of different reactions of muscle tone, resulting in rigidity or looseness of muscles, called *spasticity* and *hypotonia*, respectively or fluctuating tone, called *athetosis*, or mixes of these.

Many cerebral palsied infants are of normal intelligence, but their motor performance and perhaps language skills are delayed because of their inability to gain control over their musculature. While cerebral palsy can be viewed as nonprogressive; i.e., the brain will not become more damaged, muscle problems can become more involved because of the lack of normal movement. The role of therapists with these infants is to help parents learn how to best handle and feed the baby with motor problems, to promote normal motor development, and to prevent the tightening of muscles, called *contractures*, that are not fully exercised as during normal movement.

While cerebral palsy has been typically assumed to originate because of anoxia during the birth process, Milani Comparetti (12) has suggested that some forms may occur earlier in fetal development. Thus prenatal brain damage may result in abnormal development, reduced motor interaction before and during the birth process, and other perinatal complications.

## Preterm or Low-Birth-Weight Infants

Numerous books have been written about the topic of prematurity. Since it is such a broad topic, it will only be covered briefly here. For those who are interested in learning more about the topic, many of the references listed in this chapter may be consulted.

## Clarification of Terms

For many years, the term "prematurity" was used to refer to a collective category of infants who were small (i.e., less than 2500 grams) at the time of delivery. Such infants, however, may be small for different reasons. Tanner (21) notes that the average length of gestation measured from the 1st day of mother's last menstrual period is 280 days, or 40 weeks. Considerable variability in found in this time, however, so a range has been set for what is considered a normal term of pregnancy, i.e., from 37 to 42 (or 38–41) completed weeks. Babies born within these limits are considered "term" or "full-term," whereas those born earlier are "preterm" and later, "postterm."

Years ago, infants classified as premature may, in fact, have come from full-term or preterm categories. For example, a preterm infant may develop at a normal rate but still be small because of early birth. In contrast, a full-term infant with an impaired rate of growth would experience a normal duration of pregnancy but be small at delivery. This latter infant is now considered small for gestational age (SGA). Thus, many different terms are used now—those referring to length of term and those referring to birth weight; i.e., small for gestational age (SGA), appropriate for gestational age (AGA), or large for gestational age (LGA) (21, 90, 93, 94). In addition, the terms "low birthweight" or "small for dates" may also be used synonymously with SGA.

As our understanding of these different groups of infants and our awareness of their unique problems has increased, it has become necessary to differentiate between groups of premature infants. Indeed, their needs may differ. For example, preterm infants may require a supportive environment until their immature systems mature sufficiently for independent control. In contrast, SGA status may be an index of some failure to develop normally, or what is referred to as *failure to thrive*. There may be attendant neurological or metabolic deficits that must be corrected before normal developmental processes are gained. Thus, these infants may require some special diagnostics to determine the reasons for their delays as well as interventions to correct identified problems.

Infants who are preterm, or SGA, are often collectively regarded as "high risk," meaning that they have been subject to conditions that may contribute to central nervous system dysfunction and to later developmental delays (93). Postterm infants may also be subject to delivery risk, or their delayed status may be an index of some interference with the normal impetus for birth. Many potential short- and long-range problems exist for the high-risk infant. The general rule is that the smaller the birth weight or the shorter the term, the greater the risks for the infant.

## Short-Range Problems Associated with Risk Status

Since high-risk neonates are typically small and immature, internal organs and systems are not yet sufficiently developed for independent existence out of the mother's womb. As was explained in the earlier sections of this chapter, during the last 2 months of fetal development (and postnatally), the brain and other body systems continue to mature. If, however, an infant is born before these systems are prepared to work on their own, then the life and development of the infant may be compromised. For example, as discussed under "Development of the Brain", it is thought that early fetal movement patterns (such as sucking, hand-to-mouth, coordinated limb-trunk movements) provide a foundation for subsequent, organized neonatal behavior (11, 12). As Humphrey (11, p. 84) notes, "I believe that even good prenatal development of the muscles of the infant is dependent upon their movement. If the fetus did not kick around a bit it would not develop strong muscles before birth."

Thus, the preterm infant is not only born without full-term neuromuscular maturation but also without the opportunity to have moved and coordinated some basic sensorimotor patterns in an environment where suspended movement is clearly much easier than on a gravitationally secure surface out of the womb. In fact, the placement of these infants on a flat surface out of the womb, as will be discussed shortly, presents additional compromises to the small or premature infant's health and stability. Clearly, the uterus offers a special environment; it is considered a "natural incubator" where the fetus is suspended in fluid; maintained at a stable temperature; cushioned; and provided with oxygen, nutrients, and waste removal.

Many developmental and health-related problems are associated with at-risk status, but some are more serious than others and require immediate attention and medical intervention. Clearly, the risks of anoxia and asphyxia must be addressed. There is some concern that the infant may aspirate fetal waste, called *meconium*, into the lungs, thereby resulting in hypoxia or *respiratory distress syndrome*. Inefficient breathing as a result of respiratory distress can cause subsequent brain damage or can lead to other health complications such as pneumonia. In addition, the immature lungs have insufficient surface lining (*surfactant*) that may result in *hyaline membrane disease* or other breathing disorders (93, 94).

Related to respiratory problems is the potential for cardiovascular complications. The fetal capillaries are small and fragile as well as deficient in certain clotting factors. If compromised, these vessels may collapse or rupture, leading to intracranial hemorrhage (bleeding inside the cranium, which is the skull surrounding the brain). In addition, some hemorrhaging may occur as a result of compression of the premature, soft skull during the birth process or due to poor regulation of blood pressure during routine handling and stress in the high-risk nursery (95). Other serious health problems have to do with metabolic deficiencies, immature sucking and swallowing responses, difficulty maintaining adequate temperature, small size and lack of musculature and tissue, resulting in reduced joint protection and mobility as well as possibility of skin breakdown (93, 94).

This points to the actual hazards that exist for neonates who are prematurely detached from their natural, supportive prenatal uterine environment. At present, neonatal intensive care units (NICUs) attempt to provide a protective environment for such infants while sustaining a vig-

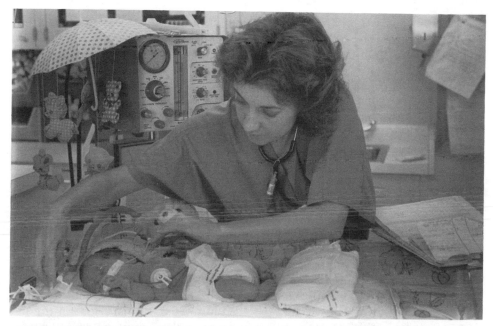

**Figure 7.12.** High-risk infants often require temporary medical interventions until their immature systems adequately develop. While the interventions characteristic of special-care nurseries are keeping alive infants who once would not have survived, the ecology of the nurseries and their potential long-term effects on development are being scrutinized. Compare this to Figure 7.13. (Courtesy of the March of Dimes, Birth Defects Foundation).

orous medical intervention aimed at life or health maintenance (Fig. 7.12). At times, this necessitates isolating the neonates in incubators that are equipped with numerous monitors as well as tubes and electrode attachments for feeding or monitoring metabolic and cardiorespiratory status; heat shields for temperature regulation; oxygen, which may be essential for respiratory distress syndrome but that may compromise development of the visual system; and placement on a surface in a position which the weak, undeveloped infant may be unable to counteract.

These NICUs are geared to address the critical needs of the high-risk infant, but considerable concern has been expressed about the various forms of intervention found on the units. Numerous studies have attributed various behavioral and motor deficits to low birth weight or preterm status; however, some recent research indicates that some of these deficits may be associated with the ecology of the neonatal or special care nursery. Studies have been looking into the characteristics of NICUs and the various types of ongoing stimulation bombarding the infant placed there.

## Complications Associated with NICUs

Unfortunately, it appears that some aspects of the special care units may be detrimental to the health and development of the neonate. For example,

excessively loud sounds (e.g., from beepers, garbage cans, monitors) penetrate the isolettes, potentially causing auditory damage and causing startles
in infants who have minimal amounts of energy to devote to such random
activity. Such energy expenditure may also result in oxygen consumption
in a neonate whose respiratory system is already challenged. In addition,
noises in the special care unit are often ongoing during 24-hour periods;
sounds are unrhythmic and uncoordinated with other sensory information.
This is atypical of the sensory stimulation that normal infants receive within
a caretaking context where smiling, talking, cuddling (i.e., visual, auditory,
tactile, and movement stimulation) are all coordinated (96–98).

Another important aspect of the NICU is the relationship of the parents
with their newborn. Parents are often overwhelmed by the technology
involved in special care nurseries, and they may feel intimidated or frightened about handling a high-risk, small infant who has various tubes and
monitors attached to the head, face, or body. Thus, the infant may be
handled infrequently, and handling that is received may be primarily for
routine medical care. The infant may come to associate handling with
negative events such as pain that occurs during medical interventions. In
addition, since monitors are often used, and since such infants often have
very little motor control and poorly developed musculature, they are often
unable to move, or to stimulate, themselves. Such lack of self-produced
movement can cause further delays in sensorimotor development as well
as pose other health and developmental risks (see chapter 9, "Active Experience is Necessary ... "). With inactivity there is always concern about
skin breakdown around the joints as well as asymmetries of later movement
patterns or abnormal muscle development due to poor positioning.

Such concerns about adequate position and warmth have led to the use
of tiny waterbeds in neonatal isolettes. The waterbeds provide containment
and therefore promote normal neonatal flexed postures as well as providing
a source of warmth for temperature control. In addition, in some cases
where the waterbed is flexible, it provides an opportunity for the kinds of
self-produced movement that a small or immature infant would not receive
on a traditional flat mattress (99). Numerous other interventions such as
this have been developed or promoted by health professionals concerned
about providing a therapeutic and a naturalistic environment for such high-
risk infants.

## Roles for Therapists in Special-Care Nurseries

Physical and occupational therapists, in particular, have been filling important roles in special-care nurseries (see 93, 94, 99–105). They perform a
number of functions, ranging from consulting with nursing staff about
adequate positioning of small or preterm neonates to promoting family-
infant interactions by educating and demonstrating handling procedures to
parents. One area of great concern to all disciplines is the potential interruption of the parent-infant emotional attachment. As this book has already
discussed, the isolated neonate is often prevented from free interaction with

family members, and it is thought that this isolation may have negative consequences for the parents and for the infant. Thus, demonstrations of neonatal responsiveness, demonstrations of proper handling, and education about medical interventions and equipment serve to reduce parents' anxiety and promote positive feelings toward their baby. (This is discussed further under "Parent-Infant Interaction.")

While some therapists fill primarily consultative functions, others evaluate and intervene directly with infants. For example, some at-risk neonates may exhibit paralysis of limbs, whereupon therapists may design splints and monitor the infants' range of motion, muscle strength, and functional, developmental progress while on the unit and after the infant is discharged. For infants whose medical conditions have stabilized, other kinds of sensorimotor interventions have been utilized. The various forms of intervention have been closely scrutinized, as it is felt that for the sick or very small infant, unnecessary handling may actually compromise physiological systems (98, 100). Numerous studies or reports by physical and occupational therapists are looking at the different forms of intervention that may be used and the different roles played in special-care nurseries (94, 99–105). It is expected that with the growing technology that has enabled smaller and younger infants to survive, the roles of therapists with these populations will continue to grow and diversify.

## Long-Term Consequences for the At-Risk Infant

Nearly every possible negative consequence ranging from the full spectrum of mild to severe brain damage and developmental delays to mental illness have been at some time traced to prematurity or perinatal stress (e.g., 5, 90, 92, 106–108). Clearly, this is a difficult area for pinning down specific causes and effects. As this chapter has so far indicated, various events may lead to at-risk status in infants. Thus, preterm or SGA status cannot be easily isolated as causative events, especially in regard to long-term consequences. Other prenatal and perinatal events must be considered as well as the child's home, nutritional, cultural, sensory, and intellectual environmental stimulation in the years following birth.

At-risk status, however, does seem to place the infant in jeopardy for one or various subsequent emotional, social, language, learning, motor (or overall developmental) problems (106–110). The greatest difficulty is determining the extent to which these effects last over the individual's life span. Recent studies, for example, have found that, similar to the effects of anoxia or malnutrition, developmental delays associated with preterm or SGA status tend to wash out over time. For example, Olson and Lamb (107, p 66) note, "Many premature infants are indistinguishable from matched full-terms by the school years, thus showing little long-term effects of the early hazards they faced." Escalona (110) further notes that whether these long-term effects are maintained depend in part upon subsequent environmental stimulation or cultural deprivation associated with the social class in which the children are raised.

This is consistent with the conclusions of Sameroff and Chandler in 1975 (111), who noted that developmental risk must be viewed according to a continuum of environmental deprivation and stimulation. Thus, infants born at risk may develop normally when presented with a supportive, compensatory environment; however, similar infants may demonstrate subsequent developmental risks when reared in a stressed, deprived environment. This "continuum of caretaking casualty," as Sameroff and Chandler called it, is similar to Horowitz's (112) three-dimensional depiction of how environment interacts with an organism's level of vulnerability to effect the high-risk neonate's ultimate developmental outcome (see Fig. 3.9).

Thus, to determine long-range consequences of any perinatal event, it is essential to examine multiple factors in the child's environment and how they all interact over time. As Sameroff (92) notes, such analyses require transactional models of development, where it is recognized that the child is actively engaged in attempts to organize and structure his or her world, and where the changing environment responds back. Sameroff points out that the key to the concept of development "transactions" in comparison to "interactions" is the notion of dynamism, or change over time. As he explains:

> Exceptional outcomes from this organismic or transactional point-of-view are not seen simply as a function of an inborn inability to respond appropriately, but rather as a function of some continuous malfunction in the organism-environment transaction across time which prevents the child from organizing his world adaptively. Forces preventing the child's normal integration with his environment act not just at one traumatic point but must operate throughout his development (92, pp 281–282).

Numerous factors may contribute to interrupted environmental-infant transactions. Some of these can be found in the infant (e.g., biological/genetic deficits), whereas other environmental factors interact with biology. One significant environmental factor is the child's social class, although poverty alone does not account for the problem. As Werner and Smith have concluded:

> A low standard of living, especially at birth, increased the likelihood of exposure of the infant to both early biological stress and early family instability. But it was the interaction of early biological stress and early family instability that led to a high risk of developing serious and persistent behavior and learning problems, in both lower- and middle class children (113, p 144).

Thus, one of the factors that must be examined in regard to predicting outcomes for high-risk infants is their level of quality interaction in their caretaking environments. This points to the necessity of examining parental relationships with normal and with high-risk neonates and children.

# PARENT-INFANT INTERACTION: BONDING AND ATTACHMENT

The development of an emotional connection, i.e., love, between the newborn and the parents and other family members has been of interest to researchers, clinicians, parents, and poets for centuries. Developmental research in this field has witnessed numerous changes. Initially, researchers focused on *attachment*, the development of affectional ties on the part of the infant and extended toward the family members, traditionally the mother. This was typically studied, if not acknowledged, as a one-way process; i.e. the attachment of the infant to the parent. However, changes in our knowledge of infant capabilities resulted in the recognition of the important effects that infants had on their parents. Thus, attachment studies gave way to examinations of the other side of the relationship, the parents' ties to the infant. This relationship is called *bonding*, as it was introduced by Klaus and Kennell (114) in the 1970s.

In the late 1970s and 1980s, with the increased application of systems approaches to behavior, the focus of attachment/bonding studies has been on the ongoing, reciprocal process that engages both partners (e.g., the mother and the infant) in a mutually significant and dynamic relationship— or "transaction." Thus, it is now explicitly recognized that not only does the infant become attached to the parents, but the parents as well become reciprocally involved with the child—and this relationship evolves over time. Thus, it is not possible to study one member in isolation from the other since their actions are mutually dependent and changing over time. However, since scientific investigations try to reduce variables to their simplest elements, many studies have attempted to examine bonding or attachment as distinct and separate processes. In this section, bonding, or the parents' responses to their fetus or newborn will be examined to the exclusion of infant attachment studies, inasmuch as that is possible. Later in this book in chapter 12, the focus will be redirected toward infant's and children's emotional development and social relationships with their families and friends.

# The Collaboration of Labor

We normally assume that bonding starts at birth, but actually, it may begin earlier. This chapter has already described some of the many developmental changes that occur prenatally. The normal fetus is an active organism, and at some point in her pregnancy, the mother becomes aware of these fetal movements. *Quickening* is the term used to refer to fetal movements and to the mother's awareness of the life of her baby as a result of them. This usually occurs in the 4th month, but some mothers may feel fetal movements as early as the 10th–11th week of gestation (12).

The movements felt during quickening may actually start parental bonding and may be very important as an assurance to the mother that she hosts a live, healthy baby (12). As Klaus and Kennell (115) report, one of the maternal tasks during pregnancy involves recognizing that the fetus will become an individual separate from herself. Thus, during pregnancy, the parents may begin to fantasize about their future baby, they may wonder what the baby will be like, and they may use a nickname and talk to the fetus (115, 116). Both parents become involved in this process (116), with the father's involvement being significant for the mother. Klaus and Kennell note that the mother's feelings of closeness to the fetus could be inhibited if she experienced difficulties during pregnancy or if the father were uninterested or provided her with little emotional support (115). This is corroborated by Blake's and colleagues' (117) findings that mothers of very low birth weight infants (less than 1500 grams) experienced less emotional problems about their children when their mates were sympathetic and offered them support.

Thus, quickening may be important for starting bonding. As Zeanah and colleagues (116, p. 205) note, "Our clinical experiences suggested that parents may begin to consider the distinctive characteristics of their infant even prior to birth. From the time of quickening, "the other" is actively interacting with the mother and sometimes indirectly with the father as well." In addition, early fetal movements may be useful diagnostically. Maternal reports of fetal movements or direct observation of these activities via ultrasound are used to identify normal fetal behavior or to signal potential developmental deficits (74). For example, in his discussion of Humphrey's important work on fetal movements, Robinson reports that mothers of babies with congenital forms of muscular dystrophy often claim they have felt little fetal movements (11). Thus, the lack of these movements prenatally may not only signal some neuromotor risk status for the fetus but also interfere with parental perceptions of the fetus as a live, independent organism.

As was noted earlier in this chapter, fetal movements are thought to be preparatory for later infant behaviors. Milani Comparetti (12) explains that these movements are probably important for bonding as well as essential for a normal delivery, which he describes as a "collaboration of labor." This collaboration is initiated by fetal kicking, which, when it is time for birth, promotes maternal contractions. From thereon, a reciprocal interaction occurs between mother and fetus, resulting in a collaborative birth process. Thus, as Milani Comparetti's research has found, a normal labor engages the mother and fetus so that the two work together. Problems of labor may therefore be due to some inability associated with the mother or to some fetal movement or developmental disorder, which prevents responding to maternal contractions. As Milani Comparetti notes:

> The mother alone is not responsible for the physiological mechanism of labor and delivery and therefore, she alone is not responsible for a disorder of this mechanism. Both mother and fetus are responsible. It is a collaboration, a team effort. Some fetuses show disorders of movement and therefore do not perform their part in the normal collaborative process (12, p 10).

This recognition of a reciprocal interaction between the fetus or newborn and the parent has given us a different approach to infant capabilities and to the probable causes of some birth defects. Thus, some disorders such as cerebral palsy, thought to occur as a result of a deficient birth process, may actually arise earlier in prenatal development. Difficulties experienced during labor may not be the cause of some movement disorders or birth defects but may actually result from fetal inability to motorically contribute its part in the collaborative process.

# The Issue of Bonding

The term "bonding" was introduced by Klaus and Kennell in their original book, *Mother-Infant Bonding* (114) and in the updated edition, renamed *Parent-Infant Bonding* (118). These authors contend that bonding is the process whereby parents become emotionally attached to their infant. An additional premise of these authors is that the time of birth is a sensitive period for bonding. Therefore, procedures that separate the infant from the parents at or near the time of birth may actually have permanent deleterious effects on the parent-infant relationship.

Klaus and Kennell's work raised important issues and questions that have been examined and debated during the 1970s and 1980s. One of these questions is whether or not there is a sensitive period for either the infant's attachment to the parents or for the bonding of the parents to the child. In attempting to answer this and other questions about the formation of early emotional ties, studies have been looking at short and long term consequences of neonatal separation from the family because of birth complications. Thus, clinicians and developmental researchers have been studying the various effects of newborns' placements in special-care nurseries, since these infants provide a natural subject pool for examining such separations. Of particular concern to these studies is the potential interruption, delay, or inhibition of an emotional relationship between the family members and the infant.

Unfortunately, examining these issues is difficult. Like so many of the other risk factors discussed in this chapter, long-range consequences are difficult to study. Neonatal separation and placement of an infant on a special-care nursery, as discussed previously, is associated with numerous other events ranging from the health and medical status of the infant to the ecology of the NICU. Some of the other factors that may impinge on a reciprocal child-family relationship include the parents' physical and genetic histories and attitudes toward their child; ongoing social, intellectual, sensorimotor, and other forms of stimulation available to the infant; the temperament of the infant; the presence of siblings and their ages; extended family members available for stimulation of the child or support of the parents; the amount of quality time available to the parents for continued interaction with their infant; ongoing nutritional status; type of daily care; and the spousal relationship (e.g., 119). All these could have effects on the emotional interaction between infants and their parents. Thus, bonding studies are extremely complex and difficult to analyze. This must be kept in mind when examining studies drawing conclusions about attachment or bonding (see chapter 12).

## What are the Characteristics of Bonding?

Love is a term that defies explanation. Still, as was pointed out in the

**Figure 7.13.** Fathers are increasingly participating in childbirth and child-rearing roles and responsibilities. They are an important part of the infant and child's ecology.

chapter on research methods, researchers recognize the importance of adequately operationally defining the variables examined in scientific studies. Klaus and Kennell (115) point out that the behaviors witnessed perinatally and characteristic of parental emotional attachment include fondling, kissing, cuddling, and prolonged gazing (Fig. 7.13). These behaviors serve two different functions: to express affection and to sustain an interaction between the partners, thereby keeping up contact. Klaus and Kennell point out that in a normal childbirth, such behaviors are common. Once the baby is born and the placenta expelled, the parents often hold, cuddle, and stroke the newborn:

> In sharp contrast to the woman who delivers in the hospital, the woman giving birth at home seems to be in control. Immediately after the birth she appears to be in a remarkable state of ecstasy. We have called this *ekstasis.* The exuberance is contagious, and the observers share the festive mood of unreserved elation after the birth and groom the mother. Particularly striking is the observer's intense interest in the infant, especially in the first 15 to 20 minutes of life (83, p 32).

Thus, Klaus and Kennell have promoted natural delivery methods because of the positive effects on the parent-child relationship.

## Is There a Sensitive Period for Bonding?

One of the controversial issues regarding Klaus and Kennell's premises about bonding is whether or not this period immediately following birth is, indeed, a sensitive period for the formation of emotional ties on the part of the parents. As was explained in earlier chapters of this book (e.g., Chapter 5) the notion of sensitive or critical periods in human development is controversial. Studies by Klaus and Kennell and others support their existence, whereas other researchers take issue with these premises. For example, Goldberg's (120) and Lamb's (121) critical reviews of the literature on bonding indicate that the studies conducted in this area thus far are methodically unsound and do not provide clear evidence to support Klaus and Kennell's premises. Both researchers conclude that while Klaus and Kennell have done much to promote positive changes in delivery practices, they may have also promoted a misconception that could have grave negative consequences for some parents. Goldberg notes:

> The popularization of the ideas and work behind these changes has often led to the distorted view that the first hours after birth are critical for the establishment of parent-infant bonding, that parents who miss such opportunities will be impaired in their ability to care for their infants, or that the relatively low-cost "intervention" of 15–60 min of contact in the delivery room can substitute for continuing and more expensive forms of support for disadvantaged parents.

> Finally, the notion of early contact as critical rather than potentially beneficial needs to be dispelled. The emphasis on "early bonding" has already created an expectation on the part of many parents that if they do not have this experience they have somehow failed and will never be fine parents. For all of those who cannot have these experiences because of medical interventions during delivery (e.g., Caesarean section) or problems with the infant's health (e.g., premature birth), it is important to emphasize that the parent-infant relationship is a complex system with many fail-safe or alternative routes to the same outcomes. Its success or failure does not hinge on a few brief moments in time. It would be irresponsible of us as professionals to encourage this incorrect and extreme view (120, p 1379).

Lamb (121) points out that while early bonding may be of benefit to some mothers, it has no appreciable long-term consequences. Richards (122, p. 52) places this information in a developmental context. He notes, "The postnatal period can be viewed as a sensitive period for the establishment of the social relationship with the child provided that it is realized that the length of any such period is extremely variable and is not a simple once-and-for-ever process."

Thus, it appears that the time immediately following birth can be significant for the parents but is not *critical* for their subsequent bonding with their child. During childbirth parents may begin affectional responses toward the newborn, but while this time period can be very important for the parents, it is not "critical" to subsequent emotional transactions between parents and child. Numerous factors will affect the bonding process. Some of these include the mother's actual labor, the support she receives from her partner and others during labor, her attitude toward pregnancy, the characteristics and responsiveness of her newborn, and many other environmental factors that occur during the infant and child's development. What may be more important than looking for some circumscribed critical period for bonding is determining how to support or facilitate a reciprocal transaction between family members and the infant if, for some reason, special events prevent them from occurring naturally. Some of these interventions are examined under "Parenthood and the Newborn's Environment."

## A Special Concern: Father-Infant Bonding

While studies in the 1940s and 1950s tended to concentrate on the relationship of infants to their mothers, the 1970s and 1980s have witnessed a shift to the study of father-infant and family-infant interactions (Fig. 7.13). Birth practices gradually changed and along with them brought about increased participation of fathers during the birth process. Other social changes in parenting and sex roles have brought about increased scientific concern about fathers' participation in the family (123). While this book will return to this issue in subsequent chapters, it is essential to introduce in this chapter the topic of father involvement in birth and bonding.

Because of some of the long-range implications of critical periods in bonding, some unrealistic expectations have been drawn about father participation in the birth process. In a critical review of studies that examined the consequences of father's birth attendance and their early contact with their newborns, Palkovitz (124) draws conclusions which he compares to Goldberg's examination of maternal-infant bonding studies. Palkovitz notes that the literature examining fathers' birth attendance as well as their early and extended contact with newborns is inconclusive. He and others (115, 117) have found that in some cases father participation in the birth process and their attention to their partners enhances the marital relationship and can thereby have indirect positive consequences for the children. But, in general, father's birth attendance—or absence—is not critical to subsequent father-child interactions:

It appears that contemporary middle-class fathers believe that their involvement during all phases of pregnancy, delivery, and "bonding" is extremely important (if not critical) for enhancing relationships with their children. What is required is a balanced approach where fathers, in consultation with their partners, choose in all stages of the parenting process levels of involvement that are consistent with their skills, desires, and perceived roles.

Researchers should study fathers' early histories with their infants in the larger context of triadic interactions with their partners. By focusing longitudinally on family interaction, attitudinal shifts, and parental role assignments during the transition to parenthood, researchers can come to a better understanding of the processes underlying family development (124, p 405).

Thus father's roles are important, but their presence at the birth event is not critical. They play significant supportive roles in the spousal relationship as well as being a crucial part of the infant's interactive environment. The father's attitudes and level of involvement with his family can directly and indirectly affect the infant over the long term regardless of his early presence during a brief time period at birth.

### Conclusions About Bonding

Regardless of the academic debate about the existence of sensitive periods of bonding, Klaus and Kennell's and others' work has reminded us of the multiple positive and nonmedical aspects of the birth process. Their goal, and that of others, has been to promote pregnancy and childbirth as a natural, positive event; they have succeeded to some extent in altering many attitudes and delivery practices. They have also pointed out the value and significance of positive, reciprocal parent-infant interactions that occur perinatally (118, 119).

This focus on parent-infant interactions has enabled developmental researchers and clinicians to take a closer look at the family unit and to make analyses of the necessary components of a warm, supportive environment that sustains and promotes positive parent-infant reciprocity. This information is of use clinically and has resulted in recommendations regarding the family that may be at risk for supportive, developmental care. Thus, the focus on bonding has also promoted a positive attempt to intervene with families that may be at risk because of inadequate parenting abilities or because of the stress of delivering and learning to care for a child with special needs. This has resulted in increased involvement for clinicians such as occupational and physical therapists in work with infants and their families and is especially important for clinicians who work with the families of infants and children with special needs.

# Parenthood and the Newborn's Environment

This chapter has looked at many of the factors associated with risk status for the mother and her term of pregnancy, for delivery, and for the fetus or newborn. Numerous studies, some of which are cited here, have examined the potential long-range consequences of these risk factors on the subsequent development of the newborn. Consistent with current ecological, lifespan approaches to development, these studies have pointed to the significance of the infant/child's family environment, which may either improve or aggravate developmental problems associated with pregnancy and birth. This range of potential for the caretaking environment to positively or negatively affect an infant's development is what Sameroff and Chandler (111) refer to as "the continuum of caretaking casualty."

The infant's environment is so essential to his or her developmental prognosis that it is important in this chapter on prenatal development to look not only at pregnancy-related risk factors, but to also explore some of the characteristics essential for adequate family and parent care of the newborn. As Sameroff (92, p. 283) notes, "It is apparent that if developmental processes are to be understood, it will not be through continuous assessment of the child alone, but through a continuous assessment of the transactions between the child and his environment to determine how these transactions facilitate or hinder adaptive integration as both the child and his surroundings change and evolve."

## Preparing Families for Child Rearing

Bronfenbrenner (125) has concluded that the family is the most economical system for fostering and sustaining the development of the child. While his analyses were directed primarily at early cognitive stimulation programs, his conclusions about family-based interventions are relevant to child development in general. He recommends that families be supported so that they, in turn, will be prepared to provide an environment that nurtures and promotes the development of their children. The elements of this ecological-based intervention approach are basic. They are designed to promote the health, integrity, and well-being of the family. Economic security for the family, Bronfenbrenner points out, is essential both before and after pregnancy, so that the family is prepared and able to meet its basic needs. Economic security will enable the family to have adequate health care, nutrition, housing, employment, and opportunity and status for parenthood. Once these basic needs are met, then a program of intervention can be directed at individuals at different stages in family evolution (Table 7.5).

### Educating Children and Teens about Pregnancy and Parenthood

Bronfenbrenner notes that the first stage in family-centered intervention should begin early, while children are still in school. There they should receive practicum experiences in child care as well as knowledge about adequate prenatal care. As was mentioned earlier in this chapter, one of the ways of reducing many of the prenatal and pregnancy-related risks is through education. Thus, prospective parents should be informed about the potential effects of such prenatal risk factors as drugs, smoking, and alcohol

Table 7.5
Stages of Family-Centered Intervention[a]

| Stage | Intervention |
|-------|--------------|
| 1 | Preparation for parenthood |
| 2 | Before children come |
| 3 | The first 3 years of life |
| 4 | Ages 4–6 |
| 5 | Ages 6–12 |

[a] Adapted from Bronfenbrenner U: Is early intervention effective? Facts and principles of early intervention: a summary. In Clarke AM, Clarke ABD (eds): *Early Experience: Myth and Evidence.* New York, The Free Press, 1976, pp 253–255.

as well as the need for adequate maternal and preventive medical care during pregnancy.

With the growing concern over teen pregnancy in the 1970s and 1980s, school systems are finding the need to incorporate educational programs to address parent-related issues. For example, in 1979 at a high school in Wisconsin, a parenting program was initiated to cope with teenage pregnancy. Since students were bringing their babies to classes, the school system decided to set up a day-care center in the school. In this program, the staff provides support for the mothers and helps them to learn proper care for their infants. This "Infant/Child Learning Laboratory" project, as it is called, serves many purposes. It supports pregnant students, provides day-care for their children, and educates other students about the responsibilities of child rearing. A report in Newsweek magazine (126, p 69) noted, "The school's other students also find the infant lab instructive. About 30 a day visit it to observe the unglamorous realities of parenthood."

The author was similarly involved in the mid-1970s with a parent-infant support program at the Zale Infant Center in Dallas, Texas. This laboratory was located directly behind a high school, situated in a low-income housing project. Since most programs at this time did not offer support for infants, many of the teenage mothers found it necessary to drop out of school to take care of their babies. The infant laboratory provided them with local day-care until their infants were old enough to transfer to a preschool program. This enabled the mothers, who could drop off their babies before school and visit them during free time, to complete high school as well as to learn from skilled child-care workers about normal child development and care.

Supportive programs such as these have been designed primarily for disadvantaged or at-risk parents; i.e., those with low socioeconomic status, teen-aged, or unmarried; however, there is a growing tendency to make these available to all parents (127). The success of the programs seems, as Bronfenbrenner's ecological approach would predict, to depend upon the duration of education and support offered to the families. For example, Larson (128) reported that a combined prenatal and postnatal home-based program for low-income families is more effective in regard to the child's health than a program administered solely during the postnatal period.

Ferland and Piper (127) examined the effectiveness of a training program given to middle-income expectant parents. In their study, both treatment and control families received prenatal information from a nurse. The treatment group, in addition, received from an occupational therapist, child development information about infant sensorimotor development and play activities and materials. The researchers found that this additional training had significant effects on several home environment measures at 3 months postnatally. By 8 months, however, no significant differences were obtained in the development scores of the infants in the treatment and control groups. The authors concluded that their findings are limited in scope but suggest that a sensory motor training program can be effective in altering a child's subsequent home environment. Ferland and Piper (127) further suggest that their program may have been more successful if it had extended into the postnatal period. Their program is important because it illustrates how education about normal development may be of value for prospective parents and also illustrates an important role for developmental therapists in education and in the possible prevention of developmental and health-related problems among infants and children through ecological intervention.

## Prenatal Care

Bronfenbrenner (125) suggested that initial family-oriented intervention be aimed at preparing children and teenagers for parenthood. Subsequent intervention should be directed at families as they are formed and before children are born. Again, once the families' basic nutritional, housing, economic, and health care needs are met, then other relevant issues can be addressed. For example, Bronfenbrenner notes that this is an optimum time to start parent intervention programs (such as those discussed above). In addition, this is an important time for emphasizing prenatal care, creating an awareness of environmental and genetic risk factors, prenatal genetic counseling if warranted, and preventive medical care.

A report regarding the National Conference on Prenatal Care in 1986 (129) notes that good prenatal care is the most cost-effective way to reduce infant deaths. This conference report noted that one of every 100 infants born in the United States does not survive past 1 month after delivery and that black babies are twice as likely to die as white infants. A report in the New York Times in 1984 (130) noted that the United States has one of the highest infant mortality rates among industrialized nations and that current trends are toward reduced prenatal care, especially among low-income, nonwhite families. The National Conference on Prenatal Care [129, p. 4] concludes, "the United States can either spend more money on neonatal intensive care for low birthweight babies or deliver good prenatal care. To save the same number of additional babies, neonatal intensive care would cost $50 for every $1 spent on prenatal care."

**Figure 7.14.** The ecological approach to child development recognizes that prior to pregnancy, families need adequate education about childbirth and child rearing, genetic counseling if necessary, awareness of birthing options, and knowledge of prenatal care. (Courtesy of March of Dimes, Birth Defects Foundation.)

### Support for Families at Delivery and Beyond

Bronfenbrenner (125) notes that an adequate ecological, family-based intervention continues support for families during early and late childhood (Fig. 7.15). For the first 3 years of life, intervention is aimed at promoting family relationships (i.e., attachment and bonding) and facilitating the transactions between infant and parent. Bronfenbrenner (125, p 254) notes that the result will be to "improve the parent's effectiveness as a teacher for the child, further the latter's learning, and, in due course, establish a stable interpersonal system capable of fostering and sustaining the child's development in the future." Continued intervention, according to Bronfenbrenner, extends to adolescence.

As Klaus and Kennell and others (89, 118) have pointed out, one of the first ways to initiate positive parent-infant interaction is to actively involve the mother and her partner in the birth experience. Many childbirth options are presently available to healthy mothers, and prospective parents need to be educated about the choices available in their communities. Childbirth can be a positive emotional experience for the mother and her family, and this experience plus the mother's initial impressions of her infant, can set a tone for subsequent spousal and parent-child interactions. For example,

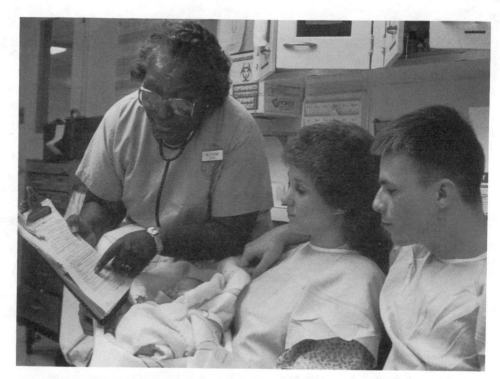

**Figure 7.15.**  The ecological approach to child development recognizes that once the baby has arrived, continued support and education needs to be directed toward the family as a child-rearing unit. (Courtesy of the March of Dimes, Birth Defects Foundation.)

Murray and colleagues (87) compared infants born to women who had been administered various levels of, or no, perinatal medications. Compared with infants from nonmedicated mothers, the babies born to the most highly medicated mothers scored poorly on an assessment of neonatal behaviors, called the Brazelton Neonatal Behavior Assessment Scale (131), frequently referred to as the Brazelton or the NBAS. Some general differences between babies were still apparent by 5 days but eliminated by 1 month postpartum.

Yet, despite the fact that at 1 month the babies scored similarly on the NBAS, the mothers still perceived them as different. Murray and colleagues (87, p 78) note, "By 1 month, although examiners noted few differences between babies in the groups, mothers of medicated babies seemed to view them less favorably and in general found them more difficult to care for compared with mothers of unmedicated babies." The researchers (87, p 81) explain these findings: "Although direct biochemical drug effects may wear off in the first days . . . , the mother's early impressions may remain to influence how rewarding she finds her baby and the manner in which she responds to her baby's initiatives at 1 month of age." Thus, maternal perceptions of the birth experience, her initial impressions and interactions with her newborn, and the support she receives during pregnancy and delivery may all affect longer-term family transactions. This points to an important role for health care professionals who are involved in obstetrics and neonatal care to ease families' anxieties and facilitate their preparenthood experiences.

Hulme and associates (132) note that physical therapists are increasingly participating in the area of obstetrics and childbirth education. For example, some specific traditional roles played by physical therapists include massage and exercise with postpartum women as well as providing therapy to maintain adequate circulation for bedridden pregnant women. These clinical researchers predict that physical therapists' roles will continue to expand in obstetrics as evidenced by the increasing focus on obstetric education in physical therapy curricula. Specific content areas emphasized include information about fetal and child development, birthing alternatives, relaxation techniques, body mechanics, nutrition, and breathing techniques (132). Other specific roles for developmental therapists in neonatal care are found in special care nurseries and in follow-up of at-risk infants and families.

## Special Kinds of Support for Families with Disabled Infants

The goal of most parent intervention programs is to promote positive, ongoing interactions among family members. For families with disabled infants, an additional challenge is faced. Many of these families must first address the issues associated with giving birth to a preterm infant or to one that is currently or may eventually end up with a physical or mental disability or delay. These families often must cope with ongoing or periodic feelings of guilt and grief (see 133) as well as the anxiety and demands associated with providing care for an infant that may be frail or have a variety of special needs (e.g., 134)

As this chapter has already addressed, many high-risk infants are often placed immediately after birth in special-care nurseries, where they stay until their health stabilizes or until their physiological systems mature. Families of these infants will often need to cope with feelings of loss of these infants as well as fears of and intimidation by medical equipment and the apparent frailty of their small or different-appearing infant. Thus, the initial contact the parents receive with their infants may be emotionally turbulent and negative; therefore, it is an important role for health professionals to intervene with such families, to ease their fears and to facilitate positive interactions between the special infants and their families (see 135).

### Interaction Coaching

Developmental therapists are specifically trained in such areas as sensorimotor development and intervention, oral-motor control, positioning, and handling. Thus, they can often offer advice to nursery staff and parents regarding optimum methods for positioning, holding, stroking, and cuddling infants that may have sensory deficits or motor problems. In addition, parents need to be counseled in regard to what kinds of signs to look for when these small or frail infants are overhandled or stressed. Therapists may suggest certain simple activities that parents can engage in with their hospitalized newborns, such as activities to promote movement or responsiveness, or to coordinate various forms of sensory stimulation within a caretaking context (94, 101, 105, 136). For example, if the neonate is placed on a waterbed, the parents may be shown how to produce rhythmic stimulation and what facial and body responses to look for in the baby to determine if he or she is responding positively or negatively (see 99). Similarly, therapists may make suggestions about positioning and interventions while feeding infants who have undeveloped oral-motor musculature. Such suggestions show the parents how to "feel in control" and show them what to look for in terms of their infants' responses. This enables them to more successfully interact and to set up a series of positive reciprocal transactions.

Such intervention is particularly helpful for parents of infants who have sensory deficits (e.g., blindness or hearing impairments), paralysis of limbs, or altered muscle tone. The preterm neonate or infant with a motor deficit, for example, may be unable to respond as a parent expects when cuddling is offered. Thus, the parent may stop cuddling the infant and may feel confused, afraid, or guilty. Similarly, a blind infant will not respond to many of the visual caretaking cues that parents take for granted, e.g., smiling, imitating, holding out toys for visual inspection, eye contact. Thus, therapists can often help parents by demonstrating successful interactions, by focusing on the baby's strengths, and by giving the parents' feedback about their interactions (101). This type of intervention is what Field (137) has termed *interaction coaching*. It is designed to provide parents with helpful demonstrations and feedback so that they can eliminate inadequate interactions and gain successful socialization experiences with their special needs infants.

While such parent intervention has proven successful with infants with sensorimotor problems, it also may be useful with parents of healthy neonates. For example, some mothers of healthy preterm infants have benefited from demonstrations of their newborn's capabilities (138). This is particularly true for first-time or inexperienced mothers, for women who are unfamiliar with infant development, or for women who may give birth to an infant that is "different," i.e., early (preterm), or small. Widmayer and Field (139) illustrated this point with teen-age, economically deprived mothers to whom they demonstrated their preterm infants' abilities. Administering the Brazelton Neonatal Behavior Assessment Scale to the neonates in the presence of the mothers resulted in increased interactions among the mothers and infants and higher scores in mental development that were still evident 4 months later.

Thus, some relatively simple interventions planned for parents before or around the time of birth (such as Widmayer and Field's [139] work illustrates) seem to be successful, but it appears that long-range parent-infant interactions may require ongoing or periodic support over time. As Bronfenbrenner (125) has noted, for some economically disadvantaged families, this intervention may take the form of economic, medical, educational, and community support. For families with newborns who have medical or developmental problems, follow-up is customary.

## Follow-up

Developmental follow-up may take place in many different stages. Follow-up after discharge from a neonatal care nursery is an integral component of any comprehensive program (101, 136). This follow-up may involve facilitating an infant's transition from the hospital to the home (140), where the parents may require reassurance, special equipment, and training in caretaking procedures. In addition, continuous assessment of the infant's health and developmental status is often recommended as well as home visits by various specialists. Later, an early-intervention program or special therapies may be recommended for the infant or child, and various support services may be useful for the parents. These may range from parent support networks (e.g., 141) to respite care, a form of temporary in- or out-of-home care offered to give temporary relief to families with handicapped members (142).

While many of these parent intervention measures have typically been aimed at mothers as the primary caretakers, the current focus in research and outreach is expanding to include more members of the infant's environment. This includes fathers, siblings, and grandparents (143–147) of handicapped infants. Thus, developmental therapists may find themselves placed in new roles in interaction with family systems, as current clinical research explores ways to improve the ecologies of able-bodied and disabled, healthy and unhealthy, infants so that they are given all possible opportunities for potential development and growth. Tinsley and Parke note:

> The ecologically sensitive expansion beyond the mother-infant to include the father-infant relationship is only a first step; an understanding of the full set of relationships among mother, father, and infant, who are recognized as part of a family system, is necessary as well. Finally, recognition of the embeddedness of the family in formal and informal social networks and of the specific direct and indirect ways in which these extrafamilial social systems can alter family functioning is necessary to improve our knowledge concerning the interplay between the individual and the environment, and to better understand the social environment of the premature infant (148, p 107).

Parental support should not end at the neonatal period. Sameroff and Chandler (111) have pointed out how important it is to provide a positive, nurturing environment that will serve to reduce caretaking casualty for at-risk infants and children. While supportive, compensatory intervention may eliminate some of the effects of early complications for infants with special needs, other problems may arise for them or for children not at risk if their family environment is not prepared, over the course of infancy and childhood, to meet their changing needs. It is important, therefore, for therapists as well as parents to be familiar with the basic principles and trends in human development. The remaining chapters of this text will cover many of these trends in infant, child, and adolescent development.

## REFERENCES

1. Luria SE, Gould SJ, Singer S: *A View of Life*. Menlo Park, CA, Benjamin/Cummings, 1981.
2. Mussen PH, Conger JJ, Kagan J: *Child Development and Personality*, ed 3. New York, Harper & Row, 1969.
3. Hetherington EM, Parke RS: *Child Psychology: A Contemporary Viewpoint*. New York, McGraw Hill, 1979
4. Gardner H: *Development Psychology. An Introduction*. Boston, Little Brown & Co, 1978.
5. Kopp C, Parmelee AH: Prenatal and Perinatal Influences on Infant Behavior. In Osofsky JD (ed): *Handbook of Infant Development*. New York, John Wiley & Sons, 1979.
6. Scarr S, Weinberg RA, Levine A: *Understanding Development*. New York, Harcourt Brace Jovanovich, 1986.
7. Friedman SL, Jacobs BS, Werthman MW: Sensory processing in pre- and full-term infants in the neonatal period. In Friedman SL, Sigman M (eds): *Preterm Birth and Psychological Development*. New York, Academic Press, 1981.
8. Denenberg VH (ed): *The Development of Behavior*. Stamford, CT, Sinauer, 1972.
9. Powledge TM: Windows on the womb. *Psychology Today* May:37–42, 1983.
10. Rosenblith JF, Sims-Knight JE: *In the Beginning. Development in the First Two Years*. Belmont, CA, Brooks-Cole, 1985.
11. Humphrey T: Postnatal repetition of human prenatal activity sequences with some suggestions of their neuroanatomical basis. In Robinson RJ (ed): *Brain and Early Behaviour. Development in the Fetus and Infant*. New York, Academic Press, 1969.
12. Milani Comparetti A: Pattern analysis of normal and abnormal development: The fetus, the newborn, the child. In Slaton DS (ed): *Development of Movement in Infancy*. Chapel Hill, University of North Carolina at Chapel Hill, 1981.
13. Timor-Trisch I, Zador I, Hertz RH, Rosen MG: Classification of human fetal movement. *Am J Obstet Gynecol* 126:70–77, 1976.
14. Bower TGR: Repetitive processes in child development. *Scient Am* 235:38–47, 1976.
15. LeCours AR: Correlates of developmental behavior in brain maturation. In Bever TG (ed): *Regressions in Mental Development*. Hillsdale, NJ, Lawrence Erlbaum, 1982.
16. Strauss S: Ancestral and descendant behaviors: the case of U-shaped behavioral growth. In Bever TG (ed): *Regressions in Mental Development*. Hillsdale, NJ, Lawrence Erlbaum, 1982.
17. Bever TG (ed): *Regressions in Mental Development*. Hillsdale, NJ, Lawrence Erlbaum, 1982.
18. Campbell SK: Book review. *Phys Occup Ther Pediatr* 4:116, 1984.
19. Willemsen E: *Understanding Infancy*. San Francisco, WH Freeman, 1979.
20. Cowan WM: The development of the brain. *Scient Am* 241:112–133, 1979.
21. Tanner JM: *Fetus into Man. Physical Growth from Conception to Maturity*. Cambridge, MA, Harvard University Press, 1978.
22. Hubel DH: The brain. *Scient Am* 241:44–53, 1979.
23. Zaichkowsky LD, Zaichkowsky LB, Martinek TJ: *Growth and Development. The Child and Physical Activity*. St. Louis, CV Mosby, 1980.
24. Rodier PM: Chronology of neuron development: Animal studies and their clinical implications. *Dev Med Child Neurol* 22:525–545, 1980.
25. Dobbing J, Sands J: Quantitative growth and development of.human brain. *Arch Dis Childhood* 48:757–767, 1973.
26. Ayres AJ: *Sensory Integration and the Child*. Los Angeles, Western Psychological Services, 1979.
27. Teyler TJ, Chiaia N: Brain structure and development. In Frank M (ed): *A Child's Brain. The Impact of Advanced Research on Cognitive and Social Behaviors*. New York, Haworth, 1984.
28. Blakemore C, Cooper GF: Development of the brain depends on the visual environment. *Nature* 228:477–478, 1970.
29. Hubel DH, Weisel TN: The period of susceptibility to the physiological effects of unilateral eye closure in kittens. *J. Physiol* 206:419–436, 1970.

30. Rice FP: *The Adolescent. Development, Relationships, and Culture.* Boston, Allyn & Bacon, 1984.
31. Chan GM, Ronald N, Slater P, Hollis J, Thomas MR: Decreased bone mineral status in lactating adolescent mothers. *J Pediatr* 101:767–770, 1982.
32. Adams GR, Gullotta T: *Adolescent Life Experiences.* Belmont, CA, Brooks Cole, 1983.
33. Elster AB, McAnaraney ER, Lamb ME: Parental behavior of adolescent mothers. *Pediatrics* 71:494–503, 1983.
34. Lester BM, Garcia-Coll C, Valcarcel M, Hoffman J, Brazelton TB: Effects of atypical patterns of fetal growth on newborn (NBAS) behavior. *Child Dev* 57:11–19, 1986.
35. Fay FC, Smith KS: *Childbearing After 35. The Risks and the Rewards.* New York, Rutledge Books, 1985.
36. Zamenhof S: Nutrition and brain development. *Brain Res Inst Bull* 3:4–11, 1979.
37. Altman J, Das GD, Sudarshan K: The influence of nutrition on neural and behavioral development. I. Critical review of some data on the growth of the body and the brain following dietary deprivation during gestation and lactation. *Dev Psychobiol* 3:281–301, 1970.
38. Lewin R: Starved brains. *Psychol Today* Sept:29–33, 1975.
39. Zimmerman RR, Steere RL, Strobel DA, Hom HL: Abnormal social development of protein-malnourished rhesus monkeys. *J Abnorm Psychol* 80:123–131, 1972.
40. Stechler G, Halton A: Prenatal influences on human development. In Wolman BB (ed): *Handbook of Developmental Psychology.* Englewood Cliffs, NJ, Prentice-Hall, 1982.
41. Grantham-McGregor SM, Stewart M, Desai P: A new look at the assessment of mental development in young children recovering from severe malnutrition. *Dev Med Child Neurol* 20:773–778, 1978.
42. Levitsky DA: Ill-nourished brains. *Natural History* October:6–20, 1976.
43. Levitsky DA, Barnes RH: Nutritional and environmental interactions in the behavioral development of the rat: long-term effects. *Science* 176:68–71, 1972.
44. Ottinger DR, Simmons JE: Behavior of human neonates and prenatal maternal anxiety. In Denenberg V (ed): *The Development of Behavior.* Stamford, CT, Sinauer, 1972.
45. Gunderson VB, Sackett GP: Paternal effects on reproductive outcome and developmental risk. In Lamb ME, Brown AL (eds): *Advances in Developmental Psychology.* Hillsdale, NJ, Lawrence Erlbaum, 1982, vol 2.
46. Levine S, Lewis GW: Critical period for effects of infantile experience on maturation of stress response. In Denenberg VH (ed): *The Development of Behavior.* Stamford, CT, Sinauer, 1972.
47. Levine S: Stimulation in infancy. In *The Nature and Nurture of Behavior. Developmental Psychobiology.* San Francisco, WH Freeman, 1973.
48. Ader R, Conklin PM: Handling of pregnant rats: effects on emotionality of their offspring. *Science* 142:411–412, 1963.
49. Denenberg VH, Whimbey AE: Behavior of adult rats is modified by the experiences their mothers had as infants. In Denenberg VH (ed): *The Development of Behavior.* Stamford, CT, Sinauer, 1972, p 92.
50. Denenberg VH, Rosenberg KM: Nongenetic transmission of information. In Denenberg VH (ed): *The Development of Behavior.* Stamford, CT, Sinauer, 1972, p 96.
51. Denenberg VH, Zarrow MX: Effects of handling in infancy upon adult behavior and adrenocortical activity: Suggestions for a neuroendocrine mechanism. In Walcher DN, Peters, DL (eds): *The Development of Self-Regulatory Mechanisms.* New York, Academic Press, 1971.
52. Fride E, Dan Y, Gavish M, Weinstock: Prenatal stress impairs maternal behavior in a conflict situation and reduces hippocampal benzodiazepine receptors. *Life Sci* 36:2103–2109, 1985.
53. Davidson DA, Short MA: Developmental effects of perinatal heroin and methadone addiction. *Phys Occup Ther Pediatr* 2:1–10, 1982.
54. Streissguth AP, Landesman-Dwyer S, Martin JC, Smith DW: Teratogenic effects of alcohol in humans and laboratory animals. *Science* 209:353–362, 1980.

55. Householder J, Hatcher R, Burns W, & Chasnoff I: Infants born to narcotic-addicted mothers. *Psychol Bull* 92:453–468, 1982.
56. Dishotsky NI, Loughman WD, Mogar RE, Lipscomb WR: LSD and genetic damage. *Science* 172:431–440, 1971.
57. Seliger DL: Effect of prenatal maternal administration of d-amphetamine on rat offspring activity and passive avoidance learning. *Physiol Psychol* 1:273–280, 1973.
58. Mole RH: Conference Report. Radiation risks to the individual in utero. Report of a scientific symposium: radiation risks to the developing nervous system. *Int J Radiat Biol* 49:183–189, 1986.
59. Spezzano C: Prenatal psychology: pregnant with questions. *Psychol Today* May:49–57, 1981.
60. Howell L: Birth defects; Genetic counseling A. Disabling birth defects. In Goldenson RM (ed): *Disability and Rehabilitation Handbook.* New York, McGraw-Hill, 1978, chap 19, pp 235–240.
61. Lampton Christopher: *DNA and the Creation of New Life.* New York, ARCO Publishing, 1983.
62. Zellweger H, Simpson J: *Chromosomes of Man.* Philadelphia, JB Lippincott, 1977.
63. Milunsky A: The prenatal diagnosis of chromosomal disorders. In Milunsky A (ed): *Genetic Disorders and the Fetus.* New York, Plenum, 1979, chap 5.
64. Hughes JG: *Synopsis of Pediatrics.* St. Louis, CV Mosby, 1971.
65. Lydic JS, Windsor MM, Short MA, Ellis TA: Effects of controlled rotary vestibular stimulation on the motor performance of infants with Down syndrome. In Ottenbacher K, Short MA (eds): *Vestibular Processing Dysfunction in Children.* New York, Haworth, 1985.
66. Milunsky A. Sex chromosomes and X-linked disorders. In Milunsky A (ed): *Genetic Disorders and the Fetus. Diagnosis, Prevention, and Treatment.* New York, Plenum, 1979, ch 6.
67. Pashayan HM, Feingold M: Genetic evaluation. In Gabel S, Erickson MT (eds): *Child Development and Development Disabilities.* Boston, Little Brown & Co, 1980, chap 9.
68. Goldenson RM: Muscular Dystrophy. In Goldenson RM (ed): *Disability and Rehabilitation Handbook.* New York, McGraw-Hill, 1978, chap 41.
69. Goldenson RM: Other disorders. In Goldenson RM (ed): *Disability and Rehabilitation Handbook.* New York, McGraw-Hill, 1978, chap 54.
70. Shah SA, Borgaonkar DS: The XYY chromosomal abnormality: some "facts" and some "fantasies." *Am Psychol* 29:357–359, 1974.
71. Milunsky A. Amniocentesis. In Milunsky A (ed): *Genetic Disorders and the Fetus. Diagnosis, Prevention, and Treatment.* New York, Plenum, 1979, chap 2.
72. Campbell S. Diagnosis of fetal abnormalities by ultrasound. In Milunsky A (ed): *Genetic Disorders and the Fetus. Diagnosis, Prevention, and Treatment.* New York, Plenum, 1979, chap 10.
73. Hollander HJ: Historical review and clinical relevance of real-time observations of fetal movement. *Contrib Gynecol Obstet* 6:26–28, 1979.
74. Wittmann BK, Davison BM, Lyons E, Frohlich H, Towell ME: Real-time ultrasound observations of fetal activity in labor. *Br J Obstet Gynaecol* 86(4):278–281, 1979.
75. Mahoney MJ, Hobbins JC: Fetoscopy and fetal blood sampling. In Milunsky A (ed): *Genetic Disorders and the Fetus. Diagnosis, Prevention, and Treatment.* New York, Plenum, 1979.
76. Milunsky A. Genetic counseling. Prelude to prenatal diagnosis. In Milunsky A (ed): *Genetic Disorders and the Fetus. Diagnosis, Prevention, and Treatment.* New York, Plenum, 1979, chap 1.
77. Jorgensen C, Uddenberg N, Ursing I: Ultrasound diagnosis of fetal malformation in the second trimester. The psychological reactions of the women. *J Psychosom Obstet Gynaecol* 4:31–40, 1985.
78. Grobstein C: External human fertilization. *Scient Am* 240:57–67, 1979.
79. Marx JL: Embryology: out of the womb—into the test tube. *Science* 182:811–814, 1973.
80. Andrews LB: Yours, mine and theirs. *Psychol Today* Dec:20–29, 1984.

81. Gillespie SA: Childbirth in the 1980s: what are the options? *Issues Health Care Women* 3:101–128, 1981.
82. Shepperson B: Home or hospital birth? A study of women's attitudes. *Health Visitor* 11:405–406, 1983.
83. Klaus MH, Kennell JH: Labor, birth, and bonding. In Klaus MH, Kennell JH (eds): *Parent-Infant Bonding*. St. Louis, CV Mosby, 1982.
84. Harrison M: *A Woman in Residence*. New York, Penguin, 1982.
85. Ernst EKM: Pioneering interdependence in a system of care for childbearing families. *J Nurs-Midwif* 29:296–299, 1984.
86. Romalis S (ed): *Childbirth. Alternatives to Medical Control*. Austin, University of Texas Press, 1981.
87. Murray AD, Dolby RM, Nation RL, Thomas DB: Effects of epidural anesthesia on newborns and their mothers. *Child Dev* 52:71–82, 1981.
88. Blank JJ: Electronic fetal monitoring. Nursing management defined. *J Obstet Gynecol Neonat Nurs* 14:463–467, 1985.
89. Brackbill Y: Obstetrical medication and infant behavior. In Osofsky JD (ed): *Handbook of Infant Development*. New York, John Wiley & Sons, 1979.
90. Robertson EG: Prenatal factors contributing to high-risk offspring. In Field M (ed): *Infants Born at Risk*. New York, SP Medical & Scientific Books, 1979.
91. Gottfried AW: Intellectual consequences of perinatal anoxia. *Psychol Bull* 80:231–242, 1973.
92. Sameroff AJ: Early influences on development: fact or fancy? *Merrill-Palmer Q* 21:267–293, 1975.
93. Pelletier JM, Palmeri A: High-risk infants. In Clark PN, Allen AS (eds): *Occup Ther Children*. St. Louis, CV Mosby, 1985.
94. Connolly BH: Neonatal assessment: An overview. *Phys Ther* 65:1505–1513. 1985.
95. Demilio PA: Periventricular-intraventricular hemorrhage in the neonate: a review. *Phys Occup Ther Pediatr* 3:45–55, 1983.
96. Newman LF: Social and sensory environment of low birth weight infants in a special care nursery. An anthropological investigation. *J Nerv Ment Dis* 169:448–455, 1981.
97. Gottfried AW, Wallace-Lande P, Sherman-Brown S, King J, Coen C, Hodgman JE: Physical and social environment of newborn infants in special care units. *Science* 214:673–675, 1981.
98. Speidel PB: Adverse effects of routine procedures on preterm infants. *Lancet* 6:864, 1978.
99. Pelletier JM, Short MA, Nelson DL: Immediate effects of waterbed flotation on approach and avoidance behaviors of premature infants. In Ottenbacher K, Short MA (eds): *Vestibular Processing Dysfunction in Children*. New York, Haworth, 1985.
100. Campbell SK: Effects of development intervention in the special care nursery. In Wolraich ML, Routh D (eds): *Advances in Developmental and Behavioral Pediatrics*. Greenwich, CT, JAI Press, 1984, vol 6.
101. Anderson J, Auster-Liebhaber J: Development therapy in the neonatal intensive care unit. *Phys Occup Ther Pediatr* 4:89–106, 1984.
102. Kirschbaum MJ: General principles of intervention for the preterm infant. *AOTA Dev Disabil Special Interest Section Newslett* 8:1,4, 1985.
103. Case J: Positioning guidelines for the premature infant. *AOTA Dev Disabil Special Interest Section Newslett* 8:1,2, 1985.
104. Redditi JS: Occupational and physical therapy treatment components for infant intervention programs. *Phys Occup Ther Pediatr* 3:33–44, 1983.
105. Stern FM: Physical and occupational therapy on a newborn intensive care unit. *Rehabil Nurs* Jan–Feb:26–27, 1986.
106. Knobloch H, Pasamanick B: Environmental factors affecting human development, before and after birth. *Pediatrics* Aug:210–218, 1960.
107. Olson GM, Lamb ME: Premature infants: cognitive and social development in the first year of life. In Stack JM (ed): *The Special Infant. An Interdisciplinary Approach to the Optimal Development of Infants*. New York, Human Sciences Press, 1982.

108. Schaefer M, Hatcher RP, Barglow PD: Prematurity and infant stimulation: A review of research. *Child Psychiatr Human Dev* 10:198–212, 1980.
109. Denenberg VH, Zarrow MX: Effects of handling in infancy upon adult behavior and adrenocortical activity: suggestions for a neuroendocrine mechanism. Special remarks of Dr. Lewis Lipsitt. In Walcher DN, Peters DL (eds): *The Development of Self-Regulatory Mechanisms.* New York, Acadmic Press, 1971.
110. Escalona SK: Babies at double hazard: early development of infants at biologic and social risk. *Pediatrics* 70:670–675, 1982.
111. Sameroff A, Chandler MJ: Reproductive risk and the continuum of caretaking casualty. In Horowitz FD, Hetherington M, Scarr-Salapatek S, Siegel G (eds): *Review of Child Development Research.* Chicago, University of Chicago Press, 1975, vol 4.
112. Horowitz FD: Toward a model of early infant development. In Brown CC (ed): *Infants at Risk. Assessment and Intervention. An Update for Health-Care Professionals and Parents.* New Brunswick, NJ, Johnson & Johnson, 1981.
113. Werner EE, Smith RS: An epidemiologic perspective on some antecedents and consequences of childhood mental health problems and learning disabilities. *J. Am Acad Child Psychiatr* 2:292–306, 1979.
114. Klaus MH, Kennell JH: *Maternal-Infant Bonding.* St. Louis, CV Mosby, 1976.
115. Klaus MH, Kennell JH: The family during pregnancy. In Klaus MH, Kennell JH (eds): *Parent-Infant Bonding.* St. Louis, CV Mosby, 1982.
116. Zeanah CH, Kenner MA, Steward L, Anders TF: Prenatal perception of infant personality: A preliminary investigation. *J Am Acad Child Psychiatr* 24:204–210, 1985.
117. Blake A, Stewart A, Turcan D: Parents of babies of very low birth weight: long-term follow-up. In *Parent-Infant Interaction.* New York, Elsevier, 1975.
118. Klaus MH, Kennell JH (eds): *Parent-Infant Bonding.* St. Louis, CV Mosby, 1982.
119. Belsky J, Tolan WJ: Infants as producers of their own development: an ecological analysis. In Lerner RM, Busch-Rossnagel NA (eds): *Individuals as Producers of Their Development.* New York, Academic Press, 1981.
120. Goldberg S: Parent infant bonding: Another look. *Child Dev* 54:1355–1382, 1983.
121. Lamb ME: Early contact and maternal-infant bonding. One decade later. *Pediatrics* 70:763–767, 1982.
122. Richards MPM: Effects on development of medical interventions and the separation of newborns from their parents. In Schaffer D, Dunn J (eds): *The First Year of Life. Psychological and Medical Implications of Early Experience.* New York, John Wiley, 1979.
123. Yogman MW: Development of the father-infant relationship. In Fitzgerald HE, Lester BM, Yogman MW (eds): *Theory and Research in Behavioral Pediatrics.* New York, Plenum, 1982.
124. Palkovitz R: Father's birth attendance, early contact, and extended contact with their newborns: a critical review. *Child Dev* 56:392–406, 1985.
125. Bronfenbrenner U: Is early intervention effective: Facts and principles of early intervention: a summary. In Clarke AM, Clarke ABD (eds): *Early Experience: Myth and Evidence.* New York, Free Press, 1976.
126. Williams D, Foote D: Bringing babies to school. *Newsweek* Feb 1:69, 1982.
127. Ferland F, Piper C: Evaluation of a sensory-motor education programme for "parents-to-be". *Child Care Health Dev* 7:145–154, 1981.
128. Larson CP: Efficacy of prenatal and postpartum home visits on child health and development. *Pediatrics* 66:191–197, 1980.
129. Inside Developments. Need for prenatal care. *Dev Child Dev Inst Univ North Carolina Chapel Hill* 11(3):4, 1986.
130. Prenatal care is declining, agency finds. *New York Times* Jan 8:E5, 1984.
131. Brazelton TB: *Neonatal Behavioral Assessment Scale.* Philadelphia, JP Lippincott, 1973.
132. Hulme JB, Nieman K, Miller K: Obstetrics in the physical therapy curriculum. *Phys Ther* 65:51–53, 1985.
133. Wikler L, Waslow J, Hatfield E: Chronic sorrow revisited: parent vs. professional depiction of the adjustment of parents of mentally retarded children. *Am J Orthopsychiatr* 51:63, 1981.

134. Petersen P, Wikoff RL: Home environment and adjustment in families with handicapped children. A canonical correlation study, *Occup Ther J Res* 7: 67–82, 1987.
135. Goldberg S: Premature birth: consequences for the parent-infant relationship. *Am Scient* 67:214–220, 1979.
136. Dickson JM: A model for the physical therapist in the intensive care nursery. *Phys Ther* 61:45–48, 1981.
137. Field T: Interaction coaching for high risk infants and their parents. In Moss HA, Hess R, Swift C (eds): *Early Intervention Programs for Infants*. New York, Haworth Press, 1982.
138. Brazelton TB: Demonstrating infants' behavior. *Children Today* 10:5, 1981.
139. Widmayer SM, Field TM: Effects of Brazelton demonstrations for mothers on the development of preterm infants. *Pediatrics* 67:711–714, 1981.
140. Affleck G, Tennen H, Allen DA, Gershman K: Perceived social support and maternal adaptation during the transition from hospital to home care of high-risk infants. *Infant Ment Health J* 7:6–18, 1986.
141. Reed D: Partners in growth: a parent support network. *AOTA Dev Disabil Special Interest Section Newslett* 10:3–7, 1987.
142. Short-DeGraff M. Kologinsky E: Respite care: Roles for therapists in support of families with handicapped children. *Phys Occup Ther Pediatr* in press.
143. Meyer DJ, Vadasy PF, Fewell RR, Schell G: Involving fathers of handicapped infants: translating research into program goals. *J Div Early Childhood* 5:64–72, 1982.
144. Gallagher JJ, Vietze PM: *Families of Handicapped Persons. Research Programs and Policy Issues*. Baltimore, MD, Paul H. Brookes, 1986.
145. Meyer DJ, Vadasy PF, Fewell RR: *Living with a Brother or Sister with Special Needs*. Seattle, University of Washington Press, 1985.
146. Powell TH, Ogle PA: Brothers and sisters: addressing unique needs through respite care services. In Salisbury CL, Intagliata J (eds): *Respite Care. Support for Persons with Developmental Disabilities and their Families*. Baltimore, Paul H Brooks, 1986.
147. Vadasy PF, Fewell RR. Meyer DJ: Grandparents of children with special needs: insights into their experiences and concerns. *J Div Early Childhood* 10:36–44, 1986.
148. Tinsley BR, Parke RD: The person-environment relationship: Lessons from families with preterm infants. In Magnusson D, Allen VL (eds): *Human Development. An Interactional Perspective*. New York, Academic Press, 1983.

# 4

# SENSORY, PERCEPTUAL, AND MOTOR DEVELOPMENT

# Section 4

Section 4 looks at sensation and perception and motor development, which are of particular interest to occupational and physical therapists. Chapter 8 examines the development of each of the sensory systems and looks at the implications of their emergence on behavioral organization. Also examined are developmental changes in learning abilities, as they are affected by sensory changes and as they are used in clinical assessment of sensory function. The coordination of and significance of movement to sensory and perceptual development are examined in the context of early experience research, which is reported in chapter 9. The implications of this research are related to the clinical fields of early intervention and infant stimulation.

Chapter 10 continues the topic of sensorimotor development but changes focus from sensation and perception to gross and fine motor development. This chapter, written by a physical therapist, examines the major developmental motor milestones from birth through childhood and adolescence and illustrates how skill acquisition is based on prior sensorimotor accomplishments.

# 8

# Sensory and Perceptual Development and Behavioral Organization

All living, moving organisms interact with the world around them, find meaning in it, and respond appropriately; i.e., they "make sense" of their environments. Birds correctly identify twigs and brush for nests, recognize and respond to the colors and calls of other birds, and correct their flight for mild or strong gusts of wind. Frogs recognize and catch insects for food, migrate distances to mating territories, and hide from humans seeking them in ponds. Humans recognize and react to visual patterns and shapes and colors as well as distinguishing between such auditory stimuli as rock-and-roll or classical music, the telephone, and train whistles in the distance. We recognize visual shapes in nature and classify similar objects into categories such as houses, factories, and barns; trees, shrubs, and plants; cars, trucks, and buses; infants, children, and adults. We also classify objects that we touch as smooth, rough, hot, or cold, and we classify tastes and odors. All this involves making sense out of the world and has to do with the topics of this chapter, sensory and perceptual development.

# SENSATION AND PERCEPTION AS PARTS OF A COMPLEX SYSTEM

Making sense out of the world requires receiving, adequately processing, and interpreting sensory information, Thus, we interpret a spherical object approaching us as a beach ball tossed in our direction, or we interpret hexagonal red signs with the four letters "S-T-O-P" as a cue to stop a car at an intersection. Golfers hearing the word "Fore!" automatically attribute correct meaning to the sound and respond appropriately. This ability to receive and interpret information involves the two complementary processes, sensation and perception.

*Sensation* is the reception of sensory information, and *perception* is the attribution of meaning to it. An integrated study of sensation and perception takes into consideration an entire complex of factors involving the nature, context, and meaning of environmental/sensory stimulation; the history and biopsychosocial nature of the individual; and the characteristics and meaning of the individual's environment (Table 8.1).

## Interaction of Sensation and Movement

The interaction of sensation and movement has been previously discussed in different parts of this text. Various sensory-motor, stimulus-response, organism-environment models (e.g., see Figs. 5.4–5.6) and their significance to occupational and physical therapy have been described. Consistent factors in these models are the concepts of input (sensation), output (response), and feedback.

The input, or information we receive about our environments, comes to us in the form of energy that is picked up by specialized cells in our bodies, *sense receptors*, and is then relayed to our brains. This neural

Table 8.1
Factors Studied in an Integrated Approach to Sensation and Perception

Nature and dimensions of sensory stimuli
Physiology of sensation (sensory processing)
How meaning is assigned to events in the environment (perception and cognition)
Organism's genetic and environmental history and how those affect the meaning of sensory events (the interaction of past and present genetics, species specific behaviors, and learning)
Significance and variability of environmental contexts (ecology)
Organism's reactions (behavior, motor responses)

"handling" of information is called *sensory processing*. It involves not only reception but also relay and transmission of stimulation. Sensory information is relayed along *afferent* neurons to the brain, which itself is a complex sensory receiving system. Once the brain receives sensory information and processes it, two functions occur: The brain directs the motor system to make a response and monitors and provides feedback to the original sensory receptors. The information that is relayed away from the brain is carried by *efferent neurons*, which provide neural information to peripheral sense receptors as well as to muscles. The muscles, in turn, are directed to respond according to the sensory message received. In most cases, this occurs instantaneously and involves more than one sensory system; for example, consider the following:

You are riding on a crowded bus, standing in the center of the aisle, holding onto an upright support, and watching ahead. The bus brakes suddenly, and you are thrown off balance. At that time, your body shifts and you receive pertinent sensory information from the bottoms of your feet, from numerous muscles and joints that begin to shift weight, and from your eyes as your position changes. In addition, as your equilibrium shifts, you may also hear brakes squeal, as well as feel other people into whom you bump. In an instant, all this sensory information is relayed to the brain, processed, coordinated; your body, seemingly automatically, responds by directing your hand to hold tighter to the support, shifting weight, shuffling feet, regaining an upright position, and mumbling an apology to the person you bumped into.

Living organisms do not passively receive sensory information; they process it through a complex chain of neuroanatomical events, then interpret the information, and respond to it with muscular reactions. In many cases, the only clear evidence that sensory information has been received, accurately processed, and interpreted is the resultant motor response. Thus, the golfer hears "FORE!", looks around, and gets out of the way. Amphibians sense light and heat and sit on rocks in the sun to absorb the warmth. Babies become hungry or overheated and wriggle and cry. Cats see a leaf flutter across the lawn and scurry to chase it. We quickly withdraw our hands from a hot stove, scratch at a spot on our bodies that is tickled by a piece of thread, or readjust our footing and thrust out our arms if we are about to fall.

The importance of most sensory stimulation, especially that received from the environment, is that it provides an impetus for action. Most of the sensory information organisms receive is significant for their survival, and that survival depends upon some kind of action. For example, an animal sees, hears, and smells prey and therefore runs to catch it; the organism feels excessive cold and seeks shelter; the organism tastes something very bitter and spits it out; or the organism hears the wingbeat or hoofbeat of a predator and runs to escape. Sensation, for the most part, begets movement, hence the common term "sensorimotor," which recognizes this unity (Fig. 8.1).

A full understanding of sensory processing and motor responses necessitates looking at the connection between brain and muscle systems, at how muscles develop and regulate movement, and how and when the brain is included or excluded from reactions to sensory information. The study of perception additionally involves examining the capacity to handle and process information, which in turn involves attention, memory, reasoning, and problem solving. This is even further complicated when attempting to examine how all these systems change and interact over time during the course of human development. Clearly, all this information is beyond the

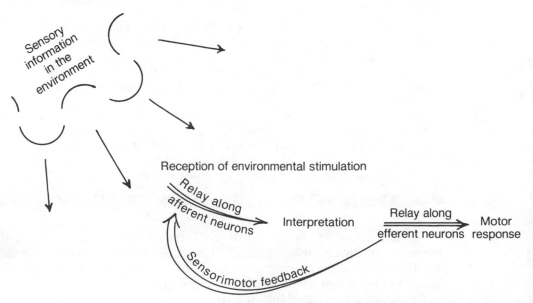

**Figure 8.1.** Simple model of sensorimotor interaction, including sensorimotor feedback, also known as "reafference."

scope of this chapter, although some basic discussion is essential for explaining normal sensory and perceptual development. Readers who want more details about specific sensory systems and neural processing of sensory information may consult a textbook in anatomy and physiology, physiological psychology, or sensory processing.

# Cognitive-Perceptual-Motor Interactions

In the occupational therapy and physical therapy professions, sensation, perception, cognition, and motor development are often simultaneously studied under the rubric "cognitive-perceptual-motor development" (1). While this integrated approach provides a more appropriate and holistic view of behavior, it also poses a challenge to the clinical educator as well as the student because of the difficulty of studying all these systems at once. To examine each of these systems in detail, one must arbitrarily be selected as a starting point for analysis. In this text, the chosen starting point is the topic of sensory and perceptual development. While cognition and motor skills must be considered as they relate to sensation and perception, a detailed description of motor and cognitive changes during development will not be elaborated until chapters 9, 10, and 11.

An interesting question to ponder is, How do infants think? If, as most of us believe, thinking is some kind of internal dialogue, then how do infants, without language, process information? Analyzing this issue, Piaget concluded that early thinking is really sensorimotor in nature. Thus, early sensory and motor functions are considered the bases of cognition. This is further illustrated by the common definition of both cognition and perception as "information processing." Much of adult information processing is in the form of words or thoughts that may be based on abstract representations. Adults, with massive stores of previous information and experiences, process information in a sophisticated fashion. Infants, however, do not have words to organize their thoughts and therefore must use some other way of coding information. That method of coding knowledge of the world without words is with schemata.

The schema is a nonverbal mental representation of experience and is the first form of the infant's knowledge (2). Psychologist Jerome Kagan explains:

> Schemata exist in all modalities—auditory, olfactory, tactile, taste, and visual. The ability to recognize a melody, the fragrance of a rose, the feel of velvet, or the taste of an apple is possible because one has created a schema for each of these experiences. In all of these examples, however, the schema represents the pattern of physical qualities in the original event; the schema is not a symbolic or linguistic representation. Further, the schema is to be contrasted with an image, which is a consciously elaborated representation created from the schema. Because conscious mental effort is required to generate an image from the more abstract schemata, the young infant probably has no images, only schemata. (2, p. 206)

Thus, the initial forms of knowledge (also known as cognition) are sensorimotor or perceptual in nature. This illustrates the close tie between sensation, perception, and cognition.

This chapter's description of the development of sensory and perceptual abilities focuses primarily on changes that occur very early in life. This is due to two major factors. First, many of the sensory and perceptual systems become functionally adult-like during infancy, so there is relatively little change to describe beyond that time. Second, in childhood, with increased sophistication of and integration of sensory/perceptual abilities and the acquisition of language and symbolic representations of thought, the traditional schemata of childhood become much more abstract. They change from simple sensorimotor to abstract schemata such as concepts and complex intersensory interpretative skills. The focus of study therefore shifts away from a sensorimotor to a cognitive emphasis. This emphasis does not represent an exclusion of the sensory and perceptual abilities of the child; it only means that in childhood these abilities become more representational in nature and are given a different name. They are therefore studied in Chapter 11, which discusses cognition.

# SENSORY RECEPTION AND PROCESSING

Sensory systems have been classified in different ways, such as according to their location in or on the body (Table 8.2) or by the nature or type of stimulation they receive (Table 8.3). The actual numbers and types of sensory systems are subject to some debate. The most commonly recognized modalities are the basic five (Table 8.3). The other less commonly recognized, but no less important, is the proprioceptive system, which deals with movement and position of the body in space. It processes information generated from the muscles and joints as well as from the vestibular system, which regulates balance. Some recent accounts identify many more than these basic sensory systems. Rivlin and Gravelle (3), for example, claim

**Table 8.2**
**Classification of Sense Receptors**

| | |
|---|---|
| Exteroceptors | — Receive information from sources external to the body |
| Interoceptors | — Receive information from within the body (viscera) |
| Proprioceptors | — Receive information from the muscles, joints, and tendons |

**Table 8.3**
**The Five Most Commonly Recognized Senses**

Vision (seeing)
Audition (hearing)
Somesthesis (feeling)
Gustation (tasting)
Olfaction (smelling)

there may be as many as 17 interrelated senses. These include, in addition to the basic five senses: the ability to pick up chemical signals in the environment through a form of nasal perception; nociception or pain perception; visual/contour and contrast, which is separate from color perception; response and synchrony to the sun; and other undiscovered extrasensory abilities such as the detection of magnetic fields and auras (3).

While many researchers do not recognize as many as 17 senses, they concur that sensory processing is complex and not fully understood. The anatomy and physiology of each sensory system as well as the neuroanatomy necessary for conducting sensory information through the body are complex. To further understand sensory development requires examining how each system evolves prenatally and over the life span. Clearly, in-depth analyses of all these topics cannot be handled in one chapter; however, a brief introduction is provided regarding sensory reception and processing, and brain development is described as it relates to sensation and perception. Following a discussion of the measurement of sensory and perceptual abilities, the developmental changes characteristic of individual sensory modalities are explored. In many cases, generalizations about the sensory systems have been drawn from studies on vision, the most highly studied and understood in regard to neuroanatomy and perception.

# Characteristics of Environmental Stimulation

One of the first steps in understanding sensation and sensory development involves analyzing the characteristics of sensory stimulation and the corresponding receptors. Physical stimulation from the environment comes in a variety of forms such as light, sound, pressure, heat, and molecules, all of which vary in characteristics. For example, light waves are a different form of energy than sound waves, and each is received by systems that are quite different from one another. In addition to differences between the various forms of physical energy, there is considerable variability in the characteristics of each form of physical energy. Light waves, for example, vary in amplitude, frequency, and duration, necessitating that visual receptors must accommodate these different dimensions.

The study and analysis of the dimensions of sensory stimulation is one of the oldest fields in psychology. In the 1830s, both Weber and Fechner formulated psychophysical laws regarding the nature and relationship of sensory stimulation (physical energy) and the different sensory modalities. This information about the characteristics of different forms of sensation has added to our understanding of the physiology of sensory reception. However, traditional studies of sensation tended to focus primarily on exteroceptive, to the exclusion of organism-generated, information. Such interoceptive or proprioceptive stimulation, which tells us about pain, movement, equilibrium, alignment and position of body parts, and the relation of the body in space, are all less understood.

Further, traditional classifications of sensory stimulation tended to focus on specific forms of stimulation in isolation rather than upon the interaction of sensory modalities. Walsh (4) points out in regard to the ecology of the brain that numerous dimensions of environmental stimulation exist, and practitioners and researchers are challenged to regard sensory development in terms of the meaning and context of various forms of stimulation. He notes, for example, that to clearly understand the effects of environmental stimulation one must consider the organism's learning history, the variety of stimulation that occurs along with each stimulus input, and the entire ecology of the organism (Table 4.7).

# Mechanics of Sensory Reception

Before each sensory system is examined individually, some general characteristics of sensory reception and processing need to be introduced. Sensory receptors vary according to the nature of the stimulation they receive. Exteroceptors are geared to receive stimulation from the environment, whereas interoceptors or proprioceptors (Table 8.2) receive information from within the body. Proprioceptors located in the muscles, tendons, and joints, will respond as the limbs move and change position, i.e., from organism-generated stimulation. In contrast, visual, auditory, or olfactory receptors receive different forms of environmentally generated stimulation. Thus, each sensory system must have its own mechanism for receiving and processing corresponding sensory information. The visual system, for example, which responds to light waves, is quite different from the skin sense, which responds to mechanical stimulation in the form of touch, pressure, or vibration.

## Peripheral Reception

The actual activation of sense receptors is sensation, which involves not only responding to the energy but directing attention toward the source of stimulation (5). Once a specialized receptor receives environmental stimulation in the form of physical energy, it must change it into a form that can be relayed along sensory neurons. This process of altering environmental stimulation into neural energy is called *transduction*. This is often assisted by additional processes performed by *accessory structures*, which amplify or modulate the initial energy so that it can be more effectively transduced and then relayed to the brain (6). Sensory receptors also interact with one another and change their sensitivity levels (*thresholds*) after repeated stimulation. A characteristic of most sensory systems is that the receptors undergo *adaptation*, which is a diminution of responsiveness following repeated stimulation. How this process occurs—i.e., whether receptors become less sensitive or fatigued is unclear—but the existence of adaptation suggests that sensory systems are geared to process stimulus change. The significance of this will be discussed further in Chapter 9.

The sensory receptors of different species are specialized to receive or process the forms and characteristics of physical stimulation that are important for their survival. One of the determinants of whether or not stimulation is processed by a particular species is the sensory threshold. A threshold is the minimum amount of stimulation that is needed to effect a response, and thresholds vary among sensory receptors as well as across species. For example, the visual system of humans receives and processes only certain wavelengths of light that are quite different from some of the visual stimulation processed by insects or birds, and these species process visual information that is beyond the normal human visual threshold. Similarly, the olfactory system of humans is relatively ineffective and undeveloped when compared with that of canines, which require a refined sense of olfaction for survival. Thus, the species-specific significance of sensory stimulation and the physiology of each sensory system will determine in part which and how each system develops.

Specific sensory processing systems change over the course of an organism's life span depending upon the growth and myelination of the central and peripheral (afferent and efferent) nerves, the ontogeny of specific sense receptors and their accessory structures, and the health and integrity of each organism's internal anatomy and physiology. Understanding the complete development of each individual sensory system is complicated. It requires an examination of the prenatal and postnatal changes of each system, as well as changes in the central nervous and the motor effector systems. In addition, since sensory information is generally interpreted (i.e., perceived) according to some context, sensation and perception are closely related. In developmental terms, sensation tends to precede perception. Sensory systems respond to "energy" and direct the organism's attention to the various forms of environmental stimulation. In contrast, perceptual systems respond to "information" and are geared toward analyzing patterns of sensory information. Thus, in normal development, sensation precedes perception, as it takes longer for the organism to develop a capacity to handle and process "patterns of information" than it does to receive and process "energy" (5). Information processing involves cognition, which in any organism, necessitates interactive experiences to mature. Neural maturational changes during the course of development affect the capacity of the organism to handle sensory/perceptual information.

## Central Sensory Processing

The brain is a sophisticated and competent sensory-processing center. Some regions of the brain such as the thalamus (Fig. 8.2) are located centrally within the brain and are geared to receive all sensory information and to relay it to other specialized regions. These other regions, such as the cerebral cortex of each brain hemisphere, are specialized to process information only from certain sensory systems. Thus, particular sensory cortical areas are associated with specific sensory functions. Other cortical regions are intersensory, meaning that they combine, process, and coordinate information from different sensory systems as well as motor systems, accounting in part for why we tend to look in the direction of sound.

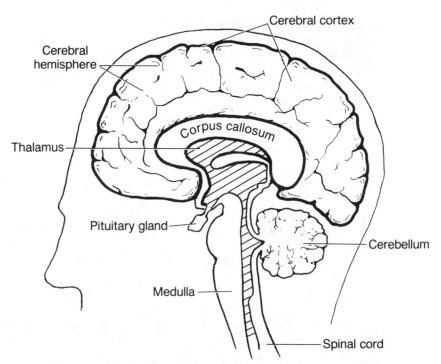

**Figure 8.2.** A section through the center of the brain. The large cerebral hemisphere with its outer surface, the cerebral cortex, occupies the outer layer of the brain. Deeper within the brain are the corpus callosum, which connects the two cerebral hemispheres, and the thalamus, a relay center for sensory information. The medulla controls heartbeat and basic regulatory functions, and the cerebellum regulates balance. The pituitary gland, located at the base of the brain, is part of the endocrine system.

### Intersensory and Sensorimotor Integration

Objects in the environment are typically not unidimensional. They have multiple sensory characteristics, and when we encounter them, we process those characteristics all at once. Thus, a squealing, happy puppy not only makes noise but also looks and feels soft and has a distinctive odor. A meal served at an elegant restaurant is made not only to taste good but also to smell and look appealing. We humans, and other organisms, are intersensory and sensorimotor in nature. We process a myriad of sensory stimulation simultaneously, and then we respond according to the messages received. The coordination of sensory and motor information is important to survival. Threatening sensory information often requires immediate motor response, which if it is not forthcoming, may cause harm or death to the organism.

During sensory processing, the brain is simultaneously functionally specific and integrated as a whole. Thus, certain neural regions process information specific to a sensory function (i.e., functional specificity) while at the same time other regions pull that information together. This pulling together of information is known generally as sensory integration.

Sensory integration is also used in occupational and physical therapy to refer to a specific form of intervention, which according to its originator, Ayres (7), is based on the premise that in certain clinical populations specific activities may facilitate the integrative capacities of the brain. For a review of the relationship of this form of intervention to neural change, the reader may consult Ayres work (7, 8) as well as critical reviews by Cratty (9) and by Ottenbacher and Short (10).

Integration means coordination, of which there are different kinds ongoing. The coordination of sensory modalities with one another is known as *intersensory* or *cross-modal communication*. The coordination of sensory and motor information is known as *sensorimotor integration*. The brain is structured for this type of intercommunication through its special anatomy and physiology. The human brain comprises two hemispheres that generally receive sensory information regarding opposite sides of the body. Thus, the right hemisphere receives sensory information and regulates motor effector systems on the left side of the body, and vice versa. Neither of these hemispheres, however, works in isolation. Information from each side of the body is relayed throughout the brain via a wide bridge of fibers called the *corpus callosum*, which connects the two cerebral hemispheres (Fig. 8.3). The corpus callosum allows interhemisphere communication—which is necessary for both sides of the body to work together—in what is termed *bilateral coordination*. As chapter 10 will examine in more detail, bilateral coordination is essential for such gross motor activities as running and catching a ball as well as for such fine motor tasks that require using both hands together such as buttoning or zipping, or skilled activity such as needlework, holding and hammering a nail, or playing a piano.

With all the intersensory, sensorimotor, and interhemisphere communication going on, it is easy to see that the brain functions as a holistic, sensory and motor integrating system that coordinates organism-generated and environmentally generated sensory information. Thus, the development of coordinated sensory abilities (and, as we will discuss in chapter 10) is dependent upon the level of maturity and the integrity of the brain.

## Brain Development

A complete understanding of sensory development cannot exist without simultaneously considering the development of the brain. Since each hemisphere of the brain receives sensory input primarily from one side of the body, an important function of the corpus callosum is to transfer sensory information from one side of the brain to the other. This interhemisphere communication, Galin and colleagues (11, p. 1330) note, "makes possible a world picture integrated across the midline" of the body. The corpus callosum, while functional and intact at birth, still continues to myelinate after birth. With the maturation of the callosum, the "cross talk" in the brain becomes more sophisticated and integrated, within and between sensory modalities, throughout childhood (11). This does not mean, however, that communication between sensory modalities does not occur early in life, only that it becomes more efficient and evident with maturation.

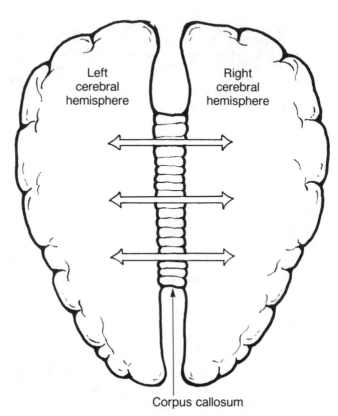

**Figure 8.3.** The corpus callosum is a wide band of fibers that connects the two cerebral hemispheres and enables cross-communication and intersensory integration.

The implications of this gradual development of neural structure connecting brain hemispheres extends beyond sensorimotor development. As Galin and his colleagues (11, p. 1331) note, "The childhood development of anatomical connections between brain subsystems may in fact be related to the parallel development of integration between mental subsystems, such as parts of the personality, leading to unity of the self." Thus, sensory and neural development are interdependent and affect other important aspects of the developing person.

Another developmental feature of the brain, which is often linked with hierarchical models of brain function (see chapter 6), is the gradual emergence of neural control by the slowly maturing, phylogenetically newer cortex. Brain activity, while constant and ongoing, is conveniently portrayed as functioning in two active, opposing fashions. Individual neurons, or entire neural regions, are viewed as having one of two effects on one another. One effect is *excitatory* in nature and is comparable to pushing the gas pedal in a car. The other effect is *inhibitory* in nature and is comparable to stepping on a brake pedal.

Hierarchical models of brain development view the central nervous system as developing gradual inhibitory control as the neural tissue matures upward and outward. Thus, phylogenetically newer regions such as the cortex are assumed, in some developmental models, to exert gradual inhibitory influence over older regions of the brain. This model of gradual neural inhibitory control is used to explain how behavior becomes progressively more regulated and controlled during development. Thus, for example, paying attention, becoming quiet, and organizing and controlling behavior, all of which are closely related to the reception and interpretation of sensory information, have been linked to gradual development and inhibition of neural tissue. The further significance of attentional mechanisms and behavioral organization in sensory development is discussed in the next section of this chapter.

# MEASURING SENSATION AND PERCEPTION: EVIDENCE OF ORGANIZED BEHAVIOR

Understanding the development of the different sensory systems provides important information about the capacities of organisms, yet it is a difficult area to study, especially with nonverbal subjects such as animals, infants, or clients with impaired communication skills. With verbal subjects, assessing sensory function can be accomplished by asking them whether they hear, see, feel, or taste different stimuli. In addition, they can be asked to *discriminate* (i.e., differentiate between) various dimensions of stimuli, by inquiring of them, for example, which stimulus is heavier, lighter, colder, louder, sweeter, brighter, etc. With noncommunicative subjects, alternative procedures need to be used.

The determination of the sensory abilities of animals and of human fetuses and infants has posed a creative challenge to which developmental researchers, particularly in the field of vision, have responded with unique and innovative procedures for "communicating" with their subjects. Many of these specific procedures will be discussed in the following sections as they are encountered in regard to their corresponding sensory systems. Other procedures that are used with many different modalities are described in general here so that they will not have to be explained separately for each system. Understanding some of these procedures may prove useful for clinicians to gain insight into the specific methods as well as their potential applications for clinical evaluation of sensory modalities. Sensory testing is an integral component of physical and occupational therapy practice. Clinicians need to know what kinds of sensory information their patients are receiving so that they can effectively communicate with them as well as plan treatment strategies geared toward normalizing sensory and motor abilities. Sensation is a complementary process to movement and to adaptation to the environment, thus it is an area of significance in both physical therapy and occupational therapy evaluation and treatment.

# Neural Recordings

Sensory responsiveness in nonverbal, nonalert, or noncooperative subjects can often only be "inferred" by neuroanatomical or behavioral responses, and sensory discrimination is often difficult, if not impossible, to accurately assess. Neural recordings are obtained by using surface electrodes to gauge the electrical discharge of a brain area or by using microelectrodes (which are approximately 1/10,000 mm in diameter) for measuring the output of single neural units (12). The recordings are often used to infer sensation by monitoring the neural response while a corresponding sensory stimulus is presented. If, as the stimulus is administered or varied, the neural output accordingly reacts, then it is assumed that the sensory system is functional, at least in regard to the particular stimulus presented. Gottlieb (13), however, points out that the activity of a sensory neuron does not necessarily imply sensory processing or perception. Thus, the presence or absence or the nature of a behavioral response is often used in conjunction with neural recordings.

# Behavioral Considerations in Sensory Testing

Behavioral responses are commonly observed during sensory testing (Table 8.4) but problems often occur when interpreting behaviors, especially those of infants. For example, infants who turn in the direction of sound or who

**Table 8.4**
**Behaviors Observed in Studies of Infant Sensory Processes[a]**

|  |
| --- |
| Motor Responses of face |
|    Sucking |
|    Yawning |
|    Crying |
|    Visual fixation |
| Movement of the trunk and extremities |
|    Head turning |
|    Arm or leg flexion or extension |
|    Finger and toe movements |
|    Rolling |
|    Arching |
|    Whole body startles |
| Neuromuscular changes |
|    Muscle tone |
|    Electromyogram |
| Change in respiration |
| Flushing of the skin |
| Psychophysiological responses |
|    Heart rate acceleration or deceleration |
|    Blood pressure change |
|    Pupillary dilation or constriction |

[a] Adapted from Kaye H: Sensory processes. In Reese HW, Lipsitt LP (eds): *Experimental Child Psychology.* New York, Academic Press, 1970, chap 2, p 34.

withdraw their feet from cold are presumed to hear and to feel cold, respectively. But infants who do *not* turn toward sound nor withdraw their feet from cold are not necessarily lacking sensation. The lack of behavioral responsiveness may imply any number of normal or abnormal circumstances, ranging from motor deficits and lack of motor control to wavering attention. Infants move rapidly from one *behavioral state* to another, vary immensely in their attention and distractibility, and fatigue easily. In addition, as previously described, sensory receptors themselves "fatigue" with use, causing variability in responsiveness over time and making it difficult to conduct repeated assessments of the same subjects during a circumscribed period of time. Different forms of sensation will have different kinds of meaning for organisms depending upon their past learning histories, the environmental context in which the stimulation is delivered, and other motivational (and developmental) factors associated with the organism. Thus, to adequately conduct sensory tests, various characteristics of the organism must be taken into consideration.

## Characteristics of the Organism

An important consideration in sensory testing is the nature of the behavior that is expected from the subject. Some subjects, for example, may sense or perceive stimuli but be unable to respond (as in the case of individuals lacking motor control), or they may respond in a way in which we are not expecting. This is very evident when assessing preterm infants or clinical patients, whose behavioral repertoires we may not fully understand. For example, the natural environment for preterm infants is the womb. When they are prematurely born, they are placed in an environment that challenges their immature motor systems, restricted energy levels, and inadequate controls against gravity. Thus, when they do respond, it may be in a very subtle or disorganized way, necessitating that observers who work closely with them stay attentive to these subtle responses that are often quite different from the well-organized, expressive behaviors exhibited by full-term babies (14).

### Approach/Withdrawal

Schnierla's (15) theory of biphasic approach-withdrawal (A-W) processes gives us an excellent perspective for understanding the nature of behavioral responses. According to Schnierla, all species engage in behaviors that increase or decrease the amount of distance between them and stimulus sources in their environments. Such A-W behaviors are crucial for the survival of all animals, as they underlie such behavioral outcomes as reaching food, mating, and avoiding predators or noxious stimulation. Approach and withdrawal are dependent upon the ability to sense and to respond to sensory stimulation, thus they are useful gauges of the integrity of sensorimotor systems.

In the rationale for the assessment of premature infant behaviors, Als and colleagues (14) have incorporated Schnierla's theory. Approach behaviors are those that bring the infant (or any organism) in closer contact with stimulation and may include actual movement of the body or arms in the direction of sensation (as in reaching toward a caretaker) or more subtle actions such as gazing or smiling in the direction of the stimulation source. In contrast, withdrawal responses are those that tend to increase the distance between the organism and the source of stimulation, and may include actual movement away or more subtle indices of "shutting down" or "turning off."

Preterm infants often lack strength and energy but also are unable to control motor responses, thus their A-W behaviors may be either very subtle or very gross and undifferentiated. Thus, they may perform subtle withdrawal movements such as tightly closing their eyes, or turning their heads away from stimulation; or they may react with an entire complex of avoidance and stress responses, including defecating, urinating, skin flushing, flailing the arms. In the latter, the arms may appear to reach toward stimulation, but when the whole body response is observed, it can be seen that the movements are either uncontrolled or form an attempt to interpose the arms between the infant and the stimulation source. Some of these subtle or whole body behaviors may be comparable to much more energetic and organized responses exhibited by older, stronger infants and children. Thus, when testing the sensory abilities of different-aged infants, it may be necessary to look for different behaviors or different degrees or intensities of behavior. Consider the following example:

Suppose you want to determine whether infants of different ages can discriminate between sweet and sour substances. As your subject pool, you select preterm, 2-day-old, and 2-month-old infants and present them with milk solutions of different tastes, one sweet and one sour. Your hypothesis is that the infants will discriminate between the two tastes and consume more of the sweet than of the sour solution. After testing all these different infants, your results indicate that the last two groups seem to spend more time sucking and have ingested more of the sweet than of the sour solution. The preterm group, however, did not suck nor ingest much of either solution. Is this because the infants could not taste the solutions or because they had oral-motor problems that interfered with strong suck-swallow responses?

Suppose you also noted that when the older infants were presented with the sour solution, they tended to "act fussy;" i.e., they would retract their heads, curl their lips, wriggle their noses, and briefly close their eyes as they moved their mouths away from the nipple. In contrast, in reaction to the sweet solution, these older infants sucked repeatedly and "appeared contented." Since these additional behaviors accompanied your other measures (sucking and ingestion), it may be informative to see if they were also evident with the preterm group. Were you, for example, able to observe in the preterm infants any of, or fragments of, these approach-withdrawal responses; e.g., turning of the head toward or away, whole-body movements, facial expressions, limb responses? These behavioral fragments may be rudimentary and, given maturation, will develop into much more organized and robust behaviors. Thus, although they are different from the responses observed in the older infants, they may be comparable and used to indicate a level of taste sensation and discrimination on the part of the preterm infants.

## *Behavioral Organization*

The ability to respond with approach or withdrawal actions is partly due to behavioral organization, which is the coordination of body parts in performing appropriate responses following stimulation. Thus, in the previous example, the older infant who purses the lips and turns the head away from sour taste displays a well-differentiated and organized response; whereas the preterm infant who, upon tasting the sour solution flails the arms, flushes, and defecates, is less well-controlled, less organized, and displays a much less differentiated response.

In the previous chapter in the section on prenatal behavioral development, Humphrey's (10) and Milani Comparetti's (17) fetal observations were described. Both observers noted that, in the course of motor development and behavioral organization, whole-body movements precede well-controlled and differentiated movements. Thus, older children display more "refined" or "sophisticated" movements than do preterm infants. For example, a child whose finger is pricked may withdraw the finger, hyperextend the wrist, and say "Ow!". An infant with less behavioral control, may make a less organized behavior and may respond by withdrawing the entire limb, extending the trunk, perhaps rotating away from the direction of stimulation, and crying.

Als and associates (14) provide an excellent description of behavioral organization exhibited by a preterm infant who was delivered at 32 weeks gestational age and assessed at 2 weeks of age. Als and colleagues describe the components of behavioral organization as the infant's ability to achieve continuous feedback from various subsystems along with input from the environment and the ability to coordinate these actions for survival and future development:

> As the examiner attempts to engage the infant in eye-to-eye contact, the infant actively avoids her, closing his eyes and then moving his head in the opposite direction. There is mild color change periorally. The infant's arm movements during this sequence, however are smooth, and there is some flexion at the elbow and wrist. With increased input from the examiner, the infant moves into a tonic neck reflex and fixates on an environmental stimulus, still avoiding the examiner's face. When the examiner speaks softly but makes her face unavailable, there is a noticeable relaxation of the infant's tone. . ., his hands open and close rhythmically, his face softens, his mouth begins to purse, and eventually he turns his head smoothly in the direction of the sounds. . .

> When the infant is then presented with an inanimate sound (rattle) his response is quite different. He turns with a large startle in the opposing direction, flinging his right hand over his face. He frowns and pales. Softening the sound results in his relaxing and now turning to the sound. With great effort, he locks onto it. The response is costly for the infant, as his color change and facial expression indicate. His movements, however, remain smooth.

This performance indicates that this infant's ability to shut out, actively to avoid stimulation that is too taxing for him, is quite sophisticated. Clearly, he processes information and can communicate whether he is able to deal with it. (14, pp. 47–48)

Als and her colleagues also have described less organized behavioral patterns among hyperreactive or lethargic infants. At one extreme, they note, is the infant who overresponds to sensory stimulation. This infant is "continuously at the mercy of environmental and internal stimuli," reacting to stimuli, "without being able to shut them out in order to protect his own regulation" (14, p. 49). These continuous responses are costly for the infant in terms of energy expenditure and potentially cause the infant delays in future behavioral consolidation and development. At the other extreme is the infant who rarely responds to stimulation, who appears lethargic and depressed, and who is noncommunicative because of the lack of behavioral responsiveness. Both infants lack behavioral organization and may be at risk for subsequent developmental and social interactional abilities.

## Species-Specific Response

An additional consideration during sensory testing is the presence of species-typical response patterns. Humans and other organisms will predictably respond to certain stimuli and not to others. Thus, the human voice, face, or language may be particularly salient for humans and evoke strong responses in other members of our species, whereas certain odors, colors, tastes that elicit strong responses in birds, dogs, or caterpillars may go either unnoticed or not responded to by humans. This may not mean that humans do not sense the stimuli but that the stimuli do not have sufficient meaning to elicit a strong response.

As ethological studies have demonstrated, certain instinctive behaviors that have adaptive significance are made in response to very specific, predictable stimulus patterns in the environment. These stimulus patterns, or releasers, will elicit simple survival-oriented instinctive responses or even behavioral *displays*, which are ritualistic responses often associated with mating or territoriality. Thus, stimuli of a certain size, color, frequency, or timing, which act as signals or releasers, may be particularly attention getting and arousing when compared with other neutral stimuli (Fig. 8.4). This fact must be considered when testing any organism's sensory abilities, as it should not be inferred that just because an organism responds to one kind of stimulus and not to another that it senses the one and does not sense the other. The organism may sense both but respond only to the "apparent meaning" of one. This apparent meaning may be directed by species specificity as well as by past learning.

## Learning History

The analysis of how meaning is attributed to environmental stimuli involves perception and cognition. Perception, as will be elaborated further in this chapter, is clearly affected by genetics, past learning, language, and culture.

**Figure 8.4.** Certain environmental stimuli elicit species-specific responses in immature organisms. For example, some hatchling birds are disturbed by hawk-like stimuli (*B,D,E*) but display no fear to goose-shaped (*C*) or neutral stimuli (*A*). Schnierla claims that it is not the specific shape of the stimulus but the component of sudden, swooping motion toward the hatchlings that produces an adaptive, protective avoidance and withdrawal response. (Modified from Schnierla TC: Aspects of stimulation and organization in approach-withdrawal processes underlying vertebrate behavioral development. In Aronson LR, Tobach E, Rosenblatt JS, Lehrman DS (eds): *Selected Writings of T. C. Schnierla* San Francisco, WH Freeman, 1972, p 357.)

For example, to some individuals, all trees are the same. But to a naturalist, every tree may be different—in species, health, age, location, etc. To many folks, snow is merely white, cold precipitation. To skiers or to individuals such as Eskimos, snow can be classified according to many dimensions, which are reflected in the different words used to describe its various forms and nature. Classical music may sound the same to an inexperienced listener, but the experienced first violinist in a prominent symphony orchestra may be able to not only identify who wrote, adapted, conducted, and performed the musical piece but also may be able to discern from among all of the music, the sound of another first violinist. Thus, certain stimuli "catch our attention", as we say, because of previous associations during experience with the world. Those environmental stimuli that catch our attention are the ones to which we are most apt to make a behavioral response.

### State and Sleep

The fact that we respond to some events and not to others is not necessarily a reflection of whether the stimuli have been sensed or not. Different states of the organism will determine when and whether an overt behavioral response occurs. Behavioral state, which has recently received considerable clinical and research attention, refers to a condition of consciousness or alertness that can be represented on a continuum from deep sleep to active wakefulness (Table 8.5). Behavioral state is affected by physiological status such as hunger, thirst, bowel and bladder functions, and intrinsic rhythms that affect sleep and other processes (18).

Behavioral state is used as an index of developmental level or dysfunction. State actually reflects the individual's "availability" for contact with the environment (19) thus controlling what input is received and processed as well as regulating infant participation in social interactions (20). For example, consider the different responses an infant may make toward a special toy when the infant is alert compared with when she or he is crying or asleep (Fig. 8.5). Clearly, state must be taken into consideration when evaluating any kind of sensory, motor, or behavioral response.

The significance of behavioral state is underscored by its inclusion in infant assessments and its implications regarding the integrity of the nervous system. State functions as an important adaptive mechanism, and fluid movement from one state to another reflects a certain intrinsic rhythmicity of behavior and is evidence of a certain level of neural maturation and behavioral organization. For example, the infant who moves smoothly from a state of crying to sleep and then later to a gradual playful alertness is displaying more organized behavior than the infant who moves unevenly from light sleep to irritability.

The intrinsic rhythmicity of behavior is not something that is evident only after birth. Ultrasound studies indicate that fetuses alternate rhyth-

**Table 8.5**
**Behavioral States**[a]

Sleep States
  Deep sleep - Regular breathing, eyes closed, no eye movements
  Light sleep - Irregular breathing, rapid eye movements with lids closed, periodic sucking
Awake States
  Drowsy or semidozing - eyes open or closed, eyelids fluttering; reacts to sensory simuli but with possible delays, smooth movements
  Alert - focuses on stimulus but possible delay, minimal motor activity
  Eyes open - high level of motor activity, startles, difficult to test sensory reactions due to high movement levels

[a] Adapted from Brazelton, TB: *Neonatal Behavioral Assessment Scale.* Philadelphia, JB Lippincott, 1973.

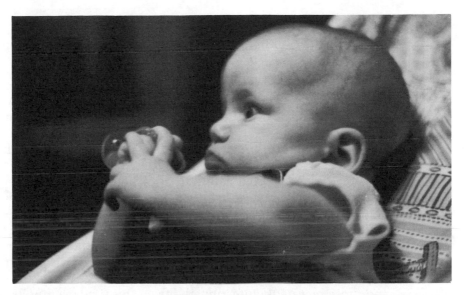

**Figure 8.5.** Behavioral state affects attention, receptivity, and responsiveness to environmental stimuli. An alert infant is much more attentive to all forms of sensory stimulation than is one who is drowsy or asleep. Attention is directed toward novel stimuli in an orienting response and toward intense stimuli in a defensive reaction preparatory for self-protection.

mically between quite and active phases (21, 22). The significance of this is illustrated by a statement from Arduini and associates (22, p. 275): "It can therefore be concluded that dynamic biophysical activities in the fetus are not random phenomena, but part of a complex mechanism of integration. The development of the central nervous system permits a harmonious correlation between these events."

The relationship of state, behavioral organization, sensory reception, and life-span development is addressed by Lipsitt as follows:

> Evidence is available, then to indicate that within the first few days of life, infants adapt to or lock in on a temporally dictated feeding schedule, they stop crying when subjected to a rocking experience, they delay crying when subjected to vestibular stimulation, and they go quickly from quiet sleep to boisterous crying when suddenly startled. While it is important to document the waxing and waning of "state" conditions under naturalistic or noninterfering laboratory conditions, it is equally necessary for the infant organism to be understood as a creature who is, even in his state fluctuations, responding to both intrinsic and extrinsic stimulation. The drums to which the newborn is listening moreover, may not always coincide with the observer's idiosyncratic attention preferences; when the observer thinks the infant is merely lying quietly and without stimulation, his internal organs might be supplying stimulation which produces a "spontaneous" or "uninstigated" startle response.

> The concept of state has special pertinence to considerations of life-span developmental psychology inasmuch as states are often regarded, erroneously or otherwise, as relatively stable attributes of the human organism. It is tempting to conceive of the child who is hyperactive at birth as the child who will

become the overactive, obstreperous adolescent or hypersensitive adult. That there are such "life plans" congenitally given is a realistic possibility, but to date there are no data of which this writer is aware that would strongly support the contention (23, p. 302).

Sleep, while a reflection of state, is also significant physiologically and pyschologically. In most individuals, sleep is a rhythmic, predictable process that, if it is repeatedly disturbed, can lead to profound negative effects on the organism. Sleep is essential as a restorative process, guaranteeing physical and mental health. It, like most body functions, displays fairly predictable changes on a daily basis as well as over the life span. For example, newborns spend approximately 16 of 24 hours in various sleep states, with approximate 4-hour sleep/wake cycles. Their seven or eight naps during 24 hours become gradually longer at night, so that by about 28 weeks they tend to sleep throughout the night (24). Different sleep stages exist (Table 8.4) with the most characteristic and studied rapid-eye-movement (REM) sleep, a time when the closed eyes exhibit rapid eye movements. Adults wakened during REM sleep often report having recently been dreaming, and those repeatedly interrupted during this time eventually become irritable and anxious. Thus, it is thought that REM sleep is physiologically and psychologically important.

Developmental changes in REM sleep have been reported, indicating that newborns and children up to about 5 years of age engage in about 8 of 16 hours of REM sleep. After age 5, this figure then drops to the adult average of 20%. Premature infants, however, spend even greater amounts of time than do full-term infants in REM sleep, and rapid eye movements have also been identified as occurring frequently in fetuses approximately 38 weeks gestational age (22, 24).

Like sleep states, awake states also vascillate. Infants, children, and adults all display varying levels of wakefulness and attentiveness. Because sleep-wake states are so important, especially when testing infants, wakefulness has been categorized on developmental tests such as the Neonatal Behavioral Assessment Scale (NBAS); most clinicians recognize that to adequately assess an infant's reactivity to various environmental stimuli, the infant should be tested when alert, rather than fussy, crying, or asleep. The degree of alertness increases with age, with newborns alert only about 11%, increasing by about 4 weeks of age, to 21% of their day.

The NBAS (18), which is in part a test of infant interactiveness with the environment, highlights sleep and wakefulness (i.e., state). Brazelton, who designed the NBAS, notes that the evaluation is geared to test an infant's "best" behavior, thus the infant's state must be taken into consideration when presenting test items. As he reports in the test manual, "An important consideration throughout the tests is the state of consciousness or 'state' of the infant. Reactions to stimuli must be interpreted within the context of the presenting state of consciousness, as reactions may vary markedly as the infant passes from one state to another" (18, p. 5). Behavioral states are also common to other evaluations, such as Prechtl and Beintema's

"Neurological Examination of the Full Term Newborn Infant" (25) and the Assessment of Preterm Infants' Behaviors (26).

### Attention

Behavioral state affects and is affected by attention, which is sometimes used to gauge sensory abilities. Attention refers to the degree to which an organism selectively orients toward some stimulus to the exclusion of other stimulation (27). Thus, infants who look up at their mother as she enters the room are attending to her as opposed to their toys, bottle, the dampness of their diaper, etc. In many cases, attention is inferred by a behavioral response such as increased eye opening, head turning in the direction of stimulation, or eye contact with the stimulus. Thus, if an infant attends to a certain stimulus, then we assume that the infant senses it. For example, if a toy music box is wound up and placed in an infant's crib, and if the infant turns the head in the direction of the toy and continues to visually scan it while the music plays, then we assume that the youngster sufficiently receives and processes auditory and visual stimulation.

Other methods for assessing attention include *psychophysiological responses*, which are physiological reactions associated with the *autonomic nervous system* and coordinated with psychological processes such as orientation, arousal, or different emotional states. The autonomic nervous system consists of peripheral nerves that regulate internal functions within the body, such as constriction and dilation of the blood vessels, sweating, and other involuntary behaviors. It is so named because it appears to operate autonomously and outside of conscious control. When an organism receives various forms of environmental stimulation, autonomic changes often occur when overt behavioral responses are not evident. Thus, psychophysiological measures are often used as an index of responsiveness to various sensory stimuli, as they may demonstrate sensory reception when other gross measures do not. Some of the responses that are examined include measures of change in heartbeat (*via an electrocardiogram [ECG]*), brain wave activity (*via electroencephalogram [EEG]*), pupil reactions, flushing of the skin, measures of sweating, and other reactions.

Attention is important in several different ways in regard to sensory testing. First, attention to some stimulus may be used as a gauge of sensory reception. Second, attention is often a prerequisite in sensory testing. Thus, if we want to see whether an infant responds differently to a red ball than to a yellow one, we must first direct the infant's attention toward the different stimuli. Here, the close relationship between attention and state is emphasized. An infant who is asleep is not paying attention to external sensory stimuli; whereas one who is awake may respond to a wide variety of events going on in the environment (Fig. 8.5). Finally, the degree of attention paid to a certain stimulus is often used as an indication of the meaningfulness or importance of that form of stimulation. We would expect that some stimuli have more meaning than others and thus are more likely to both gain as well as sustain our subjects' attention. The use of these attentional measures in sensory testing will become more apparent as each sensory modality is explored.

### Attention and the Orienting Reaction

A close relationship exists between orienting and attention. As defined in the previous section, attention is the degree to which an organism "orients" toward and selectively responds to a particular stimulus. Orientation toward a stimulus is called the *orienting response* (or reflex) and is defined as a reaction to stimulus change. Stimulus change is what often "attracts our attention," thus the close connection between attention and orientation becomes obvious. While definitions of these terms are controversial, some researchers have made a reasonable distinction between them. Reese and Lipsitt (27) describe the relationship:

> In attempting to relate the orientation reaction to attention, some writers have gone so far as to equate the two, using them interchangeably. However, if the definition of attention is limited to some active selective process, then the orientation reaction must be classified as different from, though obviously related to, attention. In keeping with this position, the registering of stimuli as evidenced by the orientation reaction can be considered a first and necessary condition of attention. However, the presence of an orientation reaction does not necessarily imply that the subject has acted selectively upon the stimulus input nor, more importantly, that he has obtained specific information of knowledge about the nature of the stimulus. The selective scanning of information has been defined here as attention, while the identification of the selected information, as to what it is or is not, would seem to be the function of perception (p. 310).

In regard to visual behavior, Cohen (28) has distinguished between attention getting and attention holding. His categories seem appropriate for distinguishing between attention and orienting. Cohen's concept of attention getting is equivalent to Reese and Lipsitt's view of the orienting response, an initial, brief reaction to some stimulus change. Attention holding refers to the prolonged process of selectively responding to one stimulus over another.

The presence of an orienting response (OR) is often inferred by changes in many of the same behavioral or autonomic reactions used as indices of attention (Table 8.6). The orienting response has been studied widely in developmental and psychophysiological fields, and implications that have been drawn regarding its presence or absence in immature organisms are controversial. For example, Sokolov (29) claims that the OR is not a "purely" autonomic-regulated function but is subject to cortical control. Thus, he infers, that its emergence in infancy is an indication of brain maturity. While this claim is not universally accepted, the absence of an OR or its perseveration in an individual is presumed to be an index of neural damage or immaturity. Like attention, orientation toward a sensory stimulus is a measure of stimulus perception and intact function of that sensory modality. The implications of the existence and clinical measurement of attention and orienting are explored further under "Habituation."

Table 8.6
Indications of Orienting Response[a]

| |
|---|
| Pupil dilation |
| Turning head toward stimulus source |
| Momentary arrest of ongoing activity |
| Increase in muscle tone |
| Change in electroencephalographic activity |
| Change in respiration |
| Change in heart rate |

[a] Adapted from Reese HW, Lipsitt LP: Attention. In Reese HW, Lipsitt LP: *Experimental Child Psychology*. New York, Academic Press, chap 9, 1970.

### Defensive and Orienting Reactions

Some researchers distinguish between orienting and defensive reactions (DR), claiming that they are different processes that prepare the organism for qualitatively different responses. The orienting reaction, which directs attention toward a stimulus, is made in reaction to stimuli of low or moderate intensity, and its purpose is to increase the organism's sensitivity to sensory input. The DR, in contrast, is made in response to intense stimulation and is protective in nature; its function is to limit or restrict stimulus input. These two responses, the DR and OR, are associated with different components of heart rate. Cardiac deceleration is associated with the OR, whereas acceleration is associated with, or a component of, the DR (30).

In the 1960s and 1970s, studies of newborns and young infants indicated that cardiac acceleration was the predominant response to a variety of forms of sensory stimulation. While the DR was present at or before birth, the OR did not emerge consistently until around 3 to 6 months of age (30, 31). This led some researchers to claim that there is a maturational shift in the onset of the OR that, if Sokolov's theory (29) is considered, is indicative of gradual cortical maturation and control. Subsequent studies, however, have demonstrated that orienting responses, measured by cardiac deceleration, can be obtained in newborn infants (32, 33). Studies of OR need to be attentive to the totality of stimulation used (32) and to the characteristics of the infants studied; i.e., to the stimulus context, the state of the infant, and the infant's past history with similar stimuli. An infant's previous experience with various sensory events can alter the meaning of those stimuli and affect responsiveness (orienting and attention) during sensory testing.

# Meaning of Environmental Stimulation in Sensory Testing

Although this topic was covered in the previous section, it needs to be highlighted here because of its close connection with attention and orienting. The nature of stimuli used during sensory testing will affect the nature of the responses observed. According to Schnierla's theory of approach/

withdrawal processes (15) as well as theories about orienting and defensive responses (30), different intensities of the same stimulus will provoke different types of responses. For example, very intense stimulation such as loud noise, bright lights, quick spinning, or heavy pressure may be painful or threatening and initiate cardiac acceleration as well as avoidance responses. This is in contrast to approach behaviors or state changes that may be exhibited in response to soothing stimuli such as soft noise and lights, rhythmic rocking, and stroking.

In addition, stimuli may have different meanings for people or species. For example, a human infant will tend to pay more attention to another infant than to a flat piece of paper. A bone from a beef roast may attract the attention of a dog, whereas the bone may be ignored by a bird or a butterfly. Similarly, small, fast darting movements by water insects may be meaningful for frogs, fish, or other insects but may be relatively ignored by humans. In contrast, social stimuli such as laughing, language, smiling, or glaring are salient to humans and other primates but may be insignificant to some other species. Organisms' different responses to stimuli are attributable to variations in learning history as well as to species-specific characteristics.

Thus, certain stimuli (especially painful, intense, and threatening events) and those to which conditioning has occurred will attract immediate attention and response (defense or orienting reactions) as will those that act as releasers for particular species-specific actions (Fig. 8.4). These factors will also interact with chronological age, causing some stimuli and events to have meaning only during certain time periods in an organism's life. In our culture, many sexual symbols (such as women's breasts) have profoundly different meaning for adolescents and adults than for infants. For organisms that have a very circumscribed mating period, specific stimuli are salient only during that time. For example, during mating season many male birds are colorful and territorial and behave aggressively toward other similarly colored birds. In contrast, during migration seasons the birds are duller in color and their behaviors are nonaggressive and cooperative, as they flock together in preparation for migratory flights. Thus, environmental stimuli take on different values and meanings, and this fact must be taken into consideration when comparing the behavioral responses of different species as well as the developmental differences of members of the same species.

## Experimental Procedures for Evaluating Sensory Processing

In Reese and Lipsitt's classic text on infant and child psychology, Kaye (34) has delineated three primary methods for studying infant sensory abilities. These are the reflex, habituation, and conditioning procedures. In the reflex procedure, a single stimulus is presented and the presence or absence of elicited reflex is noted; e.g., a nipple is placed in an infant's mouth and the presence or absence of a sucking reflex is observed. This reflex procedure is also used to assess the integrity of the motor system by physical and occupational therapists; however, rather than merely observe the presence or absence of response, they also examine the quality of movement. It is important to note that many of the tests used to explore sensory functions are similarly used to explore motor abilities. This illustrates the close interrelationship between sensory and motor systems. Additional reflexive behaviors will be examined in this chapter and in chapters 9 and 10.

## Habituation

*Habituation* refers to a response decrement with repeated presentation of the same stimulus. Because of habituation, we often no longer seem to "hear" or respond to a ticking clock, a dripping faucet, or to music playing in the background. When the habituation procedure is used to test sensory abilities, a stimulus is presented two or more times, and the researcher attends to the change in the subject's responses after each presentation. The normal reaction is a diminution of responses over time.

Though subject to some debate, habituation is considered by some as a learning process. If learning is considered a change in behavior, then habituation qualifies as a simple form of learning. Even though it appears to be relatively simple, the implications of the presence of habituation are profound. True habituation (rather than fatigue) indicates that a subject recognizes a stimulus when it is repeatedly presented. This recognition that an object stays the same over a period of time is an index of memory as well as an indicator of central nervous system integrity. Subjects with little or no memory will respond to the same stimulus as if it is novel each time it is introduced. As Lipsitt (35) explains, habituation is a normal process that enables individuals to become accustomed to, or to shut out, sensory stimulation. Some individuals with neurological deficits, however, lack the inhibitory mechanisms that enable them to habituate. Thus, they continue to repeatedly respond (i.e., *perseverate*) at every presentation of the same stimulus.

Consider for example the previous description (under Behavioral Organization) of behavioral organization in preterm infants that was presented by Als and colleagues (14). The hyperreactive infant who startles to every stimulus displays little or no habituation. Without habituation, excess energy is continually expended on motor responses and in attending to environmental stimuli. With habituation, attentional mechanisms can be readied and motor responsivity can be conserved for situations in which they may be most needed or relevant. Thus habituation is an important adaptive function that enables the organism to respond only when necessary and to shut out stimuli when appropriate. Brazelton (18, p. 13) notes that one of the most impressive mechanisms in neonates is their capacity to decrease responses to repeated disturbing stimuli; "response decrement" (as Brazelton calls habituation) to a variety of sensory stimuli (sound, pain, and vision) forms a major component of the NBAS.

The habituation procedure used for sensory testing has been expanded to include a habituation/dishabituation method. This involves application of one kind of stimulation until a response decrement is achieved, at which time a slightly different stimulus is presented. The normal response for an individual during this procedure is to exhibit habituation during presentation of the first stimulus and to demonstrate response recovery upon presentation of the second stimulus. Thus, if a bell is repeatedly sounded at the side of an infant's head, the infant will at first turn the head in the direction of the sound. Over time, however, as a result of habituation, the infant will stop turning the head and appear to lose interest in the bell. Once a new tone is introduced—and *if the infant is able to discriminate the difference* between the two sounds—the child will once again turn the head in the direction of the sound. Essentially, the normal habituation process would be undone, i.e., the infant would "dishabituate." This procedure can continue as a variety of stimuli are presented or until the subject fatigues. An illustration of this process of habituation and dishabituation is presented in Figure 8.6.

The orienting response (OR) is often tested with habituation. For example, behavioral responses and heart rate are measured as an audible bell is sounded. The normal infant will initially respond with an OR, maybe by turning and looking at the bell and by exhibiting cardiac deceleration. if the stimulus is repeatedly sounded, these heart rate and behavioral measures will gradually decline, i.e., habituate. Thus, a different bell is then used. If the infant can discriminate the sounds, then he or she will orient

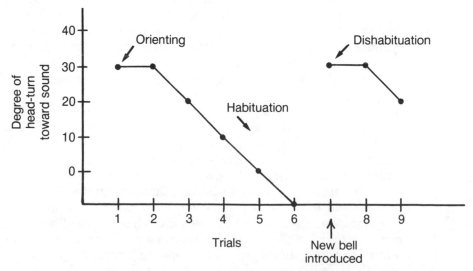

**Figure 8.6.**  Habituation of an orienting response to an auditory stimulus. On trial 1, a bell is sounded to the side of the infant's head, and it elicits a large degree of head turning toward the sound. With repeated presentations of the bell, response decrement typical of habituation occurs. But, when a new bell is introduced on trial 7, dishabituation occurs, characterized by the recovery of head turning and orienting toward the new sound.

to the new stimulus (i.e., dishabituate), once again, by turning and looking and displaying cardiac deceleration.

Since the OR is, by definition, a response to stimulus change, then it makes sense that the OR habituates with repeated presentation of the "same" stimulus. Stimulus change is an important characteristic to most organisms, including humans; thus habituation is equally important as it enables organisms to effectively "shut down" until additional responses are necessary. Testing habituation of the orienting response to different sensory stimuli may be a simple clinical assessment. If habituation does not occur, it may be an index of immaturity or of central nervous system damage. However, caution should be exercised in drawing conclusions based solely on habituation responses. For example, in contrast to the OR, the DR, which is protective in nature and which is a reaction to intense possibly threatening stimuli, does not habituate, as it serves to protect the organism and to prepare the organism for potential threat. This means that in clinical testing, the nature of the stimulus, past associations, and species-specific significance must be taken into consideration. Habituation will occur with some stimuli, whereas it will not occur with others. The fact that some infants do not display habituation to certain stimuli may be due to a feature of the stimulation rather than to some developmental or neurological deficit.

Other problems are encountered with the habituation procedure. For example, if some kind of taste reception is being tested with repeated presentations of food, then the subject may become satiated; or if the stimulus elicits a fairly gross response, then the subject may tire. Habituation can also be confused with the fatigue or adaptation of a sense receptor. As previously noted in this chapter, sensory receptors adapt after repeated stimulation and require a certain degree of recovery time before becoming responsive again. Thus, a response decrement following repeated presentation of a stimulus may occur as a result of receptor fatigue or habituation.

Clinical and developmental assessments need to take this ambiguity into account. Numerous experimental procedures exist for discriminating receptor fatigue and habituation. One procedure is habituation/dishabituation. For example, if an infant is repeatedly presented with a sweet taste and stops responding, the lack of response could be due to receptor fatigue or to habituation. If a new stimulus is introduced, however, and if the infant responds again, then receptor fatigue is ruled out. (Other characteristics of habituation are listed in Table 8.7.) It is a useful procedure for evaluating sensory and perceptual processes, and its potential developmental and clinical uses (35) should become clearer with examples that will be given later in this chapter.

## Conditioning

The conditioning procedures that were introduced in chapter 5 have proven very useful for communicating with nonverbal subjects. With classical conditioning procedures, a conditioned stimulus (CS) and an unconditioned stimulus (UCS) are presented. For example, presentation of a bottle (and

Table 8.7
Characteristics of Habituation[a]

---

Repeated application of a sensory stimulus elicits a response decrement (habituation).

If the stimulus is withheld for a period of time, the response will recover.

Habituation becomes more rapid after several instances of habituation and recovery.

The more rapid the frequency of stimulation, the more rapid the habituation.

The weaker the stimulus, the more rapid habituation will occur.

Strong stimuli may not habituate.

Habituation to one stimulus may generalize to a similar one.

Presentation of another stimulus results in recovery of habituation (dishabituation.)

---

[a] Modified from Thompson RF, Spencer WA: Habituation. A model phenomenon for the study of neuronal substrates of behavior. *Psychol Rev* 73:16–43, 1966.

nipple) with which the infant is familiar (CS) is repeatedly paired with a novel bell tone (UCS). If the infant can hear the sound, then classical conditioning should occur; i.e., the infant should come to associate the tone with the bottle so that eventually the presentation of the tone alone should elicit a suck response. Lipsitt (35) has described how numerous developmental behaviors are affected by classical conditioning and how these can be assessed using classical conditioning procedures. For example, infants have been found to make anticipatory sucking movements when placed in a familiar feeding position, and they have been conditioned to perform other motor reflexes to a variety of sensory stimuli.

In operant conditioning, a rewarding or punishing stimulus is used to strengthen or weaken, respectively, the probability of some behavior in the child's repertoire. Thus, sucking or vocalizing may be increased by presentation of food, cuddling, or a combination of social stimuli. Further developmental and clinical implications regarding conditioning procedures with sensory testing will be evident as each sensory modality is explored later in this chapter.

## Developmental Changes in Learning Capacities

One of the issues encountered when using these experimental learning procedures in testing sensation and perception is the potential for developmental changes in learning capacities. At one time, infants were considered unable to learn, therefore such procedures were not used with this age group, let alone with preterm infants or fetuses. The use of learning procedures is therefore relatively recent. It is expected that, with the increased use of sophisticated technology in fetal diagnostics and behavioral observation, additional information will be forthcoming about the habituation and conditionability of fetal sensory and motor responses.

Denenberg (36) notes that prior to the 1960s—with the exception of some work by Marquis (37) in the 1930s—studies of infant learning in the United States were practically nonexistent. It was believed that infants were essentially passive, lacked sophisticated sensory perceptual abilities, and were subject much more to maturational than to learning influences. While mothers knew better of their infants' capacities, it took some time for experimenters to consider infants as appropriate candidates for experimental study of learning skills. Initial experiments of infant conditionability were unsuccessful, in part because of the nature of stimuli that were used. Intense, aversive stimuli tended to provoke behavioral disorganization rather than learning (38).

Now, however, though subject to some controversy, findings by Western researchers indicate that classical conditioning does not necessarily occur more readily in older than in younger infants. When classical conditioning procedures are used with infants, researchers must consider the nature of the stimuli (whether autonomic or behavioral responses are conditioned), the characteristics of the conditioning procedure (intervals between stimulus presentations), and the sensory modality being assessed (38). Lack of conditionability may not be a reflection of the modality assessed or the developmental level of the infant but may reflect some insensitivity of the assessor toward infant behaviors, states, and sensorimotor capacities.

Jeffrey and Cohen (39) draw similar conclusions in regard to habituation. Developmental trends in habituation abilities have been reported in some cases. The most consistent finding is that habituation emerges at about 2 to 3 months of age. Lipsitt, however, provides numerous descriptions of neonatal habituation (35). The discrepancy between these findings may be attributed to the nature of the stimuli observed and the nature of the responses expected from the newborn. Other factors such as the state of the infant, the length of exposure to stimuli, and the complexity of stimuli must be considered. Jeffrey and Cohen (35) suggest that the habituation procedure continue to be tested because of its significance in revealing information about the nature of infant behavioral processes. All these learning processes may be useful in generating information about developmental change across the life span.

Studies of infant learning in the past 20 years (35) have provided excellent "prove" of the existence of various infant sensory and perceptual abilities—which is why this topic is covered in this chapter—as well as illustrating the interconnection between sensory, motor, and mental skills. The following study illustrates this point.

In 1964, Lipsitt and Kaye (40) demonstrated that human neonates could be classically conditioned. They used 3- to 4-day-old neonates as subjects and divided them into experimental (E) and control (C) groups. The E subjects were conditioned to associate a loud tone with presentation of a nonnutritive nipple that was inserted into their mouths. The C infants also heard the same sound and received a nipple, but these events were not paired. After about 20 conditioning trials, the E infants made frequent and significantly more suck responses when the tone was presented than did the C infants. These are the implications of this study: (*a*) three- to 4-day-old neonates can hear; (*b*) they can learn, i.e., their behavior can be altered by practice; (*c*) they can exhibit motor control of sucking behavior; (*d*) they have memory; and (*e*) they display behavioral organization as evidenced by their integration of sensory (auditory), motor (sucking), and cognitive (memory, conditioning) skills.

Numerous other examples of infant learning will be used in this chapter to illustrate infant and childhood sensory and perceptual abilities. Other developmental changes in learning abilities will be discussed in subsequent chapters of this text, e.g., social learning (chapter 12), learning and cognition (chapter 11), and gender differences in cognition and learning abilities (chapter 12).

# DEVELOPMENT OF SENSORY MODALITIES

This section briefly describes and examines specific sensory modalities from a developmental and a clinical point of view. While this will not be an exhaustive analysis, it should give the reader a basic understanding of how sensory systems change over life from prenatal time to adolescence. As stated before, most developmental change occurs during infancy at which time sensory systems become adult-like. Additional information about perceptual motor and cognitive changes in childhood and adolescence is provided in chapters 10 and 11. It is not within the scope of this book to examine sensory physiology nor to describe the anatomy and physiology of each sensory system. Students unfamiliar with the basic biology of such organs as the eye, skin, ear, or vestibular system may want to preview an elementary biology or physiology textbook for this information.

## Somesthesis

Somesthesis, also known as *cutaneous* (skin) or *tactile* (touch) sensitivity, is based on information received through the body surface, the skin. Cutaneous receptors, which respond to mechanical, thermal, electrical, and chemical stimulation, tell us about what is adjacent to our bodies, thereby providing information about the environment as well as protecting us from harm or exposure (41). Somesthesis is a critical sensory system because of its scope as well as its essential protective quality. As Montagu (42, p. 1) in his book *Touching, The Human Significance of the Skin*, notes "[The skin] like a cloak covers us all over, the oldest and the most sensitive of our organs, our first medium of communication, and our most efficient of protectors."

## Different Skin Senses

As far back as the time of Aristotle, it was believed that all skin senses could be considered together as "touch", but now many different classifications are recognized. These classifications are a mix of at least four incompletely understood skin senses: pressure, warmth, cold, and pain. While the function of each system is not clearly understood, it is thought that there may be separate kinds of receptors for each of these different sensations (6). This, however, does not explain other somesthetic sensations such as tickle, itch, tingle, or wetness, which result from stimulation of separate subsystems or combinations of other skin sensations. The sensation of pain particularly, is controversial. Different theories of pain suggest that it is due to separate somesthetic receptors, to overstimulation of receptors, or to a special pattern of activity in cutaneous receptors (41).

One reason why a variety of cutaneous receptors are thought to exist is that different afferent systems have been found relevant to somesthetic information. Below the level of the neck, cutaneous information is transferred to the brain via the spinal cord by two separate but interdependent afferent systems. These are the *lemniscal* and *spinothalamic* (or *extralemniscal*) systems, which follow different pathways to different cortical projection areas of the brain. The adaptive significance of these two systems is unclear. Some researchers have suggested that these systems carry two different kinds of information, with the lemniscal system carrying primarily touch information, and the spinothalamic dealing primarily with pain and temperature (43). Others have contended that the major difference between these systems is anatomical, resulting in the speed with which they conduct information to the brain. The lemniscal system is phylogenetically newer, has few synapses, and is therefore faster than the older, multisynaptic, spinothalamic system.

## Development of Somesthesis

Montagu (42) refers to touch as "the mother of the senses" because it is the earliest sensory modality that gives rise to all other systems. Since the cutaneous system is the earliest to develop and is so critical for survival, it is thought to be the earliest to become functional and responsive to sensory input. Responsiveness is particularly acute in parts of the body that are used for exploring the world, e.g., the mouth and fingers. Thus, as expected, sensory development occurs early in proximal areas such as the mouth, lips, tongue, and snout regions of vertebrates (13). This is clearly evident in human infants, whose perioral area is well developed from a sensory as well as a motor perspective. For example, newborns respond to a relatively light touch at the corners of the mouth by turning their heads toward the stimulation. This familiar response, *the rooting reflex*, is well developed at birth and thought to be an adaptive, approach behavior for orienting toward nourishment. The sophistication of oral development is illustrated by Lipsitt's (35, p. 499) observation: "[T]he importance of such tactual stimulation, and the low threshold of the newborn for response to it, can be demonstrated by rotating the finger completely around the lips in a circle, and noting the precise following of such stimulation which many newborns can demonstrate."

## Difficulties Studying the Development of Touch

Somesthetic development is actually poorly understood. It is a complex and difficult area to study. For example, adults have difficulty discriminating among pressure, touch, and pain; thus measuring cutaneous sensation, discrimination, and perception among nonverbal subjects is especially problematic. In an analysis of cutaneous sensory development, Rosenblith and Sims-Knight (19) note that tactile development is poorly understood because of the confounds of many studies that have used different forms of stimulation, examined different areas of the body, tested infants of various ages and during various behavioral states, and combined male and female subjects, who may display different thresholds for varying types of somesthetic stimulation.

Rosenblith and Sims-Knight (19) note that in most studies of infants, tactile stimulation can be classified as aversive or nonaversive. Those studies using aversive stimulation, many of which were conducted in the 1930s, applied pinpricks and mild electric shock to different body surfaces of infants. These studies did not support the old myth that neonates do not feel pain. While it is difficult to know exactly what they do feel when cutaneously stimulated, infants respond with avoidance reactions such as limb retraction and crying, indicating that they certainly sense and attempt to avoid such stimulation. Another form of aversive stimulation includes placing an object over an infant's mouth and nose and observing the nature and vigor of attempts to remove the object. This is not a "pure" measure of tactile sensitivity since it also occludes breathing passages, but it is an informative index of sensorimotor responsiveness as well as behavioral organization. Since it is an informative measure, it is included as a test item on various neonatal assessments such as the NBAS (18) and the Graham-Rosenblith Behavioral Test for Neonates (44).

Other relatively nonaversive procedures used in cutaneous sensory testing include stimulating various body parts with air puffs or with an aesthesiometer. The latter consists of various grades (sizes and tension strength) of filaments which are pressed against the skin until they bend. These are particularly useful because they enable researchers to quantify the degree of cutaneous stimulation that is administered. The habituation/dishabituation procedure (described previously) has also proved useful for testing cutaneous sensation. With this procedure, Kisilevsky and Muir (45) demonstrated that neonates are able to discriminate between tactile stimulation on different body parts as well as between tactile and auditory stimulation. Thus, although intersensory communication is thought to mature gradually during infancy, there is apparently some cross-communication between senses during the first few days of life.

### General Developmental Findings

Infant responses to cutaneous stimulation can be categorized as reflex, approach, or withdrawal behaviors. For example, reflexes are made to a variety of forms of cutaneous stimulation ranging from light stroking, which elicits rooting, to palmar pressure, which elicits the Babkin response. (Other reflexes are described in Table 8.8.) Avoidance responses include actual head, body or limb withdrawal, scratching or rubbing, and facial grimace. Approach responses to tactile stimulation are common and are easily illustrated by children's receptiveness to hugs or by gentle giggling or smiling in reaction to pokes and tickles.

The general conclusions from studies on cutaneous sensation are that neonates do respond to touch, that they feel pain, and that the somesthetic system is intact and functional at birth or earlier. For example, human embryos at 2 months have shown responses to tactile stimulation, especially when it is located around the head (46); the fetal observations conducted by Humphrey (16) and discussed in the previous chapter demonstrate that even before birth, fetuses make differentiated approach and avoidance responses to various types of tactile stimulation. Conclusive claims about the early development of the somesthetic system, however, must take into consideration such factors as the type of stimulation, the location on the body, the size and amount of fat on the infant, and the infant's behavioral state. Rose and her colleagues have demonstrated that although pre-term and full-term infants respond to tactile stimulation, the responses of the preterm group are particularly affected by behavioral state (47). In addition, tactile sensitivity on the body varies, with maximal sensitivity in areas such as the face and hands.

With development over the life span, touch receptors change and so therefore does touch perception. The numbers of touch receptors, especially those within a specific skin surface area, are thought to gradually decline

**Table 8.8**
**Early reflexes dependent upon tactile stimulation**

| Reflex | Description of Stimulation and Response |
|---|---|
| Rooting | Light stroking near mouth produces turning of the head toward source |
| Babkin | Pressure on palms produces opening of the mouth |
| Babinski | Light stroke on the sole of the foot produces fanning of the toes |
| Grasp | Placing an object in the hand produces grip of object |
| Crossed extension | Pressure on one leg produces extension in the other leg |

with age. Part of this is due to the increase in skin surface with growth. Along with declining receptors is a gradual diminution of touch and pain sensitivity, which either decline or remain comparatively stable throughout childhood and adulthood (46).

### Interconnection Between Touch and Other Senses

Since the brain continues maturation beyond birth, and myelination continues along peripheral nerves and in the corpus callosum, changes in tactile acuity and discrimination are expected over time. Indeed, the tactile system becomes progressively more differentiated during infancy and childhood due to maturational and life experiences. The somesthetic system is neurally interconnected with other sensory modalities as well as associated with emotional and social events in life. For example, tactile input is essentially for exploration of the environment. Since early development around mouth and snout areas is sophisticated in vertebrates, human infant exploration is often centered on the mouth (Fig. 8.7). This has theoretical implications; e.g., in regard to the oral stage of development in psychoanalytic theory (see chapters 6 and 12), as well as practical applications. Occupational therapist Ayres notes:

> Much of the young child's early information about the world and about himself comes from contact between his skin or mucous membranes and the environment. The area within the mouth, richly endowed with receptors, is an important source of sensory stimulation of mechanoreceptors and, if the stimuli are interpreted temporally and spatially, a source of information about small objects and small segments of anatomy (7, p. 167).

**Figure 8.7.** In early development, exploration of the environment often centers on the mouth, an area richly endowed with sensory receptors and motor control.

Later, this exploration shifts from oral to manual exploration, and the child depends in greater part on eye-hand coordination and visual-tactile exploration (48).

An illustration of the integration of sensory modalities is seen with use of touch, movement, and vision in manual exploration and dexterity. As touch becomes more refined in the hands and fingers, sensory discrimination becomes more acute. Objects placed in the hands can be identified with use of the sensory information received from the fingers and hands alone, which is called *stereognosis*. This involves the ability to identify objects based on texture, size, and shape. Tyler (49) reports that stereognostic abilities increase with age, and that for children 2 years of age and under, this ability is neither reliable nor well developed. By ages 3 and 4, however, stereognostic abilities have developed sufficiently to be reliably tested, and Tyler recommends that additional clinical tests explore children's abilities in differentiating shape, size, and texture for a clearer profile of cutaneous development.

The sense of touch is important not only for exploration of objects in the environment but also in terms of understanding oneself. As Mason notes:

> Simply speaking, touch is used to construct the individual's world picture and his world view . . . At first the child attempts to discover what there is in the world about him, in other words what constitutes reality. This is his world picture. He will use his sense of touch through hands, mouth, and skin surface to bring parts of the world close enough to examine them and to separate them from each other and from himself (50, p. 167).

The sensory information infants receive from the skin and other senses will provide them with an "impression" of their body and a "sense" of self. This forms the basis of the body schema, an internal mental representation of one's own body and its relation to space. All of us are informed through skin senses where our bodies stop and the rest of the world begins, thus the skin is essential in early cognitive development and ongoing environmental/ spatial relations.

Clearly cognitive factors alter touch perception, and the acquisition of language may actually "direct" tactile discrimination as the infant or child learns the meanings of such words as hot, cold, itchy, painful, scratch, and tickle. The tactile system is also important in human social interactions (Fig. 8.8). Touch is an essential element of cuddling, stroking, rocking, lifting, diapering, and dressing, as well as spanking, slapping, and other forms of pain. It is interesting that for the curious child, an impatient parental response is slapping the child's hand and saying, "No! Don't touch!" Such imperatives and early handling may affect the child's subsequent exploratory behavior and may also influence the child's learning and associations with certain people, as both punishment and rewards are often tied closely with cutaneous sensation. Lipsitt's observations of early conditioning among neonates cautions us to attend to the types of tactile experiences repeatedly administered to children:

**Figure 8.8.** Touch is an important and poorly understood sense, playing a significant role in social and emotional development and in relationships between people.

The extent to which the newborn is capable of retaining memories of experienced pain is, or should be, a matter of clinical import . . . That we do not recall 20 years later, or even five years later, painful experiences in the first few days of life does not mean that painful neonatal experiences have no effects . . . This is not to say that unremembered painful experiences are necessarily deleterious, for indeed it is possible that early stress and the recruitment of physiological and psychological resources for coping with that stress may even have a beneficial effect upon the child's later ability to handle stress . . . Very little is known about this sort of thing, first because pain perception and its significance have not been studied much in the very young infant, and second, because investigations would be loath to administer unnecessary and intensive pain stimulation. Nonetheless, in the natural course of caring for newborns, unusual levels of aversive stimulation sometimes are administered, as in circumcision, venipuncture, or excision of a supernumerary digit. Researchers would do well to capitalize upon these rather routine hospital procedures to explore aspects of pain perception in the newborn and the possible effects which such procedures may have on the infant's reciprocal relationship with caretakers in the immediately subsequent hours and days (35, p. 506).

In the 1950s and 1960s, noted psychologist Harry Harlow claimed that touch, or contact comfort, was essential for normal social and subsequent sexual development in primates (51). While the significance of the tactile system in normal development was overemphasized by Harlow and others (see chapter 12), touch is an important aspect of the attachment-bonding process as well as an essential ingredient in human sexual interactions and intimacy. Culture certainly affects the degree of touch that is tolerated within a society and therefore its emphasis in social relations. In some parts of the United States such as the southeast, hugging and touching are more common and more tolerated among adults than in some sections of the northeast. This also varies across families according to their cultural heritage and learning experiences. In the United States, touching, in general, begins to decline in early childhood, with even less touch being directed toward boys than girls. At adolescence, however, touch seems to reawaken, during which time it forms an essential ingredient in sexual awareness and exploration of self and others (52).

Cutaneous sensitivity can be highly developed to compensate for other deficient modalities, as in the case of people with visual impairments who use tactual cues for enormous amounts of sensory-cognitive input. The somesthetic system also seems to be quite subject to individual differences; some children and adults, due to genetic and learned factors, show considerable variability in their tolerance of heat, sensitivity to air currents, water temperature, drafts, clothing textures and tightness. Response to touch is linked to irritability or temperament, as some children like to be touched and cuddled, whereas other more irritable infants react to cuddling by fidgeting and withdrawing. Ayres (7, 8) has posed that some children with various neurological immaturity or deficits display *tactile defensiveness*, which is an aversion to various forms of tactile input. Thus, Ayres claims that responsivity to touch can be used clinically to evaluate the integrity of the nervous system, and tactile discrimination test items are included in her initial evaluation of sensory integration, the Southern California Sensory Integration Tests (SCSIT) (53). That disorders of tactile functions may indicate deficits in CNS integrity is supported by findings that various clinical populations engage in forms of self-abuse such as biting, burning their skin, or pulling out clumps of hair. Such individuals may not process or perceive pain in the way that others do.

## Clinical Significance of the Tactile System

The numerous clinical implications of the skin senses are vast, and describing them in detail is well beyond the scope of this chapter. Skin breakdown is one life-threatening implication of the lack of skin integrity that concerns all health professionals. When the body rarely moves, as a result of paralysis, weakness, reduced sensation, or inadequate motor control, the skin can break down, especially over bony prominences. This can result in skin ulcers, called *decubiti*, which when they occur, can lead to serious infections and medical complications. For individuals who lack pain sensation, the body may be at risk for decubiti. These individuals must be trained to use other senses to compensate for the lack of cutaneous sensitivity and pain perception. While pain is often regarded as negative, it is an important mechanism for informing us of infection or potential harm or compromise to our bodies. Pain reception and the skin itself serve significant preventative and informative functions.

Madison and colleagues (54) have proposed the clinical use of the habituation procedure with somesthetic stimulation delivered to fetuses. These researchers demonstrated habituation of fetal motor responses to vibrotactile stimulation (i.e., vibration of the mother's abdomen). Fetal movements, which were monitored with ultrasound, declined with repeated application of the stimulation and recovered upon presentation of a novel stimulus. In addition, the recovered response also habituated, and as habituation studies predict (see "Habituation") the motor responses habituated more rapidly when the original stimulus was presented again. Thus, this procedure, proven useful for sensory testing of infants, is also useful for assessing tactile sensitivity of fetuses and as a clinical assessment of fetal behavioral organization.

The somesthetic system is important also in physical therapy and occupational therapy treatment. Numerous sensory and motor interventions are based on the stimulation of cutaneous receptors. This includes the application of vibration, pressure, massage, brushing, and icing as well as the variety of sensory stimulation found with neurodevelopmental treatment (NDT) and sensory integration therapy (SIT). NDT involves an interaction between the therapist and client, with an emphasis on touch, movement, and positioning, all the while stimulating various cutaneous receptors and proprioceptors. In some forms of SIT, the client is encouraged to self-administer tactile stimulation by rubbing the body parts. Tactile stimulation forms a significant part of treatments aimed at individuals with neuromotor damage, as individuals recovering from partial paralysis or weakness may be encouraged to experience a wide variety of tactile experiences in order to restore somesthetic reception.

Consistent with her recapitulation view of development and treatment, Ayres (7) claims that the maturity of the tactile system is fundamental for development of other intellectual and cognitive functions. Ayres theorizes that deficits in tactile functions may indicate the presence of other neurological or developmental problems such as some forms of learning disabilities, and considerable research has been oriented toward exploring this relationship. For example, Finlayson and Reitan (55) reported that children and adolescents with poor tactile perception tended to perform less well on other cognitive and reading abilities than did children without tactile problems. Similarly, Haron and Henderson found that boys with poor motor planning skills displayed poorer active and passive touch skills than did a control group without motor planning deficits (56).

Thus, touch may be used therapeutically either as a developmental intervention (such as with NDT or SIT) or as a means of facilitating communication between client and therapist. Tactile stimulation is associated with a variety of developmental gains when administered to immature organisms of many species, including humans. Although touch cannot be separated from other forms of proprioceptive stimulation, "handling" immature organisms has resulted in developmental gains over nonhandled controls. [For a review of this issue see Ross (57) or Montagu (42).] While many of these studies are compelling, it is difficult to discern whether the stimulation is tactile, vestibular, proprioceptive, social, or an intricate interaction of them all. The significance of the interaction of sensory and motor stimulation will be discussed further in Chapter 9.

Touch has also been reported to have an almost "mystical" component. "Therapeutic touch" has been advanced in nursing, allied health, holistic health, and in some religions as a potential source of healing as well as a special means of communicating between members of the same or different species. Touch is special for parent-child and therapist-client interactions (42, 49, 52). As Montagu notes in regard to touching and physicians:

> In every branch of the practice of medicine touching should be considered an indispensable part of the doctor's art. As a member of a family the doctor should know what the human touch is capable of achieving in soothing ruffled feelings, in assuaging pain, in relieving distress, in giving reassurance, in making, in short, all the difference in the world. The world of humanity is the family writ large, and on a smaller scale the relationship seen in the family holds true between patient and doctor.

> What the patient expects from the doctor is a human touch and healing effect. Touch always enhances the doctor's therapeutic abilities and the patient's recuperative capabilities. The laying on of hands has for centuries been well understood in religious communion. It would be well if it were similarly understood within the healing community (42, p. 224).

Montagu's observations about touch and physicians apply equally well to occupational and physical therapists and their relationships to their patients or clients.

# Proprioception

Proprioception is regulated by special receptors located in different parts of the body: the joints, muscles, tendons, and the inner ear. These all work together in an intricate control system that regulates different degrees of tension (muscle tone) in the musculature and keeps the organism informed about the position of any part of the skeleton relative to another (41). Thus, proprioception involves awareness of skeletal movement, position in space, alignment of body parts and of the head in relation to the body and to gravity, movement in space, and equilibrium. The interaction of these abilities is essential to the activity and survival of the organism. Propriocep-tion essentially keeps us informed about what our bodies are doing so that we can prepare what to do next (58).

While the terms "proprioception" and "kinesthesia" are often used synonymously, a distinction, especially in physical and occupational ther-apy, is made between the two. Proprioception is the broader term referring to reception of information about movement and position in space. That information may be conscious or unconscious. For example, until you read this, you may not have been consciously aware of the position of your right foot, your left leg, your neck, or your 5th finger on your left hand. Generally, the body makes adjustments and readjustments without our conscious awareness, as we shift weight, fidget, stretch, and make minor skeletal adjustments. Proprioception consists of those unconscious actions as well as kinesthesia, which is the actual conscious awareness of joint position and movement. Kinesthesia is "conscious proprioception," as Scott calls it (59).

In addition to their location in the muscles and joints, proprioceptors are also located in the inner ear. These special receptors are part of the vestibular system, a component of proprioception, which responds to gravity and informs the body and head of their alignment and movement in space. Ayres explains it this way:

> The vestibular system enables the organism to detect motion, especially accel-eration and deceleration and the earth's gravitational pull. The system helps the organism to know whether any given sensory input—visual, tactile, or proprioceptive—is associated with movement of the body or is a function of the external environment. For example, it tells the person whether he is moving within the room or the room is moving about him (7, p. 57).

The vestibular system is closely connected to the musculature, affecting the body's muscle tone and the muscles of the eye, i.e., oculomotor func-tions. Since it also regulates equilibrium and maintains the body and its relation to space (i.e., spatial orientation), the vestibular system senses and processes information related to the three dimensions in which human functions occur. This three-dimensional characteristic of the vestibular system is reflected in its complex anatomy and physiology, which is located within the inner ears (Fig. 8.9). [For an excellent description of the anatomy and function of this system, the reader is referred to Clark (60).] The vestibular system actually comprises two subsystems. One consists of the otolith organs (the utricle and saccule) and tells us primarily about motion and the position of the head. The other includes the semicircular canals, each of which is geared to one of the three dimensions and which respond to gravity, motion, and rotation of the head (7, 41).

A close relationship exists between the ocular musculature, the visual system, and the vestibular system. For example, the measure often used as an index of intact vestibular functions is the presence or the duration of the vestibular-ocular-reflex, which is manifested by rapid ocular move-ments called *nystagmus*. These movements are normal and may sometimes occur spontaneously or after specific kinds of stimulation of the inner ear.

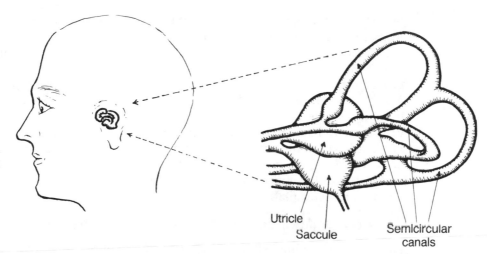

Utricle

Saccule

Semicircular
canals

**Figure 8.9.** The vestibular apparatus, located in the inner ear, consists of two subsystems: the semicircular canals and the otolith organs (the utricle and saccule).

For example, after the body is rotated, as in rapid spinning, the eyes move rapidly back and forth in *postrotary nystagmus*.

Another significant aspect of the relationship between the visual and vestibular systems is the latter's role in providing a foundation for vision. For example, as humans walk or move, their heads also move. To compensate for this, the vestibular system (specifically the semicircular canals) relay information about the position of the head and adjust the eyes accordingly. Thus, compensatory ocular movements keep the visual system stable, enabling organisms to receive fairly constant and steady visual information while locomoting (41).

## Development of the Vestibular System

In most vertebrates, the somesthetic is usually the first sensory system to develop functional capability. The vestibular system begins to react as the embryo makes whole-body adjustments. Over time, the fetal vestibular apparatus becomes more developed, and responses gradually become more differentiated (13). Vestibular development has not been studied as much as other sensory systems; however, some general information has been obtained. The system does function prenatally and is thought to be quite mature by the time of birth. Fetal head and body righting—i.e., keeping the head in line with the body and keeping segments of the body in line with one another, respectively—have been observed. Additionally, fetuses have displayed ocular movements that have been identified as either nystagmus or other types of eye deviations indicative of vestibular activity (13, 61). Whether the vestibular system is fully mature at birth is unclear. While nystagmus is present at this time, neural inhibitory influences may continue to mature, thereby inhibiting neonatal ocular movements (61) that mature, by one account (62), by 16 weeks postnatally.

Since the vestibular system is fairly well developed in the normal infant, and since vestibular responses go through a predictable maturational process, this system is often clinically evaluated for indications of central nervous system integrity. Rossi and colleagues (63) have suggested that in normal development important maturational changes occur in the vestibular system in the last weeks of intrauterine life. This has important implications for the infant whose intrauterine conditions are unfavorable or who is born prematurely. Such infants may be deprived of important maturational influences and, as a result, may have subsequent developmental problems (63).

### Clinical Significance of Proprioception and the Vestibular System

Lawrence and Feind (64), who observed nystagmus in healthy infants, noted the significance of this response for confirming the presence of other neural injuries, assessing the integrity of cranial nerves (i.e., peripheral nerves around the head), and as a potential sign of disease. Vestibular dysfunction is associated with numerous deficits; these range from hearing disorders to cerebral palsy or mental retardation, though the influence of vestibular dysfunction may be overrated in some cases (65). Reports of the relationship of vestibular functions with numerous other systems and abilities such as language and reading skills, ocular fixation, emotionality, some forms of mental illness, motor development, muscle tone, reflexes, arousal, audition, and with the development of a body schema and a conscious body image confirm the potential significance of the vestibular system (66–69). The significance of the vestibular system has been highlighted by occupational therapist Ayres and her colleagues, who note the importance of this modality in sensory integration therapy. Ayres (7, p. 57) reports, "Considerable opportunity exists for the vestibular system to exercise influence over all other ongoing sensory systems." She and others (70–73) have established various age-related developmental norms for postrotary nystagmus responses using a standardized test called the Southern California Postrotary Nystagmus Test (SCPNT) (71).

As noted under "Somesthesis": separating the proprioceptive from the somesthetic systems is difficult. When movement occurs in the body, both systems are called into play. For example, as a joint is moved, the skin is stretched and skin receptors are stimulated. As people walk, they not only maintain equilibrium, move their skeleton, muscles and joints, and keep in alignment but they also feel the air touch their skin, feel clothing touch their bodies, and sense and respond to the changing surface where they walk. Despite the close interaction of sensory systems, some systematic research studies have attempted to experimentally control for and separate the effects of visual, vestibular, and tactile stimulation. Such studies indicate that vestibular stimulation, even in isolation from other forms of sensory input, seems to be a powerful influence (67, 74) that is used therapeutically in a wide variety of settings and for several purposes. Vestibular input has been presumed to facilitate maturation, especially of the motor system; this is achieved either by neural alterations or by modifying muscle tone and thereby providing a more stable base of support for subsequent development (10, 69, 75).

Vestibular stimulation affects arousal and behavioral state. In infants, for example, gentle, rhythmic rocking has a soothing and calming influence; whereas, fast, random movements may be unsettling and can increase arousal and enhance attention or defensive reactions while also often reducing (or interrupting) crying. The developmental application and degree of control associated with vestibular stimulation is displayed by parents who often naturally rock their infants to sleep or who toss and jiggle their children during play. The youngsters in turn respond with laughter, increased arousal, and vocalization.

During the 1970s and 1980s, numerous studies by Anneliese Korner and colleagues demonstrated the powerful clinical impact of such stimulation. Korner and Thoman (74) suggested that the most effective soothers are rhythms and types of motions that are normally experienced "in utero" (i.e., while the fetus is still in the uterus). Based on this and other premises, Korner and various colleagues (76, 77) have used waterbed mattresses as therapeutic stimulation for preterm infants. The advantages of these mattresses are that they provide a source of containment and warmth similar to in utero conditions, plus they react when the infant moves. Thus, they also provide a form of feedback for self-generated movements, thereby providing natural proprioceptive feedback and stimulation.

Korner's research has found that the use of waterbeds alters behavioral organization by enhancing quiet sleep, REM sleep, and effecting fewer state changes in preterm infants (76). In addition, in premature infants without severe respiratory problems, waterbeds reduced *sleep apnea*, which is the interruption of respirations often associated with sudden infant death syndrome (77). A study in which the author was involved supports the positive effects of waterbeds on behavioral organization. In this study of healthy preterm infants, clinical researchers Pelletier, Short, and Nelson (78) found that, after an invasive feeding procedure, infants placed on waterbed mattresses displayed more approach behaviors and fewer avoidance behaviors than did controls who were placed on conventional mattresses (Table 8.9). These data indicate that the varied stimulation from the waterbeds may be calming as well as facilitating organized approach responses. The advantage of such stimulation is that it may be used clinically as a tool to elicit in preterm infants low energy behaviors, which may in turn facilitate subsequent development as well as support reciprocal interactions between the infant and his or her caretakers (66).

Although little research is available to explain how it mediates so many behaviors, vestibular stimulation has been attributed (and in some cases perhaps overrated) with numerous potent developmental influences (10, 68). The vestibular system is phylogenetically old, located in anatomically lower regions of the brain, and interconnected with many other sensory systems. Thus, recapitulation models of brain and behavioral development

Table 8.9
Approach and Avoidance Responses of Preterm Infants Placed on
Waterbeds or Conventional Mattresses[a]

| | Mean No. of Responses | |
| --- | --- | --- |
| | Waterbed | Conventional |
| Approach | | |
| Hand to mouth | 23 | 9* |
| Avoidance | | |
| Grimace | 23 | 54* |
| Startle | 3 | 11* |
| Trunkal arch | 14 | 36* |
| Finger splay | 20 | 14 |
| Salute (sudden limb extension) | 23 | 71* |

[a] Adapted from Pelletier JM, Short MA, Nelson DL: Immediate effects of waterbed flotation on approach and avoidance behaviors of premature infants. In Ottenbacher K, Short MA (eds): *Vestibular Processing Dysfunction in Children*. New York, Haworth, 1985, p 81.
* Statistically significant differences

claim that adequate vestibular function is essential before phylogenetically newer brain regions reach their full potential. This premise is supported by findings that indicate that vestibular dysfunction is associated with deficits in language and writing abilities and other skills associated with cortical regions of the brain. Considerable research in occupational and physical therapy is directed at studying the remediative effects of vestibular stimulation on some forms of learning disabilities, developmental delays, and other sensory deficits (7, 10, 68, 69, 79, 80).

Although clinicians frequently refer to the myriad effects of "vestibular stimulation", in normal development this modality cannot be viewed as an isolated sense. In everyday life, vestibular stimulation is only one component of a complex sensory interaction. Thus, the normal child who rolls, runs, falls, and climbs is bombarded by a variety of visual, auditory, tactile, and proprioceptive input all in a context of play. Similarly, the caretaking context involves lifting, stroking, rocking, jiggling, as well as cooing, smiling, feeding, singing, talking, changing diapers, etc. The ecological perspective informs us that vestibular stimulation occurs within a changing system of sensory interactions that have no real meaning when separated from one another. To understand this point, read the following description and while doing so, consider the numerous sensory systems that are involved.

Mandy is an 8-month-old infant who has just awakened from a nap. She is hungry, and her diapers are soiled so she begins to fidget and then to cry. As her crying intensifies, her father comes to her crib, all the while talking to her. As he approaches her, he reaches down, lifts her and holds her in front of his face. He furrows his brow as she cries, and he continues to talk: "Oh no, poor Mandy. What's wrong, baby? Is Mandy hungry?" He starts to cuddle and rock her, and Mandy, in turn, reduces her crying. As her father notes the wetness of the diaper, he gently sets her back on the bed and goes to get some materials to clean her up. As he moves around the room, she follows his voice and move- ments with her eyes but then begins to wail when he does not pick her up again. She continues to fuss while he changes her diaper and then stops when he lifts her to his face and gently jiggles her up and down.

At this point, her eyes widen and she looks directly at him. In turn, he makes a face and mimics her and then makes a funny noise. She smiles and giggles as he nudges her face, and he makes funny sounds as he moves her back and forth. Then, cradling her in his arm, he reaches for a rattle. This he puts near Mandy's face while he begins to walk out of the room, carrying her toward the kitchen. All the while he continues to talk to Mandy and move her slightly while she is still cradled in his arm. Once in the kitchen, he grabs a bottle, which he holds near her. He watches her face as he holds the bottle, and he asks, "Is Mandy hungry? Want some food?" As she responds with facial expressions, arm movements, and sounds, he places the nipple in her mouth. Mandy, in turn, begins to suck, grunts a little and heartily ingests the formula while her father continues to hold her, walking around the house, talking to her.

---

Clearly, in social interactions, movement, equilibrium, sound, visual stimulation, food, and so many other forms of stimulation are all intercon- nected within a special context. Thus, from a developmental perspective, vestibular changes are best understood when examined along with other proprioceptive as well as other sensory alterations. Some of these will be discussed in the following sections of this chapter, while proprioception is a significant influence examined as a component of motor development explored in greater detail in chapter 10.

## Gustation and Olfaction

Gustation and olfaction are often studied together as the body's chemore- ceptor system. While a close relationship exists between the taste and smell receptors that respond to chemical substances, the two modalities are different. Taste is not fully understood but involves "near" receptors located in the taste buds in the mouth. These are sensitive to chemicals that are placed in the mouth and that are dissolved in water (saliva). While the perception of more subtle taste qualities is not fully understood, it is thought that they may result from combinations of the most commonly recognized taste qualities: bitter, sour, sweet, and salt.

In contrast to taste, olfaction is a "distant" sense, which is specialized to receive chemical information that is sometimes highly diluted and transported a considerable distance through the air. Taste and smell are both considered primitive senses, although they can be trained to be quite acute (as in the case of wine tasters or perfume experts). Taste and smell are significant inasmuch as both may warn us about the palatability or safety of substances we ingest. In many other species, olfaction is quite sophisticated; in humans it is a fairly rudimentary sense, relied upon and consciously used less than the acute distant senses, vision and audition.

## Development of Gustation

The human infant has a wide distribution of taste buds, but it is unclear if the number increases during the neonatal period or not. The developmental tendency is, from childhood on, for a gradual reduction in the number of taste buds and, presumably, of taste sensation and discrimination throughout life. Taste is apparently functional in utero, as the fetus ingests amniotic fluid, which changes in chemical composition during pregnancy. These chemical changes in the fluid presumably reflect taste alterations, which may or may not be sensed by the fetus (81). Kaye (34, p. 56) notes, "The child is perhaps no more sensitive to taste stimuli at any time in his life than at birth, although this type of qualitative estimate is most difficult to define, much less test."

Gustation in infants and nonverbal individuals is typically tested by observing facial expressions when substances are placed on the tongue or by measuring the amount of substance that is ingested when it is introduced to the mouth. With both of these measures, newborns tend to perform similarly to adults, i.e., grimacing to bitter, puckering to sour, and ingesting very sweet substances more than less sweet substances. For example, Desor and associates (81) noted that neonates of 1–4 days of age discriminate between water and a sweet solution, and when given four different sugar solutions, prefer the sweeter, more concentrated ones. In a separate study, Desor and colleagues (82) also noted neonates discriminated between sweet and sour solutions but expressed little reaction to bitter and salt solutions. In addition, these 1- to 4-day-old neonates, similar to other newborn species, tended to display an aversion to water and a preference for milk over water. Other factors that may affect gustation in newborns include sex differences, birth weight, and wide individual differences among children (81).

In regard to more recent work, Lipsitt (35) points out that increased information about taste has been forthcoming because of new techniques used to measure oral responses. These techniques have enabled researchers to record tongue movements and to measure fine discriminations in sucking reactions. In addition, this work illustrates the close connection between sensory (taste) and motor (suck, swallow) abilities. For example, Lipsitt notes that studies of taste discrimination have indicated a close relationship between sucking and swallowing, with the latter often affecting the rate of sucking of different solutions. With preferred solutions, infants tend to suck more slowly but to swallow more often. It is of clinical interest that newborns exhibit preferences for certain substances (e.g., sweet solutions) long before these infants could be conditioned by experience. Thus, clinicians may find that infants with oral-motor deficits prefer some substances and may therefore work harder to obtain some substances than others.

## Olfaction

Olfaction operates through the process of breathing air into the nasal cavities. Substances in the air are then "picked up" by chemoreceptors that may react selectively to certain kinds of molecules, thereby accounting for perception and discrimination of different odors. Whether fetuses can smell amniotic fluid is unknown, but newborns do appear to have a functional sense of smell. They withdraw from noxious or strong odors just as do adults. Whether acuity is as great immediately following birth as it is at about 1 month is unclear. Different studies have had different results with testing newborns as well as with assessing developmental changes in acuity. These varied results may be affected by the presence of amniotic fluid and mucus in the nasal cavities immediately after birth, the crude measures that are used, the fact that infant respiratory capabilities change over time, the poorly understood nature of the olfactory system, state changes in infants, and the difficulty in presenting and testing olfactory stimuli. Presenting gradations of olfactory stimuli is difficult, and once a response is elicited, it is unclear to what the infant has responded. For example, infants presented with stimuli soaked in a wet solution may sense humidity rather than odor.

Olfactory abilities during development are tested in a variety of ways. For example, odorants are often presented on swabs and then behavioral or psychophysiological responses are observed. Behavioral responses may include avoidance responses such as head retraction, grimace, or head turning. Psychophysiological responses may include changes in heart rate or respiration following presentation of the stimulus. Another measure includes the habituation/dishabituation procedure described earlier in this chapter. In this procedure, an odor is presented to an infant until habituation occurs. Then a second odor is presented. If the infant responds to the new odor, i.e., dishabituates, then the researcher assumes that the infant can discriminate between the two stimuli (34).

Olfaction is a poorly understood sensation, whose subtleties have been studied much more fully in species other than humans. In some species a sensory system that is separate from, but closely linked to, olfaction has been examined. This *vomeronasal system*, as it is called, responds to chemical external hormones (called *pheromones*), which are produced in certain glands of different animals and released into and transported through the air. As found with insects and some mammals, miniscule amounts of these substances have powerful effects, attracting other members of their species from as much as miles away. Pheromones are thought to communicate fear, territorial marking, identification of species members, and sexual receptivity; they are closely linked with maternal and with sexual behavior (83). In humans and other species, pheromones are thought to affect sexual attraction; they differ in males and females and fluctuate in females in accordance with the menstrual cycle (3).

Whether pheromone perception increases in late childhood or adolescence in accordance with sexual maturity is not known. A functional olfactory-related system (which may be pheromone sensitive) seems significant, however, to other developmental social interactions. This was demonstrated by MacFarlane (84) and others (85) who have reported that human infants or adults can recognize one another by olfactory cues. For example, parents can discriminate their children by odor, and mothers who have recently given birth can identify their neonates by the odors of their garments (85). As MacFarlane demonstrated, 6-day-old breastfeeding infants can discriminate between the breast pads worn by their lactating mothers when compared with those worn by nursing strangers. In addition, infants may actually "mark" their own clothes and blankets with their own odors. This would support the validity of what we commonly refer to as a "security blanket", where an infant or child displays less anxiety in an unfamiliar location by carrying something familiar (and recognizable by scent) from home (3).

Olfaction does seem to play an important role in human social interactions and in safety. Olfactory cues can signal danger (e.g., by smelling smoke) and are certainly important in food preparation and consumption. Olfaction and taste interact, informing or warning us of the palatability of food and are clearly affected by learning and culture. Olfaction is considered a rudimentary sense inasmuch as it is one of the oldest senses in evolutionary terms. It may be that, in connection with pheromone sensitivity, this system affects us in many subtle and unconscious ways. Clearly, olfaction has an important function in human sexual and social interactions, as witnessed by this society's emphasis on perfumes and fragrances.

Because olfaction seems to operate for both newborns and their mothers, it is thought to be an important sense in early infant-caretaking interactions, thereby facilitating both bonding and attachment responses as well as kin recognition. As Porter and colleagues (86) have demonstrated, humans seem to be able to recognize their own kin based on subtle olfactory cues. These researchers claim that although subtle, the influence of biologically produced odors in human social behavior may be of greater import than is typically assumed (86, p. 448).

# The Distant Senses

Visual and auditory systems contain receptors geared to process information received from a distance. While these are important and the most extensively studied of the sensory modalities, vision and audition are not subject to as much focus in occupational and physical therapy as are the near receptor and proprioceptor systems, to which many of our sensorimotor interventions are directed. This is not to say that audition and vision are insignificant but that many therapeutic modalities are directed at musculoskeletal systems and therefore rely more heavily on somesthetic, proprioceptive, and motor input and feedback than upon the distant senses. Despite this, the greatest amount of research information is available regarding the development of auditory and visual sensation and perception in a wide variety of species. Part of this large body of information results from an understanding of the characteristics of light and sound wave stimulation and the nature and function of the visual and auditory receptors and corresponding structural and neural systems.

At birth, the peripheral neural development of both systems is relatively more mature than central auditory and visual cortical areas, which continue myelination postnatally. Complete auditory development follows the somesthetic and vestibular modalities but precedes the visual system, which continues extensive developmental change after birth (87). All the sensory systems, as will be explained later, interact closely with one another and with the musculoskeletal system (88).

## Auditory Development

Hearing is affected by the structure of the ear, which undergoes numerous changes during development. The ear consists of three major divisions that are all connected. One division, the external ear structures, which will eventually act to funnel sound waves from the environment to the eardrum, begins development early in embryonic life but continues growth postnatally. The middle ear, which consists of bony ossicles that transmit sound wave vibrations from the eardrum to the inner ear, appears at about 7 weeks of gestation and reaches final adult size by 6–8 months of gestation. The middle ear cavity, however, continues to elongate for at least the 1st year; developmental changes in it and in the size of the external auditory canal alter the resonance of sound waves, thereby potentially affecting auditory sensation and interpretation (89, 90).

The auditory system provides an excellent example of energy transformation for eventual neural processing. Auditory stimulation in the form of sound waves is picked up by the external ear and relayed to the eardrum, which begins to vibrate. These vibrations then cause the bony ossicles to vibrate. The inner ear, which is structurally developed by birth (88), transforms the movements from the bony ossicles into fluid vibrations, which are then transduced into neural activity. The sensory receptors in the cochlea of the inner ear are then stimulated, and neural information is transmitted along fibers of the acoustic or auditory nerve, which carries the output from the receptor cells to various parts of the central nervous system (89). Because of structural developmental change at each stage of energy transfer, auditory sensation and perception may be altered during early infancy and possibly childhood. For example, the acoustic nerve that connects the ear to the central nervous system is well myelinated at birth, but auditory cortical areas continue myelination postnatally. The fact that there are so many different components in the auditory apparatus and that they develop at different rates prenatally and postnatally makes it difficult to conclude with surety about the overall maturation of this system (89, 90).

### Auditory Stimulation

Gottlieb (13) theorizes that the auditory system in most vertebrates becomes operational after the first signs of vestibular function. The fetus, which is provided with rhythmic auditory stimulation by maternal vessels, reacts to sound with body movement and respiratory changes. These responses tend to be greater to sounds in the low- to mid-frequency regions, whereas responsiveness to higher frequencies occurs later in infancy. The development of electronic monitors to record fetal cardiac activity (through the mother's abdomen) and brain wave activity (via fetal scalp electrodes) indicate that these immature organisms do respond to sound in a manner similar to neonates. One potential confound in interpreting these studies, however, is the inability to distinguish whether fetuses are responding to sound transmitted through maternal tissues and amniotic fluid or whether their responses are secondary to maternal reactions. Thus, the fetuses may not actually "hear" the sound but may be responding to maternal emotional, hormonal changes. The rapidity of the responses, however, indicates that fetuses are actually hearing, rather than responding secondarily (91).

One method for eliminating this confound about testing fetal audition involves examining the auditory responses of premature infants who are able to respond independent of maternal reactions. While studies of this population are also confounded by the use of subjects who may have a variety of problems associated with (or responsible for) their prematurity, they do show that neonates and preterm infants consistently respond with heart rate changes to auditory stimulation (31, 33). Als and colleagues demonstrated that a 34-week-old preterm infant could display relatively sophisticated responses to auditory stimulation administered during different behavioral states. The infant displayed response decrement (habituation) to auditory stimulation when asleep, turned toward a soft noise when awake, and stared at and startled to a loud sound. The researchers who conducted this study concluded that with better observational methods and sensitivity to the behavioral repertoire of preterm infants, researchers may find that prenatal auditory responsiveness is greater than originally believed (92).

Studies of the auditory responses of full-term infants have been conducted for more than a century. Researchers in the late 1800s and early 1900s tended to look at the effects of auditory stimuli on various infant behaviors such as startles, eye blinks, body movements, reflexes, facial changes, and awakening from sleep. These early studies tended to use complex sounds such as rattles and human voices rather than pure tones, and they did not make distinctions between different qualities of sound (Table 8.10). Currently, studies of auditory developmental changes fall into two different categories: those geared toward determining developmental

Table 8.10
Characteristics of Sound[a]

| Loudness | — | Intensity of sound (loud vs. soft) |
| Pitch | — | Frequency of sound wave (high vs. low tones) |
| Duration | — | Length of time stimulation occurs |
| Location | — | Source of origin of sound |

[a] Adapted from Rosenblith JF, Sims-Knight JE: *In the Beginning. Development in the First Two Years.* Belmont, CA, Brooks Cole, 1985.

changes in auditory sensation and discrimination and those attempting to identify *absolute auditory thresholds*; i.e. the softest sounds to which the infant or child responds.

Studies of newborns and older infants indicate that they do discriminate different intensities of auditory stimulation; they startle more to louder than to softer sounds. Loud complex sounds often produce startles and whole-body movements, Moro reflex, or facial responses such as the auropalpebral reflex, which is the closing of previously opened eyes or tightening of eyes that are already closed (as when stimuli are presented during sleep states) (89). When compared with preterm infant or fetal responses, full-term neonates and older infants display an auditory competence characterized by greater variety and differentiation of behavioral responsiveness (90). Within a few months after birth, the normal infant habituates rapidly to repeated sounds and displays auditory discrimination. There are marked maturational changes reflected in neural recordings of auditory cortical responses as well as heart rate responses to auditory stimulation. The absolute threshold for neonates is higher than in adults, which means that more stimulation is required of neonates before a response is made. While this auditory threshold tends to decrease rapidly over the first few days of life and decreases additionally between 3 and 8 months of age, the threshold for speech sounds still remains higher during the first year of life than it is for adults (19, 90).

Infants respond to and can discriminate different frequencies of sound. While low- and middle-frequency sounds produce greater responsiveness in preterm infants and neonates, this tendency shifts in older infants, who tend to be more sensitive to high-frequency sounds than are adults. Neonates seem to show preferences for, and seem more responsive to, human voice and complex sounds rather than pure tones, such as that produced by a flute. High-frequency, intense sounds, which for adults communicate alarm, also appear to distress infants. Other developmental trends are a tendency for a shorter latency to respond (i.e., reduction in the amount of time it takes to respond) to acoustic stimuli and for sensory fatigue to increase with maturation (19, 87, 89).

Like vestibular stimulation, different qualities of auditory stimulation can affect behavioral state or psychophysiological response. Kearsley (93) demonstrated that infants are differentially responsive to auditory stimuli. While some controversy exists regarding premature infants or neonates' abilities to produce orienting, as opposed to defensive, reactions (31, 33), Kearsley demonstrated that neonates do respond with either reaction depending upon the nature of the auditory stimulus. Intense, arrhythmic sounds can be distressing and produce defensive or avoidance responses, whereas sustained and rhythmic auditory stimulation can be soothing, causing decreases in heart rate and spontaneous motor activity and producing more regular heart rate and respiration in both preterm and full-term infants. Heart beat sound may enhance sleep as well as affect state and organization of behaviors. This was demonstrated by Schmidt and colleagues (94), who found that the therapeutic application of heart beat sound during the first hour of sleep had a major impact on the sleep states, motility, and cardiac responsiveness of preterm infants, which brought preterm infants closer in responsiveness to that of full-term infants. Another characteristic of auditory stimulation that may have special meaning is whether sound is speech related.

### Significance of Early Auditory Development

Different types of studies of infants' responses to auditory stimulation support the assumption that human infants may respond selectively to certain sounds that are likely to have adaptive, social meaning. Various ecological, behavioral, and anatomical studies have collectively demonstrated that human infants are geared to selectively perceive human speech sounds.

The human being, a species highly dependent upon language and speech for communication, socialization, and for record keeping and history, displays neuroanatomical specialization for speech and language. The two hemispheres of the brain, while often considered symmetrical, are functionally different. This functional asymmetry is known as *lateralization*. Electrophysiological studies and clinical information obtained from patients suffering brain damage have informed us that in most humans (i.e., 90% of the population who are right handed), the left hemisphere of the brain is specialized for receiving and interpreting language and for the motor control required for speech. Anatomical studies indicate that even at birth lateralization is evident, with the left hemisphere of the brain more heavily involved than the right in processing sound patterns characteristic of human speech (89). These anatomical studies indicate that infants process speech and nonspeech sounds differently, and behavioral studies concur. Infants react differently to patterned and non-patterned acoustic stimuli. For example, patterned signals and speech sounds elicit differentiated movements, vocalizing, facial reactions, and visual search; whereas nonpatterned stimuli tend to produce undifferentiated whole-body movements. Acoustic discrimination studies indicate that infants in the first 3 months can differentiate a wide variety of similar speech sounds, such as "da" and "ba" (89).

Other behavioral and ecological studies indicate that speech sounds are significant early in development. Infants seem to be especially responsive to sounds in the frequency range of the human voice, and newborns become restless when they hear the sound of another newborn crying. Infants also tend to pay attention more to the human female voice than the male voice even though both make the same sounds (35, 89). A study by Condon and Sander (95) indicates that infants synchronize their motor behaviors to fluctuations in speech sounds. Condon and Sander believe this interaction between the communication that infants hear and the infants' motor responses is evidence of an entrainment mechanism, whereby infant motor responses are essentially "driven" involuntarily by the verbal stimulation occurring in their environments. This entrainment may serve as a facilitator for subsequent language development:

This study reveals a complex interaction system in which the organization of the neonate's motor behavior is entrained by and synchronized with the organized speech behavior of adults in his environment. If the infant, from the beginning, moves in precise, shared rhythm with the organization of the speech structure of his culture, then he participates developmentally through complex, sociobiological entrainment processes in millions of repetitions of linguisitic forms long before he later uses them in speaking and communicating (95, p. 101).

Ecological approaches to early auditory development point out the significance of speech in social interactions. The proponents of parent-infant bonding, researchers Klaus and Kennell (96), indicate that the early synchrony between adult speech and infant motor responses may serve to facilitate reciprocal interactions between infants and their caretakers. Thus, the infant's motor responses may reinforce the parents' verbal reactions. In turn, the infant continues to respond, thereby establishing a reciprocal cycle in which the parents' verbalizations are reinforced by the infant's responses. As a consequence, the infant receives ongoing verbal stimulation and attention.

An implication of the ecological study of early auditory stimulation can be seen with preterm infants placed in neonatal nurseries (Fig. 8.10). Such infants are often deprived of speech-related sounds, especially within a

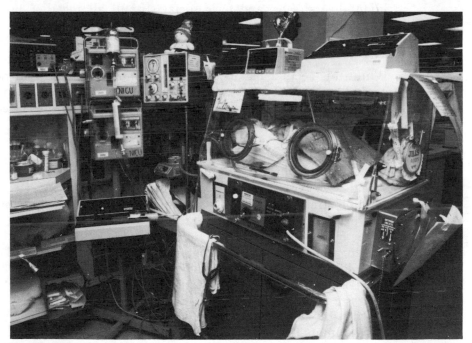

**Figure 8.10.**   Ecological studies of preterm infants take into consideration their sensory-perceptual environments. (Courtesy of Neonatal Intensive Care Unit, University of Connecticut Health Center's John Dempsey Hospital, Farmington, CT.)

caretaking context. While they are positioned in isolettes, they may hear conversations taking place around them but lack the related somesthetic and visual stimulation that normally accompanies speech. In addition, these infants may be bombarded by a wide array of non-speech-related, arrhythmic, and intense auditory stimuli (which we know from previous studies) often produce whole-body startes and defensive reactions, which may use up energy reserves required for self-maintenance. It is not known if the noise pollution from continuous incubator and air circulation motors and equipment, the clatter of doors and trash cans, and the sound of beepers that occurs on special care nurseries and that infants hear early in life will affect later auditory perception and development. Some studies have indicated that when compared with full-term infants, preterm infants display early hearing losses and language difficulties when they are older. Whether these are due to early extraneous and continuous auditory stimulation associated with the neonatal nursery is not known. Infections, drugs, hypoxia, and other complications associated with preterm delivery may be the contributors to subsequent auditory deficits, or they may all interact with the noise pollution. Additional studies will need to be conducted to investigate and isolate the offending variables (90, 97). In the meantime, health professionals working with this population must be sensitive to the need for delivering verbal stimulation during gentle caretaking, so that the infant hears, sees, and feels social stimuli all together in a caretaking context.

### Integration of Audition with Other Sensory and Motor Abilities

While this section has focused on auditory sensation and perception, it should not detract from the close coordination between this modality and other sensory abilities. For example, neonatal auditory localization, or turning in the direction of the source of stimulation, has been demonstrated (98). While localization may appear to be a simple ability, it requires sophisticated auditory perception and integration with motor (and in this case, ocular) abilities. To localize an auditory stimulus requires hearing the sound with both ears and recognizing the direction of the source. If the sound is straight ahead, then the sound waves hit both ears simultaneously. However, if the sound is to one side, then the sound waves hit one ear prior to the other. A newborn who turns in the direction of sound demonstrates auditory localization, motor control of the head and neck, and coordination of auditory, visual, and motor abilities.

Localization is often assessed in neonatal tests in which the examiner shakes a rattle or a bell to the side of an infant and observes the response. A study by Turner and MacFarlane (99) indicated that as young as 4 days of age, infants were able to correctly localize and therefore discriminate between the sound of a human voice delivered to their right, left, and at the midline. This ability was better in infants whose mothers did not receive anesthesia during labor than it was in neonates exposed to the drug. Rosenblith and Sims-Knight (19) point out that the localization behaviors observed in newborns may be quite different from those displayed by infants over 3 months of age. Localization in newborns is often inconsistent and rudimentary when compared with that of older infants. While newborns may turn their heads and visually scan, this does not indicate that the infants are actually looking precisely toward the source of sound. This precise ability may require further maturation of the nervous system as well as experience with objects and events in the environment, which facilitates further cognitive development and intermodal communication. One of the implications of precise localization, Rosenblith and Sims-Knight point out, is the "expectation" that an auditory event is associated with an object that has visual characteristics. Otherwise, why turn to "look at" the source of auditory stimulation? Thus infants less than 3 or 4 months may be displaying orienting reactions rather than accurate localization responses, which require an understanding of the nature of objects in the environment as well as a close connection between auditory and visual modes.

In addition to localization, other auditory perceptual abilities are included in clinical and developmental assessments. In the 1930s, in one of the first experimental demonstrations of auditory sensation, Dorothy Marquis (37) demonstrated that newborn infants during the first 10 days of life can be classically conditioned to a sound. These infants were presented with an auditory stimulus at the same time they received a nipple placed in their mouths. After many trials, the sound alone elicited sucking responses. In a review of the clinical applications of learning procedures, Lipsitt (35) notes the relevance of classical conditioning for assessing deafness. Infants and children who cannot be classically conditioned to auditory stimuli, especially when sounds are paired with salient stimuli, such as a pinprick to the sole of the foot, may show evidence of hearing impairments or the inability to learn; i.e., to make an association between the two stimuli. Since learning requires basic memory abilities, infant tests of conditionability to auditory stimuli may be used as evidence of CNS damage (37).

Lipsitt also described the clinical benefit of the close connections between audition and sucking. Suck suppression, sometimes called the Brohnstein effect, is a cessation of ongoing sucking behavior when an extraneous stimulus (e.g., an auditory, visual, or olfactory event) is presented (100). In a review of research about this effect, Lipsitt notes that in normal infants, responses during suck suppression habituate over time; e.g., a sound will interrupt sucking less and less as the sound is repeatedly presented. Since suck suppression is thought to occur as a result of arousal or of cognitive processes that interfere with ongoing behavior, habituation of suck suppression responses is an indicator of sophisticated memory and inhibitory abilities. Lipsitt notes research that has shown that when compared with normal infants, those born at risk take many more trials to habituate in a suck suppression procedure. In addition, if suck suppression occurs in response to an auditory stimulus, it is an indicator that the child or nonverbal client could hear the sound. As such, it can be used as a test for the absence of deafness.

With age and throughout childhood, auditory abilities become more refined and more integrated with other motor skills and stimulus modalities. With maturation of neural structures and with specific learning experiences, children develop capacities to attend to and process the sounds (words) of their culture. Discriminative abilities become more acute, and with numerous associations, auditory events (emergency alarms, school bells, train whistles, sirens, telephone bells, alarm clocks, music) take on special properties. Audition in childhood and during subsequent development can be affected by environmental events. For example, temporary or permanent structural hearing loss may occur as a result of continuous exposure to very loud music played at concerts or through headphones. Auditory and linguistic skills are affected by intelligence, vocabulary, and verbal contexts (101).

## Vision

Most research in the area of infant perception has been conducted in the field of vision. The resulting information has been used to generate models for how sensory and perceptual development occurs in the other modalities (102). The structural development of the visual system and its level of immaturity at birth, however, are quite different from those of other sensory modalities. With the possible exception of the pupillary reflex, which is present at birth, visual system maturity follows that of the auditory (and other sensory) system and displays numerous changes postnatally. Despite the scope of research in visual perception, existing knowledge about physiological maturation of the visual system in humans is still limited.

### Structural Changes in the Visual System

Development of the eye is first evident at about the 4th week of gestation, at which time the ectoderm layer thickens and forms what will be the lens and other parts of the eye. The retina, which lines the back of the eye and which functions similarly to film in a camera, comprises three layers that differentiate by about 7 months. At this time, the eye is sensitive to light, although parts do not completely differentiate until months or years postnatally. For example, all cortical neurons are present at birth, although the visual cortex is still quite underdeveloped at this time (41, 88, 89).

Many different stages are involved in the transfer of visual information, which is received by the eye, processed by the retina, and relayed to and by each eye's optic nerve, which consists of about 1,000,000 axons. The optic nerves extend outward from the back of each eye and then meet at a location at the base of the brain over the roof of the mouth. This is a crossover point, called the *optic chiasm*, where optical information from each eye is split up. From this point, visual information from the right half of each eye is carried by optic tracts to one hemisphere, and information from the left half of each eye is carried to the other hemisphere. The "optic tracts," as they are now called, leave the optic chiasm and carry sensory information first to the lateral geniculate bodies of the thalamus and then to the visual cortical areas of each hemisphere, where they are processed (Fig. 8.11) (43).

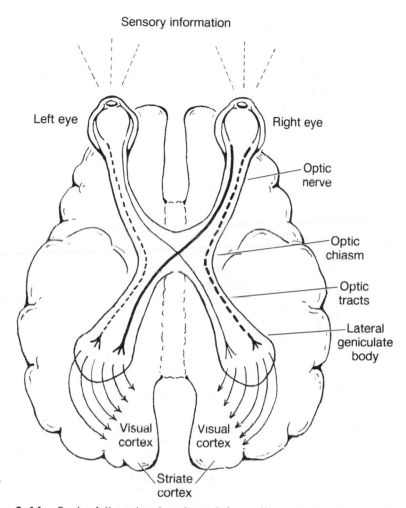

**Figure 8.11.** Part of the visual system. Information received by each eye is relayed by the optic nerves to the optic chiasm. At this point, nerves from the right half of each eye are relayed to the left hemisphere, and nerves from the left half of each eye are relayed to the right hemisphere. From the optic chiasm, optic tracts relay information to the lateral geniculate bodies, then to the visual cortex, where some visual-neural information is processed and relayed to other regions of the brain. (Modified from Ludel J: *Introduction to Sensory Processes.* San Francisco, WH Freeman, 1978 and Hubel DH: The visual cortex of normal and deprived monkeys. *Am Scient* 67: 532, 1979.)

Each of these optical neural processing and relay areas develops at different rates; peripheral components of the visual system usually mature earlier than more central ones. The optic nerve is incompletely myelinated at birth; it is fully myelinated at about 3 weeks to 4 months postnatally. In contrast, cortical areas associated with vision do not myelinate until well after birth. The cortical neurons additionally seem to undergo numerous changes; there are alterations in the cell bodies, changes in the length of axons, and gradual myelination that continues at least through the first 10 years of life (89).

Researchers disagree about the maturity of the visual system at birth and about its development early in life. One major area of disagreement is in regard to the structure of the infant retina, the lining of the eye. The retina consists of two types of receptors: the cones, which are sensitive to color, and the rods, which are sensitive to all lightwaves. Central on the retina and behind the lens is the fovea, which is densely packed with cones; these normally decrease in concentration as the distance from the fovea increases. At 23 weeks of gestation, rods and cones can be identified, although their distribution prenatally may be dissimilar to that of adults. It is thought that at birth the cones are shorter and less concentrated at the fovea and that not until around the 4th month does the concentration become similar to that of adults. Although the retina is functional at birth, the degree of functionality is unclear. While often inconsistent, current studies of infant visual perception indicate that this modality may be more advanced developmentally than originally thought (88, 89). However, as Boothe and colleagues (103, p. 512) note, infant visual perception is immature: "Development is more than just growth, and infants are not just little adults. Different visual functions emerge in different age ranges, and with different time courses."

### Measuring Visual Discrimination and Perception

A commonly used procedure for assessing visual sensation and perception is *fixation time*; i.e., the duration an individual looks at a visual stimulus (102). If an infant fixates longer on one stimulus than on another, then it is assumed that visual discrimination exists. Further, it is assumed that novel stimuli or those that are most intriguing, appealing, or interesting are associated with longer fixation times. Although fixation appears to be a simple response, observing one stimulus longer than another involves a series of responses including visual detection, attention, retention, and generation of a visual-motor response (104). Fixation, however, is only one of many methods used to explore visual sensation and perception. Other methods used include behavioral and psychophysiological measures, such as those used to test other sensory modalities, suck suppression to visual stimuli, habituation, and various forms of learning or conditioning procedures.

Numerous sophisticated and novel procedures have been developed to test postnatal anatomical changes and behavioral changes associated with the development of visual perception. Entire books have been written on this topic, as it is an area of extensive study and interest. Clearly, many procedures and data generated about this area cannot be reviewed here, and the reader is referred to reviews included as references in this section (e.g., 88, 89, 102, 103).

### Visual Ocular Movements

As described earlier, most of the axons from the optic nerve are enlisted in visual perceptual capacities; however, some travel to neural areas other than the sensory processing areas of the lateral geniculate bodies and the visual cortex. For example, some axons from the optic nerve travel to neural regions that regulate the degree to which light enters through the iris of the eye, and other neurons relay information regarding the extraocular muscles that move and rotate the eyes (43). These latter muscles are involved in nystagmus, *saccadic eye movements*, and in ocular *tracking* and scanning.

Nystagmus, which has already been discussed in this chapter, is the rapid movement of the eyes following specific kinds of visual or vestibular stimulation. *Optokinetic nystagmus* is elicited by presentation of repetitive patterns such as vertical black-and-white stripes. Although this response is evident in newborns, it is immature for at least the first few months of life (103). *Saccadic eye movements* are also evident at birth but immature. Saccades are rapid movements that bring stimuli from peripheral vision into view. They have significant adaptive value, as moving stimuli approaching from the periphery may be threatening, thus it is significant that humans and other organisms rapidly process and respond to such events. In humans, saccadic eye movements are also important in processing visual information ranging from pictures to reading material, which are scanned by a series of visual saccades and fixations.

Saccadic eye movements are found in newborns although they are characterized by less accuracy and longer latencies than at later ages, and they occur more commonly at this age for nearer than for farther stimuli. Although the mature form of saccadic movements may not occur until at least 7 weeks of age, their presence indicates that newborns can detect peripheral information and are born with the ability to locate visual events in their environments (44, 103). In addition to saccadic eye movements, other ocular actions are necessary for visually exploring the environment. As noted under "Integration of Audition with Other Sensory Motor Abilities", whether newborns actually "scan" the environment in search of objects is unclear; however, they do visually track moving stimuli and display developmental trends in visually scanning stationary objects. *Tracking*, or *visual pursuit* of an object that is moved (or moves) in front of the eyes occurs for a wide distance in neonates, although it is immature and not smoothly developed for some events until the first 6 months of life. With age, smooth pursuit movements become interspersed with saccades during tracking, and the proportion of smooth pursuit movements increases.

Visual scanning of stationary objects has been studied using equipment that can discern where the individual focuses. Developmental trends in scanning indicate that newborns tend to look at some single feature on the outside of stimuli. Thus, when presented with an image of a triangle or a face, they tend to look at one corner of the geometric shape, or the external portions of the head, respectively. By 2 months of age, however, scanning becomes broader and includes the entire triangle as well as both the outside and inside facial features (44). As will be discussed later, these visual ocular movements and preferences may have important social and cognitive consequences.

### Visual-Motor Abilities

An important aspect of visual ability is found in its relationship to motor skill acquisition and to manipulation and exploration of objects. Abravanel indicates that early in development a preference emerges in regard to visual instead of tactual exploration of objects (48). Predictable developmental trends throughout infancy and childhood are found in regard to visual-motor skills, and these changing abilities are often examined in developmental assessments. As motor abilities improve, visually guided reaching becomes more refined. While young infants do not respond consistently, older infants display predictable responses. For example, the abilities to pick up a raisin and place it in a small bottle and to make a tower of cubes involve coordinated efforts between motor development and visual guidance. Both of these are included in various assessments of adaptive motor abilities, which will be described in more detail in chapter 10, which discusses motor development. The general trend is for orienting motor responses and coarse sensory capacities to emerge earliest and for finer sensory processing and visually guided and regulated reaching and placing to emerge later (103).

### Visual Acuity and Accommodation

At one time it was generally assumed that infants were blind at birth, but studies in the 1950s and 1960s indicated that neonates not only see but also exhibit visual preferences. Many of these studies of visual acuity, i.e., the ability to distinguish detail in patterns (pattern discrimination), were originally conducted by Robert Fantz (105). Since Fantz' classic studies with the looking chamber (a device designed to hold an infant, present visual stimuli, and then to assess and gauge infant visual fixation) many sophisticated procedures, which have been developed to study central and peripheral vision and acuity, have supported Fantz' original findings which showed that infants have visual capabilities beyond what was originally believed.

For example, Fantz assessed infants' abilities to discriminate gray from black-and-white patterns of equal brightness. Since individuals with poor acuity will see no difference between solid gray and small black-and-white patterns, fixation on one over the other indicates acuity and discriminative abilities. Fantz demonstrated that infants less than 1 month of age were able to perceive ⅛-inch stripes at a distance of 10 inches, whereas by 6 months they were able see stripes ¹⁄₆₄-inch wide (105). Although visual acuity is evident in infancy, it is not comparable to that of adults. Acuity improves rapidly during the first year of life and appears to be within the range of normal adult vision by 6 months to 1 year (88). Developmental trends in pattern perception continue as the infant grows, and preferences are shown for more complex and intense stimuli with time (89).

Studies of visual acuity and accommodation are still clarifying specific developmental trends. Most of the early studies of infant acuity presented stimuli at a set distance from the infants' faces because of the belief that in early life visual accommodation does not occur. Accommodation is the change in the shape of the lens, which is necessary to bring objects at different distances into focus. Current studies indicate that after about 1 month of age, infants adjust their focus for farther or closer distances and that this ability becomes gradually refined over time. By 3 to 4 months, accommodation is roughly appropriate for the distance of the object that is viewed (103).

# Visual Perception

Numerous types of visual perception exist, and how each operates is difficult to fully understand, let alone to assess and to generate developmental trends. Pattern perception, for example, which is assessed with studies of visual acuity, has been shown to exist early in life. Recent sophisticated procedures indicate that both acuity and focus in infants is better than previously thought, but continued investigation will be needed to resolve inconsistent data gathered over the years and to provide a complete developmental picture of these visual abilities (88, 89). Pattern discrimination, however, is only one of many examples of visual perception (Table 8.11); each type of visual perception has its own structural explanation, developmental trends, and conflicting theories as to how and why it occurs as it does. Many of these, such as the perception of color or all of the elements of spatial discrimination, are not fully understood. They involve cognitive, interpretive capacities that are difficult to examine, especially in preverbal organisms such as infants, and therefore it is difficult to identify developmental trends.

## Color Perception

Our understanding of how people actually perceive color is incomplete and therefore a clear picture of the development of color perception is unavailable. Studies of infants' perception of color have been confounded by using visual stimuli that are not equated in terms of brightness. Thus, infants who respond to differently colored stimuli may be responding not specifically to differences in hue but to changes in intensity or brightness of the stimuli.

**Table 8.11**
**Types of Visual Perception**

| |
|---|
| Color perception (hue vs. brightness) |
| Spatial perception |
|   Depth perception |
|   Picture perception |
| Perceptual constancies |
|   Size |
|   Shape |
| Perception of form/pattern/contour/complexity |
| Perception of contrast |
| Face perception |

Despite discrepancies in studies and confounds in research methodologies, general conclusions have been made about infant color perception.

It appears that color perception exists early in life. Fantz' fixation studies in the 1960s indicated that while infants prefer patterns such as faces and bull's-eyes over solid colors, they do fixate longer on colors than on noncolored objects (105). Subsequent studies have indicated that infants as young as 10 weeks of age can discriminate colors but that this ability may not be identical to adult's perception of color (102). As Bornstein and colleagues have noted, infants of 4 months categorize colors in the same way as adults and they do this long before they have the words for such discriminative abilities (106). Thus, although it is unclear how such perceptual abilities arise, they do appear to be operative early in infancy.

## Face Perception

Another area of visual development that has received considerable attention is the emergence of the perception of human faces. Fantz' early studies indicated that infants preferred faces to bull's-eyes, and when presented with either a model of a face or a model with facial features jumbled up, the infants preferred the regular face (105). Such findings have led some scientists to conclude that since face perception emerges very early in development, it must be innate. However, this aspect of visual perception is subject to considerable controversy, partly because of the difficulty establishing an accurate and useful procedure for testing face perception and because of the confounds due to the numerous variables that are involved. For example, some studies examine infants' responses to pictures of faces, whereas others use three-dimensional models. Some studies have used familiar faces such as those of infants' mothers compared with stranger's faces, whereas others have studied infants' responses to their mothers making different facial expressions. Other variables that are included and that may confound the issue are the presence or absence of color and the degree of contrast in faces, in control, or comparison stimuli. In addition, a series of studies have compared infants' responses to faces with the features intact and to other faces with the features upside down or jumbled up. All these studies may also be confounded by infants' state and attention and by cognitive variables such as memory and schema development.

The development of face perception has been extensively reviewed, and often, to the dismay of the individual attempting to obtain developmental trends, conclusions are based upon the theoretical biases of the authors. From an ethological point of view, the innateness of face perception is apparent and logical. Early attention of the infant to his or her caretakers is seen as an adaptive function that ensures attachment of the infant, that provides security, and that promotes reciprocal interchange that is significant for the infant's survival. However, scientists from nonethological perspectives point out that actual recognition of individual faces cannot occur until after some exposure to different faces, schema construction, the development of visual discriminative abilities, and memory. These abilities, some scientists argue, take time—both for neural maturation and experience—to emerge. Thus, these scientists contend, face perception cannot be innate and findings like Fantz' report of infant fixation on faces must be explainable in other terms. For example, as noted earlier, visual scanning initially starts on the outside of objects and then, with development, gradually begins to take in features and detail. Visual attention in infants, as in adults, is attracted to areas of high contrast and to movement. Thus, very young infants may be highly attracted to the human face, not because it is a face, per se, but because the face is characterized by variability in movement (expression, movement of eyes and mouth) and in contrast (e.g., the eyes).

The consensus of reviews regarding face perception do not support its existence at birth (88, 89, 102). This capacity appears to emerge gradually, much as do other visual perceptual abilities. Developmental trends in regard to face perception and recognition indicate that 3-month-old infants can identify their mothers and by 4 months respond to whole configurations of facial features. Thus, it appears that by 4 months of age, infants have acquired an "expectation" or schema of how a normal face is to appear in regard to arrangement of features. This age is important because many studies of infant face perception reveal that 4-month-old infants pay particular attention to facial models with the features jumbled up. As Kagan explains:

> As the child acquires schemata, the relation between them and immediate experience competes with the original power of contour, movement, color, and curvature to attract and to hold the child's attention. Specifically, events that are a partial transformation of existing schemata begin to dominate the infant's attention. These events are called "discrepant" (2, p. 37).

Thus, once a child establishes a schema for "human face," he or she is then particularly attracted to discrepant schema. This illustrates the close connection between sensorimotor and cognitive events. From 5 months on, infants become increasingly able to respond to individual features on faces. At this time, they are able to distinguish one face from another and to discern the same individual in a variety of poses (102, 107). This general developmental trend, as will be discussed further in the next section, does not eliminate the ethological view regarding face perception, it means only that innateness, neural maturation, and environmental experiences all combine as interactive forces that affect development in different ways and at different times in life.

## Nativism and Empiricism

The study of infant visual perception has a long history with profound implications regarding the study of human development, in general. It was originally thought by many researchers and lay people that humans, who are clearly immature at birth, could not, so early in development, possess sophisticated visual perceptual abilities that are necessary for discriminating depth and color or for understanding the nature and relationships of objects in space. Since infants were originally believed to be blind, it followed that visual perceptual skills only emerged gradually according to maturation of neural structures and experience with objects in the environment. Evidence of early perceptual abilities, however, contrasted with these initial expectations of infant "inabilities." Thus, in the scientific study of human development, a controversy arose over the issue of whether perceptual skills were actually innate or whether they required time and experience to emerge. This nature vs. nurture debate in the field of visual perception was championed by the nativists on one side and the empiricists on the other. The nativists believed that, as Bower (108, p. 36) explains, "the ability to perceive the world is as much a part of man's genetic endowment as the ability to breathe." In contrast, the empiricists believed that such complex perceptual abilities could only be acquired through experience, learning about one's self, objects, and their nature and relationships in the world.

The nativism-empiricism debate resulted in many imaginative experiments designed to elucidate what and how infants really see. For example, the challenges and the complexity of this issue can be found in the study of depth and space perception. The ability to perceive depth is necessary for successful interaction in a three-dimensional world. At issue in developmental psychology was whether infants are born with an innate awareness of depth or whether they acquire it as a result of various experiences with near and far objects, with body parts that move close and far away, and with dropping objects or falling down. In the 1950s and 1960s numerous studies attempted to explore this issue, and conclusions based on their results swung back and forth on the nativist vs. empiricist scale.

### Visual Cliff

Data from psychologist Eleanor Gibson's classic visual cliff experiments were used to support the nativist position. The visual cliff consists of a large sheet of glass suspended off the floor. In the center of the glass is a small board for support, and on one side of the board, a patterned material, like a checkerboard table cloth, is placed flush against the underside of the glass. On the other side of the board, the cloth is laid on the floor some distance underneath the glass. Thus, on one side of the center support, there appears to be a cliff, whereas the other side appears to be close (Fig. 8.12). This visual cliff device was used to test the abilities of human and animal infants to discriminate the shallow from the deep side. For example, if an infant is placed on the center support and called by the mother from either end, the infant should, if he or she has depth perception, venture out on the shallow side but avoid the apparent cliff. Gibson and Walk describe their findings with this procedure:

> We tested 36 infants ranging in age from six months to 14 months on the visual cliff. Each child was placed upon the center board, and his mother called him to her from the cliff side and the shallow side successively. All of the 27 infants who moved off the board crawled out on the shallow side at least once; only

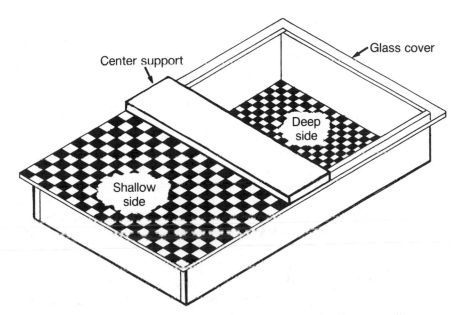

**Figure 8.12.** The visual cliff used to measure depth perception.

three of them crept off the brink onto the glass suspended above the pattern on the floor. Many of the infants crawled away from the mother when she called to them from the cliff side; others cried when she stood there, because they could not come to her without crossing an apparent chasm. The experiment thus demonstrated that most human infants can discriminate depth as soon as they can crawl (109, p. 19).

Gibson and Walk's findings were used in support of nativists' claims that depth perception was an innate capacity. However, empiricists argued that by the time infants are able to crawl, they may have had sufficient experience with their bodies, with falling down, and with close and distant objects to have acquired the perception of depth. More difficult to explain for the empiricists, however, were Gibson and Walk's findings that of other animal species such as kids, lambs, chicks, and kittens, some of whom can ambulate the day they are born, all shied away from the deep but ventured out onto the shallow side of the visual cliff.

### Spatial Organization, Depth, and Constancies

Depth perception is only one component of spatial perception, concerning the nature and organization of objects in space. Spatial organizational abilities arise gradually from a number of sources: understanding one's own body, the relationship of the body and its parts in space, movement through space, the position of the body in relation to objects in the environment, and the relationship and nature of objects in the environment. Spatial perception is, therefore, visual, cognitive, and proprioceptive; it concerns understanding the nature of objects in the world and an understanding of one's body and its parts and their relationship to one another and to the world around one's self. Clearly, developmental changes in spatial perception affect and are affected by concurrent shifts in maturation of the nervous system, experience with the world, sensory and motor changes, and cognition.

Another characteristic of spatial perception is picture perception; i.e. the ability to discriminate objects in a picture. This ability is related to the perception of patterns, color, brightness, form, contrast, and shading. Such skills are enhanced with maturation of nervous system and with experience. In addition, it appears that some developmental spatial abilities that emerge during childhood are different for males than for females.

A component of spatial perception is what is referred to as *constancy*. Constancy is the notion that objects stay the same, (i.e., are constant) despite perceptual information to the contrary. Thus, in our experiences with the everyday world, we know that objects retain their size and their shape, despite physical information to the contrary. Read the following for an example of constancy.

---

Extend your arms out in front of you. Supinate your arms (i.e., so that your palms are facing upward). Now, flex your wrists so that your palms are facing you and are perpendicular to the floor. Bring your left hand about 3 inches from your face, while keeping your other arm extended. Look back and forth at both palms. The palm of your left hand appears much larger because it is closer to you. Now, gradually move your left hand back. As it moves, it becomes smaller. As you move your left hand forward and backward in front of your face, the hand appears to grow or to diminish in size. You recognize, however, that your hand does not keep changing size. What, in fact, occurs is that the retinal image your hand projects changes as the hand becomes closer or farther away. The recognition that your hand retains its size despite the perceptual information from the retinal image is size constancy (Fig. 8.13). Shape constancy is similar and can be demonstrated by slowly rotating your arm as you look at your hand. Your hand appears to change shape, as you see first the palm, then one side, and then the dorsum (back) of the hand. Because of shape constancy, you recognize that your hand retains a constant shape despite its "apparent" changes.

---

Constancies are important components of depth perception, but they are only one of many. Sorting out which components, if any, of spatial perception (depth, constancies) exist early in life is very difficult. Gibson altered the patterns that were used with her visual cliff in order to discern what types of cues infants used in perceiving depth (109). Fantz' fixation studies indicated that infants as young as 2 months of age preferred three-dimensional to two-dimensional objects and therefore possessed depth perception even before they were able to crawl (105). Perceptual researcher

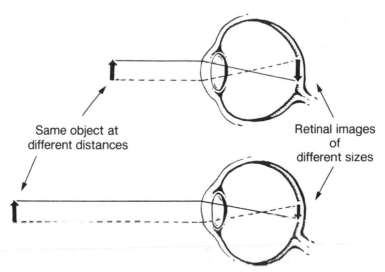

**Figure 8.13.**  The same object at different distances will project retinal images of different sizes. Perceptual size constancy enables us to recognize that despite retinal information to the contrary, as an object moves closer or farther, it still remains a constant size.

Bower, on the other hand, designed a conditioning procedure to test infant depth and constancy abilities (108). These and other studies of infant visual perception collectively, revealed a sophistication of infants not thought possible years ago. As Bower concludes:

> The overall picture of perceptual development that is emerging is very different from traditional ones. It has long been assumed that perceptual development is a process of construction—that at birth infants receive through their senses fragmentary information that is elaborated and built on to produce the ordered perceptual world of the adult. The theory emerging from our studies and others . . . is based on evidence that infants can in fact register most of the information an adult can register but can handle less of the information than adults can. Through maturation they presumably develop the requisite information-processing capacity (108, p. 44).

The logical next step for analysis, once it had been determined that perceptual abilities are existent early in life, was to explore what early anatomical-maturational as well as experiential conditions were necessary for visual perception to emerge and to continue normal development throughout life. These different areas have all been extensively studied, and because visual anatomy was quite well understood in animals and in adult humans, and because so much study was conducted in regard to infant visual perception, conclusions from this area were extended to other perceptual abilities. This research fits within the field of development known as early experience, because it involves understanding the prerequisite experiences necessary for the occurrence of normal developmental abilities early in life. This topic is examined in the next chapter.

# REFERENCES

1. Llorens LA, Rubin EZ, Braun J, Beck G, Mottley N, Beall D: Cognitive-perceptual-motor functions. *Am J Occup Ther* 5:202–208, 1964
2. Kagan J: *The Nature of the Child*. New York, Basic Books, 1984.
3. Rivlin R, Gravelle K: *Deciphering the Senses. The Expanding World of Human Perception*. New York, Simon & Schuster, 1984.
4. Walsh R: *Towards an Ecology of the Brain*. New York, SP Medical and Scientific Books, 1981.
5. Munsinger H: Light detection and pattern recognition: Some comments on the growth of visual sensation and perception. In Goulet LR, Baltes PB (eds): *Life-Span Developmental Psychology. Research and Theory*. New York, AP, 1970, chap 9, pp 227–246.
6. Thompson RF: *Foundations of Physiological Psychology*. New York, Harper & Row, 1967.
7. Ayres AJ: *Sensory Integration and Learning Disorders*. Los Angeles, Western Psychological Services, 1973.
8. Ayres AJ: *Sensory Integration and the Child*. Los Angeles, Western Psychological Services, 1979.
9. Cratty BJ: Sensory-motor and perceptual-motor theories and practices: An overview and evaluation. In Walk RD, Pick HL (eds): *Intersensory Perception and Sensory Integration*. New York, Plenum, 1981, p 345.
10. Ottenbacher K, Short MA: Sensory integrative dysfunction in children. A review of theory and treatment. In Wolraich M, Routh DK (eds): *Advances in Developmental and Behavioral Pediatrics*. Greenwich, CT, JAI Press, 1985, vol 6, p 287.
11. Galin DM, Johnstone J, Nekell L, Herron J: Development of the capacity for tactile information transfer between hemispheres in normal children. *Science* 204:1330–1331, 1979.
12. Houston JP, Bee H, Rimm DC: *Essentials of Psychology*, ed 2. New York, Academic Press, 1985.
13. Gottlieb G: Introduction to behavioral embryology. In Gottlieb G (ed): *Studies on the Development of Behavior and the Nervous System*. New York, Academic Press, 1973.
14. Als H, Lester BM, Tronick EZ, Brazelton TB: Toward a Research Instrument for the Assessment of Preterm Infants' Behavior (APIB). In Fitzgerald HE, Lester BM, Yogman MW (eds): *Theory and Research in Behavioral Pediatrics*. New York, Plenum, 1982, vol 1, p 35.
15. Schnierla TC: Aspects of stimulation and organization in approach-withdrawal processes underlying vertebrate behavioral development. In Aronson LR, Tobach E, Rosenblatt JS, Lehrman DS (eds): *Selected Writings of T. C. Schnierla*. San Francisco, WH Freeman, 1972, p 344.
16. Humphrey T: Postnatal repetition of human prenatal activity sequences with some suggestions of their neuroanatomical basis. In Robinson RJ (ed): *Brain and Early Behavior. Development in the Fetus and Infant*. New York, Academic Press, 1969.
17. Milani Comparetti AM: Pattern analysis of normal and abnormal development: The fetus, the newborn, the child. In Slaton DS (ed): *Development of Movement in Infancy*. Chapel Hill, University of North Carolina at Chapel Hill, 1981, p 1.
18. Brazelton TB: *Neonatal Behavioral Assessment Scale*. Philadelphia, JB Lippincott, 1973.
19. Rosenblith JF, Sims-Knight JE: *In the Beginning. Development in the First Two Years*. Monterey, CA, Brooks/Cole, 1985.
20. Thoman EB, Tynan D: Sleep states and wakefulness in human infants: Profiles from motility monitoring. *Physiol Behav* 23:519–525, 1979.
21. Timor-Trisch I, Zador I, Hertz RH, Rosen MG: Classification of human fetal movement. *Am J Obstet Gynecol* 126:70–77, 1976.
22. Arduini D, Rizzo G, Giorlandino C, Vizzone A, Nava S, Dell'Acqua S, Valensis H, Romanini C: The fetal behavioural states: An ultrasonic study. *Prenat Diag* 5:269–276, 1985.
23. Lipsitt LP: The experiential origins of human behavior. In Goulet LR, Baltes PB (eds): *Life-Span Developmental Psychology. Research and Theory*. New York, Academic Press, 1970.
24. Hall E, Lamb ME, Perlmutter M: *Child Psychology Today*, ed 2. New York, Random House, 1986.

25. Prechtl H, Beintema D: *The Neurological Examination of the Full Term Newborn Infant. Clinics in Developmental Medicine No. 12.* London, Spastics Society and Heinemann Medical, 1964.
26. Als H, Lester BM, Tronick EZ, Brazelton TB: Manual for the assessment of preterm infants' behavior (APIB). In Fitzgerald HE, Lester BM, Yogman MW (eds): *Theory and Research in Behavioral Pediatrics.* New York, Plenum, 1982, vol 1.
27. Reese HW, Lipsitt LP. Attentional Processes. In Reese HW, Lipsitt LP (eds): *Experimental Child Psychology.* New York, Academic Press, 1970.
28. Cohen LB: Attention-getting and attention-holding processes of infant visual preferences. *Child Dev* 43:869–879, 1972.
29. Sokolov EN: Higher nervous functions: The orienting reflex. *Annu Rev Physiol* 25:545–580, 1963.
30. Graham FK, Clifton RK: Heart-rate change as a component of the orienting response. *Psychol Bull* 65:305–320, 1966.
31. Berkson G, Wasserman GA, Behrman RE: Heart rate response to an auditory stimulus in premature infants. *Psychophysiology.* 11:244–246, 1974.
32. Kearsley RB: The newborn's response to auditory stimulation: A demonstration of orienting and defensive behavior. *Child Dev* 44:582–590, 1973.
33. Schachter J, Williams TA, Khachaturian Z, Tobin M, Kruger R, Kerr J: Heart rate responses to auditory clicks in neonates. *Psychophysiology.* 8:163–179, 1971.
34. Kaye H: Sensory Processes. In Reese HW, Lipsitt LP (eds): *Experimental Child Psychology.* New York, Academic Press, 1970.
35. Lipsitt LP: Sensory and learning processes of newborns: implications for behavioral disabilities. *Allied Health Behav Sci* 1:493–522, 1978.
36. Denenberg VH. Learning. In Denenberg VH (ed): *The Development of Behavior.* Stamford, CT, Sinauer, 1972, p 195.
37. Marquis DP: Can conditioned responses be established in the newborn infant? *J Genet Psychol* 39:479–492, 1931.
38. Fitzgerald HE, Brackbill Y: Classical conditioning in infancy: development and constraints. *Psychol Bull* 83:353–376, 1976.
39. Jeffrey WE, Cohen LB: Habituation in the human infant. In Reese HW (ed): *Advances in Child Development and Behavior.* New York, Academic Press, 1971, vol 6, p 63.
40. Lipsitt LP, Kaye H: Conditioned sucking in the human newborn. *Psychonom Sci* 1:20–30, 1964.
41. Gleitman H: *Psychology.* New York, WW Norton, 1981.
42. Montagu A. *Touching. The Human Significance of the Skin,* ed 2. New York, Harper & Row, 1978.
43. Ludel J: *Introduction to Sensory Processes.* San Francisco, WH Freeman, 1978.
44. Rosenblith JF: The Modified Graham Behavior Test for Neonates: test-retest reliability, normative data and hypotheses for future work. *Biol Neonate* 3:174–192, 1961.
45. Kisilevsky BS, Muir W. Neonatal habituation and dishabituation to tactile stimulation during sleep. *Dev Psychol* 20:367–373, 1984.
46. Birren JE, Kinney DK, Schaie KW, Woodruff DS: *Developmental Psychology. A Life-Span Approach.* Boston, Houghton Mifflin, 1981.
47. Rose SA, Schmidt K, Riese ML, Bridger WH: Effects of prematurity and early intervention on responsivity to tactual stimuli: a comparison of preterm and full-term infants. *Child Dev* 51:416–425, 1980.
48. Abravanel E: How children combine vision and touch when perceiving the shape of objects. *Percept Psychophys* 12:171–173, 1972.
49. Tyler NB: A stereognostic test for screening tactile sensation. *Am J Occup Ther* 26:256–260, 1972.
50. Mason A: Something to do with touch. *Physiotherapy.* 71:167–169, 1985.
51. Harlow HF, Mears C: *The Human Model: Primate Perspectives.* New York, John Wiley, 1979.
52. Huss AJ: Touch with care or a caring touch? *Am J Occup Ther* 31:11–18, 1977.
53. Ayres AJ: *Southern California Sensory Integration Tests.* Los Angeles, Western Psychological Services, 1972.

54. Madison LS, Adubato SA, Madison JK, Nelson RM, Anderson JC, Erickson J, Kuss LM, Goodlin RC: Fetal response decrement: True habituation? *Dev Behav Pediatr* 7:14–20, 1986.

55. Finlayson MAJ, Reitan RM: Tactile-perceptual functioning in relation to intellectual, cognitive, and reading skills in younger and older normal children. *Dev Med Child Neurol* 18:442–446, 1976.

56. Haron M, Henderson A: Active and passive touch in developmentally dyspraxic and normal boys. *Occup Ther J Res* 5:101–112, 1985.

57. Ross EF: Review and critique of research on the use of tactile and kinesthetic stimulation with premature infants. *Phys Occup Ther Pediatr* 4:35–49, 1984.

58. Myers DG: *Psychology.* New York, Worth, 1986.

59. Scott AD: Evaluation and treatment of sensation. In Trombly CA: *Occupational Therapy for Physical Dysfunction,* ed 2. Baltimore, Williams & Wilkins, 1983.

60. Clark DL: The vestibular system: An overview of structure and function. In Ottenbacher K, Short MA (eds): *Vestibular Processing Dysfunction in Children.* New York, Haworth, 1985.

61. Groen JJ: Postnatal changes in vestibular reactions. *Acta Otolaryng.* 56:390–396, 1963.

62. Mitchell T, Cambon K: Vestibular response in the neonate and infant. *Arch Otolaryng* 90:40–41, 1969.

63. Rossi LN, Pignataro O, Nino LM, Gaini R, Sambataro G, Oldini C: Maturation of vestibular responses: Preliminary report. *Dev Med Child Neurol* 21:217–224, 1979.

64. Lawrence MM, Feind C: Vestibular responses to rotation in the newborn infant. *Pediatrics.* 12:300–306, 1953.

65. Rabin I: Hypoactive labyrinths and motor development. *Clin Pediatr* 13:922–937, 1974.

66. Short MA: Vestibular stimulation as early experience: Historical perspectives and research implications. In Ottenbacher K, Short MA (eds): *Vestibular Processing Dysfunction in Children.* New York, Haworth Press, 1985, pp 135–152.

67. Ottenbacher KJ, Petersen P: A meta-analysis of applied vestibular stimulation research. In Ottenbacher K, Short MA (eds): *Vestibular Processing Dysfunction in Children.* New York, Haworth Press, 1985, p 119–134.

68. Ottenbacher D. Developmental implications of clinically applied vestibular stimulation. *Phys Ther* 63:338–342, 1983.

69. Weeks ZR: Effects of the vestibular system on human development. Part 1. *Am J Occup Ther* 33:376–381, 1979.

70. Ayres AJ: *Southern California Post Rotary Nystagmus Test.* Los Angeles, Western Psychological Services, 1975.

71. Kimball JG: Normative comparison of the Southern California Postrotary Nystagmus Test: Los Angeles vs. Syracuse data. *Am J Occup Ther* 35:21–25, 1981.

72. Punwar A: Expanded normative data: Southern California Postrotary Test. *Am J Occup Ther* 36:183–187, 1982.

73. Short MA, Watson PJ, Ottenbacher K, Rogers C: Vestibular proprioceptive functions in 4 year olds: Normative and regression analyses. *Am J Occup Ther* 37:102–109, 1983.

74. Korner AF, Thoman EB: The relative efficacy of contact and vestibular-proprioceptive stimulation in soothing neonates. *Child Dev* 43:443–453, 1972.

75. Ottenbacher K, Short MA (eds): *Vestibular-Processing Dysfunction in Children.* New York, Haworth, 1985.

76. Korner AF, Ruppel EM, Rho JM: Effects of water beds on the sleep and motility of theophylline-treated preterm infants. *Pediatrics* 70:864–869, 1982.

77. Korner AF, Guilleminault C, Van den Hoed J, Baldwin RB: Reduction of sleep apnea and bradycardia in preterm infants on oscillating water beds: a controlled polygraphic study. *Pediatrics* 61:528–533, 1978.

78. Pelletier JM, Short MA, Nelson DL: Immediate effects of waterbed flotation on approach and avoidance behaviors of premature infants. In Ottenbacher K, Short MA (ed): *Vestibular Processing Dysfunction in Children.* New York, Haworth Press, 1985, p 81–92.

79. Short-DeGraff MA, Ottenbacher K: *Collaborative Research in Developmental Therapy: A Model with Studies of Learning Disabled Children.* New York, Haworth, 1986.

80. DeQuiros JB, Schrager OL: *Neuropsychological Fundamentals in Learning Disabilities.* San Rafael, CA, Academic Therapy, 1978.

81. Desor JA, Maller O, Turner RE: Taste in acceptance of sugars by human infants. *J Comparat Physiol Psychol* 84:496–501, 1973.

82. Desor JA, Maller O, Andrews K: Ingestive responses of human newborns to salty, sour, and bitter stimuli. *J Comparat Physiol Psychol* 89:966–970, 1975.

83. Whittaker RH, Feeny PP: Allelochemics: chemical interactions between species. *Science* 171:757–768, 1971.

84. MacFarlane A. Olfaction in the development of social preferences in the human neonate. In *Parent-Infant Interaction.* New York, Elsevier, 1975, pp 103–113, Ciba Foundation Symposium 33.

85. Porter RH, Cernoch JM, McLaughlin FJ: Maternal recognition of neonates through olfactory cues. *Physiol Behav* 30:151–154, 1983.

86. Porter RH, Cernock JM, Balogh RD: Odor signatures and kin recognition. *Physiol Behav* 34:445–558, 1985.

87. Friedman SL, Jacobs BS, Werthmann MW: Sensory processing in pre- and full-term infants in the neonatal period. In Friedman SL, Sigman M (eds): *Preterm Birth and Psychological Development.* New York, Academic Press, 1981, p 159.

88. Berg WK, Berg KM: Psychophysiological development in infancy: state, sensory function, and attention. In Osofsky J (ed): *Handbook of Infant Development.* New York, John Wiley & Sons, 1979, p 283.

89. Acredolo LP, Hake JL: Infant perception. In Wolman BB (ed): *Handbook of Developmental Psychology.* Englewood Cliffs, NJ, Prentice Hall, 1982, p 244.

90. Parmelee AH: Auditory function and neurological maturation in preterm infants. In Friedman SL, Sigman M (eds): *Preterm Birth and Psychological Development.* New York, Academic Press, 1981, p 127.

91. Sontag LW, Steele WG, Lewis M: The fetal and maternal cardiac response to environmental stress. *Hum Dev* 12:1–9, 1969.

92. Als H, Lester MB, Brazelton TB: Dynamics of the behavioral organization of the premature infant. A theoretical perspective. In Field T, Goldberg S, Sostek A, Shuman HH (eds): *The High-Risk Newborn.* New York, Spectrum, 1979.

93. Kearsley RB: The newborn's response to auditory stimulation: a demonstration of orienting and defensive behavior. *Child Dev* 44:582–590, 1972.

94. Schmidt K, Rose SA, Bridger WH: Effect of heartbeat sound on the cardiac and behavioral responsiveness to tactual stimulation in sleeping preterm infants. *Dev Psychol* 16:175–184, 1980.

95. Condon WS, Sander LW. Neonate movement is synchronized with adult speech. *Science.* 183:99–101, 1974.

96. Klaus MH, Kennell JH: Labor, birth, and bonding. In Klaus MH, Kennell JH (ed): *Parent-Infant Bonding.* St. Louis, CV Mosby, 1982.

97. Newman LF: Social and sensory environment for low birth weight infants in a special care nursery. *J Nerv Ment Dis* 169:448–455, 1981.

98. Wertheimer M: Psychomotor coordination of auditory and visual space at birth. *Science.* 134:1692, 1961.

99. Turner S, MacFarlane A: Localisation of human speech by the newborn baby and the effects of pethidine ('Meperidine'). *Dev Med Child Neurol* 20:727–734, 1978.

100. Lipsitt LP: Learning capacities of the human infant. In Robinson RJ (ed): *Brain and Early Behaviour. Development in the Fetus and Infant.* New York, Academic Press, 1969.

101. Bromley DB: *The Psychology of Human Aging*, ed 2. Baltimore, Penguin, 1974.

102. Cohen LB, DeLoach JS, Strauss MS: Infant visual perception. In Osofsky J (ed): *Handbook of Infant Development.* New York, John Wiley & Sons, 1979, p 393.

103. Boothe RG, Dobson V, Teller DY: Postnatal development of vision in human and nonhuman primates. *Annu Rev Neurosci* 8:495–545, 1985.

104. White KD, Brackbill Y: Visual development in pre- and full-term infants: A review of chapters 12–15. In Friedman SL, Sigman M (eds): *Preterm Birth and Psychological Development.* New York, Academic Press, 1981, p 289.

105. Fantz RL: The origin of form perception. In *The Nature and Nurture of Behavior. Developmental Psychobiology*. San Francisco, WH Freeman, 1973, p 66.
106. Bornstein MH, Kessen W, Weiskopf S: The categories of hue in infancy. *Science*. 191:201–202, 1976.
107. Fagan JF: The origin of facial pattern perception. In Bornstein MH, Kessen W (eds): *Psychological Development from Infancy. Image to Intention*. Hillsdale, NJ, Erlbaum, 1979.
108. Bower TGR: The visual world of infants. In *The Nature and Nurture of Behavior. Developmental Psychobiology*. San Francisco, WH Freeman, 1973, p 36.
109. Gibson EJ, Walk RD: The visual cliff. In *The Nature and Nurture of Behavior. Developmental Psychobiology*. San Francisco, WH Freeman, 1973, p 19.

# 9

# Perceptual-Motor Development and Early Experience

Early experience refers to changes in the normal range of environmental influences that have potential effects on the later behaviors of organisms (1). These influences may affect one or any combinations of the following areas: sensory-perceptual, emotional-social, language, cognitive, motor, and neural; however, it is important to point out that all experience comes to us in the form of sensory information.

Research into early experience examines the effects of stimulation or deprivation that may occur naturally or under experimental conditions. Early stimulation research aims to determine whether or not developmental processes can be accelerated or whether some forms of stimulation can be used to compensate for various forms of environmental deprivation or developmental delay. Early deprivation studies focus on organisms that are deprived of experiences they would normally encounter in the course of development (2). Subjects of such deprivation studies include organisms with sensory or motor impairments; infants or children removed from their homes and adopted, institutionalized, or placed in foster homes; children or animals exposed to various levels of environmental deprivation, such as neglect or unusual rearing conditions; or various types of animals experimentally deprived through behavioral or surgical manipulations. Such experiments look at physiological or behavioral effects of early deprivation and look at whether these effects are permanent or temporary. The purpose of these deprivation studies, though the methods used may appear cruel in some cases, is to elucidate the factors that are essential for normal development; this is done by determining what factors, if absent, seriously interrupt or interfere with normal developmental processes.

In the past several decades, the field of early experience has grown vastly. For convenience, it can be divided into different research areas depending upon the developmental domain investigated and upon the clinical or theoretical background of the investigator. This author has previously (3) distinguished between two major divisions or approaches within the field of early experience (Table 9.1). While considerable crossover exists between empirical and applied approaches, fundamental differences divide them. The empirical approach, also referred to here as the "scientific study of early experience," is associated with research exploring how and when various developmental abilities emerge and whether the various contributions of nature or nurture on different aspects of development can be sorted out. This empirical field, while having numerous applications, tends to focus on basic research, to generate and test theories, and to use a variety of species to generate models and principles of development. Although his reference is primarily in regard to the cognitive domain, Denenberg (4) aptly describes the premises of the empirical researcher in the field of early experience. In the following quote, Denenberg refers to animal research in contrast to human research; this is equivalent to the distinction made here between empirical and applied research, respectively:

> Animal researchers also have shown interest in the development of cognitive processes. This interest has not stemmed from a philosophy of education or an

Table 9.1
Major Divisions in the Field of Early Experience[a]

| Division and Approach | Subjects | Topics | Goals |
|---|---|---|---|
| Empirical<br>  Comparative<br>  Theoretical | Variety of species including humans | Early stimulation and deprivation<br>Maternal deprivation<br>Early learning<br>Environmental enrichment<br>Perceptual deprivation | Generate theories and principles of normal development<br>Generate animal models |
| Applied<br>  Medical<br>  Rehabilitative<br>  Educational | Human normal or clinical populations | Early intervention<br>Infant stimulation | Test efficacy; generate principles of intervention; facilitate or accelerate development |

[a] Adapted from Short MA: Vestibular stimulation as early experience: historical perspectives and research implications. In Ottenbacher KJ, Short MA (eds): Vestibular Processing Dysfunction in Children. New York, Haworth, 1985.

attempt to develop better methods of rearing or educating children. Instead, the basic motivation has been to find out how the brain develops and how this development is affected by various sorts of early experiences.... Even though the stimulation for the animal studies is quite different from the motivation for human research, we know from the history of science that lines of thought and experimentation which develop independently may ultimately converge for the benefit of all. It is possible that animal studies on the effects of experience upon brain development and research on procedures to improve the cognitive capabilities of underprivileged children are nearing convergence. (4, p. 312)

The empirical approach in early experience has focused on, among other areas, environmental enrichment and visual perception, and it is these areas that this part of the chapter will initially review. The applied approach in early experience will be examined later. The applied approach, which designs and tests the efficacy of different intervention strategies with various special populations, includes the fields of early intervention and infant stimulation.

# SCIENCE OF EARLY EXPERIENCE

Although philosophical assumptions about early experience can be traced to Aristotle, Plato, Descartes, Locke, Rousseau, Freud and many others over the centuries; the scientific study of early experience has its origins in the work of D.O. Hebb. As Denenberg (4) explains, Hebb (5) initiated the first series of systematic experimental investigations in the field of early experience. In the 1940s and 1950s, Hebb and colleagues made important experimental contributions and also generated theories that shaped the field of early experience for decades. Hebb was interested in determining if early stimulation provided to infant rats could somehow alter their cognitive abilities. Thus, he designed what is called an *enriched environment*, a virtual rat playground that was a precedent for many of the pet-store ramps, wheels, and tubes that are sold for gerbils, mice, or other rodents. The enriched environment, Denenberg points out, was one of Hebb's important methodological contributions to the field of early experience. The other was a test of "problem solving" abilities, actually a maze, which was used to test the effects of exposure to enriched environments.

## Environmental Enrichment

According to Hebb, experience in enriched environments should result in higher "intellectual" or problem-solving capacities in stimulated than in control rats. This hypothesis was expanded to a range of stimulation opportunities for a variety of animal species. The following questions resulted from this research. Does environmental enrichment work; i.e., does it cause positive change? If so, in what species of animals and in what clinical populations? Via which sensory modalities, and how? Many of these questions still do not have answers, despite decades of research by prominent scientists in the field of early experience. The difficulty with understanding the cause(s) of environmental enrichment is, as Henderson (6, p. 49) notes, "that the enriched rearing conditions allow a variety of unspecified experiences that can have many sensory, motor, cognitive, and emotional effects."

## Research on the Development of Visual Perception

In a review of 25 years of research in the field of early experience, psychologist Norman Henderson (6, p. 46) notes that while steady, modest progress has been made in the field of environmental enrichment, another area has exhibited rapid growth and advancement. This area is "the study of visual deprivation or enhancement and their effects on later visual discrimination/ information processing and on peripheral and central nervous system development, brain chemistry, and the response of single neurons to stimuli." Over the decades, research in this field has been advanced by work conducted in many different laboratories in the world and from scientists from many different disciplinary and theoretical perspectives. In the 1940s, Riesen (7) demonstrated that behavioral deficits resulted when chimpanzees were reared under conditions of light deprivation. Others, such as Nobel Prize winners Hubel and Weisel (8–11), explored the anatomical development of the cat and rhesus monkey visual cortex and the cellular responses to various types and durations of visual deprivation. Others, such as Blakemore and Cooper (12), examined the electrophysiological and behavioral responses of animals reared under specific conditions of visual stimulation.

These studies of animal visual sensation and perception were also combined with data from a variety of excellent studies by individuals, e.g., Fantz (13), Gibson and Walk (14), Bower (15) and others who were working with human infants (see chapter 8). Henderson (6) points out that this multidisciplinary, cross-species, experimental approach contributed in part to the steady advances made in the field of visual sensation and perception. Thus, since there was such a breadth of information from research examining this specific modality, generalizations were applied to other sensory fields.

# Nobel Prize Winning Research

A considerable body of information in regard to the visual system was prompted by work conducted by neurophysiologists David Hubel and Torsten Weisel (8–11). These two researchers, along with Richard Sperry, won the 1981 Nobel Prize in Physiology and Medicine for their contributions to the understanding of the function of the visual system and development of the brain. Hubel and Weisel were recognized for their numerous decades of research in regard to the information-processing abilities and the effects of visual experience and postnatal influences on specific aspects of the visual system. Since their work has been often applied to behavioral and clinical studies in development, it will be described in detail here.

Working together for many years, Hubel and Weisel initially set out to understand the normal neuroanatomical development of the visual system. Once this was well understood, they embarked on an exploration of the effects of various postnatal experiences on neural development. Using both kittens and rhesus monkeys as subjects in their experiments, the researchers explored the effects of visual deprivation, early and later in life, on cells in specific regions of the visual cortex. The research focused on one small region of the visual cortex, the striate cortex, which is special because of its association with certain visual sensory functions and because it receives visual sensory information from both eyes.

If you recall from chapter 8's description of the development of visual anatomy (see Fig. 8.10), as information from each eye leaves the optic chiasm (the cross-over part of the visual system), each brain hemisphere receives sensory information from a portion of each eye. Thus, the left hemisphere deals with information from the right half-field of vision, and the right hemisphere processes information from the left-half field. The striate cortex is only one area of the visual cortex that receives fibers from the lateral geniculate bodies, and sensory information relayed to the striate cortex continues to other cortical processing regions. The striate cortex, which is the area scrutinized by Hubel and Weisel, consists of different types of cells geared to process certains features of sensory visual information. An important characteristic of some individual cells is that information from both eyes converges on them. This property, called *binocular convergence*, proved important to Hubel's and Weisel's work (10).

Contrary to other studies, such as Reisen's (7), which looked at the effects of light deprivation in general, Hubel and Weisel's study looked specifically at visual deprivation. These two researchers set up experiments in which individual animals could serve as their own controls by visually depriving one eye and leaving the other intact. In their animal subjects, Hubel and Weisel left one eye normal and sutured the other one closed, a temporary and reversible procedure for subsequent studies. Thus, the researchers were able to compare the anatomy and development of the two different eyes and to look at the effects of monocular deprivation upon the connections of the other, nondeprived, eye.

# Hubel and Weisel's Findings

The results of decades of work led to important findings about the function of the visual system during normal development and during deprivation. These conclusions have been used to build a model of early development and were generalized to other areas of the visual cortex and to the field of early experience in general. While this work was important in many fields and for many reasons, three primary generalizations are discussed here. These three findings concern (a) the sophistication of the visual system at birth, (b) the demonstrated plasticity of neural systems, and (c) the evidence of plasticity only during critical periods in the development of the species.

What these researchers specifically found was that the brain cells from the striate cortex that normally received afferent input from both eyes changed as a result of monocular deprivation. Thus, as a result of closing one eye, the brain cells that typically displayed binocular convergence now only received input from the one, the open, eye. Apparently input from the closed eye dropped out. As Hubel (10) points out, there are two explanations for these findings: One explanation is that because of the lack of visual experience, neural connections actually failed to develop after birth. An alternative explanation is that the cortical machinery is present and intact at birth, but it drops out if it is not used. This latter explanation, however, was initially disregarded by many psychologists because of the prejudice against the maturity of sensory systems early in life.

Many experimenters in the field of perceptual development tended to lean toward the empiricist side of the nature-nurture debate and therefore possessed a general expectation that sensory and other systems were relatively immature and required environmental stimulation to mature. Thus, as Hubel (10, p. 539) notes, the maturity of the visual system at birth "wasn't considered because of psychologists tendency to view the newborn brain as a tabula rasa on which postnatal experience writes its message". Continued research by Hubel and Weisel and others in the field, however, forced a reconsideration of the development of perception. The work indicated that cells in the striate cortex of newborn monkeys were functionally adult-like. Therefore, the visual system at birth was quite sophisticated. The prominent researchers concluded that "experience has little to do with the formation of the connections responsible for the high degree of specificity in that part of the brain" (10, p. 539). The defects that were produced by eye closure occurred as a result of the deterioration of neural connections that were already present at birth.

Thus, Hubel and Weisel had established that certain visual-perceptual functions were present in cats and monkeys that had had no prior visual experience. The differences between the immature and adult animals' systems was one of precision and interaction; i.e., lack of experience and fine-tuning of perceptual abilities, but not one of differences in maturity of the neural machinery (11). They established that, contrary to empiricists' expectations and consistent with nativists' claims, the infant is born with certain innate abilities.

The next issue, once it was established that neurophysiology was mature at birth, was to determine what happened to the striate cortical cells to cause them to change and to cause the afferent input from the unused eye to drop out. Hubel points out that at first they thought that the deterioration of connections from the unused eye was due to lack of use, but as their work went on, they realized that disuse was probably not the main cause. The main reason why connections dropped out from the deprived eye was because the other eye began to take over the shared cells. As Hubel (10, p. 539) describes, "It was as though the connections from the open eye had somehow taken advantage of their rivals from the closed eye in their competition for space." The significance of this is the demonstration of brain plasticity; i.e., that neural connections already present early in life can be modified by subsequent visual experiences.

The third major finding was the presence of time constraints on the modifiability of the neural connections. Hubel and Weisel found that neural modification, which is referred to as *plasticity* of brain tissue, was possible only during certain months after birth; its duration depended on the species under examination. In the human, for example, cataracts late in life may have no effect on visual anatomy, and sight can be restored after their removal. In contrast, the presence of cataracts during infancy can cause permanent changes in cortical anatomy and visual perception. Furthermore, if the damaging effects of visual deprivation are to be reversed, the eye must receive stimulation before the end of the critical period of development (9, 11).

# Behavioral and Neuroanatomical Effects of Early Rearing

Hubel and Weisel's work focused primarily on basic anatomical changes and neural functions and not at how those changes affect behavior. Other pioneering research conducted in the 1970s provided an additional view about rearing conditions on both neuroanatomy and behavior. Neurophysiologists Blakemore and Cooper's work (12) combines some of the hypotheses drawn from environmental enrichment studies (i.e., that exposure to an environment may affect development) as well as from the visual-perceptual research (i.e., that visual experiences may alter visual cortical anatomy). The especially important aspect of Blakemore and Cooper's study is their examination of behavioral (rather than just neural) effects of rearing conditions.

In their study (12), two groups of cats were raised in the dark except for the 5 hr/day when they were exposed to one of two different visual environments. One group was exposed to an environment in which only vertical stripes were present, and the other group was exposed only to horizontal stripes. Following these experiences during the first 5 months of life, the cats were then tested for behaviors indicative of visual perceptual functions. These behavioral studies found that the kittens reared in the vertical environments were virtually blind to horizontal input. Similarly, the kittens raised in horizontal environments were blind to vertical stimuli. Thus, the latter kittens could jump to a horizontal chair seat but bump into vertical chair legs, as if they could not see them.

Hubel and Weisel's findings predicted that the neuroanatomy of these vertically or horizontally reared cats would be changed as a result of their different visual experiences. Blakemore and Cooper actually tested this. They measured the neural responses of these cats to stimuli presented in vertical or in horizontal orientations and discovered that the neural changes corresponded to the cats' rearing experiences. Thus, the neurons of the cats reared in vertical environments responded to vertical, but not to horizontal, stimuli. These neuranatomical data supported the behavioral findings. Blakemore and Cooper (12, p. 478) conclude, "It seems ... that the visual cortex may adjust itself during maturation to the nature of its visual experience. Cells may even change their preferred orientation towards that of the commonest type of stimulus; so perhaps the nervous system adapts to match the probability of occurrence of features in its visual input." Blakemore and Cooper's findings are important because they provide a behavioral and neuroanatomical confirmation of Hubel and Weisel's work.

# Drawing Conclusions from Empirical Research in Early Experience

Hebb initiated systematic study in the field of early experience and generated theories about, and tests of the effects of, environmental enrichment. Hebb's initial theories were important for setting the tone for and prompting subsequent research in the field. Beyond the initial assumption that stipulated that exposure to enriched environments would produce gains in problem-solving abilities, he also theorized that the effects would be (a) permanent, and that in order for them to be maximally effective, (b) exposure must occur early in life (4, 5). One must keep in mind, however, that Hebb's work was initiated in the 1940s. Therefore, how applicable are his findings now, given the tremendous amount of research in the field of environmental enrichment and in human development? In regard to the visual system, Hebb's findings are corroborated by Hubel and Weisel's work (8–11). Their findings can be summarized as follows: (a) Very specific innate perceptual mechanisms are present at birth, but *experience is necessary* for their maintenance and full development. (b) In order for that experience to have maximal effects, it must occur during some *critical period early in life*. Thus, Hubel and Weisel confirmed Hebb's theory that environmental stimulation has effects, *if* it occurs early in life. Hubel and Weisel's work also pointed out the presence of sophisticated and plastic neural apparatus that was unanticipated in Hebb's time.

An understanding of the significance of these findings to the science of human development necessitates looking at some of the issues surrounding Hebb's theory and from Hubel and Weisel's work. These are the issues that must be explored: (a) drawing conclusions about anatomy and generalizing to behavior; (b) generalizing from work with animals to humans; (c) the existence and role of critical periods in development; (d) the notion of brain changes as a result of experience; and (e) coming to some conclusions from the science of early experience that may be generalizable to human development.

## Generalizing from Anatomy to Behavior

In describing the work that earned Hubel and Weisel the Nobel Prize, Whitteridge (16) notes that this research is forcing a reconsideration of the development and functions of all other cortical areas: As Weisel notes:

> Innate mechanisms endow the visual system with highly specific connections, but visual experience early in life is necessary for their maintenance and full development. Deprivation experiments demonstrate that neural connections can be modulated by environmental influences during a critical period of postnatal development. *We have studied this process in detail in one set of functional properties of the nervous system,* [my italics] but it may well be that other aspects of brain function, such as language, complex perceptual tasks, learning, memory, and personality, have different programs of development. Such sensitivity of the nervous system to the effects of experience may represent the fundamental mechanism by which the organism adapts to its environment during the period of growth and development. (9, p. 373)

Whether or not developmentalists can generalize from these findings remains to be seen. As Weisel notes, he and his colleague have studied only "one functional property of the nervous system." While appealing, such a generalization from one specific anatomical system to much more complex, multifactor and elusive functions such as personality, learning, and some complex perceptual abilities may be inappropriate. This will be discussed more fully in the following sections regarding brain changes and critical periods in development.

## Animal Models

An issue that often arises in clinical fields is the usefulness of experimental information generated from studies of species other than humans. Many of the studies in the scientific study of early experience, and of development in general, have used nonhuman species. Some of the reasons for this are obvious. First, life-time effects of experimental manipulations are more easily observed in species that have shorter life spans. Second, the effects of deprivation or experimental manipulations can be explored with animals when ethical and humane constraints prevent testing humans. Third, as Denenberg (4) points out, many developmental researchers are interested in general principles of development and behavior that apply to most species including, but not limited to, humans. These scientists may use one particular species as a model of behavior and then use the information they obtain to generate hypotheses about behavior in general.

A rule of thumb about the application of these findings is, the more species studied with similar findings obtained, the greater the application to humans or to broad principles of behavior or function. This is illustrated with Hubel and Weisel's work. Hubel and Weisel's animal studies are easily integrated with other studies in the field of human infant perception. For example, as noted in chapter 8, the work of Bower (15), Gibson (14), Fantz (13) and many others revealed that the visual system of human infants is, like Hubel and Weisel's kitten and monkey studies indicated, quite sophisticated at birth. Furthermore, data that are obtained with animal subjects but that bridge contemporary theories often have a wide appeal and tend to be widely adopted and applied. Whether or not the research findings are accurate remains to be tested, but their "fit" with contemporary thinking lends them to adoption in models of development. For example, one of the appealing aspects of Hubel and Weisel's work is that it bridges the nature-nurture dichotomy. They found that the visual perceptual system is relatively sophisticated early in life (nature) but that it is changeable depending upon experience (nurture). This interactionist position fits conveniently with contemporary positions about developmental influences (see chapter 2).

The cross-species model that was generated from Hubel and Weisel's and others' research with many species, including humans, appears to be widely acceptable to scientists from a variety of theoretical positions. The following interpretation of Hubel and Weisel's work regarding visual perceptual development may be compatible with learning, maturational, ethological, and ecological theories: The discovery of certain early perceptual preferences and predispositions to respond in certain ways has led to the conclusion that different species are innately geared (or geared early in development) to react to specific salient features of their environments. What those features are depends upon their adaptive significance for the species, and whether or not those innate abilities develop depends upon subsequent experiences with their worlds. The nature of those experiences are examined more fully in the next sections.

## Critical Periods and Brain Changes as a Result of Experience

Hebb's theory posed that environmental enrichment effects would only occur during a critical period of development early in life. Much of the subsequent research in the field of environmental enrichment adopted Hebb's theory and focused on determining what specific changes were induced by various early experiences. Many of these studies looked specifically at changes in the brain. For example, several decades of work at the University of California at Berkeley by prominent researchers Rosenzweig, Bennett, Diamond and their colleagues consistently demonstrated neurochemical and neuroanatomical changes in the brains of immature rats exposed to enriched environments (17). While other findings such as accelerated physical growth and improvements in other complex cognitive, problem solving, and social abilities have also been reported, they are less consistently found than, and are unlikely to correspond to neural changes (18). This points to two important points: (a) the issue of critical periods in brain development, and (b) the response of the brain to environmental experiences.

## Critical Periods

The critical period hypothesis is very important in the scientific study of development and is discussed in numerous chapters of this text. As discussed earlier in chapter 6, the notion of permanent change during a circumscribed critical period of development is controversial. While it may be applicable to some structural embryonic changes, the application of the critical period concept is less acceptable in regard to postnatal neural changes and to behavior, especially in regard to complex, multidimensional behaviors such as cognition, problem-solving, social-emotional interactions, and learning (19).

Since the concept is controversial, how do Hubel and Weisel's findings of critical periods in the development of striate cortex fit in? Their report of a critical period in the development of a particular cortical area is not surprising given that research (already discussed in this chapter) indicates that visual cortical structures continue maturation (myelination) following birth. Whether or not these findings generalize to other sensory modalities, however, is still under investigation. It is possible that Hubel and Weisel's findings are specific to the visual system. For example, Gottlieb (20) and Friedman and associates (21) have speculated that the visual system is the least mature sensory system at birth. Thus, as Friedman and colleagues demonstrated, the visual system is more subject to perinatal influences than, for example, the tactile system, which is functionally mature long before birth. Postnatal visual experiences may therefore affect neural changes in the visual system, whereas other sensory systems may be less susceptible to influences after birth (21).

Hebb's (4) initial premises have been revised as cumulative information gathered from environmental enrichment studies do not lend support to the notion of critical period effects. Many of the initial studies in this field, as would be expected in the study of "early" experiences, focused on immature organisms; however, as data were collected from mature subjects, it was found that they, too, displayed neural responses to environmental enrichment. This information began to force a reconsideration of the critical period concept about perceptual experiences—that the neural changes once thought to occur only during critical periods of early development are also found in mature subjects. In a review of environmental enrichment effects, Henderson (6, p. 49) concludes, "It is generally acknowledged that most of these effects are not due to experience unique to early life" and that most gross anatomical changes that occur as a result of environmental enrichment may be "age independent." These data, and others discussed in this text, are gradually reshaping perspectives of critical periods in development.

### Does the Brain Change as a Result of Experience?

An issue that has plagued scientists for centuries is whether or not the brain changes as a result of experience. The early empiricists such as John Locke (see chapter 3) assumed that experiences were "etched" onto the blank slate of the brain. In addition, because of biases in regard to immature subjects, it was assumed that these effects occurred primarily early in life. Now, with investigations of mature subjects, it appears that the bias toward neural change in early development is suspended. But the issue of brain change as a result of experience still remains, and this field of study is a quagmire. A current issue is not whether the brain changes, but whether neural change actually has any special meaning.

For example, studies of environmental enrichment effects have regularly reported neural change in a variety of species. The question of interest is not whether change occurs but whether or not it is significant. Enrichment specifically means opportunities for variety in quality and quantity of environmental, i.e., sensory-perceptual, motor, and social stimulation. The implications of positive findings from environmental enrichment studies are important for if, indeed, specific experiences can enhance some aspect of development, then these same forms of stimulation may be effective in remediative programs for individuals who have been environmentally deprived, or, for that matter, for accelerating different aspects of development in normal individuals. This is a topic of interest to empirical researchers as well as to clinicians. In fact, the finding of neural change with environmental enrichment has led to an overly enthusiastic reaction on the part of many researchers and clinicians who have assumed that they could induce neural change (and therefore rehabilitation and recovery) in their patients through various forms of therapeutic stimulation (19, 22). While improved sensory and motor functions often occur during therapeutic intervention, such recovery is often untraceable to corresponding changes in the brain.

In reality, the significance of neural change as a result of experience is not understood. Walsh (18, 23) and other eminent brain researchers have concluded that the brain is in constant activity. Thus, every stimulus and every experience, especially those that are multisensory and sensorimotor in nature, may possibly alter the brain. In addition, it is also known that behavior can change independently of brain alterations (24). Thus, while it is appealing to attribute developmental or behavioral changes to corresponding changes in the brain, this may be a wholly inaccurate and misleading approach to neural function (19, 22–24).

How, then, do we integrate Hubel and Weisel's findings regarding specific functional characteristics and plasticity of neural tissue? As noted earlier, Weisel (9) explains that their research focused on one set of functional properties of the nervous system. They looked at one specific aspect of a specialized tissue in the nervous system, the cortex. The cortex, while unique and necessary for conscious interaction with the world, is not the whole brain. It is only one aspect of the most poorly understood organ of the body. And while we do not know exactly how this organ works, we do know that parts of it are specialized and parts are not, and all of these parts work together in a way that is not yet understood.

The cortex has specialized areas associated with certain sensory and motor functions. This was discovered by different researchers over the years. In the late 1800s, German physiologists Fritsch and Hitzig found that when they delivered mild amounts of electrical current to certain cortical areas, their dog subjects moved specific body parts. Similar studies of humans in the mid-1900s by neurosurgeon Wilder Penfield (25) demonstrated a corresponding effect in humans. These patients not only displayed motor reactions upon cortical stimulation but also reported specific sensory experiences that were associated with stimulation of certain cortical regions. The association of these specific sensory and motor functions with specific neural regions is known as "localization of function." Discoveries by Penfield and others have led to the construction of actual maps of the cortex, with localization of specific sensory or motor functions identified with certain regions. This is consistent with Hubel and Weisel's finding of specific, localized visual functions within the striate cortex.

However, as Hubel and Weisel have pointed out, the striate cortex is only one small component of the visual system and an even smaller component of the whole brain. Visual sensation and complex visual perceptual abilities that involve memory, reasoning, associating, and coordinating many different types and levels of knowledge cannot be pinpointed to one, or even a few, areas of the brain. Thus, while specific sensory and motor abilities can be localized in the cortex, this does not hold true for other integrative functions or complex behaviors such as processing information, making a judgment, and effecting a subsequent motor response.

Thus, while cortical studies by Hubel and Weisel and others may continue to isolate very specific functional areas of the brain controlling sensory or motor functions, this does not mean that all behaviors are controlled by specific areas of the brain or that all experiences significantly change the brain. A balance between localization of function and holistic brain function must occur, but how it occurs is controversial and unclear. For example, neurophysiologists have sought, in vain for decades, to locate where memory is stored in the brain. As a result of this research, neurophysiologists speculate that complex processes such as learning and memory may involve the whole brain and therefore cannot be localized. In addition, given the plastic nature of the brain, functions may shift according to external and internal demands.

Brain researcher Roger Walsh (23, p. 7) points out that environmental enrichment studies have demonstrated positive effects of neural change, but "Whether the effects will be detectable or lost forever in neural noise is another question." Rather than pursuing what may be an elusive goal of determining precisely "whether" environmental stimulation works in the brain, it may be more important to determine the essential elements of stimulation and under what conditions it affects development. Knowing this will provide us with important clinical information regarding the use of therapeutic stimulation.

### Active Experience is Necessary for Environmental Enrichment and Visual Perceptual Effects

Research into environmental enrichment and visual perception has pointed to the specific nature of experiences necessary to produce positive behavioral and developmental effects. For example, one such study was designed by Richard Held and Alan Hein and is widely known in the field of visual perception (26). This is a study of the active and passive cats. These cats, like those in the study by Blakemore and Cooper (12), were raised in the dark. When they were old enough to be tested in the experiment, they were hitched up to a merry-go-round device for about 3 hr/day. Otherwise, they stayed with their mothers and littermates in unlighted cages. When in the merry-go-round device, the active cat was harnessed and connected to a center post. This, in turn, connected to the passive cat. The passive cat was placed in a little compartment (gondola) that fit around its neck. Thus, the passive cat was unable to see its paws or the rest of its body, but it could move its limbs inside the gondola. The active cat was able to move freely and could see its limbs while moving. As the active cat moved, the passive cat was pulled.

After about 30 hours of exposure in this apparatus, the active and passive kittens were assessed for visual-motor or visually guided functions, such as visual placing when brought to a table top (i.e., preparing the paws for supporting body weight), blinking at an approaching object, and avoiding the deep side of the visual cliff. The passive kittens, in contrast, did not display these behaviors until after they had the opportunity for a couple of days to move in a normal environment. Held concludes from this and from many other experiments, that visual-motor interactions are important for perceptual and motor behavior.

These findings complement those of Rosenzweig and colleagues (17), who were looking into the effective elements of environmental enrichment. Ferchmin, Bennett, and Rosenzweig (27) compared the effects of "passive" exposure, i.e., observation, vs. active movement throughout enriched environments. Thus, similar to Held and Hein's finding, these researchers discovered that mere exposure to an enriched environment is not sufficient to produce positive effects. Rosenzweig concludes, "Apparently, active and direct interaction with the enriched environment is necessary for production of the cerebral changes; passive and indirect contact is not enough [and] no particular sensory modality appears to be required." (17, p. 39) These findings were corroborated by other researchers in the field such as Walsh and Cummins. In a review of neural responses to therapeutic environments, these researchers reported that subjects must interact *physically* with the stimuli and that *reafferent* stimulation may be particularly important (18).

Reafferent stimulation is a force often recognized and studied in occupational and physical therapy. As described at the beginning of chapter 8 (see Fig. 8.1), sensory information comes from different sources and is carried toward the brain by afferent neurons. Researcher Richard Held (26) describes that the terms "afference," "reafference," and "exafference" come from neurophysiologists von Holst and Mittelstadt. They describe *afference* as excitation of afferent neurons in general. Afferent stimulation, however, may be organism or environmentally generated. *Reafference* is described as organism-generated sensory stimulation. It is typically labeled "feedback" in models of sensorimotor function such as in Figure 8.1. Reafference is sensory feedback that comes from self-produced movement, or as Held (26, p. 74) explains, "neural excitation following sensory stimulation that is dependent on movements initiated by the sensing animal." In contrast, environmentally generated stimulation is exafference or neural excitation independent of self-produced movement.

In the scientific study of early experience, reafferent stimulation (i.e., the active participation or self-produced movement of the organism) has been isolated as the effective element in bringing about change as a result of environmental stimulation. Although this is an important experimental finding, its implications are even more important when applied to developmental and clinical situations. As Dru (28, p. 266) and colleagues have noted, "Self-produced locomotion must accompany exposure to a patterned visual environment for recovery to occur. These observations have implications for the treatment of patients after stroke or other trauma affecting the visual system." The significance of reafferent stimulation, or as it is more commonly called, feedback, is widely recognized in therapy. Walsh and Cummins (18, p. 189) note that reafferent stimulation, "Certainly . . . is one of the major functions of a therapist during clinical rehabilitation."

# EARLY INTERVENTION

The systematic therapeutic intervention with infants or young children is known as *early intervention*. Early intervention is a broad term that is commonly used, but rarely defined. Recent reviews (29, 30) of the field point out the different forms of early intervention that exist, and classifications can be developed according to the population addressed or the type of intervention offered (Table 9.2). For example, Browder (29) distinguishes between intervention for children who are environmentally deprived or biologically impaired. Environmentally or culturally deprived children are typically approached with cognitive, psychoeducational programs that address intellectual goals and parent-child relationships. Children with biological impairments make up a very heterogeneous group such as those

Table 9.2
Classifications of Early Intervention

| Populations Addressed | Treatment Approaches |
|---|---|
| Environmental Risk | Cognitive |
|    Cultural/social deprivation | Psychoeducational |
| | Ecological |
| |    Parent-child |
| |    Family systems |
| |    Community |
| Biological Impairments | |
|    Sensory impairments | Health-medical |
|    Brain damage | Rehabilitative |
|    Developmental delay | Broad-based |
|    Physical-motor deficits | Transdisciplinary |
|    Health problems | Ecological |
| |    Parent-child |
| |    Family systems |

with orthopedic problems as well as neuromotor deficits and developmental delays. Such infants or children may face a variety of interventions including occupational, physical, or speech therapy or combinations of these in what is called a *transdisciplinary approach*.

# Interdisciplinary and Transdisciplinary Approaches to Early Intervention

The transdisciplinary approach arose from the more common and traditional interdisciplinary team approach. Harris explains:

> Within the interdisciplinary team, each of the team members has a defined role and provides specific intervention based on his own medical or educational expertise. Each team member provides feedback to the infant's parents and to other team members based on his assessment of the infant. Although the parents may be actively involved in goal setting and decision making, they are primarily recipients of the advice and instructions offered by the interdisciplinary team members. (31, p. 74)

The transdisciplinary model, Harris points out, was developed to reduce the fragmentary effects of the interdisciplinary approach. The transdisciplinary model is based upon an assumed level of active transactions occurring between team members and family. It recognizes the parents as the primary care givers and includes them in team goal setting and intervention. Each team member functioning under the transdisciplinary model is aware of the theory, aims, and approaches of the other team members and can incorporate these varying goals within treatment. Thus, intervention is provided by one, rather than an assortment of, individuals, on the assumption that this approach will provide greater continuity and consistency of care. The team member selected to provide intervention is the one whose discipline is most closely associated with the major needs of the child (31).

The transdisciplinary model is only one intervention approach and is not applied solely in the field of early intervention. In treatment programs aimed at various client groups, and where administrative and staff philosophies and intercommunication enable it, this model has been successfully adopted.

# Infant Stimulation

A recent subarea within the field of early intervention is infant stimulation, which itself can be subdivided into four topical areas based on the primary purpose or aim of stimulation (Table 9.3). The medical subdivision is aimed primarily at infants who fail to thrive, need immediate medical attention during the perinatal period, or are high-risk status. Intervention is generally associated with acute care and high technology specialized to stabilize and maintain the neonate's functions and health (Fig. 9.1). Some such infants, once stabilized, may then be treated with the habilitative approach, which is aimed at infants whose health is stabilized and who exhibit specific biological impairments or who are either at risk for, or demonstrate some kind of, developmental delay. The habilitative approach also addresses parent-infant relationships that may have been interrupted as a result of hospitalization and separation or as a result of infant sensory or motor impairments that interfere with effective parent-child interactions. While parent-child intervention has been a consistent part of cultural deprivation programs, it was less frequently incorporated into treatment for children with biological impairments. Now, however, with the recent acknowledgment of the long-range value of positive family relations on the development of all children, family intervention is a focus of many infant- and child-oriented programs.

Table 9.3
**Categories of Infant Stimulation**

| Type | Purpose |
|---|---|
| Medical | To maintain and/or stabilize health of at-risk or ill infants |
| Habilitative | To compensate for delays or interruption in various developmental domains as well as infant-family relationships |
| Educational | |
|   Remediative | To compensate for cognitive delays or interruptions |
|   Accelerative | To speed up development of normal or gifted children |
| Research | To systematically investigate factors that affect infants' behavior and development |

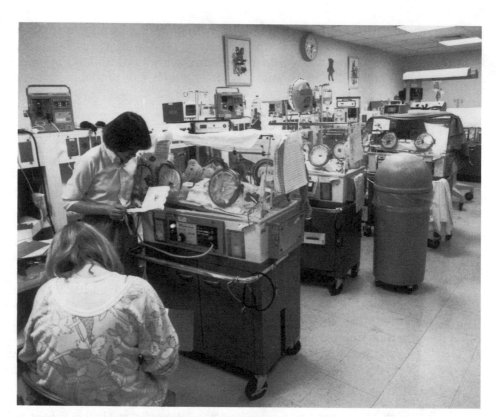

**Figure 9.1.** Neonatal intensive care or special-care nurseries are geared to stabilize the infant's medical status, physiological functions, and health. (Courtesy of the Neonatal Intensive Care Unit, University of Connecticut Health Center's John Dempsey Hospital, Farmington, CT)

The third category of infant stimulation includes a variety of cognitive and psychoeducational approaches that are habilitative in nature or that are accelerative and aimed at advancing the intellectual skills of normal or gifted infants. The latter area is receiving considerable attention and controversy in the fields of early child development and education. At the heart of the issue is the concern on the part of some parents that, with the growing emphasis on technology and advanced education, children must receive as much early training as possible so that they are competitive in the narrowing preschool, private school, and college "markets" (Fig. 9.2). The other side of the issue is the growing concern about the potential stress such training may place on the child as well as the overemphasis on information and intellectual development to the exclusion of play, socialization, sensorimotor, emotional, and other developmental domains.

The fourth area of infant stimulation is that of research. This area is broad based and may involve studies of the efficacy of interventions used in the other three approaches (Table 9.3). Both applied and basic research are considered, as one finds that even the most basic research in the field

*"Attention, please. At 8:45 A.M. on Tuesday, July 29, 2008, you are all scheduled to take the New York State bar exam."*

**Figure 9.2.**   Drawing by Leo Cullum; © 1984. The New Yorker Magazine, Inc.

of infant stimulation may be used in clinical or educational applications. Examples of the types of research studies including topics already discussed in this book are infant state and behavioral organization; the development of reliable and valid assessments of term and preterm infants; studies of normal fetal behavior or predictors of risk conditions; ecological studies of newborn or preterm infants; studies of bonding and consequences of its interruption; effects of adverse stimulation (prenatal and pregnancy-related risk conditions, perceptual overstimulation or deprivation); and potential short- and long-range consequences of different forms of sensory-perceptual stimulation.

Research in infant stimulation is relatively recent. Many researchers and lay people in the first half of this century generally believed that infants were relatively passive beings, with little active perceptual processing abilities. Not until data from the scientific study of early experience (e.g., from Fantz, Bower, Hubel and Weisel, and many others) indicated that infants are relatively sophisticated in perceptual apparatus and reactivity did research begin to explore infant learning and the effects of perceptual stimulation. Additionally, advances in medical technology paralleled the growing body of empirical information about infants. As a result, prenatal diagnostics and viable preterm infants provide us with a growing body of knowledge about the sensory and motor abilities of fetuses and preterm infants. Thus, information is constantly forthcoming about the capabilities and responsiveness of younger and younger infants.

# Research in Early Intervention

Research in the field of early intervention has been advanced by information accumulated from the scientific study of early experience. For example, many of Hebb's original premises and generalizations from environmental enrichment research have been applied to other fields of early experience, such as infants' responses to stress, maternal deprivation and institutional-ization, malnutrition, and a wide variety and combinations of sensory stimulation experiences used in the design of therapeutic environments. In addition, specific forms of early (sensory) stimulation that were examined in animals were subsequently applied to humans. This research makes up an enormous field of study that examined the effects of various forms of stimulation on an even larger variety of outcome measures. Some of these issues will be examined here, primarily as they relate to sensory stimulation and development. Other topics will be covered in later chapters, e.g., maternal deprivation and institutionalization and their effects on social-emotional development and early hormonal effects on sexual differentiation of the brain and behavior (chapter 12).

The developmental effects of stimulation of every sensory modality was, and is currently, being investigated by various researchers across the country (Table 9.4). This includes tactile stimulation and handling, rocking, stroking; various forms of auditory experiences using rhythmic music or heartbeat sounds; and vestibular, olfactory, and combinations of sensory experiences in sensorimotor and sensory integration programs (3, 18, 19, 22–24, 32–37). The most important finding, from a therapeutic point of view, is that different forms of sensory stimulation facilitate various aspects of development. For, if development can be advanced, then such stimulation can be used in treatment programs geared to habilitate those who are delayed or to rehabilitate those whose development has been interrupted (See chapter 3). Early intervention research has focused on comparing and contrasting various forms of environmental stimulation programs in a desire to determine which produces the greatest amount of gains, in which developmental domains, and in which specific client populations.

Studies of programs aimed at children subject to environmental depri-vation have a longer history, are more common, and have produced more consistent results than studies of programs for children with biologic im-pairments. As Browder (29, p. 42) notes, "There are considerable psychologic and developmental data to provide a sound rationale for early intervention in children subjected to environmental deprivation . . Furthermore, studies support the concept that the mental growth of deprived children is favorably

**Table 9.4**
**Examples of Studies Exploring Sensory Stimulation Effects on Premature Infants**[a]

| Type of Stimulation | Outcome |
| --- | --- |
| Handling | Quieting, reduced crying |
| Handling and stroking | Reduction in apnea |
| Handling and colored mobiles | Improved interactive behaviors |
| Oscillating waterbed | Reduction in apnea |
| Rocking in hammock | Visual orientation change but no difference in tactile or auditory responses |
| Rocking bed and recording of heartbeat | Early development of sleep patterns, faster motor development |
| Recording of woman's voice | Increased alerting, motor maturation |

[a] Modified from Schaefer M, Hatcher RP, Barglow PD: Prematurity and infant stimulation: a review of research. *Child Psychiat Hum Dev* 10: 199–212, 1980.

influenced by infant stimulation programs with strong provision for improving parenting skills and parent-child attachment." As this text has already explained in regard to the ecological-systems model of Bronfenbrenner (37) and others, support of the parent-child relationship and of the family at large provides continued positive environmental intervention for the child and prevents further effects of environmental deprivation.

In comparison to studies of environmental deprivation, the study of programs for biologically impaired infants and children is less consistent and complete (29). As Ferry (22, p. 40) notes, "Large-scale, systematic, and scientifically controlled studies with rigorous methodology are sorely needed." Many of the various studies in existence, which comprise a large and inconsistent body of information, have found various short-term developmental gains, primarily in psychosocial areas rather than in motor behaviors and skills (29). As is usual in developmental studies, long-range gains are less frequently investigated and therefore less commonly observed. While it is not possible to thoroughly review this enormous field, it is possible to generate some basic premises about the nature of therapeutic stimulation. These premises are based on the previous sections of this chapter regarding normal sensory-perceptual development and environmental enrichment and visual perceptual research (Table 9.5).

## Developmentally Appropriate and Therapeutic Stimulation

Therapeutic environments are designed with the goal of providing appropriate sensory-perceptual stimulation, for it is the sensory-perceptual mode through which we gain information about our worlds. Understanding and knowing what to expect in terms of normal sensory-perceptual development and motor responsiveness enables us to design sensitive intervention strategies for those whose development may not be proceeding normally. For example, the following statement from Ramey and colleagues, while applied to infant stimulation programs, applies equally well to intervention programs in general: "The goal of intervention is to provide an optimal set of conditions in which the infant [child, client] and environment can adapt to one another. Thus, we are concerned with the nature of the relation between the infant's and the environment's ability to respond to each other." (38, p. 411) Understanding the characteristics of the child or the client, the nature of the stimulation, and the child's available motor responses informs the therapist how the stimulation is being received. Knowing this, the therapist can re-adjust the stimulation accordingly (Fig. 9.3).

Thus, the primary consideration in developing effective intervention is an understanding of the nature of the individual being addressed. This is accomplished by diagnostic evaluations, by a thorough understanding of normal development, and by understanding the nature of the normal human and the characteristics of the biological impairment involved. While this text is not written to include biological and medical conditions, it does address normal development; some general principles of intervention can be created based on this knowledge alone (Table 9.5). For example, the research already discussed in this chapter (e.g., from Hubel and Weisel, Gibson, Bower, Fantz and others) showed us that humans, across developmental levels, are perceptually quite similar, with the exception that infants can handle less information than can more experienced children and adults. Thus, stimuli or events that are important and potentially therapeutic for children and adults are often also important for

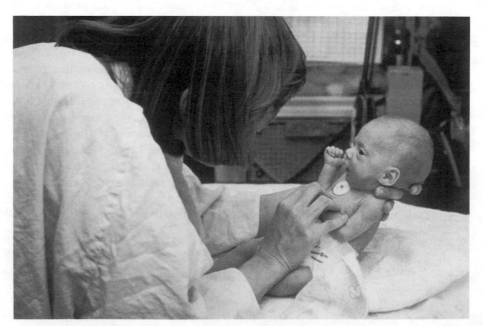

**Figure 9.3.** The goal of effective intervention is to set up conditions so that the environment and the infant (or client) can adapt to one another. (Courtesy of the Neonatal Intensive Care Unit, University of Connecticut Health Center's John Dempsey Hospital, Farmington, CT)

**Table 9.5**
**General Principles for Developmental and Therapeutic Stimulation**

Be sensitive to early predispositions to respond to (approach/withdraw from) specific forms of stimulation

Facilitate active participation thereby generating self-produced movement and reafferent stimulation

Recognize the significance of contexts that satisfy the need for long-range supportive systems of reciprocal interaction (families, caretakers, institutions, communities)

infants, with the exception that infants may become more easily overloaded or overwhelmed by too much, or too intense, stimulation from one particular sensory mode or from combinations of multisensory stimulation. Particular attention to their state changes and arousal levels, as well as subtle or overt motor responses indicative of attentiveness, approach, and avoidance reactions, will inform us how that sensory information is being received. As Ramey and colleagues note,

> Blindly providing an extensively stimulating environment to an infant with poor state control may prohibit the infant's existing capacities from taking advantage of environmental attributes by overly arousing the infant into a state that is not optimal for orienting to the environment. Conversely, reduced amounts of response-contingent stimulation may also result in nonoptimal development. (38, p. 411)

Data from the scientific study of visual perception indicates that infants (like adults) are attracted to movement, faces, high contrast (like black-and-white bull's-eyes patterns), contour, and medium complexity. Thus, visual stimuli and toys with these elements should attract and sustain attention more than other types of visual stimuli (e.g., soft, flat, two-dimensional pastels) (Fig. 9.4). We know from the science of early experience that humans, as a group, are predisposed to seek out (i.e., pay attention or respond to) certain stimulus qualities in the environment and that they are specifically attracted to, and (within certain limits) will respond positively to, social stimuli or to stimuli presented in social contexts. Humans habituate to repetitive stimuli, respond to stimulus change and contrast, and are soothed by slow rhythmic stimuli in a variety of modalities (e.g., tactile, auditory, vestibular). Furthermore, specific forms of stimulation attract our immediate attention. Those that are threatening, i.e., intense or arrhythmic in any modality, provoke defensive reactions and potential avoidance responses. Because of the lack of habituation and the tendency to continually respond to such stimuli, repeated exposure can mean a great cost in terms of energy expenditure and prolonged emotionality.

We also know that stimulation that actively engages the infant, child, or patient produces the greatest degree of interest and the greatest potential for sensorimotor gain because of the positive effects of reafferent stimulation. Thus, toys and environmental stimuli that encourage or require self-produced movement and feedback may be most developmentally significant. Finally, from an ecological and a systems point of view,

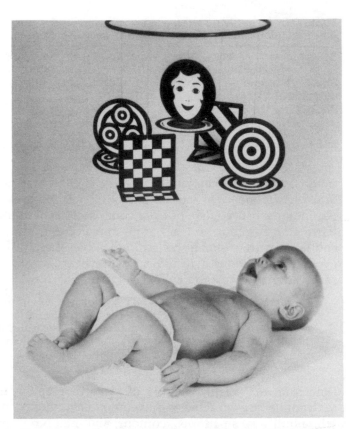

**Figure 9.4.** This mobile takes advantage of research indicating that infants have sophisticated visual perception systems and that they fixate longer on visual stimuli of faces and on stimuli that have high contour, medium complexity, and high contrast. (Courtesy of Infant Stim-Mobile, Wimmer-Ferguson Child Products.)

sensorimotor environments that are reciprocally responsive will set up their own self-sustaining feedback systems with long-range opportunities for continued stimulation and promotion of development.

## Caretaking as Reciprocal Perceptual Stimulation

While the previous description sounds quite esoteric and academic, it actually describes the developmental processes of a normal infant within a responsive caretaking context. As previously discussed in this text, the normal caretaking situation naturally provides sensory stimulation in every modality; the newborn and infant are geared to take advantage of it. With sophisticated and predisposed sensory responsiveness, infants quickly identify family members through olfaction, attend to the perceptual change and contrast in the eyes and the face, and synchronize their movements with social stimulation. As infants respond to these various stimuli, so, too, does the caretaker, who also is attracted to and reacts to perceptual contrast and multistimulus change. The parent responds to the infant with eye contact, stroking, and conversation, and the infant responds, in turn; this establishes a mutually satisfying reciprocal relationship that will be self-promoting for the members involved.

The violation of that reciprocal relationship is particularly revealing regarding the infant's perceptual and motor readiness to respond socially. This has been demonstrated by experiments with mothers and their infants who have been brought together in an observational setting and their reciprocal interactions recorded. Once this baseline of interaction is established, the mothers are instructed to withdraw and to maintain a consistent, nonresponsive posture (blank face) toward the infant. The response on the part of the infants is startling. As Brazelton describes it:

> [The mother] is instructed to follow a three-minute play period with a second three minutes in which she sits in front of the baby but remains unresponsive, staring at him with her face perfectly still. When this interactive system is violated by the parent's nonreciprocity, the infant will respond in an expectable manner, indicating how powerfully he is affected by the violation of his expectation for playful, interactive responses. . .

> For instance, a three-month-old infant began reacting to the still face by showing the characteristic wary pattern of behavior. About a minute and a half into the interaction he looked at his mother and laughed briefly. After this brief tense laugh, he paused, looked at her soberly, and then laughed again, loud and long, throwing his head back as he did so. At this point the mother became unable to maintain an unresponsive still face, broke into laughter and proceeded to engage in normal interactional behavior. The intentions and emotions of the older infant are similar to those of a younger infant. The richness and skill in reestablishing a reciprocal interaction, however, are greater. (39, p. 11)

It is interesting that a chapter devoted to the topic of sensory and perceptual development ends up emphasizing "social" encounters. This movement from sensory to social and emotional aspects of the child demonstrates the difficulty of examining isolated developmental domains. All aspects of development are interwoven and related. Thus, the sensory, perceptual, and motor capacities of infants gear them to becoming social beings, and in so doing, set up social contexts for continued perceptual-motor stimulation. As chapter 11 will examine, this forms the basis for further cognitive development; and as chapter 12 explores, our important human social interactions, which involve the sensory functions of touch, smell, synchronized movements, looking, and listening are fundamentally sensorimotor in nature.

# Sensorimotor Deficits and Development

Intervention in physical and occupational therapy has been traditionally aimed at restoring or normalizing peoples' sensorimotor systems (Figs. 9.5 and 9.6). Numerous sensory stimulation and sensorimotor programs have been created, and as discussed in chapter 4, investigation of the efficacy of these varying interventions is a primary focus of contemporary research in occupational and physical therapy. An additional and fairly recent focus of

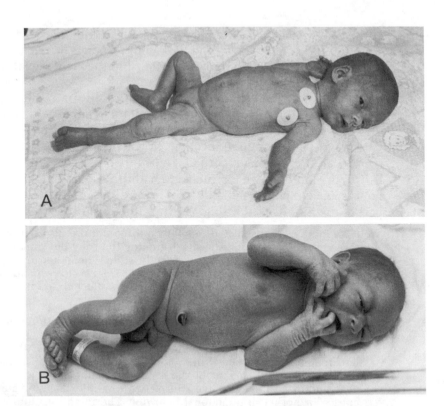

**Figure 9.5.** *A,* An infant who is weak or lacks muscle tone may be unable to move out of a posture such as the asymmetrical tonic neck reflex. Such an infant may be less likely to engage in effective social interaction and self-stimulation than a child in the normal flexed posture typical of most newborns (*B*). (See Fig. 9.6.) (Courtesy of the Neonatal Intensive Care Unit, University of Connecticut Health Center's John Dempsey Hospital, Farmington, CT)

therapy is based on the recognition of effects that interrupted sensorimotor functions in the developing infant and child can have on the family-child relationship. The parent-child relationship can be interrupted because of a variety of reasons such as stress, emotional or psychiatric disturbances, or lack of knowledge and fear on the part of the parents. Interruption can also occur because of some child-related deficit such as a sensory or motor impairment that interferes with the infant's responsiveness. Consider the following examples:

A blind infant may not smile when her mother smiles and may not make the facial responses her mother expects from her. Feeling disappointed and slightly rejected, the mother begins to respond less to her baby. Without really realizing it, she provides less social stimulation, talking, touching, and interacting less because she feels her infant does not respond. The infant, now lacking sensory stimulation in numerous modalities, begins to become listless and withdrawn; while the mother becomes more tense and anxious, feeling incompetent as a mother and angry that her baby is not normal. The infant, sensitive to the mother's tenseness and abrupt handling, fusses and cries even more, thereby frustrating the mother more.

An infant with immature oral musculature has a difficult time feeding. He is the third child in this family, and his mother has nursed his other two siblings. The mother has a part-time job, and feels constrained by the time demands with three children and work. She expected that feeding her baby would be pleasant and fun, but it takes longer than expected and is tedious and messy. She begins to resent the fact that her newborn does not behave like her first two children. She has begun to label him a "difficult child," and she finds herself picking him up and cuddling him less. The baby, in turn, is hungry. He gets less food at each meal, and he is uncomfortable and cries more often. This concerns and frustrates his mother even more, and her tendency is to withdraw from her baby even more.

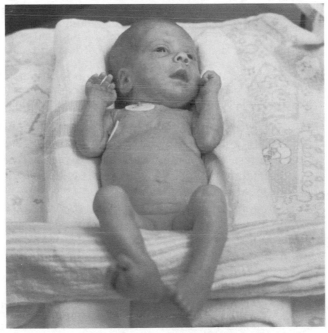

**Figure 9.6.** Preterm infants and older clients who are delayed or lack muscle tone are often bolstered to improve their posture and alignment and to help them with self-regulation. Compared with other postures typical of infants with low muscle tone (see Fig. 9.5A), a flexion posture enables such infants to receive and to respond to sensory and social stimulation. (Courtesy of the Neonatal Intensive Care Unit, University of Connecticut Health Center's John Dempsey Hospital, Farmington, CT)

In each of these cases, the natural reciprocal mother-infant relationship is being eroded initially because of a sensory or motor deficit. Many parents do not know how perceptually sophisticated their infants and children are and are often unaware of their infant's potential for responsiveness. They do not know what cues to look for, in terms of state changes, attention, approach, withdrawal, and they are even more overwhelmed when an infant or child experiences some sensorimotor deficits. Such deficits, early intervention research shows us, can interrupt the delivery or reception of normal social interactional cues, thereby interfering with, and perhaps preventing, positive caretaking behavior. Thus, therapists—recognizing that the long-range care of their young clients is contingent upon parental care and understanding—are incorporating parent intervention into their treatment programs. Therapists are educating parents how their children may compensate for deficits in sensory and motor systems and how to "read" the behaviors of these youngsters (Fig. 9.3). Oehler describes this in regard to parental responses to preterm infants:

> Although further research is needed, parents can now be informed that the premature infant—even the youngest and sickest one—responds to social stimuli that parents can easily provide: talking and touching. We can enhance parents' understanding of their child's interactive capabilities by demonstrating the specific behaviors that infants exhibit in response to these stimuli. More specifically, the finding that the youngest preterm infants engage in both stabilizing activity to maintain interaction and avoidance signals to discourage interaction suggests the possibility of teaching parents to monitor these cues and use them to moderate their interaction with their infants. Finally the information that sick infants respond to the combinations of talking and touching with increased avoidance signals can be used to suggest to sick infants' parents that they provide only one mode of stimulation at a time (40, p. 32)

Field's (41) interaction coaching for high-risk infants and their parents (see chapter 7) is aimed at educating parents about the special responses of their infants. These interventions are geared primarily toward teaching parents to understand the infant's attempts at behavioral organization and self-regulation, at the feeding context, and at face-to-face encounters involving gazing at and responding to one another. As Brazelton (39) notes, the infant at risk includes premature and minimally brain damaged infants who cannot adjust as easily as well-equipped neonates to environmental deprivation and disruption; quiet, nondemanding infants who do not elicit mothering from overstressed mothers; and hyperkinetic and hypersensitive infants who overstress their parents and provoke inappropriate responsiveness. Brazelton describes the responses of these infants' parents:

> By history, they associate their failure with this child to an inability to "understand" him from the neonatal period onward, and they claim a difference from the other children in his earliest reactions to him as parents. If we are to improve the outcome for such children, assessment of the risk in early infancy must mobilize preventive efforts and programs for intervention before the neonate's problems are compounded by an environment which cannot understand him without such help. (39, p. 4)

Brazelton explains that early intervention needs to address such infants and their parents and that we need to intervene so that the systems that will support the infants' development can be put into gear as soon as possible. Intervention as early as possible will help both the infant and the family develop a sense of mastery, which expand into childhood and throughout life.

These two approaches from Field and Brazelton do not mention, however, the infants and children with specific sensory and motor impairments who make up much of the clientele of physical and occupational therapists. Some of the same parent-intervention approaches discussed by Brazelton and Field can be applied to parents of infants and children with sensorimotor deficits, so that these families can be helped to understand the characteristics of sensorimotor deficits and how to use other sensory systems to compensate for one(s) that may be impaired. Thus, for example, in the case of a child with visual impairments, parents can be taught to look for motor cues rather than changes in facial expressions or eye contact. For infants with disorganized motor responses, intervention can be directed at helping them develop motor control (Fig. 9.6) as well as demonstrating to their parents how to facilitate and to identify approach and withdrawal responses. For infants with visual and auditory impairments, somesthetic interactions can be emphasized, and parents can be taught how to "communicate" with touch, movement, changing pressure, and body interactions. Physical and occupational therapists' knowledge of movement and sensory/perceptual development provides a special role in promoting parents' understanding of their special children. Als (42) has developed a framework for regarding the special needs infant in terms of maintaining a balance between the various behavioral and physiological systems and the need for self-regulation. She illustrates how an interventionist's attentiveness to the infant's sensorimotor functions and behavioral cues can help facilitate the infant's struggle for self-control and behavioral organization (Figs. 9.5 and 9.6).

---

Vestibular-proprioceptive stimulation has been explored extensively in human and nonhuman species, and a considerable body of research indicates that positive effects, such as gains in motor and adaptive behaviors, are associated with programs of vestibular stimulation (3, 32, 33). While vestibular stimulation has been "applied" to children and various client populations (e.g., in the form of rocking or spinning), it may be even more meaningful to look at the uses of such stimulation as a mechanism for enhancing interactions between children with sensorimotor deficits and their environments (3). Such an approach has been adopted with the therapeutic use of waterbeds. Consider the numerous sensory and clinical advantages of such stimulation.

Waterbeds often help control body temperature and prevent skin breakdown. They also respond as the individual responds. This may be particularly important for those with little motor control or with weakness. Such individuals can easily move and then benefit from the feedback generated from self-produced movements. Such activity may, in turn, facilitate a low level of arousal and self-interest in hypoactive infants or clients. For individuals with poor motor control, including infants, waterbeds provide a form of containment that may help behavioral organization by holding the body in a more normal, flexed posture. In addition, in some infants who cannot be "reached" through other modalities, rhythmic movements may evoke approach responses, to which parents can be trained to respond. This, in turn, may enable parents to find a way to gently interact with their infants, to gain some feelings of confidence and control, and to provide an effective entry to a more relaxed and effective parent-child relationship. Such effective social interchange may set up a system for continued perceptual stimulation and development for the infant or child.

# CONCLUSION

The conclusions from this and the previous chapter are important and compelling. Collectively, they form a tidy picture that is consistent with an interactional approach to human development. These are conclusions from the science of early experience and from the study of infant sensory and perceptual development: At birth, infants are relatively sophisticated beings who are also affected by their environmental experiences. While their perceptual capacities are not identical to those of adults and while they cannot handle all the information adults can, infants possess many basic perceptual abilities that are then elaborated during life. As an example of this, in regard to visual development, infants, like adults, show certain preferences for high-contrast patterns rather than solid colors or solid black; solid, three-dimensional objects rather than flat objects; curved rather than linear stimuli; medium rather than high- or low-complexity stimuli; faces rather than nonsense; and stimuli that move and change rather than stationary stimuli (13).

What is the meaning of this sensory/perceptual sophistication at birth? Does it imply, in regard to nature versus nurture, that abilities are essentially inborn and therefore unchangeable? The answer to this last question is, No. Hubel's and Weisel's (8–11) and other research have illustrated that at least part of the visual system is plastic and can be modified by experience. Thus, at least in regard to visual perception, development is a result of a blend of innate, maturational capacities that are reinforced by life experiences and learning. What seems to be particularly significant is that the most important experiences are those that engage the infant in interaction in which ongoing sensorimotor feedback is obtained.

Ethologists and infant researchers would agree that infants of different species seem predisposed to respond to circumstances that are significant to their species. Given the dependent nature of the human infant, the stimuli most important to the human species are those that will bring about long-range protection and care. A mutually satisfying social interaction between the infant and the caregiver(s) will not only provide this care but also serve as a source of continued feedback that promotes further development.

Infant capabilities that support this social interaction are evident across the various sensory modalities at or around the time of birth. While ethologists would say that infants have an innate disposition to respond to faces, an equally plausible explanation in nonethological terms is that human infants are predisposed (or adapted) to respond to visual stimuli that are significant for their continued development and survival. Thus, in their attraction to areas of contrast, areas of medium complexity, and to stimulus change, they are naturally drawn to the human face. The same generalizations appear to hold for other sensory modalities and for what motor responses are available to humans at each age. Infants are multisensory organisms, and their sensory systems are well developed and fairly predirected at birth. Thus, auditory systems seem geared to the human voice and language, and sensorimotor responses are made in accordance with changes in conversation. Infants respond to touch and handling, and early in life, they can recognize family members by odor. In turn, the family members react to infant responses and provide additional stimulation. Early motor reflexes are oriented toward survival, such as rooting and sucking for locating and obtaining nourishment.

The early sensorimotor actions and repetitive movements so characteristic of infant behavior and child's play, are self-reinforcing. They provide the infant or child with feedback from his or her own self-produced movements and serve to promote continued stimulation and development. It appears that young humans are geared to promote their own perceptual enrichment through maturation, self-stimulation, and reciprocal social exchange. Sensory/perceptual development is therefore a significant part of an ongoing natural system that promotes continued motor, cognitive, and social development, which are the topics of the next chapters in this book.

## References

1. Simmel EC, Baker E: The effects of early experiences on later behavior: a critical discussion. In Simmel EC (ed): *Early Experiences and Early Behavior. Implications for Social Development*. New York, Academic Press, 1980, p 3.
2. Bronfenbrenner U: Early deprivation in monkey and man. In Bronfenbrenner U (ed): *Influences on Human Development*. Hinsdale, IL, Dryden Press, 1972, pp 256–300.
3. Short MA: Vestibular stimulation as early experience: Historical perspectives and research implications. In Ottenbacher K, Short MA (eds): *Vestibular Processing Dysfunction in Children*. New York, Haworth Press, 1985, pp 135–152.
4. Denenberg VH: Environmental enrichment. In Denenberg VH (ed): *The Development of Behavior*. Stamford, CT, Sinauer, 1972, p 312.
5. Hebb DO: *The Organization of Behavior*, New York, Wiley, 1949.
6. Henderson ND: Effects of early experience upon the behavior of animals. The second twenty-five years of research. In Simmel EC (ed): *Early Experiences and Early Behavior. Implications for Social Development*. New York, Academic Press, 1980, p 45.
7. Riesen A: Arrested vision. In *The Nature and Nurture of Behavior. Developmental Psychobiology*. San Francisco, WH Freeman, 1973, p 62.
8. Hubel DJ: The visual cortex of the brain. In *The Nature and Nurture of Behavior. Developmental Psychobiology*. San Francisco, WH Freeman, 1973, p 5.

9. Wiesel TN: The postnatal development of the visual cortex and the influence of environment. *Biosci Rep* 2:351–377, 1982.

10. Hubel DH: The visual cortex of normal and deprived monkeys. *Am Scient* 67:532–542, 1979.

11. Hubel DH, Weisel TN: The period of susceptibility to the physiological effects of unilateral eye closure in kittens. *J Physiol* 206:419–436, 1970.

12. Blakemore C, Cooper GF: Development of the brain depends on the visual environment. *Nature* 228:477–478, 1970.

13. Fantz RL: The origin of form perception. In *The Nature and Nurture of Behavior. Developmental Psychobiology.* San Francisco, WH Freeman, 1973, p 66.

14. Gibson EJ, Walk RD: The visual cliff. In *The Nature and Nurture of Behavior. Developmental Psychobiology.* San Francisco, WH Freeman, 1973, p 19.

15. Bower TGR: The visual world of infants. In *The Nature and Nurture of Behavior. Developmental Psychobiology.* San Francisco, WH Freeman, 1973, p 36.

16. Whitteridge D: Visual machinery of the brain. *Nature* 294:113–114, 1981.

17. Rosenzweig MR: Effects of environment on brain and behavior in animals. In Schaepler E, Reichler RJ (eds): *Psychopathology and Child Development.* New York, Plenum, 1973, p 33.

18. Walsh RN, Cummins RA: Neural response to therapeutic sensory environments. In Walsh RN, Greenough WT (eds): *Environments as Therapy for Brain Dysfunction.* New York, Plenum, 1976, p 171.

19. Ottenbacher K, Short MA: Sensory integrative dysfunction in children. A review of theory and treatment. In Wolraich M, Routh DK (eds): *Advances in Developmental and Behavioral Pediatrics.* Greenwich, CT, JAI Press, 1985, vol 6, p 287.

20. Gottlieb G: Introduction to behavioral embryology. In Gottlieb G (ed): *Studies on the Development of Behavior and the Nervous System.* New York, Academic Press, 1973.

21. Friedman SL, Jacobs BS, Werthmann MW: Sensory processing in pre- and full-term infants in the neonatal period. In Friedman SL, Sigman M (eds): *Preterm Birth and Psychological Development.* New York, Academic Press, 1981, p 159.

22. Ferry PC: On growing new neurons: are early intervention programs effective? *Pediatrics* 67:38–41, 1981.

23. Walsh R: *Towards an Ecology of the Brain.* New York, SP Medical and Scientific Books, 1981.

24. Thompson WR, Schaefer T: Early environmental stimulation. In Fiske DW, Maddi SR (eds): *Functions of Varied Experience.* Homewhood, IL, Dorsey Press, 1961, p 49.

25. Penfield W, Rasmussen T: *The Cerebral Cortex of Man.* New York, Macmillan, 1950.

26. Held R: Plasticity in sensory-motor systems. In *The Nature and Nurture of Behavior. Developmental Psychobiology.* San Francisco, WH Freeman, 1973, p 73.

27. Ferchmin PA, Bennett EL, Rosenzweig MR: Direct contact with enriched environment is required to alter cerebral weights in rats. *Comparat Physiol Psychol* 89:360–367, 1975.

28. Dru D, Walker JP, Walker JB: Self-produced locomotion restores visual capacity after striate lesions. *Science* 187:265–266, 1975.

29. Browder JA: The pediatrician's orientation to infant stimulation programs. *Pediatrics* 67:42–44, 1981.

30. Simeonsson RJ, Cooper DH, Scheiner AP: A review and analysis of the effectiveness of early intervention programs. *Pediatrics* 69:635–641, 1982.

31. Harris SR: Transdisciplinary therapy model for the infant with Down's syndrome. *Phys Ther* 60:420–423, 1980.

32. Ottenbacher KJ, Petersen P: A meta-analysis of applied vestibular stimulation research. In Ottenbacher K, Short MA (eds): *Vestibular Processing Dysfunction in Children.* New York, Haworth Press, 1985, pp 119–134.

33. Ottenbacher K: Developmental implications of clinically applied vestibular stimulation. *Phys Ther* 63:338–342, 1983.

34. Schaefer M, Hatcher RP, Barglow PD: Prematurity and infant stimulation: a review of research. *Child Psychiatr Hum Dev* 1064:199–212, 1980.

35. Ottenbacher KJ, Biocca Z, DeCremer G, Gevelinger M, Jedlovec KB, Johnson MB: Quanti-

tative analysis of the effectiveness of pediatric therapy. Emphasis on the neurodevelopmental treatment approach. *Phys Ther* 66:1095–1101, 1986.

36. Cornell EH, Gottfried AW: Intervention with premature human infants. *Child Dev* 47:32–39, 1976.

37. Bronfenbrenner U: Is early intervention effective? Facts and principles of early intervention: a summary. In Clarke AM, Clarke ABD (eds): *Early Experience: Myth and Evidence.* New York, Free Press, 1976.

38. Ramey CT, Zeskind PS, Hunter R: Biomedical and psychosocial interventions for preterm infants. In Friedman SL, Sigman M (eds): *Preterm Birth and Psychological Development.* New York, Academic Press, 1981, p 395.

39. Brazelton TB: Early intervention: What does it mean? In Fitzgerald HE, Lester BM, Yogman MW (eds): *Theory and Research in Behavioral Pediatrics.* New York, Plenum, 1982, vol 1, p 1.

40. Oehler JM: Examining the issue of tactile stimulation for preterm infants. *Neonat Network* 25–33, 1985.

41. Field T: Interaction coaching for high-risk infants and their parents. In Moss HA, Hess R, Swift C (eds): *Early Intervention Programs for Infants.* New York, Haworth Press, 1982.

42. Als H: A synactive model of neonatal behavioral organization: framework for the assessment of neurobehavioral development in the premature infant and for support of infants and parents in the neonatal intensive care environment. *Phys Occup Ther Pediatr* 6:3–54, 1986.

# 10

# Motor Development

Motor development provides the individual a means of independence that enhances sensory stimulation and interaction with people and physical environment, thereby facilitating all other areas of development. During infancy, movement and manipulation of objects foster cognitive and perceptual development including understanding about the nature of objects in the environment, spatial relationships, and problem solving. Motor activity is integral to childhood play, contributing to self-image and socialization. Academic performance is also affected by motor skill, particularly perceptual-motor tasks such as writing. During adolescence, motor ability influences leisure time interests and socialization with peers and may even affect eventual choice of vocation.

While it is implicit that motor development affects and is affected by all other developmental domains, this chapter will focus on the numerous and exciting changes that take place during the acquisition of motor abilities during infancy, childhood, and adolescence. As chapter 3 has noted, age-specific divisions in development are arbitrary. There is no magical and noticeable event that separates a 24- from a 25-month-old, nor is there an event that distinguishes a child who is 5 years 11 months from when he or she becomes 6. Despite this, age-specific divisions are particularly convenient for describing developmental trends, especially in an area as complex as motor development. Thus, in this chapter, specific stages are used. The reader must recognize that such divisions and their corresponding names or ages may vary across developmental references. The age-related stages referred to in this chapter are listed in Table 10.1.

Motor development is divided into gross motor and fine motor domains. Gross motor development includes reflexive behaviors, the ability to assume and maintain postures such as sitting and standing, and the ability to perform whole body movements such as creeping, walking, and jumping. Fine motor development encompasses arm and hand movements involved in reaching, grasping, manipulating, and releasing objects. Coordination of hand movement, vision, and other somesthetic and proprioceptive abilities are also necessary for many fine motor tasks. Although the term "motor development" is commonly used "sensory-motor development" is a more accurate term. Sensory input to the central nervous system is important for initiation and regulation of movement. Sensory feedback also contributes to the modification and refinement of movement, which are processes that are essential to motor learning. Consequently, development of the sensory modalities, which is described in the previous chapter, is closely involved with changes in motor abilities over time.

This chapter examines the development of skilled movement. Particular attention is devoted to infancy because it is the most dramatic period of motor development. The general principles that are covered here are important for a basic understanding of how gross and fine motor abilities change during normal development from birth through adolescence. These principles are also especially important for occupational and physical therapists because of their developmentally oriented interventions, which are geared not only for children but also for adults with movement disorders.

Table 10.1
Age-related Stages in Motor Development

| Stage | Age Range |
|---|---|
| Infancy | birth to 2 yr |
| Early childhood | 2 to 6 yr |
| Late childhood | 7 to 12 yr |
| Adolescence | 12 to about 18–21 yr |

Basic principles of motor development are integral to treatment aimed at assisting individuals to use movement in a purposeful manner to achieve maximal function and self-sufficiency.

# MOTOR DEVELOPMENT IN INFANCY

Motor development has been most extensively studied during infancy, which encompasses the first 2 years of life. The newborn infant has minimal ability to support the body against gravity and is completely dependent upon caretakers. However, in 2 short years, children are independent in locomotion and effectively use their hands to play. To provide a basis for understanding the motor development sequence, initial consideration is given to factors that influence motor development and the development of automatic movement.

## Factors That Influence Motor Development

Motor development is characterized by a predictable rate and orderly sequence. The mean age and variability of achievement of motor behaviors has been determined predominately through cross-sectional research as described in chapter 4. Progression through the motor development sequence follows a predictable time course. For example, on the Bayley Scales of Infant Development (1), the mean age for sitting is 6.6 months, and for walking alone it is 11.7 months. Equally important is the variation in achievement of each motor behavior. On the Bayley scale, 5% of the normative sample walked at 9 months, 50% by 11.7 months, and 95% by 17 months. Although developmental norms cannot be used to predict an infant's rate of development, they support the concept that motor development is age related. Motor development also follows an orderly sequence. The locomotor sequence during the first year is well documented and includes the following progression: rolling, crawling on stomach, creeping on hands and knees, cruising (lateral steps while holding onto a support), and walking. Similarly, infants initially grasp reflexively, later grasp with the palm and fingers, and eventually grasp a pellet using a pincer grasp with the tip of the index finger opposed to the tip of the thumb.

The fairly predictable rate and orderly sequence of motor development, which seem quite consistent across children and which are responsible for the ability to obtain norms like the ones listed above, are attributed to maturation of the central nervous system. Maturational theory states that the emergence of a behavior is dependent upon a sufficient level of development in the areas of the central nervous system that mediate (regulate) the behavior. As noted in chapter 7, maturation is related to anatomical changes in the central nervous system. It includes differentiation, migration, and growth of neurons; formation of synaptic connections (2); and myelination of axons and dendrites. Myelin formation, in particular, is associated with motor development in infancy; however, a direct causal relationship between myelination of neurons and functional movement has not been established (3).

In addition to maturation, movement experiences are necessary for motor development. Myrtle McGraw (4), whose descriptive studies of motor development are classic, considers the infant's overall rate of motor development to be limited by neuromaturation of the central nervous system. However, McGraw proposes that during critical periods, specific experiences augment the acquisition of emerging motor behavior. The development of walking illustrates this concept: The 7-month-old infant is able to take weight in supporting standing and able to bounce but lacks the maturation necessary for balance and reciprocal stepping. Consequently, to practice walking at 7 months is futile. In contrast, at 12 months, when the infant is beginning to stand alone, encouraging an infant to cruise or holding onto the infant's hand to assist in walking provides experience conducive to the development of independent walking.

Opportunity for movement and play provides the experience needed for motor development in infancy. Early stages of the motor development sequence provide experiences that prepare the infant for later more advanced stages; thus specific training associated with learning motor skills at older ages is not necessary. The infant is adept at interacting with people and the environment to obtain the experience necessary for motor development. As will be described more fully in the next chapter, cognitive theorist Piaget (5) conceptualizes infants as possessing an innate drive to explore and master their surroundings. The infant's spontaneous motor action and response to new play experiences are also valuable sources of feedback to parents. This feedback guides parental expectations and reinforces spontaneous interaction, further assuring that the infant receives appropriate movement experiences.

Differences in rate of maturation, socioeconomic status, and childrearing practices are possible explanations for sex and race differences that have been observed in motor development. These differences are seen primarily in infancy and are not pronounced; therefore, they are often not emphasized. The differences include an accelerated rate of physical maturation in females, which may enhance motor development. Goldberg and Lewis (6) report that males are more vigorous in play and prefer toys that require gross motor coordination, while females prefer toys that emphasize fine motor manipulation. The implications of these data and other sex differences will be discussed in chapter 12. In regard to racial differences, black infants have been shown to be advanced in gross motor development, compared with white infants (7). For example, of the infants who formed

the normative sample of the Bayley Motor Scale (8), the median age placement for black infants was at least 0.7 months earlier on 11 of the 60 items
on the test, including sitting and walking.

# Principles of Motor Development

Motor development, particularly during the first year, is characterized by
the five principles discussed in the following sections. As discussed in
chapter 5 in regard to maturational theory, these are general trends that,
though variable, are commonly recognized principles that illustrate the
changes in motor control that occur as a result of maturation of the central
nervous system.

### Cephalocaudal

Development of motor control proceeds from the head to the foot. Head
control develops first, then trunk control, and finally control of the arms
and legs. The infant develops head control in prone position prior (Fig. 10.1)

**Figure 10.1.**  While in a prone position, the infant is gradually able for longer
periods of time to keep her head held up while she is propped on her forearms.
Head righting keeps the head in line with the body, while the mouth, eyes, and
hands explore the environment. Hands often remain fisted, as precise manipulation of objects does not emerge until sitting is attained, thereby freeing the
hands for exploration.

to developing sufficient trunk control to sit, and sits prior to developing sufficient control of leg movement to walk.

### Proximal-Distal

Control of proximal body parts precedes control of distal body parts. Shoulder control for reaching develops prior to control of the hand for grasp. Hip control for rolling and creeping develops prior to ankle and foot control for walking.

### Automatic-Voluntary

Movement progresses from a reflexive, or automatic, level to a voluntary level under conscious control. For example, the newborn reflexively grasps a finger inserted into the palm. During the first 6 months, the grasp reflex is integrated, and voluntary grasp emerges. Integration of reflexes is an important concept in motor development. It refers to the gradual "disappearance" of a reflex, presumably due to the maturation of neural areas that gradually exert inhibitory control over primitive behaviors. Thus, as neural control becomes more inhibitory, actions become voluntary and less automatic. In regard to grasp, by the end of the first year, the infant has moved from a reflexive to a voluntary grasp that is part of goal-directed play. In regard to gross motor abilities, a similar progression is found in the neonate who demonstrates reflexive stepping movements compared with the 1-year-old child who voluntarily initiates stepping movements to walk.

### Stability-Mobility

Postural muscle stability to maintain a developmental position precedes independent mobility. For example, the infant is able to stay upright when placed in a sitting position prior to independently moving in and out of this position. Rocking on hands and knees is seen prior to creeping, and standing precedes walking. This is because the postural stability that is characteristic of rocking and standing is necessary before weight shift of the arms and legs occurs during the independent movements characteristic of creeping and walking.

### Unrefined-Refined

An emerging motor behavior is performed in an awkward, inefficient manner. Through repetition and neuromaturation, movement becomes progressively more refined. This process occurs at each stage of motor development. For example, the infant initially demonstrates minimal trunk rotation in sitting and holds the arms at chest level to maintain balance. Within a few months sitting is refined, the trunk freely rotates, and the hands are used to manipulate toys. After sitting is refined, walking begins. The infant initially walks with minimal trunk movement and with the arms flexed at chest level, a striking resemblance to the initial sitting posture. A few months later, the infant walks with improved balance and takes longer steps. Trunk rotation increases, and the arms rest alongside the body.

# Automatic Movement

Coordinated movement consists of both automatic and voluntary actions. Automatic movements are elicited by sensory stimuli, are generally not under conscious control, and are mediated primarily by subcortical centers of motor regulation in the brain. Automatic movements of particular relevance to motor development include primitive reflexes, righting reactions, equilibrium reactions, and protective arm extension.

### Primitive Reflexes

During the first 3 months of life, primitive reflexes are an integral part of movement. Primitive reflexes are motor behaviors (Table 10.2) that form a continuum between fetal and neonatal movement. In the normal infant, movement is not limited to primitive reflexes nor are primitive reflexes *obligatory*, meaning that once initiated, the infant cannot help but to continue the complete reflex action. Although they are recognized as innate behaviors under neural control, primitive reflexes are influenced by various environmental effects. This is an excellent illustration of the interaction between maturational and environmental factors. For example, the frequency and variability of primitive reflex responses are dependent upon such factors as body position, behavioral state, time since last feeding, and body temperature.

Hierarchical views of brain development and control (see chapter 6) regard neonatal primitive reflexes as mediated by pathways in the spinal cord and brainstem. As midbrain and cortical areas of the central nervous system mature, voluntary motor control is regarded as being superimposed upon spinal and brainstem centers. Consequently, by 4 to 6 months, primitive reflexes are modified and no longer observed in stereotyped form. Maturation of cortical centers of motor control is thought to inhibit subcortical centers, thereby modifying reflexive behavior. The term "integration of primitive reflexes" has traditionally been used to describe this process. Table 10.2 lists a representative sample of primitive reflexes. For a complete list of primitive reflexes, consult the chapter by Tower (9).

Primitive reflexes have functional and clinical significance. For example, the rooting reflex assists the infant in locating the mother's nipple. The asymmetrical tonic neck reflex (ATNR) posture (Table 10.2, Fig. 9.5A) places the infant's hand in the visual field, promoting awareness of the hand as part of the body. Since the manifestation of these reflexes is time related and under predictable neural control, testing these primitive behaviors provides a standard method for assessing the integrity of neuromotor maturity. Primitive reflexes that are obligatory or that persist in stereotyped form beyond the first year are indices of a central nervous system imma-

**Table 10.2**
**Primitive Reflexes**

---

Rooting Reflex
    Onset:        28 Weeks gestation
    Stimulus:    Stroking of the skin at the corner of the mouth with the index
                  finger.
    Receptors:  Touch receptors of skin
    Response:   The head turns toward the stimulated side and the infant at-
                  tempts to suck the examiner's finger.
    Integration: 2–3 months
Moro Reflex
    Onset:        28 Weeks gestation
    Stimulus:    Supported in supine, the infant's head is suddenly dropped
                  backward (20–30°) and caught.
    Receptor:   Labyrinths of inner ear
    Response:   Abduction and extension of arms, with spreading of fingers; fol-
                  lowed by adduction and flexion of the arms.
    Integration: 5–6 months
Palmer Grasp Reflex
    Onset:        Birth
    Stimulus:    The examiner inserts index finger into the infant's hand from the
                  ulnar side and gently presses against the palm.
    Receptor:   Touch receptors and proprioceptors of hand
    Response:   The infant's fingers flex around the examiner's finger.
    Integration: 4–9 months
Automatic Stepping
    Onset:        37 Weeks gestation
    Stimulus:    The infant is suspended upright and slightly forward, so that the
                  soles of the feet touch the surface.
    Receptor:   Proprioceptors and touch receptors of the ankle and foot.
    Response:   Rythmical reciprocal stepping movements.
    Integration: 2 months
Asymmetrical Tonic Neck Reflex (ATNR)
    Onset:        Birth
    Stimulus:    In supine, rotation of the infant's head to one side.
    Receptors:  Proprioceptors and joint receptors of cervical muscles.
    Response:   Arm and leg on face side extend. Arm and leg on skull side flex.
    Integration: 3–4 months

---

turity that may occur with developmental delays or with neuromotor disorders such as cerebral palsy. Although testing of primitive reflexes is an integral part of a neurodevelopmental evaluation, many physical and occupational therapists find that, given the variability of these early motor responses, observing how primitive reflexes effect spontaneous movement is of greater value in treatment planning than is eliciting isolated responses.

## Righting Reactions

Righting reactions are automatic movements that maintain and restore the vertical position of the head in space, the alignment of the head and trunk, and the alignment of the trunk and limbs. Righting reactions develop during the first 6 months postnatally and are elicited by one or more of the following sensory stimuli: vestibular, proprioceptive, visual, or tactile (i.e., change in body contact with supporting surface). With experience and maturation of the central nervous system, righting reactions are integrated with equilibrium reactions and continue to function in automatic control of posture and movement throughout life.

Two different kinds of righting reactions contribute to development of head control and movement around the body axis: head righting and body righting. *Head righting* maintains the position of the head in line with the body. Thus, it allows the 4-month-old infant, for example, to maintain vertical alignment of the head while infant is in a prone position, a supported-sitting position, and while being carried. The development of *body righting* results in modification of the initial log-rolling pattern in which the body moves as a unit. Body righting provides for segmental rolling, in which arm, trunk, and leg movements take place in a sequence. Head and body righting also enable the infant to assume antigravity positions using rotation around the body axis. This is exemplified by the infant who in a *supine* position (lying on his or her back) turns to one side and pushes with arms into a sitting position. Later, the infant uses the righting reactions to pivot from a sitting position onto the hands and feet and then pushes up to standing.

## Equilibrium Reactions

Equilibrium reactions function to maintain and regain balance during movement. Based primarily upon vestibular input and secondarily upon proprioception and vision, equilibrium response is a counter rotation of the head and trunk away from the direction of displacement. Arm and leg responses vary with developmental position but, in general, during displacement the arms *abduct* (move outward from the body), the leg toward the side of displacement (downside) extends, and the leg on the upside flexes. This posture is essentially what we see when people try to "catch their balance." These seemingly automatic abilities are not evident at birth, and they lag behind the ability to maintain a posture. As a result, balance and movement are initially precarious at each stage of the developmental sequence that proceeds between 6 and 18 months in the following order: prone, supine, sitting, quadruped, and standing. By about 4 years of age equilibrium reactions are mature. By this time, instead of relying on righting reactions to move from a sitting to a standing position, the child is now able to maintain balance and push straight up with the legs into a standing posture.

## Protective Arm Extension

If the speed or magnitude of displacement of the individual's center of gravity is too great, equilibrium reactions are not sufficient to regain balance, and protective extension is evoked to prevent injury. Protective extension of the arms prevents the head and face from injury when balance is lost. The response, which is elicited by vestibular stimulation, is an extension of the arm(s) and hand(s) to support the body weight and protect against a fall. Protective extension emerges between 6 and 10 months in the following order: downwards and forwards, sideways, and backwards. It is an important protective motor ability that is necessary for good sitting balance and for security when standing and moving in the upright position.

# Stages of Motor Development in Infancy

Thus far, factors that influence motor development, principles of motor development, and the development of automatic movement have been discussed. In this section, a brief description of the infant's accomplishments at each stage of motor development is presented. The descriptions do not account for the experimentation and variable movement combinations that are typical of motor play. The age levels stated in this section and the remainder of the chapter are intended to provide only a general perspective of when specific movements develop. Readers interested in normative data should consult specific motor development assessments (1, 10–13). Motor behaviors that occur before birth are discussed in chapter 7.

## Birth to Three Months

During the neonatal period (first month) motor behavior is primarily random and automatic in nature. However, despite immaturity, the neonate is capable of adaptive motor capacities. The full-term neonate postures in flexion, which is a reflection of fetal position during the last trimester of pregnancy. At this time, extensor muscle tone is insufficient for head and trunk control. Thus, when the neonate is in supine, the head is turned to one side, the arms are flexed alongside the head, the hands are fisted, and the legs are maintained in flexion. In prone, the head is lifted only momentarily to turn to one side. The arms are curled under the chest and the legs are drawn up under the buttocks. When supported in sitting, the infant's posture reflects the dominating flexion influence, as the trunk is rounded, and the head drops forward. Primitive crawling movements of the legs in prone and swiping motions of the arms in supine are spontaneously performed and are automatic and lack postural stability. Adaptive movements during this time specifically relate to flexion responses. Self-quieting responses that reflect behavioral organization include bringing hand to mouth and curling up in a flexed posture.

By the end of the first stage of development, body schema and the postural muscle control necessary for voluntary movement are emerging. The 3-month-old infant has adjusted to the extrauterine environment and is alert for longer periods of time. The infant smiles and turns to visual and auditory stimuli. Arm and leg movements are self-perpetuated and promote *body schema*, which is a gradual awareness of body parts and how they all function as a part of whole, individual self. The infant postures with the hand open or loosely closed, makes scratching finger movements, and retains an object placed in the hand. At first, the hand is observed when it happens to enter the visual field. Shortly afterwards, visual regard of hand movement begins. The infant may also initiate unilateral swiping motions of the arm toward an object in the visual field. These latter two behaviors are precursors of voluntary reach.

By 3 months, integration of primitive reflexes contributes to improved postural control and midline orientation. In prone, shoulder and upper trunk stability enable the infant to hold the head up while being actively propped on either forearms or partly extended arms. In supported sitting, the head is held upright for short periods. The presence of the ATNR previously interfered with the ability to coordinate both hands and head together. The gradual integration of this reflex allows the infant to posture in supine with the head centered and the hands clasped together on the chest. This symmetrical posture marks the onset of midline body orientation which will be very important for later prehension tasks and eye-hand coordination.

## Four to Six Months

During the second stage of motor development, sitting and independent mobility emerge along with voluntary reach and grasp. The infant vigorously engages in spontaneous movement, often to the amusement of adults. Play in prone is directed toward improving postural extension. The infant readily pushes up on forearms in prone and, by 6 months, on fully extended arms. Also in prone, first the legs and then both the arms and legs are actively extended off the floor. This prone extension posture is thought to enhance proprioceptive feedback, thereby enhancing postural muscle control (14). In supine, the infant actively kicks, brings hands to feet, and feet to mouth. This exploratory play incorporates the legs and feet into the body schema.

### Sitting

The middle of the first year is notable for the onset of sitting. Infants raised in the United States are generally exposed to supported sitting well before they can sit independently. Play in prone and supine also promotes the development of head and trunk control which is necessary for sitting. The infant initially sits with a rounded trunk while propped forwards. Development of trunk extension and protective arm extension coincides with independent sitting. While the 6-month-old infant sits alone, this posture cannot be assumed independently and is not completely functional because the arms are required to help maintain balance.

### Rolling and Crawling

Rolling and crawling on the stomach are the infant's first methods of independent mobility. While actively playing in prone, the infant log rolls to supine initially by accident. Log rolling is replaced by segmental rolling as body righting develops. The 6-month-old infant frequently rolls from prone to supine to free the hands in order to manipulate toys. In prone, as weight shifting improves, the infant pivots in a circle and crawls forward on the stomach. Supported in a standing posture, weight is held by the infant's legs, and by 6 months, the infant delights in bouncing.

### Voluntary Reach and Grasp

Incorporation of the hand into the body schema, midline orientation, and improved postural stability are all prerequisites for voluntary reach and grasp. In supine, volitional reaching emerges at about 4 months. Visual regard of an object triggers reaching; however, hand movement is not visually monitored and arm movement is fairly circuitous. By 6 months, a direct unilateral reach (with one arm) is observed. The infant visually glances between hand and object, providing feedback for coding distance. The infant will eventually demonstrate a mature reach when the hand is brought from outside the visual field directly to an object.

The development of voluntary grasp also follows an orderly sequence during the first year (Table 10.3). Overlap is demonstrated between each successive grasping pattern. The 3- to 4-month-old infant grasps only when the object contacts the palm, with the fingers moving as a unit and squeezing the object into the palms. Grasp includes no thumb involvement and is from the outer portion of the hand and arm, referred to as the *ulnar* side. At 4 to 5 months, a mass palmar grasp is displayed. The fingers press the object against the palm and thumb, which is *adducted* (moved toward the hand). At 6 months, the thumb opposes the first two fingers (in a radial-palmar position) to grasp a 1-inch cube. The 6-month-old infant also secures small objects using a raking motion of the fingers (Fig. 10.2). Unfortunately, to the infant's dismay, small food objects such as raisins, which are often used in testing, are occluded in the palm, thereby preventing self-feeding. By 5 to 6 months, release is achieved by actually being able to drop an item or by assistance with adult removal of objects from the hand.

Voluntary reach and grasp expand the horizons of infant exploration and promote bilateral hand coordination which is exemplified by transfer of an object from hand to hand and bringing two objects together at the midline of the body. The fine motor play of the 6-month-old infant involves repetition of simple actions including banging and shaking objects. The infant is beginning to visually observe the consequences of manipulation and to develop a greater understanding of the nature of objects; and toys are repeatedly brought to an organ rich in sensory receptors—the mouth.

**Table 10.3**
**General Trends in Voluntary Grasp**

Use of the hand moves from ulnar to radial control
Whole hand grasp with no thumb movement precedes selective opposition of
   the tip of the index finger and thumb
The wrist is initially flexed, then held in neutral position, and finally extended

**Figure 10.2.** The infant initially grasps with a raking motion. Eyes and hands are not yet well coordinated to produce smooth, fine motor functions.

## Seven to Twelve Months

The second half of the first year is characterized by rapid changes in locomotor ability and hand use. During the first half of this period, sitting balance improves. The development of protective extension and equilibrium reactions allow the infant to rotate the head and trunk to cross the midline of the body and to use the hands for manipulation. The infant first learns to move out of a sitting posture at about 7 to 8 months, and at 8 to 9 months uses righting reactions to pivot from either prone or quadruped into sitting. After it is independently achieved, sitting becomes the preferred posture for fine motor play.

### Creeping and Walking

Creeping on hands and knees is a continuation of the prone progression of locomotor ability. When placed in supine, the infant now prefers to roll to prone. While playing in prone, weight is shifted backwards and the infant pushes onto hands and knees. The transitional phase of rocking on hands and knees prepares the arms and legs for creeping, which begins at about 8 months. The mature creeping pattern is characterized by reciprocal arm and leg movement. Some infants also "plantigrade walk," i.e., on hands and feet, which is a variation of creeping. By 12 months, creeping is generalized, and climbing begins. These motor abilities provide the infant an effective way to explore the immediate surroundings, and the combination of independent mobility and unabated curiosity create a number of precarious situations for the infant as previously inaccessible places and objects are investigated. (Fig. 10.3)

The infant makes steady progress from creeping to walking. This progress is facilitated by the development of equilibrium reactions in quadruped that enable the infant to free the arms and pull up to kneeling (Fig. 10.4) and later to standing. Parents generally provide experiences in standing before the infant is able to assume or maintain this position, but as selective leg movement is acquired, standing is assumed independently through half-kneeling. The ability to pull to standing is an indication of maturational readiness for mobility in the upright position.

After the infant assumes standing, the progression to walking accelerates. Creeping, however, remains the most effective method of locomotion. The infant creeps throughout his or her environment and pulls to stand at various low surfaces. Play in standing is performed initially by leaning

**Figure 10.3.** The combination of creeping, climbing, and curiosity increase the infant's mobility and access to new objects and environments.

against a supporting surface with two-hand support. Standing then progresses to an upright posture with one hand free to manipulate objects. Mobility begins as weight is shifted onto one leg while the other leg is lifted off the ground. This prepares the legs for cruising. A toy just out of reach is frequently the stimulus that prompts the infant to take steps laterally while holding onto a firm surface (Fig. 10.4).

Toward the end of the first year, the infant becomes preoccupied with learning to walk, the onset of which highlights the infant's many accomplishments during the first year. Parents often sense the infant's motivation and encourage stepping movements by holding the infant's hands. While engrossed in play, the youngester is frequently observed to momentarily take both hands off the supporting surface and to stand alone. At about 12 months, the infant is taking independent steps. These occur within an immature walking pattern with short lateral strides and little trunk movement and with the arms held to the sides in a flexed "high guard" posture. This posture keeps the hands and arms in a position where they are prepared to guard against falling.

**Figure 10.4.** A toy just out of reach often encourages mobility in creeping and cruising postures.

### Grasp and Release

Rapid changes in grasp also occur between 7 and 12 months. The radial-palmar grasp of the 6-month-old infant progresses to a radial-digital grasp, in which the thumb opposes the index and middle fingertips. At about 9 months, the wrist extends while grasping a small object like a cube, and the mature grasp pattern is established. The ability to isolate finger movement coincides with maturation of the pincer grasp, which is when the distal pad of the index finger opposes the distal pad of the thumb. This mature grasp emerges only after several other immature grasps occur. The raking grasp of the 6-month-old infant is gradually replaced by a scissors grasp, in which the thumb presses against the side of the index finger. The next stage is the emergence of an inferior pincer grasp, in which the thumb opposes the side of the index fingertip; finally a pincer grasp emerges. By 12 months, the infant can grasp a raisin with a mature pincer grasp that is characterized by opposition of the index finger and thumb and elevation of the wrist.

Voluntary release develops toward the end of the first year. At first, unassisted release is awkward and characterized by exaggerated extension of all fingers. The infant's growing cognitive development and awareness of the permanence of objects in the environment reinforces practice in dropping objects and observing either the hand movement or the flight of the dropped object. By 10 months a clumsy release of a cube into a container is demonstrated, and by 12 months this action becomes smoother and more controlled. At this age, a favorite play activity, in which the infant's emerging hand skill is evidenced, involves repeatedly releasing objects into a container and then dumping them out.

## Summary

At the end of the first year, vision, reach, grasp, and release are all coordinated. The infant capitalizes on these abilities by spending more time in fine motor play. Active exploration and systematic manipulation of objects characterize play. These attributes encourage problem solving and subsequent cognitive and perceptual-motor development.

## One to Two Years

During the second year of life, walking is the primary method of mobility, release of objects is refined, and perceptual-motor experiences are expanded in play.

### Standing, Walking, and Running

Through the combination of maturation and practice, a number of changes in standing and walking occur during the second year. One of the many challenges in walking is assuming the position and maintaining balance (Fig. 10.5). After begining to walk, the infant uses righting reactions to assume a standing posture. This involves pivoting from sitting, shifting weight onto the hands and feet, and then pushing up to standing. As balance

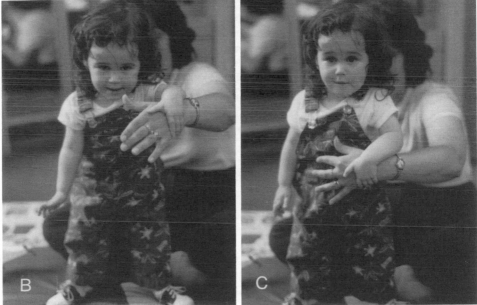

**Figure 10.5.** *A–C* Standing and walking require practice, determination, and neuromuscular maturation. A combination of vision, head-righting and balance, motivation, and motor coordination and control are involved.

improves, the infant is able to squat down from standing to pick up a toy. At this age, squatting position is often assumed during play.

Walking is at first only in the forward direction. The infant next makes turns in wide circles and by 15 months takes a few steps sideways and backwards. By 18 months the infant can take steps backwards while pulling a toy. After the onset of walking, the infant is called a "toddler," which refers to the short, unsteady steps that characterize the gait at this time. By 2 years of age, however, the infant, who is now considered a child, demonstrates mature walking by taking longer strides, controlling heel-toe patterns, and displaying coordinated trunk and pelvic rotation, with the arms positioned loosely at the sides.

Running and stair climbing are basic motor patterns that emerge during the second year. Compared with walking, they require greater unilateral weight shift and balance. A fast walk is the transition pattern between walking and running. To run, the infant must maintain a brief period of free flight with feet off the ground. In late infancy the running pattern has many of the same characteristics seen during early walking. These immature patterns include limited trunk rotation and leg flexibility as well as flexion of the arms in the high-guard posture. Stairs are initially navigated by creeping or plantigrade walking. By 18 months, first with hands held and then holding onto a railing, the infant climbs steps by placing both feet on the same step.

### Grasp, Release, and Tool Use

Basic grasp and release patterns are established by 15 months. Controlled, graded release emerges between 12 and 15 months; this is characterized by the ability to drop a raisin into a small bottle and to stack two cubes. As eye-hand coordination improves, the infant successfully plays with pegboards and "pop beads." Between 12 and 18 months, the infant can discriminate a circle, square, and triangle in succession. This discriminative skill is often demonstrated through play with shape boxes and simple form boards. Stacking blocks is also enjoyed, and it contributes to perception of vertical relationship of objects in space.

Tool use, which distinguishes human manipulation from most other species, is another important accomplishment during the second year. After being shown how, a 12-month-old infant will make a swiping stroke with a crayon across paper. The crayon is grasped with a fisted hand and flexed wrist, and the arm and hand move as a unit. By 18 months, this has changed. The infant now scribbles using a pronated grasp where the crayon is held in the fingers, the wrist is deviated laterally toward the outside of the arm (*ulnar deviation*), and the forearm is pronated. Movement of the crayon in this grasp is directed by action of the shoulder and forearm. By 24 months, visual-perceptual-motor coordination is displayed when, after demonstration, the child is able to imitate a vertical stroke.

### Summary

The transition from infancy to childhood is marked by increased independence, characterized by independent mobility (walking), some eye-hand coordination and accuracy, as well as other social skills, such as the acquisition of language, which will be discussed in other chapters. This time period is often one of consternation (as well as pleasure) for parents, for the young child now possesses an insatiable curiosity coupled with the mobility and dexterity to explore objects previously out of range (Fig. 10.6).

# MOTOR DEVELOPMENT IN EARLY CHILDHOOD

During early childhood, which ranges from about 2 to 6 years of age, the locomotor and prehension patterns characteristic of infancy are refined, and the motor developmental sequence (Table 10.4) is completed. This consists of motor patterns that are considered basic, fundamental, or primary, which are terms that are synonymous. Fine motor development is highlighted by the emergence of tool use, which includes manipulation of a pencil and scissors. These childhood motor abilities occur as a result of maturation of the central nervous system as well as specific movement experiences. Maturation of the central nervous system remains an important factor in motor development, as maturational readiness is necessary

**Figure 10.6.** Often, to the chagrin, but mostly to the delight of parents, the toddler's insatiable curiosity, mobility, and dexterity combine to enable him to explore objects and environments previously out of range.

before experience can enhance motor development. In addition, whereas in infancy opportunity for movement and exploration were sufficient for development, in early childhood, motor learning, which involves specific instruction and practice, is necessary for acquisition of certain motor abilities. Motor learning has a role in development of skills that require eye-hand coordination such as catching objects, sequencing a series of movements such as that required to tie a shoe, and combining movements in novel ways such as skipping.

Various regions of the brain such as the cortex and cerebellum are important for the motor changes seen during the early years of life. Maturation of cortical motor control centers provide a neurological substrate (basis) for such factors as strength; balance; motor planning or *praxis*, which is the ability to perform novel movements; the establishment of hand preference; and two kinds of coordination, visual-motor and bilateral (which is the coordinated interaction between both sides of the body [Fig. 10.7]). All

**Figure 10.7.** Playing with and manipulating small objects with both hands at the midline provides a foundation for self-care skills such as buttoning, zipping, and tying (see Fig. 10.12). Such early play activities reflect bimanual coordination and a level of neuromotor maturity which is not evident in infancy but that gradually emerges during childhood.

these factors, while affected by neural development, are also influenced by other ongoing, developmental changes. For example, the development of strength is also related to growth. Specialization of function in each cerebral hemisphere (lateralization) and more efficient transfer of sensory information between hemispheres contribute to the emergence of hand preference and bilateral coordination. The combined actions of the sensorimotor cortex and the cerebellum are essential for balance, coordination, and motor planning abilities. Since the cerebellum is also one of the last regions of the brain to myelinate after birth, developmental skills coordinated by this region take time to emerge. Thus, the ability to maintain balance when moving and to efficiently sequence and time a series of movements does not develop until early childhood.

Motor development is enhanced by play, which provides practice and elaboration upon basic motor patterns (Fig. 10.8). During this time of life, the child begins to pursue motor activities in which he or she is successful. These pursuits are determined by each child's specific temperament, motivation, innate abilities, and the experiences provided by adults and others in the environment. A 4-year-old child who may engage predominantly in gross motor activities will enjoy playing ball with older siblings. In contrast,

**Figure 10.8.** Motor development is enhanced by play. Younger children are challenged by the motor skills and abilities of older siblings and friends.

another 4-year-old child may prefer fine motor play and spend considerable amounts of time coloring, completing simple puzzles, and constructing with blocks. While both children would be expected to have some success throwing a ball and coloring with a crayon, the first child is apt to be advanced in throwing and the second child is apt to be advanced in grasp and manipulation. Practice is essential for refinement of basic motor patterns, and self-directed play is an important means of practice that not only enhances motor development but also contributes to differences in children's motor abilities.

The remainder of this section summarizes basic trends in gross motor and fine motor development between the ages of 2 and 6 years. Readers interested in a more in-depth account of the progression from immature to mature performance of basic motor patterns should consult Espenschade and Eckert (15) and Wickstrom (16).

# Development of Gross Motor Abilities in Ages Two to Six

The gross motor abilities that emerge or are refined during the early childhood period include walking and running, climbing up and down stairs, maintaining balance, jumping, hopping, and skipping.

## Walking and Running

The initial walking pattern that emerges toward the end of infancy is stiff-legged and characterized by lateral strides, excessive body sway, and arms that are flexed in a high-guard posture. By age 2, stride length increases and weight shift is in an anterior-posterior direction. In addition, a heel-toe pattern is demonstrated, and the arms rest at the sides. Reciprocal arm motion is incorporated into the walking pattern by age 3, and by age 4, stride length has further increased, with the parameters of the walking pattern similar to that of an adult.

Running incorporates the basic limb movements of walking, but the tempo is faster. Since there is a momentary period of free flight when both feet are off the ground, running requires greater strength and balance than walking. After the stiff-legged pattern used by the infant, an elementary pattern emerges between 3 and 4 years of age. Stride length and speed

Table 10.4
Gross Motor Developmental Sequence

|  |
|---|
| Advanced Ball Skills |
| Skipping |
| Early Ball Skills (Throwing, Catching, Kicking) |
| Jumping |
| Stair Climbing |
| Running |
| Walking |

increase at this time, and the free-flight phase is more pronounced. The elementary running pattern is also characterized by trunk rotation and arm swing. By 5 to 6 years of age, the mature running pattern develops. This is characterized by increases in knee motion, lengthening of stride, swinging the arms through a wider arc in opposition to leg movement, and a more vigorous arm backswing.

## Stair Climbing

Walking upstairs precedes walking downstairs. The latter is more difficult because the supporting leg must grade movement into flexion in order to step down. Initial climbing efforts are made with support, but by about 2 years of age, the child begins to climb upstairs without holding; by about 2¼ years the child is able to climb downstairs without support. This pattern initially involves both feet, which are brought to the same step. When the child begins to alternate feet, a railing is again used for support. The mature pattern of walking upstairs with alternating feet and without hand support develops at 3 years and is followed by walking downstairs about one-half year later.

## Balance

There are two primary measures of balance: (a) static balance, which is the ability to maintain a posture; and (b) dynamic balance, which is the ability to keep from falling while moving. Both of these are integral to coordinated movement. One-foot standing, which is a measure of static balance, goes through a regular progression. Between 15 and 24 months, the infant is able to momentarily lift one foot off the ground. At 2 years, one-foot standing balance can be held for just 1 second, whereas by 5 to 6 years, the child can maintain this posture with hands on hips and with minimal body sway for 10 seconds. One-foot standing is also performed with eyes closed, which is a more difficult task because of the absence of vision, an important modality for maintaining balance.

   Development of dynamic balance is assessed by having the child walk along a line or a 4-inch-wide wooden beam (Fig. 10.9). Between 18 and 24 months, an infant walks in the general direction of a line. At 4 years, the child begins to walk a line touching the heel of one foot to the toe of the other. This is referred to as *tandem walking*. At 5 to 6 years, the child tandem walks along a line with hands on hips and with minimal body sway.

   Walking across a balance beam also follows a developmental progression (Fig. 10.9). At 18 to 24 months, the infant takes steps with one foot on a balance beam. Between 2 and 3 years, the child first takes steps holding onto an adult's hand, then takes independent steps sideways, and finally is able to take a few independent steps forwards. By 5 to 6 years, the child is able to walk forwards, backwards, and tandem walk across a balance beam.

## Jumping and Hopping

Jumping requires not only trunk and leg strength to elevate the body off the ground but also dynamic balance to maintain body alignment and to

**Figure 10.9.** *Left* and *right*, Walking on a balance beam requires excellent balance and motor coordination of the arms, trunk, head, and legs. Balance on both feet occurs first, then tandem walking, and then balancing on one foot, a skill observed in older children. Visual, as well as proprioceptive, feedback help to maintain balance at every stage.

land without falling. At 2 years of age, children begin to jump in place. Between 2 and 3 years, children are able to jump down from a step. From 3 to 6 years, the child makes steady improvement in the ability to broad jump over a string. The initial pattern of broad jumping is characterized by vertical trunk alignment, no arm movement, and limited flexion of the legs upon landing. By 5 to 6 years, the child prepares to broad jump by swinging the arms backwards, and bending the trunk forward. Complete leg extension is demonstrated at take-off, and upon contact with the ground, weight is shifted forward to land in a crouched position.

Hopping on one foot requires greater strength and balance than jumping. To hop, the child must elevate the body off the ground with one leg and land on a narrow base of support. At 3½ to 4 years, children begin to hop in place on the preferred leg. Between 4 and 5 years, hopping is performed with either leg, and a series of hops are sequenced. By 6 years, children are able to hop along a line with hands on hips.

### Skipping

Skipping is the most difficult basic gross motor pattern as it requires motor planning abilities. Skipping is a combination of two separate movements: the step and the hop, which are reciprocally sequenced in a rhythmical pattern. A precursor to skipping is galloping, which is performed by 3- and 4-year-old children. To gallop, the child turns sideways and alternates leaping with the lead foot and stepping with the rear foot. A transitional pattern between galloping and skipping is a unilateral pattern in which one leg skips and the other steps. True skipping does not develop until at least 5 years of age, and by age 6 arm and leg movements are smoothly coordinated with the overall pattern.

## Ball Skills

The actions of throwing and catching can initially be performed with larger, than with smaller, balls. A standard developmental progression for both motor abilities has been demonstrated with a small ball such as a tennis ball.

### Throwing

Throwing a tennis ball proceeds from a gross total arm-flinging motion that is exhibited by an infant to a highly differentiated, well-controlled ability. The 3-year-old child still exhibits an immature pattern, standing perpendicular to the target with legs remaining stationary and throwing with a pushing motion and with the greatest movement occurring at the elbow. By 4 years, preparatory arm swing precedes the throwing motion, trunk rotation occurs from the throwing to the nonthrowing side, and weight is shifted forward over the front foot. Mature throwing follows and is characterized by diagonal arm motion, controlled weight shift to the rear leg, followed by a step forward and weight shift over the front leg.

### Catching

Catching requires eye-hand coordination to time arm movements to the flight of the ball. Children are first able to catch a large ball (10 inches in diameter) at about 2½ to 3 years. This precedes the ability to catch a tennis ball which occurs at 4 to 4½ years. While catching, the child initially extends the arms out in front of the body and, in a display of an avoidance response, turns the head away from the approaching ball. A ball is caught only if it lands in the hands, often much to the surprise of the child. At this time, arm adjustments to the flight of the ball are not observed. In the intermediate pattern, however, the arms are flexed and positioned in front of the body. Now, the eyes follow the ball and the hands open in anticipation. If the ball makes contact with the hands, it is clasped against the chest. The mature catching pattern begins with the arms relaxed at the sides and the elbows flexed. The arms are brought forward in synchrony with the flight of the ball and the hands cup together to catch the ball in front of the body.

### Kicking

Kicking a stationary 10-inch ball off the ground requires balance, coordination, and an interaction of visual-motor abilities to time the leg movements. While a child kicks, one leg supports body weight and maintains balance while the other leg makes contact with the ball. The 15- to 18-

month-old infant lacks sufficient balance to kick and therefore walks into the ball to propel it forward. The 2-year-old child stands with the arms resting at the sides or flexed in a high-guard position and swings the leg forward through a small arc to kick. In contrast, the 3-year-old child demonstrates an initial backswing and then swings the leg forward through a wider arc. Mature kicking incorporates reciprocal arm-leg movement, with the leg continuing to extend in follow-through after striking the ball (Fig. 10.10).

# Development of Fine Motor Abilities in Ages Two to Six

The fine motor abilities that occur during early childhood are significant. They are characterized by the appearance of hand preference, which then affects the grasp and manipulation of writing implements and scissors.

### Hand Preference

The vast majority of people are right-handed, and therefore, their left cerebral hemispheres are lateralized for fine motor control. As noted in previous chapters, the left side of the brain is also specialized for language, thereby optimizing the neural regulation of writing abilities. Heredity,

**Figure 10.10.** Mature kicking involves balance, coordination, reciprocal arm-leg movements, and follow-through. (Courtesy of the March of Dimes Birth Defects Foundation.)

cortical maturation, and experience have been proposed as factors that contribute to the establishment of hand preference. The hereditary basis is supported by the existence of a higher incidence of left-handedness in children who have a family member who is also left-handed. The interaction between early experience and inheritance in regard to handedness is suggested by the relationship between early postural attitudes and eventual hand choice. For example, infants who demonstrate a left- or right-side preference while in the ATNR posture (Fig. 9.5A) spend more time observing the hand on the side toward which the head is turned. That this may affect eventual handedness has been supported by Gesell's (17) report that a strong preference for the left ATNR posture was correlated with left-handedness. Although conclusions cannot be made from one study, these findings do suggest that preferential viewing of one hand may be an early manifestation of handedness.

Environmental factors can also influence hand preference. For example, educators used to believe that children should write with the right hand. As a result, children who were biologically left-handed were forced to learn how to write with the right hand. Fortunately, forced use of the right hand is no longer encouraged; however, because society is right-hand oriented, parents are often apt to encourage right-hand use, especially for holding a crayon.

The brain becomes progressively more specialized for fine motor control between 3 and 6 years of age. This is also the time when most children establish a hand preference for fine motor tasks. Not coincidentlly, by age 3, children spend more time using writing instruments and engaging in fine motor tasks that encourage hand choice. Interestingly, some people never develop a strong hand preference. For example, some people may write with the left hand and throw with the right, exhibiting mixed cerebral dominance for fine motor control.

Many of the fine motor accomplishments of early childhood require coordinated use of both hands, which is affected by hand preference. For example, to succeed in tasks such as unscrewing a jar, the preferred hand is used to perform the primary movement while the other hand serves as an assist or as a stabilizer. The ability to coordinate use of the preferred and assisting hands enhance fine motor planning and the acquisition of fine motor skills.

## Grasp and Manipulation of Writing Instruments

Pencil grasp and control improve through maturation and practice. The fisted crayon grasp and the whole-arm swiping movements of the 12-month-old infant and the pronated grasp and scribbling of the 18-month-old, which were previously described, provide a transition to the more functional static tripod grasp that develops between 3 and 4 years (Fig. 10.11). At this time, the child shifts from crayon to pencil grasp with the thumb opposed to the index and middle fingers and with movement directed from the forearm and wrist. The mature dynamic tripod grasp develops by 5 to 6 years of age. It is characterized by the distal phalanges of the thumb and first two fingers

**Figure 10.11.** The immature fisted grasp characteristic of infants gradually shifts to the tripod grasp which involves thumb opposition to the index and middle fingers. This grasp and hand preference both emerge by about 3 to 4 years of age. (Courtesy of the March of Dimes Birth Defects Foundation.)

in opposition to each other, the wrist slightly extended, and the pencil grasped distally with precise finger movements used to make strokes.

Writing is not mastered until middle childhood as it is a complex skill that is learned only through considerable instruction and practice. In addition to fine motor control, eye-hand coordination is a prerequisite for writing. Furthermore, writing requires the ability to discriminate letters and an understanding of the concept that letters are combined to form words. In early childhood, the ability to draw geometric figures follows a developmental progression that serves as an indicator of readiness to learn to write. Children first imitate a form after demonstration and then copy a picture of the form. A vertical line is imitated by a 2-year-old child, and within the same year a horizontal line can also be imitated. The 3-year-old child is able to copy a circle. The perception of both vertical and horizontal is integrated by the 4-year-old child, who can copy a cross and join vertical and horizontal lines at a right angle to form a square. The ability to reproduce a diagonal line develops during the 5th year as evidenced by the child's ability to copy a triangle, and then a diamond by the 6th year. In addition, by 5 to 6 years of age, visual-perceptual-motor coordination and cognition are sufficiently developed for children to learn how to print their name and copy simple words.

## Manipulation of Scissors

To cut a piece of paper with scissors, the preferred hand grasps the scissors with the thumb opposed to the index and middle fingers while the other hand holds the paper. Cutting out shapes requires manual dexterity, bilateral coordination, and eye-hand coordination. Because of the relative complexity of the task, instruction and practice are necessary to learn how to cut with scissors. The 3-year-old child unsuccessfully cuts by snipping the end of paper with alternating total hand flexion and extension movements. Through practice, however, the child learns to cut a piece of paper in half. Between 4 and 6 years, hand control is refined and the child is able to cut along a line as well as to cut out simple shapes such as a circle and square. The mature method of cutting with scissors involves precise thumb and index finger motions while the assisting hand effectively maneuvers the paper.

## Adaptive Fine Motor Abilities

During early childhood, improvement in grip strength, manual dexterity, bilateral coordination, and eye-hand coordination are responsible for a number of developmental accomplishments (Fig. 10.12). The child effectively learns to stabilize with one hand and to use the preferred hand to perform a number of tasks, including unscrewing a jar, activating a wind-up toy, and buttoning a shirt. Stringing beads, which emerges at 2 to 3 years, and tying shoes, which is an accomplishment of a 5- or 6-year-old child, are examples of tasks that require considerable bimanual coordination. During the latter part of early childhood, play with construction toys and puzzles reflects the child's expanding cognitive, perceptual, and motor competencies.

# MOTOR DEVELOPMENT IN LATE CHILDHOOD AND ADOLESCENCE

Motor development during late childhood (from approximately 7 to 12 years) and during the adolescence years involves refining primary motor patterns and combining movements in novel ways to learn motor skills. Motor skill refers to goal-directed movement performed with proficiency, which is obtained with specific practice. Motor skills are generally more complex than basic motor patterns, which when refined and performed in a goal-directed manner become skilled movements. Bruner (18) proposes that motor skill is achieved by reducing variability in how the movement is performed, increasing anticipation of the outcome of the movement, and improving the efficiency of the movement. For example, the 4-year-old child is able to throw and catch, but throwing accuracy is variable, the flight of the ball is not always anticipated when catching, and the ball is frequently dropped. The 4-year-old child has not yet attained a skilled level

**Figure 10.12.**  During childhood, manual dexterity, bilateral coordination, and eye-hand coordination are responsible for a number of developmental accomplishments and independent self-care tasks. Social interaction and play often facilitate motor skill acquisition.

of throwing and catching. In contrast, the 10-year-old child who throws and catches efficiently while playing baseball demonstrates skilled movement. Writing is another example of a task in which the basic motor pattern develops in early childhood but in which skilled performance is not attained until late childhood. Achievement of skilled writing involves refinement of grasp and coordination of hand movement to make smooth strokes, form smaller-size letters, reduce space between letters, and increase speed.

Late childhood is a period of great physical activity when games with rules emerge as a theme of motor play. Increased attention span, cognitive abilities, and cooperation allow meaningful participation in organized sports and recreation. Seefeldt (19) conceptualizes that the basic or fundamental gross motor patterns that develop in infancy and early childhood form the foundation for learning transitional motor skills, sport skills, and dances. Transitional skills include roller-skating, jump rope, and hopscotch; sport skills include gymnastics, baseball, and swimming, which are activities often performed in a group.

As chapters 11 and 12 will describe more fully, late childhood and adolescence are times when social interactions and cognitive skills expand.

While the family unit is the primary social influence during the early years, social and cognitive experiences widen with age. Thus, it is not purely coincidence that the motor experiences of late childhood and adolescence are organized around games that involve not only peer interaction but also rules that are often not fully understood by the younger child. In addition, motor abilities are socially significant. Children and adolescents often challenge each other in motor activities, as superiority in motor skills is a means of achieving peer status.

Motor learning is integral to development of skilled movement in later childhood and adolescence. Motivation, instruction, and practice have all been mentioned as factors critical to motor learning (Fig. 10.13). Fitt's (20) three-stage model is helpful in analyzing how motor skills are learned and how they interact closely with other developmental abilities. In stage 1, the *cognitive* stage, the child must first understand the requirements of the skill and then formulate a motor strategy to perform the necessary movements. Instruction, demonstration, and physical guidance accelerate this stage of learning. In stage 2, the *associative* stage, movement is practiced. This is a time when sensory and motor abilities are associated or coordinated. Thus, sensory input coupled with knowledge of success provide feedback that is used to modify movement until the desired pattern is learned. After the movement is performed correctly, efficiency is achieved through repetition. In stage 3, the *autonomous* stage, movement is refined

**Figure 10.13.** Motivation, instruction, and practice all affect skilled movement.

and skilled performance is attained. During this stage, a memory of the motor program is formed which allows the skill to be performed at will.

Learning motor skills is a continual process in childhood and adolescence. Children simultaneously practice a number of motor skills, and depending upon the nature and complexity of the skills, the youngster is frequently at different stages in the learning process. The research of Krus, Bruininks, and Robertson (21) suggests that two main dimensions underlie skilled motor performance in childhood and adolescence. One dimension is the combination of speed, strength, and precision; the other dimension is balance and coordination. The practice necessary to learn a motor skill, therefore, is likely to focus on improving ability in one or both of these two dimensions.

As the remaining chapters in this book will describe, childhood and adolescence are times when cognitive, motor, and social experiences are all closely coordinated. Infancy and early childhood involve maturity of the central nervous system and a corresponding development of the foundational skills, which are then refined in later childhood and adolescence. During these later stages of development, socialization with peers becomes increasingly more important, as the child's and adolescent's social worlds expand beyond the home to school, local community, and beyond. These experiences affect social relationships, cognition, and subsequent acquisition of skill. As the older child and adolescent become more independent of the family, peer groups take on increased importance. The motor skills that the youngster acquires and refines (e.g., playing the drums or guitar, participating in sports, driving vehicles) may be inconsistent with family values and desires. In addition, the physiological changes that also occur during puberty will affect muscle strength and motor performance. These physical and hormonal changes that are most apparent during the adolescent period highlight many of the sex differences in motor performance, cognition, and other abilities. The importance of the interrelationship of sensory-perceptual abilities (discussed in the previous two chapters) and motor achievements (discussed here) are important in socialization as well as a fundamental part of cognitive development, which will be described in detail in the next chapter.

## REFERENCES

1. Bayley N: *Bayley Scales of Infant Development.* New York, Psychological Corp, 1969.
2. Schulte F: Neurophysiological aspects of development. In *Biologic and Clinical Aspects of Brain Development.* Evansville, IL, Mead Johnson, 1975.
3. Springate JE: The neuroanatomic basis of early motor development: A review. *Behav Pediatr* 2:146–150, 1981.
4. McGraw MB: *The Neuromuscular Maturation of the Human Infant.* New York, Halfner, 1963.
5. Piaget J: *The Origins of Intelligence in Children.* New York, International University Press, 1952.
6. Goldberg S, Lewis M: Play behavior in the year-old infant: Early sex differences. *Child Dev* 40:21–31, 1969.
7. Knobloch H, Pasamanick B: Further observations of the behavioral development of Negro children. *J Genet Psychol* 83:137–157, 1953.

8. Bayley N: Comparisons of mental and motor test scores for ages 1–15 months by sex, birth order, race, geographical locations and education by parents. *Child Dev* 36:379–411, 1965.

9. Tower G: Selected developmental reflexes and reactions—A literature search. In Hopkins HL, Smith HD (eds): *Willard and Spackman's Occupational Therapy*, ed 6. Philadelphia, Lippincott, 1983.

10. Knobloch H, Stevents F, Malone AF: *Manual of Developmental Diagnosis*. Hagerstown, MD, Harper & Row, 1980.

11. Folio MR, Fewell RR: *Peabody Developmental Motor Scales and Activity Cards*. Dallas, DLM Teaching Resources, 1983.

12. Bruininks RH: *The Bruininks-Oseretsky Test of Motor Proficiency*. Circle Pines, MN, American Guidance Service, 1978.

13. Erhardt RP: *Developmental Hand Dysfunction—Theory, Assessment, Treatment*. Laurel, MD, RAMSCO Publishing Co, 1982.

14. Stockmeyer SA: A sensorimotor approach to treatment. In Pearson PH, Williams CE (eds): *Physical Therapy Services on the Developmental Disabilities*. Springfield, IL, CE Thomas, 1972.

15. Espenschade AS, Echert HM: *Motor Development*, ed 2. Columbus, OH, Charles E Merrill, 1980.

16. Wickstrom R: *Fundamental Motor Patterns*, ed 3. Philadelphia, Lea & Febiger, 1983.

17. Gesell A: The ontogenesis of infant behavior. In Carmichael L (ed): *Manual of Child Psychology*, ed 2. New York, John Wiley & Sons, 1954.

18. Bruner J: Organization of early skilled action. *Child Dev* 44:1–11, 1973.

19. Seefeldt V: Developmental motor patterns: Implications for elementary school physical education. In Nadeau CH, Halliwell WR, Newell KM, Roberts GC (eds): *Psychology of Motor Behavior and Sport, 1979*. Champaign, IL, Human Kinetics Publishing, 1980, p 314.

20. Fitts P: Perceptual-motor skill learning. In Melton AW (ed): *Categories of Human Learning*. New York, Academic Press, 1964.

21. Krus PH, Bruininks RH, Robertson G: Structure of motor abilities in children. *Percept Motor Skills* 52:119–129, 1981.

# 5

# COGNITIVE-INTELLECTUAL AND SOCIAL-EMOTIONAL DEVELOPMENT

# Section 5

Since we enter the world as helpless, dependent beings, our early emotions, thoughts, language, and personality take shape through ongoing interactions with others. All areas of development merge, and none can be fully understood without also exploring the social context. A fairly new area of research in the scientific study of development is known as social cognition. It has to do with people's awareness of themselves and others and their understanding of interpersonal relationships. This new research area illustrates the close interrelationship between the two domains covered in the fifth and final section of this text, cognitive and social development.

Chapter 11 looks at the diverse approaches to human knowledge and intelligence. Included is a description of Piaget's theory of intellectual development, which illustrates the connection between sensory, motor, and cognitive domains. In addition, the contemporary information processing approach to intelligence and other areas of cognitive science are explored.

Chapter 12 covers the many diverse topics that are a part of social behavior. Since social and emotional functions are considered partly dependent upon the attachment relationship, a historical perspective of attachment research is provided as a basis for understanding subsequent studies of social development. Some additional topics that will be examined in this chapter include the ontogeny of personality and the sense of self; the nature and nurture of temperament and language; and the numerous social factors, ranging from initial parental influences to school and peer forces, which affect the developing infant, child, and adolescent.

These many and varied topics have deliberately been left until the last chapter to illustrate how all developmental domains fit together and are interwoven within a social context. Thus, the final section of this book requires a familiarity with much of the descriptive and theoretical information already introduced in the text, since many different theories and approaches to behavior are necessary for understanding social and cognitive development. Consistent with the focus of the rest of the text, the final chapters emphasize an ecological-systems level of analysis that focuses on the interaction among the many developmental domains that make up the whole person.

# 11

# The Development of Understanding:
## Cognition and Intelligence

The study of cognitive development is the study of the acquisition of knowledge. While we cannot readily "see" the processes involved in thinking or acquiring knowledge, researchers in the field of cognitive development attempt to understand how thinking occurs and how it changes over time from infancy throughout childhood and adulthood. They do this in a variety of ways that will be examined in this chapter; e.g., they may identify fundamental mental units or structures of thought, or they may study the steps involved in mental problem solving. Theories are often then generated about how these mental units, structures, or processes change during life.

Compared with the study of sensory processing or motor behaviors discussed in the previous chapters, the study of cognition is difficult. Motor behaviors can be seen, they can be elicited and observed, and their causes can be explored by examining the anatomy and physiology of movement. Similarly, sensation can be explained in terms of the activation of a sense receptor and the relay of afferent sensory information to sensory processing areas in the brain. In contrast, cognition cannot (at least at this time) be traced to specific physical functions.

Understanding how individuals come to know about or gain knowledge of their worlds is difficult. Gaining knowledge involves such abstract and elusive processes as intelligence, reasoning, learning, problem solving, memory, and thinking. Each of these processes is not only difficult to study but also difficult to define. Thus, the study of cognition and intelligence, more than other developmental domains discussed so far, has centered around philosphical discussions of abstract phenomena, controversies regarding definitions of terms, and varied research approaches to systematic study in the field. The terms, "cognition," "intelligence," and "learning" are associated with different lines of research and traditions in the scientific study of human development (Fig. 11.1). In an attempt to illustrate these varied approaches, this chapter is divided into sections that will separately discuss learning, intelligence, and cognition. The differences between these areas should become apparent as they are discussed.

# LEARNING

Philosophical study regarding the nature of knowledge and ideas (epistemology) can be traced at least as far back as Aristotle. The 17th and 18th century British empiricists, characterized by John Locke, contended that knowledge is obtained by the acquisition of sensory impressions from the environment. The position of the early empiricists has been maintained by many contemporary learning theorists, who study how behavior is acquired by the accumulation of stimulus-response association. Such researchers do not focus on knowledge, thinking, or on mental activities and are not concerned with the nature of the human mind. What they are interested in are the products of experience (1) or changes in behavior as a result of experience, i.e., learning (see chapters 3 and 5).

Learning, according to its customary use in learning theories is the

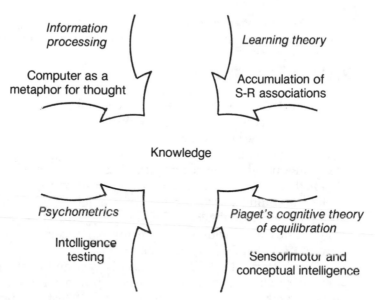

**Figure 11.1.** Diverse approaches and traditions regarding the acquisition of knowledge.

accumulation of stimulus-response (S-R) associations. Thinking, cognition, or mental processes are called *mediators* because they mediate or "go between" stimuli and responses. In conventional S-R models of learning, what is important is behavioral change, i.e., change in the "R" or response side of S-R models. Cognitive mediators are simply ignored. Thus, in the scientific study of development, the topic matter of learning theories (stimuli and responses) and of cognitive theories (what goes on between a stimulus and a response) is vastly different.

# WHAT IS INTELLIGENCE?

Historically, the concept of intelligence, like the term "learning," has evolved over the decades; along with the concept has come certain assumptions, specific to the field of psychology and quite different from its use in lay language. Sternberg (2) notes that intelligence is among the most elusive of concepts, invented by people so that they could evaluate and place other people in some kind of order based on their abilities to perform tasks that the culture deems important. Determining what those abilities are, however, has been a source of constant controversy.

Intelligence traditionally has been viewed as an innate ability or group of abilities that differentiate humans from other species. Thus, it has often been associated with human language, reasoning, and ways of processing information. In the field of psychology, intelligence is examined within the subfield of *psychometrics*, the measurement of psychological abilities, or, in this case, intelligence testing. Thus, to many, intelligence has come to be associated with some measures of performance on an intelligence test.

# Intelligence Testing

In the 1920s and 1930s, when many intelligence tests had been recently constructed, certain assumptions existed about the nature of intelligence: (a) that it was a homogeneous and measurable trait, (b) that it was innate (and therefore not environmentally alterable), and (c) that it reached a maximum by adolescence and stayed constant throughout life until senility. Given those assumptions, it was then also assumed that the scores from such tests could be used to reflect or predict the educational and vocational levels an individual would achieve in his or her lifetime. While these assumptions were widely held in the early part of this century, they are now contested and rarely accepted (3).

The history of intelligence tests is vast and controversial. These tests were originally designed to identify children with subnormal intellectual abilities and then later extended to adult tests designed to "weed out" recruits and assign jobs to members of the Armed Forces (3). As chapter 2 discusses, intelligence tests have been used for a variety of other controversial purposes, and what they assess is unclear. One assumption about what such tests assess is that there is one measurable, general factor of intelligence common to all people. This assumption, however, is not held by everyone, as some theorists claim that intelligence is made up of several distinct human abilities such as comprehension, reasoning, vocabulary, memory, imagination, and other factors. The numbers of these factors may range from a few, such as those proposed by Thurstone and listed in Table 11.1, up to 150 factors, as proposed by Guilford (2).

## Uses and Abuses of IQ Scores

An individual's performance on an intelligence test is often converted to a score that enables educators and psychologists to measure individuals relative to others in their own age group. Initially an intelligence quotient, or IQ score, was determined by taking the highest age-related test score the child received, which is called the mental age (MA) score, dividing it by the child's chronological age (CA) and multiplying by 100:

$$IQ = \frac{MA}{CA} \times 100$$

Problems, however, arise with this conversion. Mental age increases steadily up to about age 15 to the middle 20s, and then the scores begin to flatten out when adult intellectual abilities are gradually achieved. Thus, using the above formula, in which MA stays steady and chronological age continues to increase each year, gives the impression that IQ declines with age (3). As a result of this problem, contemporary tests use a variety of conversions such as standard scores to compare individuals based on normative

Table 11.1
Thurstone's Primary Factors of Intelligence

| |
|---|
| Verbal comprehension |
| Word fluency |
| Number |
| Spatial abilities |
| Memory |
| Reasoning |
| Perceptual speed |

data obtained for specific age categories. Standard scores have a mean and standard deviation (see chapter 4) and therefore enable one individual's performance to be compared to another's performance or to the mean of the standardization sample.

The overuse and abuse of IQ scores is an additional problem (see chapter 2). Given the concern that intelligence cannot readily be defined or that it reflects a conglomeration of abilities, using one score to reflect these capacities yields a much too narrow interpretation of intelligence. Thus, it may be too simplistic to use single scores to label testees and to place them into superficial hierarchies that are then used for educational placements or for broad generalizations about the testees' abilities. Some additional concern exists regarding the failure of IQ tests to assess such factors as creativity, imagination, and other important (but elusive) skills and abilities.

## Intelligence, Academic Performance, and Culture

IQ tests were originally assumed to measure some innate human ability that is stable over life (see chapter 2), but recent views of intelligence recognize that performance on IQ tests can be affected by a variety of environmental factors such as one's level of motivation, practice, educational experiences (and therefore learning of specific facts and bits of information), and culture. One recent and controversial issue revolves around the concern that IQ tests are created by and for white middle-class individuals and that these tests therefore assess information deemed important or valuable to only one section of society. This concern about the cultural bias of intelligence tests led to the elimination of intelligence testing in the 1960s in the New York City Public school systems (4) and to the creation of various tests that are designed for specific cultural groups or that attempt to be "culture fair." However, given the fact that motivation to perform on such tests may also be affected by cultural attitudes toward test-taking, teachers, and examiners, culture-fair assessment remains a challenge.

Many intelligence test questions are formed within a language framework, thus for children who do not possess verbal skills because of unfamiliarity with a language or because of reading problems, intellectual performance may be judged much lower than their real adaptive abilities indicate. In addition, for the urban minority child, adaptive capacities may mean the acquisition of an aggressive, "street smart" ability to cope. Such skills are rarely valued or assessed on educational, intelligence tests, and those items that are assessed may be of little interest to such children. Further evidence for cultural biases in intelligence testing is found in recent reports of exceptional performance by Asian-American students in educational programs and on IQ and scholastic aptitude tests. Proposed causes for these differences have, once again, resurrected the nature-nurture debate (chapter 2). Thus, some have proposed that Asian-American students are actually genetically superior to other racial groups, whereas others attribute this exceptional performance to a cultural emphasis on academic motivation among Asian-American families and to high attendance rates in school and attention to academic information (5).

These differences in academic performance of various cultural groups in the United States point to fundamental issues that go beyond concern about IQ tests and extend to the educational system in general. Meanwhile, although problems with IQ tests are recognized, they still provide a convenient, standardized method that enables educators to evaluate and to place children in various academic programs. Such tests will probably continue to be widely used until equally convenient and widespread alternative test methods are designed.

# Intelligence and Environment

The extent to which intelligence is a reflection of genetic or environmental factors is an issue that continues to resurface in regard to intellectual development. As chapter 2 discusses, this issue is socially and politically charged.

### Environmental Deprivation and Early Intervention

Studies have shown that environment can affect intellectual development, both positively and negatively. For example, children institutionalized or reared in conditions of severe deprivation were found to display severe mental retardation, but once some of these children were removed from low-stimulation environments and placed in more enriched conditions, they made subsequent gains in IQ scores (6–8).

As chapter 9 introduced, one approach in the field of early intervention is aimed at remediating cognitive deficits in children whose environments have not been as enriched as those of middle- and upper-class children. Preschool programs such as Head Start (chapter 2) were designed to involve parents and to give deprived children an extra boost with health-related, nutritional, and educational (intellectual) stimulation before entrance into the regular elementary school settings. Such programs were designed to assist culturally deprived children and their families and comprised a very diverse group of individuals, thus goals varied from program to program.

Despite the fact that the goals of these programs were so variable, initial assessments tended to focus on the academic achievements of the participants (9) and tended to use conventional IQ tests such as the Stanford Binet Intelligence Test (10). As chapter 2 points out, many of these initial assessments of the preschool programs were disappointing. Laosa (10) explains that one of the reasons for the initial perception that these programs failed is that the selection of IQ tests as a measurement instrument does not do justice to the diverse nature and goals of the Head Start programs. Despite these initial bad reports, subsequent follow-up studies indicated that important gains were achieved by many of the participants (11–13).

Bronfenbrenner's (12) review of some of these programs revealed that children do show substantial IQ gains but that these gains are not maintained over time. Eventually, the children begin to progressively decline so that by the time they are involved in elementary school, they fall back into problem categories again. Follow-Through programs were therefore designed to provide continued remedial assistance for children whose home and cultural environments were insufficiently stimulating. Bronfenbrenner reported that the children who least benefited from cognitive stimulation programs were those from the most deprived environments. His findings support his ecological approach, which takes into consideration children's environment and long-range opportunities for continued stimulation from the home and nearby community (12).

An interesting consequence of the early Head Start program and other preschool programs is that follow-up studies looked at many additional variables other than academic measures. These studies are optimistic about early intervention effects and support Laosa's (10) assertions regarding the varied approaches and program goals of Head Start and other preschool programs for the economically disadvantaged (14, 15). Brown has summarized some of these findings, which are listed in Table 11.2. It is interesting to note that many of the positive consequences are not solely academic but also include social and emotional effects on the children as well as parental effects. In addition, Lazar (15, p. 305) has noted that no special curriculum approach was more successful than another: "Our findings suggest that the hunt for 'the' best curriculum is a futile, or at least, a commercial search. Any reasonably designed, age-appropriate set of learning goals can be

**Table 11.2**
**Some Long-Range Gains Associated with Participation in Preschool Programs for Low-Income Children**[a]

Reduced number of children assigned to special education classes.
Fewer children have to repeat a grade.
Short-term academic gains.
Reduced delinquency and pregnancy leading to improved likelihood of employment.
Higher vocational aspirations for their children reported by participants' mothers.
Achievement-related reasons given by students for being proud of themselves.

[a] Adapted from Brown B: Head Start. How research changed public policy. *Young Children* 40:9–13, 1985.

achieved in a number of ways. There are many roads to Allah." Lazar has summarized a cluster of five interrelated characteristics of preschool programs that are associated with positive outcomes for the children (Table 11.3), and this is his conclusion:

> It is my belief that the basic reason these early programs had such long-lasting effects is not curricular, but rather is a result of changes in the parents' values and anticipations for their children. A single shot of preschool seems hardly enough to produce a life-long change. I believe that the increased participation of parents provided the value change that led them to encourage and reward their children's learning activities. (15, p. 305)

Low socioeconomic status is found to be consistently related to mental deficits and to other potential developmental problems in children (KE Diamond, unpublished data; 16–18). It seems a reasonable conclusion that poverty of environment can lead to poverty of stimulation that in turn may contribute to poor intellectual performance. Moynihan (19) notes that despite programs aimed at reducing the problem childhood poverty continues to be a growing concern in the United States in the 1980s. The effects of poverty are wide-ranging and include not only the intellectual climate of the child but also the entire family structure (19). Thus, the programs that address the child and family and that provide opportunities for parent involvement are those that take into consideration the long-range ecological needs of the child and that offer cognitive and other forms of stimulation for the child and siblings (12).

## Nature and Nurture of IQ

As chapter 2 has already discussed, conclusions regarding many of the early preschool programs such as Head Start were used to either support or refute assertions that IQ is primarily genetically based. Researchers such as Jensen (21), Eyseneck (22) and others used data from twin studies as evidence to contend that intelligence and other traits are primarily based on inheritance. These studies revealed that the IQs of genetically identical twins, even when they are reared apart, are closer than are the IQs of fraternal twins

**Table 11.3**
**Characteristics of Effective Preschool Programs**[a]

Age of intervention:
  The earlier the entrance to the program, the better.
Adult-child ratio:
  The fewer children to each adult, the better.
Home visits:
  The more often visits are made, the better.
Parent participation:
  The more direct participation, the better.
Ecological approach:
  The more services for the family, rather than just the child, the better.

[a] Adapted from Lazar I: Early intervention is effective. *Educ Leadership* 38: 303–305, 1981.

reared in the same home environment. On the other hand, supporters of the nurture side of the issue—i.e., those who support the environmental basis of development—point out flaws in the twin studies. Even though identical twins may be separated at birth, most still tend to be placed in similar environments. In a reanalysis of some of the twin study data, Bronfenbrenner (23) found that the IQs of separated twins who were reared in the same community are more closely related than the IQs of twins reared in different towns. Similar findings apply to twins attending the same versus different schools and to twins residing in communities with similar versus different characteristics.

The nature and nurture of intelligence continues to be examined, yet as discussed in chapter 2, the relative influence of genetics and environment on IQ is undetermined and clearly variable. It is an area of continued investigation despite claims that trying to separate out the influence of environment and genetics on development is like trying to determine the separate contributions of length and width in an oil painting. They are both important and inextricable. Genetic factors certainly play a substantial role in determining individual differences in mental abilities, but there is no single fixed figure that determines how much either genetics or environment contributes. Genetics may set a range, and environment helps to determine the extent to which an individual's genetic potential is reached (24). The degree to which intellectual abilities are "fixed" in early life and keep stable throughout development is discussed further under "Intelligence and Continuity".

## Birth Order and Intelligence

Other lines of research investigated the effects of numerous additional environmental conditions on children's intellectual development, e.g., family size, spacing between siblings, birth order, and sex of siblings. Work by Zajonc and colleagues (25) indicates that the smaller the family and the earlier a child is born in relation to siblings (i.e., firstborn vs. subsequent birth status), the greater will be his or her intellectual potential. Studies of birth order effects have consistently reported that firstborn children tend to have higher IQs, tend to be more highly represented in college and in graduate school, and tend to occupy positions of eminence more than later-born siblings (25–27). The theory behind birth order and family size effects assumes that the fewer siblings in the family, the greater the opportunity for intellectual stimulation and for interaction with parents. The firstborn child therefore has an opportunity for one-to-one interaction with parents as well as chances to educate younger siblings. In contrast, additional siblings in the family, especially if they are spaced close together, have an intellectual environment that is "diluted" by the presence of brothers and sisters who may be less stimulating than older siblings or parents. Thus it appears to be to a child's advantage, if he or she is to have older siblings, that they be spaced widely apart (26).

Research on the influence of space and birth order on intelligence is controversial and difficult to examine. Henderson (26) notes that spacing seems to interact with the sex of the child and is also complicated by the numerous variables that are involved. In a family with only two children, eight possible types of sibling structures exist: four for a male child and four for a female, each of whom may have a younger or older brother, and a younger or older sister. With three children, 24 types of family combinations may result, and this does not take into consideration other factors such as socioeconomic status, presence of both parents, the quality of interactions in the family, or the age of family members.

In separate reviews of studies examining birth-order effects, Schooler (28) and Henderson (26) point out that this area consists of controversial and inconclusive research. Henderson points out that many birth order effects, such as the greater representation of firstborns in college, are confounded by other factors such as socioeconomic status, the fact that affluent families tend to have fewer offspring, that smaller families have greater resources, and that more affluent families are more likely to send their children to college. What evidence exists regarding spacing of children, Henderson also points out, is far from conclusive. It appears that in regard to positive effects on cognitive measures, spacing of siblings has different effects on males than on females. For males with brothers, the larger the space between the firstborn and second-born male, the greater the advantage for the second-born male. For females with sisters, smaller spacing seems to be more favorable for the second-born female, indicating that for cognitive development for females, the best situation is to have an older sister who is close in age (26).

While these findings about family size and birth order effects on intelligence are interesting, work in other areas of development suggest caution until continued investigation confirms them. In the meantime, like other areas of human development, this one must be examined in regard to the many variables that can affect the numerous aspects of the child's development and his or her potential for all forms of stimulation, not just intellectual. This will depend upon the various environmental opportunities available and the quality of the interactions to which children are exposed in their homes, schools, and communities.

# Critical Periods in Intellectual Development

Many remedial programs such as Head Start and other early intervention programs have been based, in part, on the assumption that a brief intervention during early development might be sufficient to compensate for any negative consequences of deprivation. In regard to intellectual and other developmental domains, some camps believe that there is a critical period during which stimulation can have permanent compensatory benefits and, on the negative side, a critical period during which deprivation can cause permanent losses. Thus, some researchers believe that there are optimal periods in early childhood for receiving specific kinds of cognitive stimulation and that young children are able to learn certain kinds of intellectual

**Figure 11.2.** Cognitive development involves imagination and other processes difficult to measure and to define.

skills and are able to handle subject matter more efficiently than adults. An additional assumption, therefore, is that the child who fails to learn age-appropriate information will forever be handicapped in acquiring those skills (4).

As discussed previously in this text, the concepts of critical periods and readiness are controversial. While some believe in the existence of optimal periods for intellectual stimulation, others such as Ausubel and Sullivan (4, p. 592) point out, "It is quite one thing to claim that lack of adequate intellectual stimulation in the preschool years may stunt later intellectual ability and quite another to make assertions such as those stating that critical periods exist." These researchers claim that optimal readiness has never been empirically demonstrated in regard to acquiring specific kinds of intellectual abilities at particular age periods during development. For example, in several longitudinal studies children who were evaluated as severely mentally retarded, early in their development subsequently gained in intellectual status after a period of exposure to normal levels or compensatory forms of stimulation (6, 7, 29).

A more probable explanation in regard to critical periods in intellectual development is that many of the consistently observed cognitive deficits associated with early deprivation occur because of the cumulative nature of learning in our educational system (4). Thus, it may not be accurate that if children do not acquire certain forms of information, they will never be able to obtain it. What is probably more likely is that continual deprivation can lead to accumulated losses. Thus, the child who does not acquire adequate reading skills will be handicapped at analyzing literature and poetry, and the child who does not learn basic math will be handicapped when required to apply that information in subsequent algebra and geometry classes. Similarly, children who do not pay attention in class because of hunger, motivational problems, or because of emotional concerns and distractions regarding their families may incur progressive academic losses (30).

A view of intellectual development that is more optimistic than the critical period concept claims that the same degree of cognitive capacity that establishes readiness at earlier ages is probably still present at least in equal degree at some future date during development (4). As a personal example of this: as an occupational therapy consultant for our county public health department, I have been visiting a woman who, since she was a young child, has had to work and has therefore never had the opportunity to learn to read. Now, because of doctor's orders to discontinue work, she is taking the time to do some of the things she has always wanted to do. So, at 80 years of age, she is learning to read. She is, as the saying goes, having the time of her life!

# Intelligence and Continuity

The degree to which behavior is stable or discontinuous throughout life is a theme often discussed in the scientific study of development (31). Learning theory approaches to development tend to reflect development as a gradual, continuous process characterized by progressive increases in S-R associations. In contrast, stage theories view development as a sequential progression from one level to another, in which qualitatively different kinds of behaviors and capacities emerge.

Measuring the stability of intelligence is difficult because no one universally accepted gauge exists. Using intelligence tests poses problems because intellectual growth curves vary so much across the life span. Intelligence seems to grow rapidly in infancy and early childhood and then begins to level off. Thus, looking at a 3-month change in intellectual growth for a 3-year-old child is not comparable to a 3-month change for a 6-year-old or a 12-year-old youngster (4). Furthermore, studies of intelligence across life reveal different trends for different skills in adulthood and in infancy.

## Infant Intelligence

In general, the relationship between infant and adult intelligence is unpredictable. Thus, a child who scores well on an infant assessment will not necessarily score equally well on a childhood or adulthood IQ test. The most consistent prediction appears to be for the infant who scores poorly. This child is more likely, than the child who does well, to score similarly on later tests of intelligence. With age, however, predictions of adult intellectual abilities based on childhood performance gradually increase.

One proposed reason for the meager relationship between infant and adult intelligence tests is that the early measures assess different abilities than do adult scales. For example, a common infant "IQ test," the Bayley Scale of Mental and Motor Development, examines infant adaptive responses such as attention and responsiveness to the environment, not verbal reasoning and problem-solving capacities, which are examined with many adult IQ tests. At present, test results and theories such as Piaget's suggest that what are considered "intellectual abilities" or adaptive capacities for infants are different from those of adults. As psychologist Kagan explains:

> One reason for the weak support for continuity from infancy to later childhood is that the behaviors displayed during the first year or two are particularly suited to the problems of that developmental era and not to those of later childhood. It would be maladaptive for a child to retain for too long a time the hierarchical organization of cognitive and behavioral structures that are characteristic of the first year. And nature will not permit him that luxury, for maturational forces replace old behaviors with new abilities that permit, indeed demand, a different interaction with the world (32, p. 195).

At issue is whether or not there is some general, measurable factor of intelligence that is continuous throughout life and that can be effectively assessed at all the different developmental stages (16, 17). Fagan (33, p. 7) claims that continuity does exist. He notes that certain infant perceptual responses, such as visual attention to stimulus novelty, are related to later performance on intelligence tests and that "the bases of both continuity and the general factor in intelligence reside in the same small set of basic cognitive processes."

## Canalization

The continuity/discontinuity issue relates partly to the nature and nurture of intelligence. Proponents of the nature side of the debate might argue that IQ is a reflection of a stable, genetic, maturationally directed pattern and is therefore continuous throughout development. McCall (17) notes that intelligence may actually be under different degrees of genetic pressure during early versus later stages of development. He claims that mental development is strongly "canalized" early in life and that it becomes progressively less so after infancy.

*Canalization*, a concept from the field of evolution, refers to a restricted developmental course (like a canal) that is followed by all members of a species. This developmental course is called a "creod," which is assumed to provide the genetic (phenotypic) limits for a species, *if the appropriate environmental circumstances are present*. The canalization concept applied to intelligence implies that a general range of "intelligent" behavior exists for members of our species and that we will operate within that range, provided that appropriate environmental circumstances exist. Individuals exposed to atypical environments may actually return to their developmental creod once they are removed from the abnormal circumstances and experience species-appropriate stimulation. McCall (17) claims that this developmental path (creod) is more firmly established in infancy, and therefore smaller deviations in behavior are to be expected. With time, however, the creod broadens, and along with it is the display of wide individual differences and diversity in childhood.

## Zone of Proximal Development

Another approach to intellectual development has been proposed by Russian psychologist Vygotsky (34) who distinguishes between the child's actual and potential levels of development. The actual level is the one achieved by the child on tests, whereas the potential level is the one the child might achieve if the child were to receive help. This latter level is quite within the child's reach and therefore is a reflection of potential development. Vygotsky labeled the distance between this potential level and the child's actual level of performance the *zone of proximal development* (Fig. 11.3). He noted that this zone can often be bridged by the child during play activities, where help is received by more capable peers or during certain learning situations with teachers. The width of this zone varies for every child. Thus, with help, one preschool child may be able to perform a concrete operational task, whereas another may find that the reasoning is just too difficult. For children with learning problems and with mental retardation, this zone of potential development is quite broad, and therefore the child may need consistent and repeated help to bridge it. For a realistic appraisal of the child's intellectual ability, Vygotsky's theory points out, the child's potential, or zone of proximal development must be taken into consideration.

## Developmental Quotients

There are practical considerations for determining what measures can be used to predict whether a child may experience problems or display great potential in intellectual abilities. Accurate early assessment of potential deficits might help to locate the child who is at risk for later problems and

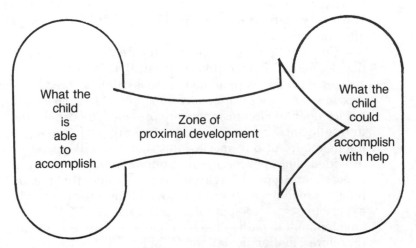

What the child is able to accomplish

Zone of proximal development

What the child could accomplish with help

**Figure 11.3.** Vygotsky's zone of proximal development.

to provide a rationale for early intervention. Most developmental therapists, if not trained to administer infant and childhood intelligence tests, still examine a variety of abilities that give information about the developmental capacity of the child (Table 11.4). Thus, instead of an intelligence quotient, a *developmental quotient* (DQ) is obtained. A DQ is a ratio like IQ but is based on the age level of performance the child attains on some test of developmental abilities, e.g., motor performance. Like IQ scores, DQ scores enable therapists to gauge how the child is performing on certain tasks relative to other children of the same age. Also like IQ scores, DQ's may be computed using the following formula:

$$DQ = \frac{Motor\ Age\ Level}{Chronological\ Age} \times 100$$

or they are based on standard scores, which allow therapists to compare the child's performance against a standardization sample.

The Peabody Developmental Motor Scales (PDMS) assess gross and fine motor abilities and enable therapists to determine whether a child is delayed in these areas. Scores from the subscales of this test can be used in conjunction with tests from other professionals to give an overall profile of the adaptive skills of an infant or child. The scores can also be used to validate the findings from other tests. For example, scores from the Fine Motor Scale on the PDMS tend to relate to scores on the Mental Scales from the Bayley Scales of Infant Development (35). Given that intelligence is difficult to define and to assess, the more information obtained about a youngster's overall capacities, the more confident the therapist will feel about making assumptions regarding the youngster's potential for subsequent development or for the need for referral for treatment.

**Table 11.4**
**General Developmental Areas[a]**

Gross motor skills:
  Postural reactions, balance, sitting, standing, creeping, walking
Fine motor skills:
  Use of hands and fingers in prehension, approach, grasp, manipulation and release of objects
Personal social:
  Reactions to people and culture in which child lives
Language:
  Facial expressions, gestures, postural movements, verbalizations, mimicking, and comprehending others in communication
Adaptive behaviors:
  Organizing stimuli, perceiving relationships, coordinating eyes and hands in reaching and manipulation, using motor equipment to solve practical problems
Functional applications:
  Using sensorimotor behaviors in play and in daily living skills such as dressing, toileting, and other goal-directed behaviors

[a] Modified from Knobloch H, Pasamanick B: *Gesell and Armatruda's Developmental Diagnosis*, ed. 3. New York, Harper & Row, 1974.

### Intelligence and Development

Performance on IQ tests tends to gradually increase throughout infancy and childhood until around adolescence and young adulthood, when it begins to level off as it reaches adult status. Different studies and theories exist, however, in regard to the nature of intellectual changes from late adolescence onward. Cross-sectional studies (see chapter 4) have led to a pessimistic view of intellectual changes over time. Results of these studies indicate that IQ scores gradually decline beginning at about age 20. While part of the reason for these findings may be due to some real changes in intellectual abilities, other reasons such as alterations in reaction times and coordination necessary to perform some IQ tests, have also been proposed. In addition, it is possible that older-age groups tend to be less well educated and less familiar with testing settings and test-taking than are younger individuals. Longitudinal studies, which tend to be better than cross-sectional studies at measuring behavioral continuity, indicate a much more optimistic view of aging and development. These studies point to continued increases, or at least stability of intelligence, across the life span. However, as Cattell (36) points out, two kinds of intelligence may exist, and each may be affected differently across life.

### Fluid and Crystallized Intelligence

One form of intelligence proposed by Horn and Cattell (37) is called *fluid intelligence*. This is a basic form of intelligence that involves logical reasoning and problem solving. It relies partly on physiological efficiency as it involves speed of thinking, reaction times (as in speed of recall of a list of numbers), and memory. In contrast, *crystallized intelligence* reflects one's learning experiences, education, and stored information.

Horn and Cattell (37) proposed that over life, fluid intelligence reaches a peak in adolescence and then gradually diminishes. Thus, like the notion of optimal readiness, some forms of intellectual ability are considered more "fluid" or more accessible early in life and become increasingly less accessible and usable over time. In contrast, since learning continues throughout life, crystallized intelligence is thought to continue to increase and, in some cases, make up for the declines in fluid intelligence. These predictions and categories of Cattell and Horn are not universally accepted, and, as will be examined shortly, there are other theories and views of human intelligence and cognition. Other longitudinal studies that have pursued Cattell and Horn's work indicate that intellectual abilities do not progressively decline but continue through life (38).

### Gender Differences

As adult levels of intelligence are gradually reached in adolescence, certain differences emerge in performance between males and females. Although intelligence tests such as the Stanford Binet Intelligence Test and the Weschler Intelligence Scale for Children have been constructed to eliminate sex differences, certain sex-related trends are sometimes found. In some cases, these have tended to favor males over females, although diverse individual differences exist. In tests of specific cognitive abilities, gender differences emerge as well. Females tend to perform better than males on language-related tasks such as word fluency and memory; whereas males tend to perform better than females on tasks requiring analysis of spatial relationships, quantitative reasoning, and problem solving (4, 39). The reasons for these findings are debatable and range from propositions of innate intellectual differences to obvious differences in sociocultural expectations and training for males and females. These and other cognitive and social gender differences and their potential causes will be examined in greater detail in chapter 12.

# Contemporary Views of Intelligence

Over the years, the field of psychometrics has expanded, and with it so has the concept of intelligence. With recent ecological approaches to human capabilities, simple IQ scores have been considered too narrow. New views have tended to look at human intelligence in terms of the relationship an individual has with, and the effectiveness of that individual on, his or her environment. Thus, in a broad sense, intelligence is the ability to adapt to or alter one's environment through the application of knowledge one obtains during life. Such a broad view does not rely solely on a single score or intelligence quotient and therefore has the potential to better predict cognitive continuity across life. Obtaining adequate measures of such a broad view of intelligence, however, poses a challenge that has yet to be met.

## Theory of Multiple Intelligences

Psychologist Howard Gardner (40) claims that human intelligence is actually composed of, not one component, but a whole set of discrete competencies or "frames of mind." His proposal, which is still incompletely formed, is based on the notion (which he points out is not new) "that there exist at least some intelligences, that these are relatively independent of one another, and that they can be fashioned and combined in a multiplicity of adaptive ways by individuals and cultures" (40). The different intelligences recognized by Gardner are included in Table 11.5.

### Triarchic Theory

Criticizing traditional views of intelligence as narrow and unrelated to the "real world," Sternberg (2) has developed a contemporary approach to human intellect that is quite different from Gardner's multiple intelligences. Sternberg recognizes that intelligent performance is affected by cognitive and motivational and affective functions; he defines intelligence as purposive adaptation to, and selecting and shaping of, real-world environments that are relevant to one's life. Thus, intelligence must be viewed as flexible and changeable over time, as the very same behavior that may be intelligent in one situation may not be intelligent in another (2).

**Table 11.5**
**Gardner's Theory of the Different Frames of Mind**[a]

Linguistic Intelligence:
  Understanding and using language systems (exemplified by the poet)
Musical Intelligence:
  Independent of physical objects, relating to an oral-auditory system
Logical-Mathematical:
  Familiarity with and use of abstractions and patterns of objects in arrays
    (different from the logic in music and linguistics)
Spatial:
  Tied to the concrete world, ability to transform objects within the environ-
    ment and to relate to objects in space
Bodily-Kinesthetic:
  Ability to use one's body and physical actions in interaction with objects in
    the world
Personal:
  Knowledge of one's self and other people, access to one's own and to
    other's feelings, motivations

[a] Adapted from Gardner H: *Frames of Mind. The Theory of Multiple Intelligences.* New York, Basic Books, 1983.

Sternberg (41) notes that various approaches have typically located intelligence in one of three domains: within the individual, within the environment, or as an interaction between the individual and environment. Within each of these are different aspects of intelligence (Table 11.6). Sternberg (2) has proposed a triarchic theory, consisting of three subtheories which recognize that intelligence should be understood in terms of the external world, the internal world, and the interrelationshp between them. Sternberg's theory reflects contemporary developmental perspectives and incorporates some elements of the ecological-systems approaches already introduced in this book. One potential criticism of this approach is that it is too broad and is therefore difficult to assess. Regardless, Sternberg's approach to intelligence has important applications to physical therapy and occupational therapy. For example, included as one aspect of intelligence is the ability to perform daily living skills. In addition, the triarchic view and clinical fields share a mutual concern regarding humans' abilities to effectively use personal capabilities and to make environmental modifications to alter individual-environment relationships.

# COGNITIVE SCIENCE

Webster's dictionary defines cognition as "the process of knowing" (42), and intelligence is the application or use of that knowledge. In the scientific study of human development, however, the term "cognition" is defined more specifically. It has its own history and specific theories, which will be explored in detail here.

Piaget's work has dominated cognitive theory for many years, but recently the field has taken a variety of turns. While Piaget was a pioneer,

Table 11.6
Locations of Intelligence[a]

---

*In the Individual*
  Biology
    Across organisms (evolution)
    Within organisms (genetics) and their interactions
  Cognition
    Knowledge
    Cognitive processes (attention, learning, reasoning, problem solving, decision making)
    Metacognition and their interactions
  Motivation
    Level of energy
    Direction of energy and their interaction
  Behavior
    Academic
    Social (within and between people)
    Practical (occupational, everyday living) and their interactions
  Biological-Cognitive-Motivational-Behavioral Interactions
*In the Environment*
  Culture/Society
    Demands
    Values
    Niche within culture/society and their interactions
*Individual-Environment Interactions*

---

[a] Modified from Sternberg RJ: A framework for understanding conceptions of intelligence. In Sternberg RJ, Detterman DK (eds): *What Is Intelligence? Contemporary Viewpoints on Its Nature and Definition.* Norwood, NY, Ablex Publishing, 1986.

his work has been expanded and revised by numerous researchers. Many scientists have been exploring the nature of thought and have moved beyond Piagetian approaches into many different realms of investigation, collectively called "cognitive science." Psychologist Howard Gardner (43, p. 6) explains that the emerging field of cognitive science is a "contemporary, empirically based effort to answer long-standing epistemological questions—particularly those concerned with the nature of knowledge, its components, its sources, its development, and its deployment. Though the term cognitive science is sometimes extended to include all forms of knowledge,—animate as well as inanimate, human as well as nonhuman—I apply the term chiefly to efforts to explain human knowledge". Cognitive science is a collective field of study that covers all the topical material in this chapter.

# Structuralism

As this text discussed in chapter 2, the nature-nurture debate provides a fundamental basis for understanding the field of human development. That debate, in regard to the nature of sensory experiences or the nature of human knowledge was dichotomized by the empiricists and the nativists.

In contrast to the *tabula rasa* of the empiricists, the nativists believed that humans are born with certain intrinsic abilities that exist independently of experience. Those abilities in regard to cognition exist as certain fundamental structures of knowledge that are either present at birth or emerge through normal maturational processes. The early investigation of these inherited mental structures led to what is today a general developmental approach called *structuralism*. This school of thought is not specific to cognition but includes a number of theories of development that share certain common themes. One of these themes is the opposition to strict mechanistic and empirical approaches to development characteristic of learning theory. Another theme is an emphasis on the existence at birth of innate capacities (or structures) that are foundational for subsequent development.

Learning theory, which is not structural, explains development as a gradual process involving the accumulation of stimulus-response associations. Thus, development, according to learning theory, is a quantitative process as the child acquires more and more associations. In comparison, structuralists often see the child as progressing through qualitatively different developmental stages as different structures become available for use. For example, intellectual development is seen as a process of restructuring, not just accumulating, knowledge and skills. As different kinds of mental structures are constructed, each new way of thinking enables the child to more capably and intelligently deal with the environment (1). Such theorists claim that understanding how children think and approach tasks and understanding what kind of mental structures children use proves valuable for teachers, parents, and therapists who want to assist children in maximizing their intellectual capacities.

Structuralism is common to many theories. Ethologists, who point to the presence of certain instinctual abilities and predispositions early in life, are considered structuralists. They contend that organisms are born with certain innate behavioral (or physiological) structures that facilitate subsequent development and adaptation to the environment. Piaget's theory of cognitive development is also considered structuralistic.

# Piaget's Cognitive Theory

Cognitive theory has often been considered synonymous with the work of Jean Piaget. Because his is such a widely recognized and influential theory, it will be examined in detail here. Part of the appeal of Piaget's work is its eclectic nature. It is simultaneously an ecological, interactive, biological, logical, and cognitive theory that has educational, developmental, and clinical applications. Its adaptability to, and incorporation into, clinical fields such as occupational and physical therapy have already been discussed in chapter 5.

Piaget's theory reflects his personal background in biology and philosophy (specifically epistemology). As Kagan (44) explains, however, Piaget described himself as a philosopher, not a child psychologist. And while

Piaget remains a significant figure in cognitive development, his theory is not universally accepted. The following section will review Piaget's theory and his proposed stages of cognitive development and then introduce some limitations of his theory as seen by other researchers in the field. (Readers may want to look back at chapter 5, as the following sections will elaborate on material already introduced there.)

Piaget is considered a structuralist because of his investigation of specific mental structures in thought. He was interested in how children learn to understand the nature of space and time and the stability of the external world (45). Although he recognized the existence of specific mental structures that facilitate those perceptions, Piaget was not a nativist. His theory (see chapter 5) synthesizes nativism and empiricism in a position referred to as "interactionism and constructionism." That is, his theory contends that each child actively "constructs" his or her own way of thinking by using innate abilities in "interaction" with the world (Fig. 11.4). Thus children construct their own internal structure of the world, representing their own unique experiences with what the world is all about (1, 46). The child does this through the process of equilibration.

## Equilibration and Intelligence

Piaget viewed intelligence as both biological and logical. Since his cognitive theory is based on biological concepts, intelligence is viewed as actions that

**Figure 11.4.** According to Piaget, action is the basis of cognitive development. Through movement the child comes to understand the nature of the world by exploring the relationship between himself and objects in the environment as well as the relation of his body parts to one another and to space.

are adaptive for the organism. As Fischer (47, p. 481) explains, "All cognition starts with action . . . Piaget has pointed out that cognition is essentially what the organism from its own point of view can do whether the doing is commonly classified as motor, perceptual, or mental." According to Piaget, just as the body seeks to maintain an equilibrium through homeostasis, so too does the mind through the process of equilibration. Intelligence is adaptation, which is maintained through equilibration, a balance of the two complementary processes of assimilation and accommodation (8).

Piaget's theory asserts that the basis of knowledge is action. Individuals act on their environments, and environments act on people. Intelligence is keeping an equilibrium between the action of the organism on the environment and vice versa. As Piaget describes it, assimilation, in its broadest sense, refers to the action of the individual on objects in the environment, providing the action involves previous behaviors with the same or similar objects. However, knowledge cannot exist via assimilation alone because things cannot be known purely by themselves. Thus, individuals must also have a way of obtaining knowledge of new objects and events never before encountered, and of modifying knowledge based on experiences with new events and information. This process of change is accommodation (8).

## Stages of Cognitive Development

According to Piaget, the form of intelligence changes over life as capacities for knowledge shift and as experiences with the environment change. What is adaptive early in life differs from what is adaptive later in life. Adaptive behaviors for young infants are sucking, reaching, looking, and responding to a variety of sensory stimuli in their environments. Early in life, adaptive behavior involves fundamental sensorimotor abilities and adjustments that enable infants to survive and to adjust their behavior according to the environment. Thus, the first form of intelligence is sensorimotor in nature, and reflexes reflect early intelligent behavior (48).

### Sensorimotor Intelligence

The earliest adaptations in life are reflexes. They enable the infant to suck to obtain nourishment, to move the head, to look, to reach, and to adjust to the environment as needed. Piaget claims that the infant gradually adjusts these early reflexes so that they become more and more refined, purposeful, and internalized. Infants do this through the process of assimilating information from similar objects and by changing their behavior and accommodating when new objects are encountered. For example, the sucking reflex enables infants to suck their mother's breast and to assimilate similar objects such as other nipples or pacifiers into their sucking repertoire. An object too large to be sucked, however, must be explored in a different way, requiring infants to accommodate to this new event by incorporating new behaviors into their repertoire.

Since infants do not yet have language, they cannot use words as mental representations for their thoughts. Instead, Piaget proposed, they form representations for their sensorimotor actions. These mental representations are termed *schemata* and form the basis of the infant's knowledge. Initially, schemata are representations of very simple acts such as a sucking response or a grasp reflex. Gradually, however, as behaviors are combined and new behavioral patterns are required, new and more complex schemas are formed. Familiar experiences use existing schemas to assimilate knowledge, but unfamiliar experiences require accommodation and change. Thus, experiences with unfamiliar objects and events in the environment require the infant to obtain new schemata.

As knowledge and environmental demands change, however, the child's adaptive capacities become less purely sensorimotor and more mental and symbolic. Gradually, through six different sensorimotor stages, cognition becomes more mature and shifts to a qualitatively different form, from sensorimotor to conceptual. Initially, actions are the most challenging and consuming behaviors of the infant, but as the infant becomes more skilled at motor abilities, acquires language, has greater exposure to varied objects and events in the environment, and is involved in more complex interactions, cognitive structures begin to shift away from action to thought. This ability to form internal, mental counterparts for objects or actions signals the beginning of conceptual thought and is an important developmental achievement. The shift, however, from action-oriented to rudimentary internal thought processes is gradual, and although it takes place around 18 to 24 months, the process continues for years. To provide the reader with an understanding of how cognition develops gradually into more mature forms of intelligence, the following section describes the six steps that make up the sensorimotor period (Table 11.7).

### Stage 1: The Use of Reflexes

Piaget explains:

> Intelligence does not by any means appear at once derived from mental development, like a higher mechanism, and radically distinct from those which

Table 11.7
**Piaget's Stages of Sensorimotor Intelligence[a]**

| Stage | Adaptive Capacity |
|-------|-------------------|
| 1 | The use of reflexes (birth to 1 month) |
| 2 | The first acquired adaptations and the primary circular reactions (1 to 4 months) |
| 3 | Secondary circular reactions (4 to 8 months) |
| 4 | Coordination of secondary schemata and their applications to new situations (8 to 12 months) |
| 5 | The tertiary circular reactions and the discovery of new means through active experimentation (12 to 18 months) |
| 6 | The invention of new means through mental combinations (18 to 24 months) |

[a] Adapted from Piaget J: *The Origins of Intelligence in Children*. New York, WW Norton, 1952.

have preceded it. Intelligence presents, on the contrary, a remarkable continuity with the acquired or even inborn processes on which it depends and at the same times makes use of. Thus, it is appropriate, before analyzing intelligence as such to find out how the formation of habits and even the exercise of the reflex prepare its appearance. (48, p. 21)

During the first month of life, early reflexive abilities such as sucking, grasping, looking, hearing, and producing some simple sounds form the infant's initial response repertoire for "handling" the environment (Fig. 11.5). These are basic unlearned responses that are necessary for the infant's survival and adaptation to the environment. Initially, these behavioral capacities are rigid and uncontrolled. For example, Piaget (48, p. 25) points out in regard to the sucking reflex, "As soon as the hands rub the lips the sucking reflex is released. The child sucks his fingers for a moment but of course does not know either how to keep them in his mouth or pursue them with his lips." During the subsequent stages, however, reflexive behaviors become more regulated and specific.

### Stage 2: The First Acquired Adaptations and the Primary Circular Reactions

From about the 2nd–5th month, reflexes, or inherited adaptations, shift to adaptations that are acquired through interaction with the environment.

**Figure 11.5.** Initial schemata involve looking, sucking, and grasping and are gradually coordinated as the infant develops.

Thus, they may also be considered conditioned reflexes or habits. For example, as Piaget points out, added to the simple inherited sucking reflex are behaviors that are obviously new. Sucking shifts from simple movements of the mouth to actions centered on the body as in thumb sucking (48). Thus, although infants may have sucked their thumb before, they have not, until this stage, possessed a set of behavioral adjustments that enable them to bring the thumb to the mouth and to keep it there (45).

During this stage, these recently acquired behaviors are repeated over and over, thus the term "circular reactions." In addition, since they are centered on the body and are performed for their own sake rather than directed at attaining a goal or object in the environment (46), circular reactions are considered primary in nature. These *primary circular reactions* do not involve just sucking but all other simple adaptations of reflexive behaviors; while they are primary, they are not centered just on the self. For example, looking shifts from simple reflexive fixation to active exploration of objects in the environment (45), and sucking involves oral exploration of various objects as well as body parts. However, even though some behaviors are directed at objects in the environment, the infant's understanding of the world is still very limited. The infant does not yet understand concepts about space, time, and the permanency of objects in the world (46).

### Stage 3: Secondary Circular Reactions

The 5th–8th month marks the onset of intentional behaviors, a concept which Piaget points out, is most difficult to define. Implicit in the concept of intention is a goal and a desire and means to reach it (48). Like primary circular reactions, the responses that emerge during this stage are repeated over and over (circularly), but they are considered secondary because the actions themselves are not as important as their consequences. Secondary circular reactions are centered on their consequences; i.e., upon maintaining events in the environment rather than on actions focused on the child's body. Thus, for example, the child shifts from the primary circular reaction of grasping for its own sake to grasping intentionally in order to hold onto something. In another example, Piaget describes how his son Laurent repeatedly shakes a chain attached to a variety of different toys over his bassinet. Compared with sucking responses, these movements are oriented around maintaining an action in the environment:

> At [4 months, 3 days] he pulls at will the chain or the string in order to shake the rattle and make it sound: the intention is clear. I now attach a paper knife to the string. The same day I attach a new toy half as high as the string (instead of the paper knife): Laurent begins by shaking himself while looking at it, then waves his arms in the air and finally takes hold of the rubber doll which he shakes while looking at the toy. The coordination is clearly intentional. (48, p. 164).

Other similar secondary circular reactions include shaking a rattle to hear its noise, striking toys to see them move, and Lucienne shaking, "her head from side to side in order to shake her bassinet, the hood, ribbons, fringes, etc." (48, p. 165). Such behaviors qualify as secondary circular reactions because they are repeated over and over, and their goal is to produce and maintain some consequence, such as hearing the rattle or seeing the toys move.

Another important achievement of this stage is beginning of the coordination of visual motor prehension, which continues refinement in stage 4. In addition, infants begin to conceive the world around them as stable. This concept of the existence and stability of objects, known as *object concept* is important in Piaget's theory. He tested the development of this concept by examining children's responses when objects were abruptly removed from them. Piaget reasoned that if a child possesses the concept that objects are stable and permanent, then he or she should be surprised when the object disappears.

As Piaget theorized, object concept is gradually but actively constructed during this and the next sensorimotor stage. Prior to this time, Piaget contended, objects apparently do not exist for children except as they interact with them. Thus, when an object is removed from a young infant, so that it can no longer be seen, felt, smelled, or heard, the youngster does not actively look for the toy nor appear distressed by its absence. Instead, attention is shifted to some other activity. With the development of *object permanence*, which is the notion that things (objects) exist even when out of view, the child gradually constructs a schema for searching for the missing object. Thus, the infant begins to have certain expectations about how the world is organized and about space and time. If an object is removed from view, and the infant begins to search for it, he or she exhibits an awareness that an object must occupy space somewhere else and that it exists over time. The infant is now able to follow the trajectory of objects that are moved or dropped and to anticipate where they may fall; the infant will reach for an object that may be partially concealed and interrupt an activity and return to it later. All these behaviors illustrate an increasingly sophisticated understanding of the nature of the world (49).

According to Piaget, the development of object concept is gradual, as the child becomes more interested in objects in the environment. Thus, during this third stage of cognitive development, object concept is still limited. The infant's search for missing objects is typically very brief and only for objects that are associated with movement, where the infant is able to follow the trajectory of the moved object. In the next stage, object concept and object permanence become more fully established.

### Stage 4: Coordination of Secondary Schemata and Their Application to New Situations

During stage 4, which lasts from about 8 to 12 months, the child becomes capable of performing more intentional behaviors directed at achieving specific goals. Thus, behavior is increasingly intentional and flexible as the child demonstrates considerable progress in using familiar schemas in new situations and combining and coordinating schemata with one another in single acts (Fig. 11.6) (48). The child no longer tries to repeat or prolong certain behaviors solely in circular reactions but actually pursues goals. Piaget describes how his son gradually acquires intentional behavior. In this circumstance, Laurent has been holding a toy in both of his hands. Piaget, however, repeatedly offers him a more attractive toy and watches how his behavior gradually shifts by using existing schemata to obtain a goal. Laurent initially accidentally drops the less preferred toy but gradually acquires the ability to intentionally discard one toy in order to hold onto another:

> At [7 months] he has in his hands a small celluloid doll when I offer him a box (more interesting). He grasps the latter with his left hand and tries to grasp it with both hands. He then knocks both objects together, at once separates them (very much surprised at the result), and recommences to knock them again . . . [7 months, 28 days] I again note the same reactions: he knocks involuntarily the two objects he is holding when he wants to grasp one of them with both hands . . . Finally at [7 months, 29 days] Laurent finds the solutions. He holds a little lamb in his left hand and a rattle in his right. A small bell is held out to him; he lets go the rattle in order to grasp the bell. The reaction is the same several times in succession but I have difficulty in discerning whether he

**Figure 11.6.** Infant intelligence involves active trial and error exploration of the environment. Simple schemata are coordinated so that visually guided reaching gradually becomes efficient.

> simply lets the rattle escape him or really discards it. While he holds the bell I offer him a big box: he grasps it with his (free) left hand and with his right (by sticking the bell against the box) but noticing the difficulty, this time he definitely discards the bell. . . .

> Henceforth Laurent knows how either to discard one object in order to grasp another or to place it, or to let it fall intentionally. (48, pp. 221–222)

Goal-directed behavior requires the coordination of sensorimotor abilities. Thus, locating a toy, grasping it and bringing it to the mouth, and then reaching out to use the toy to strike another object involve the coordination of looking, reaching, grasping, and sucking schemata.

Imitation also beings to develop at this time. While imitative actions are apparent earlier in development, during stage 4 children become able to imitate actions that they have never seen themselves perform. This ability reflects their growing capacity to separate themselves from objects in the environment and to see themselves as separate from what they observe. As object concept continues to develop, children can see that actions that take place are separate from themselves, so they begin to take an interest in novel events, assimilating and then imitating them (Fig. 11.7) (46).

Object permanence also advances so that now, when an object is placed under a cushion, the child actively looks for it and attempts to retrieve it. Still, the object concept is limited, and the child exhibits what is called "stage 4 error." That is, if, in full view of the child, a toy is hidden under one pillow, removed, and hidden under a second pillow, the infant will look where the toy was first concealed. Piaget contends that the knowledge of the object at this time is still limited, and the infant's search for the object is guided more by a misperception of the nature of the infant's actions than by an actual mental representation of the object (50). This misperception gradually changes in the last two sensorimotor stages.

### Stage 5: The Tertiary Circular Reaction and the Discovery of New Means Through Active Experimentation

During stage 5, children use new behaviors that they have never before performed. Many of these new actions are built upon secondary circular reactions. However, while primary circular reactions were centered on the self and secondary circular reactions were centered on obtaining some object in the environment, tertiary reactions are centered on variation. As circular reactions, they are repeated over and over but, contrary to previous stages, they are now varied. These variations are actual trial-and-error behaviors that represent an active form of experimentation, a "I'll-try-this-to-see-what-will-happen" behavior that leads the child to new acts of intelligence (48). For example, during this stage, the infant in a highchair often repeatedly drops food onto the floor or repeatedly bangs a spoon and cup and other utensils together, thus experimenting with falling objects, noises, and the nature of the world. Parents, however, often see this behavior as negative rather than a positive step toward learning about the

**Figure 11.7.** The ability to imitate develops gradually and is an important cognitive as well as social capacity.

world, and such experimental behaviors, combined with inquisitiveness and newly acquired motor skills (chapter 10), may lead to trouble if the child is unattended.

Piaget describes tertiary circular reactions performed by his child, Jacqueline:

> Observation 147—In her bath, Jacqueline engages in many experiments with celluloid toys floating on the water. At [1 year, 1 month, 20 days] and the days following, for example, not only does she drop her toys from a height to see the water splash or displace them with her hand in order to make them swim, but she pushes them halfway down in order to see them rise to the surface.

> At [1 year, 7 months, 20 days] she notices the drops of water which fall from the thermometer when she holds it in the air and shakes it. She then tries different combinations to splash from a distance. She brandishes the thermometer and stops suddenly, or makes it catapult.

> Between the ages of a year and a year and a half, she amuses herself by filling with water pails, flasks, watering cans, etc., and studying the falling of the water. (48, p. 273)

Stage 5 is the last purely sensorimotor stage. In this stage the child's, "sense of space is broadened, articulated, and coordinated" but there are still limitations (50, p. 59). In the next stage, the child begins to develop actual cognitive mental representations for the external world, to mentally anticipate consequences of actions, and to, therefore, begin mental problem-solving (45).

### Stage 6: The Invention of New Means Through Mental Combinations

The final sensorimotor stage beings representational, conceptual thought. As Piaget (48) explains, it is an essential moment in development of intelligence when the awareness of relationships is sufficiently advanced to permit thinking based on mental rather than on physical experiences. During this stage, the child can picture events and therefore can problem-solve mentally rather than through active trial and error. Schemata using different body parts become coordinated together during stage 6, thus sensorimotor coordination and integration are refined (see chapter 8). This enables the child to use information from all sensory modalities in understanding the nature of the world (Fig. 11.5).

The ability to mentally represent objects and actions involves an understanding of the nature of objects, how they exist in space and time, and how they relate to other objects and to one's self. The child can now see the world as a stable place and will look for misplaced objects. The ability to mentally represent objects and to hold them over time enables the child to imitate actions that occurred previously. This will affect social interactions, as the child can now observe an action or a model and imitate that behavior at some later time (deferred imitation). As the child becomes increasingly more physically mobile, so also does the child's thinking (Fig. 11.4). These abilities, in turn, as will be explored in greater detail in this and the next chapter, affect the child's social interactions.

## Preoperational Period

The preoperational period is often considered a transitional period rather than a separate stage of development. Thus, either three or four stages of cognitive development are recognized depending upon whether or not the preoperational period is included. During preoperational time, many of the initial behaviors seen in the sensorimotor period are elaborated. Preschool children display numerous examples of deferred imitation, indicating that they can indeed create and hold mental images. Their play so apparently illustrates their ability to remember and to later imitate behaviors they may have seen at home. This imitative ability coupled with their cognitive ability to make-believe lies at the basis of dress-up games, tea parties, "playing house," and making forts (Fig. 11.8). With increasing cognitive abilities and greater imagination, play becomes more complex and symbolic. They are able to act out behaviors, as well as to draw impressions from, actions they may have seen some time before.

By name, "preoperational" refers to the time period *before* the devel-

**Figure 11.8.** During the preoperational period, play becomes more imaginative.

oping child acquires operations. During this time the gradual nature of cognitive change becomes evident as the child vascillates between immature and more mature forms of thought. Thus the child is often inconsistent, at one time using logic and at other times appearing illogical [or prelogical (8)] and disorganized in thinking.

A specific trait characteristic of the preoperational period is *centering*, which is concentrating attention on one idea or focusing attention on one aspect of an event to the exclusion of other aspects. An example of centering can be found in cross-sectional research that explores children's perception of age. In these studies, children are given either actual doll models of people or drawings of people representing different stages of life. When compared with older children (age 8 or 9), preschool children have difficulty accurately separating these different stimuli into age-appropriate categories. One of the possible causes for this difficulty is that the younger children tend to focus (center) on just one dimension of the stimuli and therefore cannot use all available cues to judge the ages of the different people (51, 52). One theory is that preschool children tend to discriminate people based primarily on the characteristic of size. Thus, if they see a group of adults, all basically the same height but of varying ages, preschoolers may not see these people as noticeably different. An example of centering was recently described to me by a preschool teacher. She said that one of her 4-year-old students asked her how old she was. When she told him her age, he said, "Gee, you should be taller!"

*Egocentrism* which is characteristic of preoperational thought, is the centering of attention on events from the child's own point of view. This does not mean that the children are selfish but that they are unable to take another's point of view. Egocentrism is illustrated by children's inability to fully understand how other people are related to them, only how they are related to other people. Piaget and his followers tested egocentrism by placing children in front of a three-dimensional model of a village. Each child was then asked to draw or to pick out pictures that show how the village looks from their own perspective as well as from another point of view, say to someone standing and looking at the model from a different vantage point. Because of their egocentrism, preoperational children are unable to solve the latter part of the problem as they are unable to consider another's point of view.

With age, the thinking of the preoperational child becomes more stabilized and equilibrated. Older children are able to decenter and can therefore take all available cues into consideration. This enables them to make accurate age judgments based on facial traits, size of the stimuli, etc. (52) and to accurately determine another person's perspective when viewing a model. With the loss of centering and with the acquisition of operations, children's thinking becomes organized, and they begin to behave much more consistently and logically. When this occurs, the child has begun operational thought (45).

## Stage of Concrete Operations

Preoperational children appear illogical because they do not yet possess consistent mental operations that enable them to approach problem solving with an integrated system of thought. One of the ways to understand how preoperational children behave is to look at their behavior in comparison with children with concrete operations. This will be the approach discussed in this section. However, before this approach is introduced, it is necessary to explain what Piaget meant by operations. *Operations* are internalized mental actions that conform to certain rules and principles of logic. While schemata are like sensory and perceptual representations, operations are actual mental manipulations of information (53) and are characterized by their reversibility. For example, like simple arithmetic operations, cognitive operations can be reversed to their original state. The simple arithmetic operation $2 + 2 = 4$ can also be reversed to $4 = 2 + 2$. Similarly, boys plus girls equals children can be reversed to children minus boys equals girls. Mental operations, like arithmetic operations, can function one way and then be reversed. This should become clearer as some of the concrete operations are described.

### Conservation

One of the most highly studied of mental operations, conservation is the notion that things remain the same despite undergoing some perceptual change. One of the experiments used to test this capacity is conducted with conservation of liquid. A preoperational child is presented with two equal-sized containers with two equal levels of liquid. The child is asked about and agrees that the glasses contain the same amount. Then, in front of the child, the liquid from one glass is poured into a very wide glass, so that the level in one is now much lower. The child is then asked whether one glass has more or the same amount of liquid. At this point, the preoperational child claims that the shallow dish contains less liquid because the level is lower. If the liquid is then poured into a tall, thin glass, the child then believes that this container now holds more (Fig. 11.9). Since preoperational children cannot mentally reverse their thinking, they cannot imagine what conditions were like before the liquid was poured into a different container. Thus, they are essentially deceived and think that the amount of liquid is transformed merely by pouring it into a wider or thinner container.

Such illogical thinking gradually disappears once the operation of conservation is acquired. Once children are able to conserve, they can recognize that the liquid does not change in quantity, despite its higher or lower appearance. Conservation applies not only to liquid but also to number, volume, weight, and other properties or dimensions. When the preoperational child is confronted with two identical rows of objects, the child recognizes that they are equal. If however, the bottom row is made smaller by placing the objects closer together, then the child thinks that the top row actually contains more objects than the bottom, despite the fact that the organization but not the number of objects has changed (Fig. 11.10).

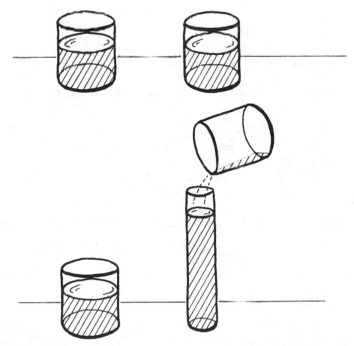

**Figure 11.9.** Conservation of liquid. The preoperational child cannot reason that the liquid remains the same despite some perceptual change.

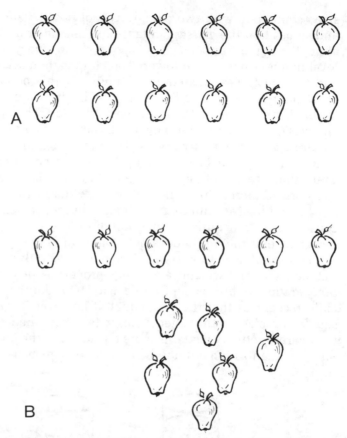

**Figure 11.10.** The preoperational child recognizes that the top rows in (*A*) are equal but reasons that the lower row in (*B*) contains fewer items. This is because the preoperational child does not yet have conservation of number.

Such inability to conserve is typical of preschool children, who may argue that their brother had more cookies because the cookie was broken into many pieces. Without conservation, the preoperational child cannot reason that the volume, number, or mass of an object does not change when some apparent perceptual alteration occurs.

### Seriation and Understanding Relationships

The ability to arrange objects along some dimension is called seriation. This is exemplified by giving a child a set of wooden sticks of various sizes and asking the child to arrange them in order. While an 8- or 9-year-old child can successfully perform this task, a preoperational child may have difficulty.

Very similar to seriation is the understanding of relational terms such as "bigger than, lighter than, brother of, next to, on top of, longer than, older and younger than, under, over". While preoperational children understand some simple relational concepts that they may learn in school and frequently practice, they have difficulty with more complex relations such as those necessary for arranging a whole series of objects. For example, preschool children tend to learn the relationship big/small fairly early and are able to classify younger and older people based on the dimension of size. Thus, preschool children can discriminate infants and children from adults, but they have difficulty ordering similarly sized people who may be 20, 30, 40, 50, and 60 years old (51, 52). In addition, because of their egocentrism, preoperational children may understand how things relate to themselves but not how they relate to someone else. The latter requires their taking on someone else's point of view and understanding relations from a different perspective. Such children are therefore often miserable storytellers because of their inability to describe information so that someone else can understand it. I can remember a preschool boy who was so excited because his aunt was visiting him at school. He saw me going up the stairs and yelled up a whole flight of stairs to tell me, "Hey! Guess what! I'm an aunt!"

## Class Inclusion

Another operation also involving the understanding of relationships is called class inclusion, which is the ability to reason about the whole and a part of the whole at the same time. This is exemplified by showing a child a plate with many brown and a few white cookies and asking if there are more brown cookies (part) or more cookies (whole). The preoperational child, unable to reason about both the part and the whole at the same time, will tend to center on the parts and claim there are more brown cookies. Another example illustrates the preoperational child's egocentrism and his lack of class inclusion and relational concepts: A family contains three boys and two girls. The youngest boy, age 4, is asked how many brothers he has. He correctly says, "two." If he is asked, however, how many brothers there are in the family, he is apt to reason that since he has two brothers, there are only two in the family. Because of his egocentrism and lack of relational concepts, he cannot reason that he, himself, is a brother to others and therefore there are three brothers in the family. Additionally, because of the lack of class inclusion, if asked if there are more brothers or more children in the family, he is likely to think there are more brothers.

In contrast, children with concrete operations are able to think through all these various relationships. They can imagine the transformations involved in pouring liquid back and forth from different containers and can look at objects in terms of their relationships to one another as well as in terms of a variety of dimensions at once. They are no longer constrained by egocentric thought, nor are they so illogical. They have gradually developed mental images for actions and for series of actions which can be represented in their mind. Because of this, their reasoning has become more logical, and they are able to perform actions not before possible. Because they can represent actions in their mind, they can, for example, give instructions to someone as well as tell meaningful stories. Such thinking, as is apparent, will have an effect on this child's social interactions and scholastic achievements. Yet, the thinking is still restricted. Concrete operations are so named because thinking is still concrete, meaning that it is limited to the here and now, rather than abstract. The following section describes some examples of concrete thought.

A physical therapist described how her 7-year-old son amuses the family by what they label a dry sense of humor and what is actually concrete thought. She explained that before they headed on vacation, she told her son that they needed to, "run over to the mall." Concerned, he said that they should probably take the car since the mall was so far away it would take them too long to run. As they were preparing to decorate their Christmas tree, David, overhearing their discussion about "trimming the tree" brought a pair of scissors out to help them do the trimming. Once the family went on vacation to Florida, David's grandfather pointed out a school of fish. David quickly and seriously asked, "If that is a school, then where is the teacher?"

A young girl singing "Rudolph the red-nosed reindeer, you'll go down in history" asked her parents where history was. They were unable to adequately convey to this child the abstract concept of history. With her concrete thought, she believed that history was an actual location, like the North Pole, where Rudolph went to live.

One of the best examples of concrete thinking was relayed to me as an anecdote but which may be an actual event. A therapist said she handed a form to her client and asked him to complete it. She gave preliminary instructions to him to write his name, skip a space, and then complete the rest of the form. Following her instructions, which he apparently interpreted very concretely, he sat down, wrote his name, then stood up, physically skipped (hopped) a space, then sat back down and completed the form.

## Stage of Formal Operations

From about the age of 11 or 12, adolescents begin to reach logical, efficient thought. Mental abilities are refined, and these enable adolescents to deal with sophisticated academic material that they would not have been able to master earlier. Thus, the adolescent becomes interested in algebra, chemistry, and philosophy and is able to apply systematic logic to problem solving. According to Piaget, formal thought involves the ability to generate hypotheses, to mentally consider all of the possible ways a problem may be solved, as well as being able to examine and evaluate one's thought processes (8).

These characteristics of formal thought explain the adolescent's interests and activities. Adolescents often become preoccupied with themselves, their beliefs, and their plans. They begin to evaluate their lives and to consider whether their beliefs are valid or not. They may take an interest in abstract games such as chess or computer simulations where all possible combinations of problems are considered and logically (mentally) worked out. In the following description, cognitive psychologist Howard Gardner points out some of the far-reaching implications of formal thought.

In contrast to the practical-minded 8- or 9-year old, many adolescents become engrossed in systematic theoretical speculation on such topics as the merits of a new religious sect, the platform of a presidential candidate, the arguments for and against legalized abortion, the quality of an actor's or athlete's performance. They become dreamers, interminably considering the varied possibilities of their lives. They revel in bull sessions, speculate endlessly, even as they continuously make inferences, predictions, and idealistic claims. Perhaps for the first time, they are able (and willing) to follow their beliefs to their ultimate conclusion, which is why they can become intensely involved in ideological matters. And, equally important, they seem for the first time inclined to care passionately about ideas and prepared to give their lives for them. It is the adolescent and young adult—not the child or the middle-aged person—who has participated centrally in the revolutionary movements of our time. (49, pp. 483–484)

# Brief Evaluation of Piaget's Theory

Piaget's is a comprehensive and influential theory, but it is not without criticism. Some of the most general criticism is that as a stage theory, Piaget's work does not accommodate individual differences (55) and that it has underestimated children's cognitive abilities. For example, chapters 8 and 9 of this book have discussed research by some scientists who believe that very early in development infants possess sophisticated perceptual capabilities that reflect an innate knowledge regarding the nature of the world. Thus, contrary to Piaget's proposal that infants must act on the world and through sensorimotor actions gradually come to understand the nature of objects, the existence of depth perception or perceptual constancies early in life argue for the presence of a fundamental, innate, perception-based orientation to the world.

The application of the sensorimotor basis of knowledge is one primary concern regarding Piaget's theory that is particularly relevant to physical and occupational therapy. While sensorimotor bases of development and treatment are widely recognized and applied in many clinical fields, Piaget's focus on action as the basis for cognition is inconsistent with findings that individuals with mild or severe sensory and motor deficits develop normal cognitive functions despite their physical limitations. Thus, Piagetian theory has difficulty accounting for individuals who, with severe sensorimotor deficits, display normal cognitive abilities. Clearly, continued research needs to investigate the significance of movement, the nature and role of reafferent stimulation (chapter 9), and the sensorimotor basis of cognition. Gratch (50) points out that to obtain a more complete view of the development of knowledge, continued naturalistic study is necessary to integrate Piaget's cognitive theory with contemporary developmental research.

Consistent with the criticism that Piaget has underestimated children's abilities are the claims that concrete and formal operational thought may occur earlier than Piaget had proposed. Such claims assert that Piaget's methods of collecting data are too strongly dependent upon logic and are not sufficiently naturalistic to obtain accurate impressions of children's capabilities (50). Thus, some young children may possess formal operations but lack sufficient language skills to display their reasoning. This criticism applies to Piaget's preoperational studies as well. For example, Gelman (56) points out that numerous studies have shown that the preschool child is not as egocentric as Piaget claimed. Gelman notes that if procedures are set up suitably for young children, then the results of many of these studies indicate that preschoolers are much more cognitively capable than Piaget proposed. Gelman is quick to point out that she does not believe that these youngsters can think like adults but that many of Piaget's tasks were not set up to optimally communicate with or to test these children's abilities.

The age of task attainment, however, was not a great concern for Piaget. As Gardner (49) points out, Piaget was indifferent to age. Piaget recognized that cognition was a reflection of maturation, learning, and direct experiences and training. Thus, more important than age-appropriate attainment of cognitive abilities is the child's ability to build upon each stage and to progress through the cumulative sequence of cognitive development. It is this sequence (and not specific ages associated with it) that seems to be most widely recognized as accurate and most consistently evaluated as Piaget's primary contribution. And although he is regarded as one of the most highly influential developmental theorists of this century, his work will continue to be evaluated for its clinical applications and for its ability to accommodate to additional ongoing research in the scientific study of development.

# COGNITION AS THINKING

Piaget's theory is only one way of organizing the body of knowledge about cognitive development. Some approaches, e.g., Fischer's skill theory of cognitive development (47), are extensions and modifications of Piaget's work, whereas other approaches look at mental processes quite differently than did Piaget. One distinct approach to human cognition and intelligence looks at the mental processes associated with thinking. Such approaches focus on attention or memory or on determining specific mental units or rules involved in solving mental problems. Some researchers consider mental problem solving the basis of thinking and reasoning, and while the terms "thinking" and "reasoning" are often used synonymously with cognition, they deal with a subject matter quite different from Piaget's theory. The term "thinking" is often used in reference to some kind of mental activity or to the processing of mental information. It has been defined as a generalized type of reflection that goes on in an individual's head and that is directed at the connections and interrelationships of events in objective reality (57, p. 100). Thus, while Piaget focused on sensorimotor, biological, and logical aspects of cognition, other approaches extend primarily to mental processes associated with the mind, not with biology.

# Models of Thought

Because thinking is something that occurs "in one's head" it is difficult to study and to conceptualize. Cognitive theorists, unable to find apt biological correlates of thinking, have used a variety of different metaphors or models for thought. Just as the mind has been conceived in many different forms, so also has the product of mind, thought. These models of the mind exist at various levels, many reflecting the social-technological status of society at the time they were originated. Some models of the brain (or mind) are simple, mechanical, reductionistic approaches typical of traditional S-R views of learning that compare the human mind to a simple telegraphic model with input/output and basic connections in between (58).

Many reductionistic approaches to human thought attempt to locate the biological basis of knowledge, but other cognitive theories point out that what is involved in thought processes is elusive and defies reduction to simple S-R or neuronal models. Whether cognition can be understood from a biological, simple S-R approach is debatable:

> It might seem that the principal contribution of biology to the study of intellectual development would be information about the anatomy and physiology of the human brain. Such is not the case. A good deal is known about various "parts" of the brain and how they work, right down to single cells or even molecules, but knowledge about the parts does not add up to knowledge about the system as a "whole". Yet the cognitive theorist is asking questions about the system as a whole. He wants to know how it works to produce intelligent behavior and how it changes so that behavior becomes more intelligent as a result of development. If these questions cannot be answered by a study of isolated parts, they must be answered by a study of the whole system in operation, that is, by observing behavior and the situation in which it occurs. In other words, intellectual development is a problem in the field of psychology, not anatomy or physiology. (1, p. 118)

Other models of the brain and of human thought processes are less mechanistic and more abstract than neuronal approaches. Because thinking has often been considered a uniquely human activity, it has often been associated with another characteristically human ability, language. Thus, at the leve of communication and symbolic interactions, models of the mind and of thought examine certain structures that are the basis of mutual understanding among people. This relationship between language and cognition will be examined further in chapter 12. In his book, *Maps of Mind*, Hampden-Turner (59) looks at a variety of approaches to human mind and thought. One approach, which he labels the "psychobiological approach", studies mind and thought in terms of the ecology of living systems and in the context of environmental relationships. More abstract models of human thought look at the nature of human intelligence and their relationship to the universe. Each of these various types and levels of models of the mind contribute to our understanding of human cognition.

# Cognition as Information Processing

Models of human thought continue to expand with contemporary holistic and systems approaches to behavior as well as advances in our awareness of information processing and computer functions. Renowned geneticist Crick (60) notes that many levels of approaches are probably necessary for understanding human functions. Reductionistic approaches are important, but so also are approaches that consider complex information processing systems:

> There are some human abilities that appear to me to defeat our present understanding. We sense there is something difficult to explain, but it seems almost impossible to state clearly and exactly what the difficulty is. This suggests that our entire way of thinking about such problems may be incorrect. In the forefront of the problems I would put perception, although here others might substitute conception, imagination, volition or emotion. All of these have in common that they are part of our subjective experience and that they probably involve large numbers of neurons interacting in intricate ways.

> To understand these higher levels of neural activity we would obviously do well to learn as much as possible about the lower levels, particularly those accessible to direct experiment. Such knowledge by itself, however, may not be enough. It seems certain that we need to consider theories dealing directly with the processing of information in large and complex systems, whether it is information coming in from the senses, instructions going out to the muscles and glands or the flow of information in the vast amount of neural activity between these two extremes. (60, p. 219)

Information-processing theories of human cognition are different from other approaches discussed so far. Kail and Bisanz (61) point out that Piaget examined human thought in terms of formal logic and verbal descriptions, and learning theory approaches are interested in specific stimulus-response associations as the basis of knowledge. In contrast, information-processing approaches use the modern digital computer as a model for human thought and are interested in the activities that intervene between stimuli and responses. Human cognition and computer functions are both viewed as manipulations of symbolic information, and such models recognize that information must be sensed, perceived, encoded, stored in memory, and then retrieved at some later time (Fig. 11.11). The following sections will examine some general developmental trends in cognitive processes: attention, memory, mental representations of information, and strategies for handling information. Although the information-processing approach to cognition is not developmental in nature, such a perspective can be applied (61).

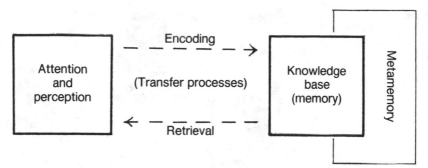

**Figure 11.11.** Memory system based on an information processing approach. (Adapted from Perlmutter M: A life-span view of memory. In Baltes PB, Featherman DL, Lerner RM (eds): *Life-Span Development and Behavior.* Hillsdale, NJ, Lawrence Erlbaum, 1986.)

## Cognitive Representation

Thinking is regarded as a form of symbolic mental activity. Objects in the environment (e.g., chairs, mother, book, cat) cannot be represented directly in the mind, so there must be a mechanism to convert objects into some form of internal mental representation that can be used for thought. Piaget, for example, proposed schemata as fundamental mental representations that are initially sensorimotor in nature and gradually become more abstract with time. Later, operations form the basis for adult thought. While Piaget's view of cognition is not universally accepted, it does appear that, as he has suggested, mental representations change with time. The most dramatic example of this change occurs with the acquisition and use of words.

### Symbols and Concepts

Once a child begins to acquire language, thinking becomes more symbolic. *Symbols* stand for or represent things. For example, the three letters C-A-T stand for (or represent) a child's pet that meows and purrs. Words do not, in and of themselves, possess meaning; meaning is attributed to words according to what they symbolize. Although words that are used as names for things are excellent examples of symbols, many symbols in everyday usage do not take the form of words (Fig. 11.12).

With development, the nature and the levels of meaning of symbols vary. Obviously, with increasing experiences and expanding vocabulary, more symbols are acquired and can be used for thought. In addition, those symbols take on different meanings. The word "cat" may initially be used by a youngster to refer specifically to a pet or toy. With time, the child begins to realize that the word also applies to other similar animals, and with even greater exposure to words and their meanings, the child will learn that the term "cat" can refer to a whole class of objects that share similar features. Similarly, the word "teacher" may be used initially to refer to one person. With experience, the word comes to stand for a whole class

**Figure 11.12.** While words are used as symbols, there are also many nonverbal symbols.

of individuals who share some common characteristics. Once children have developed this capacity to refer to classes of objects, they can use concepts.

Concepts [also called categories or ideas (44)] are names for groups or classifications of objects. Thus, while symbols represent single events, e.g., the word "cat" which stands for an individual cat, concepts represent whole classes of events, such as when the word "cat" is used to stand for the cat family, i.e., mammals with many similar attributes (62). With age, the numbers of symbols and concepts and their levels of abstraction change. Children gradually acquire new words for objects, and this expansion of vocabulary increases the numbers of symbols and concepts available to them. There is evidence, however, that concepts can exist independent of language. Kagan (44, p. 212) notes that adults may possess concepts for conditions that have no real names, e.g., "the silence of a forest, the tension of a city, or the special feeling of fatigue that follows an hour of strenuous exercise, each category being based on the qualities shared across similar experiences." Additionally, evidence exists that preverbal infants can clump together objects with similar features and therefore have rudimentary conceptual abilities before they use words for communication (63).

With development, the quantity and quality of mental representations change. As Kagan (44, p. 217) explains, "a concept is not a static element of knowledge but a dynamic and constantly changing one. The mind is continually working on its knowledge and detecting features shared by two events or ideas that were originally separate and unrelated". Initially, children tend to categorize objects according to action and function or according to some perceived similarity between the objects. Apples and oranges may be combined because they are objects that are eaten (an action) or because they are both round (perceived similarity). Later in development these same two objects may still be placed together in a more sophisticated superordinate concept, which is based on the child's recognition that objects share some common feature. In this case, apples and oranges are categorized as fruit. Clearly, superordinate classifications are affected by experience. Piaget claimed that superordinate concepts were not used until around the ages of 7 or 8 when children entered the stage of concrete operations. As children have greater knowledge of the world, experiences with different types of objects, and as they acquire various new words, superordinate categories are more numerous and accessible.

### Cognitive Rules

Another mental unit is called a cognitive rule or proposition. This is the relationship between two or more concepts: "It rains when there are clouds in the sky", or "animals have ears." Just as children are gradually able to think in terms of more than one dimension at a time, children also gradually begin to use two or more propositions in solving problems. With age, propositions become more numerous and more coordinated (44). Thus, adolescents display greater problem-solving abilities than children because of their increased flexibility in generating and in using concepts and rules.

The increasing access of concepts and of rules is important because it enables a certain economy of thinking. Rather than having to mentally represent (and keep accessible) every single object encountered, children are able to group objects together as concepts and to generate rules for the relationships of objects in the world. This enables older children to have even more bits of information available to them because of their increasing ability to clump information together. This ability is most evident in the cognitive process of memory.

### Cognitive Processes

Cognitive processes are defined by Kail and Bisanz (61) as mental activities that generate, transfer, or manipulate representations. Mental representations and mental processes form a system of knowledge or a knowledge base that is sometimes referred to as "long term memory" or "memory storage." (Kail and Bisanz resist such terminology as "memory" because of the false implication that it resides somewhere and can therefore be localized in the brain [see chapter 9]). Cognitive processes include attention, sensation, and perception, which are involved in receiving information. The next processes transfer information to memory (encoding), where it is stored and then retrieved for later use (Fig. 11.11).

### Attention

Attention, which was examined in detail in chapter 8, involves orienting toward one stimulus or event to the exclusion of others. Attention appears to be initially stimulus bound; i.e., it is determined in large part by the perceptual characteristics of the stimulus (chapter 8). Thus, infants are fairly distractable and compelled by the nature of stimuli in their environments. Around the ages of 5 to 7, however, a gradual shift occurs in the nature of attention, and the child seems less stimulus bound, is able to pay attention for longer periods of time, and seems to become more cognitively oriented. Attention in middle and late childhood gradually becomes more self-controlled, intentional, and systematic so that the youngsters are not bound by the nature of the stimuli. They can determine the salient characteristics important for specific circumstances, focus appropriately on that information, act reflectively, i.e., take time to evaluate information, and respond flexibly by imposing organizational strategies as the nature of the task and information change (66).

### Five-to-Seven Shift

This change in attentional mechanisms, which occurs at about ages of 5 to 7 is partly the basis for speculation that infant and preschool thought is primarily perceptually oriented but gradually becomes conceptual and language based. The use of language, concepts, and other complex mental units enables older children to think more flexibly and to impose their knowledge on certain forms of information. The 5-to-7 shift, as it has been termed by White (65), is important in numerous approaches to development. For example, Piaget believed that a major change in cognitive development occurs at about this time as the child shifts from prelogical thinking during the preoperational period to concrete operational thought. In addition, the observation of significant improvements in sustained attention as well as apparent physical changes that occur at this age have led to speculation that corresponding internal changes occur in the 5- or 6-year-old child's body and brain (66).

Mussen, Conger and Kagan (62) point out that, starting at about 4 to 5 years of age, a set of cognitive processes begins to emerge and gradually takes control of thinking by early adolescence. These "executive functions," as they call them, monitor and coordinate perception, memory, and reasoning so that the child gradually becomes more adept at problem solving and thinking. Improved cognitive abilities become more apparent because the older child and adolescent are increasingly able to draw on memories of past or similar experiences and to use information stored in memory for problem solving.

### Memory

Memory is a highly studied and poorly understood process that involves the retention of information over time. Kail and Hagen (67) point out that memory consists of an assortment of cognitive processes that develop at different rates. Some aspects of memory are already adult-like in childhood, while others continue to change during adolescence and adulthood. It has been speculated that information is processed in three consecutive stages in memory: sensory register, short-term memory, and long-term memory. Once information has been sensed, it is stored briefly in short-term memory for about 30 seconds; then if the information is important, relevant, familiar, or rehearsed, it will be stored for long-term access. Long-term memory is the knowledge base of information used in thought processes and problem solving. For example, we may store in short-term memory an unfamiliar telephone number just long enough to dial it, but if we reach a busy signal, we may find that the number has been forgotten and that we must look it up again. In contrast, familiar numbers that are frequently used and that we have rehearsed often are stored in long-term memory and are readily accessible when needed.

Memory research indicates that we impose certain strategies for encoding (taking in and transferring) information in memory and that these strategies change with age. Three main strategies include rehearsal, which is repeating the information over and over; elaboration, e.g., writing a list; and organization or categorization (68). For example, similar bits of information are clumped together in *chunks*. Older children, with greater vocabularies and concepts available to them, can represent more elements in memory partly because of their ability to form chunks and to include more information per chunk than younger children. The previous section of this chapter discussed how older children are able to step back from, and organize, information thereby using increasingly more effective and powerful methods for solving problems. This increased efficiency of thought, which goes on from about the age of 5 throughout adolescence, is partly due to more rapid or effective use of memory and of certain strategies for encoding and retrieving information (61).

Retrieval of information from memory occurs in different ways. Recall requires that we produce the entire stored item from memory, whereas recognition requires only that we identify information we previously learned. An essay or short answer test calls upon recall, whereas recognition is used with a multiple choice test. With age, both recognition and recall abilities improve but at all ages recall is more difficult. For example, you may be able to recognize all your classmates but are unable to recall all their names.

Developmental changes in recall abilities are difficult to assess because they are dependent upon word or language use. Thus infants who cannot yet talk or write cannot be assessed for this memory ability (69). Recognition abilities, however, have been assessed, and appear to be highly accurate throughout childhood (64). The fact that during habituation studies (chapter 8), fetuses and newborns respond differently to stimuli as they are repeatedly presented over a period of time indicates that the ability to retain stored information is functional very early in development (69). Specific developmental shifts appear to occur in regard to memory efficiency at about 5 years of age and at adolescence. Five-year-old children consistently outperform 3-year-old children on memory tasks; with age, memory abilities gradually improve, due not to changes in storage capacities but to the use of more efficient and flexible strategies for processing (i.e., encoding and retrieving) information (67, 69). Compared with older children, preschoolers rarely use strategies for remembering information, especially when that information is novel or has very little relevance to them. Such children also seem unable to respond to specific instructions to remember information (64, 67).

---

A preschool child, Emily, is shown pictures of 20 different animals and is instructed to remember them. Emily has no real understanding that the instructions necessitate her attempting to actively process information for later use.

Thus, during the 2-minute period given to her to remember the items, she makes up a little game where some of the animals talk to one another. Later, she is asked to recall what animals she had seen. She may remember a few based on their meaning to her. For example, she might remember a cat like the one she has at home and a monkey, like the one she saw at the zoo.

The same task is given to 9-year-old Alice. Alice is able to understand that the instructions mean she needs to "study" the pictures given to her. She therefore develops a specific method of organizing the pictures so that she can remember them. She may note that there are 10 farm animals, two common animals found around the house, and eight zoo animals. Within those animal categories, she might also find similarities and chunk information together; e.g., lions and tigers in one group; elephants, rhinos, and hippos in another, farm birds in another. Later, when asked to recall the pictures, she then uses the same strategy for retrieving the information. She may state to herself, "There were three different groups of animals: Dogs and cats in one group. The farm animals were cow and horse, chicken and duck and turkey, etc., and the zoo animals consisted of lions and tigers plus large, gray animals like elephants, rhinos, and hippos."

The older child's increased vocabulary (e.g., knowing "rhino" and "hippo"), understanding of concepts (farm animals, zoo animals, domestic animals) and her ability to understand what memory is and to develop strategies for encoding and retrieving make her a more efficient and effective information processor. The older child's ability to think about memory and her awareness that she should apply some kind of active process in assisting to remember involves a certain degree of self-awareness and level of cognitive development. This ability to "think about thinking" or to think about remembering is called *metacognition* or *metamemory*.

# Metacognition

The prefix *meta* is used to apply to one's awareness of mental events. Metacognition is a fairly recent and still poorly defined area of research that is often associated with the work of cognitive theorist John Flavell (70) and that refers to one's knowledge about one's own mental states, abilities, and tasks. The statement, "I'd better write that down because I'll forget it" involves metacognition (64). Another example is provided by students who start planning for a test. First they evaluate whether they know the information, plan ahead by structuring their time to prepare for the test, and periodically monitor their understanding by giving themselves brief tests while they attempt to learn the information (71).

Developmental changes appear to occur with metacognitive abilities. Preschool children are unable to grasp the full meaning of such terms as attention and memory and therefore may not apply such instructions as: "Pay attention and remember this." Additionally, such youngsters tend to overestimate their memory abilities, believing they can accomplish cognitive tasks at which they do not succeed. Around the ages 9 or 10, however, children begin to understand that they can apply specific executive processes to different tasks. They are now able to evaluate, plan, and regulate their own thinking abilities. Flavell (70) suggests that these kinds of processes can be taught to children who do not use them or to children who do not use them effectively to improve their learning skills. These training procedures, which Flavell calls "cognitive monitoring strategies", may help children to improve their memory capacities and may have subsequent effects on other cognitive and social processes.

One of the implications of training children regarding cognitive monitoring is that it may assist them with learning to monitor their own thoughts in regard to everyday social communications and pressures; e.g., in regard to social messages from media or peers regarding smoking, or drug abuse. Thus, some cognitive monitoring strategies, Flavell points out, may be used to help youngsters in formal educational settings as well as to help children, adolescents, and adults to make wise and thoughtful life decisions (70). There may be limitations to the applications of metacognitive strategies, however. Fabricius and Wellman (71) note that while some metacognitive strategies can be taught to children, the strategies tend to be effective only for specific kinds of tasks. Furthermore, young children are often unable to generalize strategies they learn regarding one set of information to different kinds of settings. Since the field of metacognition is such a recent focus of research, continued study should point out additional applications and limitations as well as developmental changes in this area.

# INTERRELATIONSHIP BETWEEN COGNITIVE AND SOCIAL DEVELOPMENT

Metacognition refers to the ability to reflect on one's own patterns of thought and therefore requires a certain degree of cognitive as well as social development underlying self-understanding. Although they are quite distinct areas with their own theoretical bases, social and cognitive development are closely interrelated. This is probably no more evident than in language acquisition. As discussed in this chapter, the acquisition of words used as symbols for objects signals a change in the structure of the child's thought processes. Language acquisition may facilitate the use of effective thinking strategies and therefore clearly interacts with the development of cognition. Language, also, is the means whereby individuals communicate their needs and interact with others. It is a powerful and significant social tool. The development of language and additional social-cognitive processes is explored further in the next and final chapter of this book.

## References

1. Rohwer WD, Ammon PR, Cramer P: *Understanding Intellectual Development*. Hinsdale, IL, Dryden Press, 1974.
2. Sternberg RJ: *Beyond IQ. A Triarchic Theory of Human Intelligence*. Cambridge, UK, Cambridge University Press, 1985.

3. Vernon PE: *Intelligence: Heredity and Environment*. San Francisco, WH Freeman, 1979.
4. Ausubel DP, Sullivan DV: *Theory and Problems of Child Development*, ed 2. New York, Grune & Stratton, 1970.
5. Butterfield F: Why Asians are going to the head of the class. *New York Times Education Life* Aug 3, 1986, pp 18–23.
6. Dennis W: *Children of the Creche*. New York, Appleton-Century-Crofts, 1973.
7. Skeels HM: Adult status of children with contrasting early life experiences: A follow-up study. In Bronfenbrenner U (ed): *Influences on Human Development*. Hinsdale, IL, Dryden, 1972, p 224.
8. Piaget J: *Psychology of Intelligence*. Paterson, NJ, Littlefield, Adams & Co, 1963.
9. Turner RR, Connell DB, Mathis A: The preschool child or the family? Changing Models of Developmental Intervention. In Turner RR, Reese HW (eds): *Life-Span Developmental Psychology. Intervention*. New York, Academic Press, 1980, p 249.
10. Laosa LM: The sociocultural context of evaluation. In Spodek B (ed): *Handbook of Research in Early Childhood Education*. New York, The Free Press, 1982, p 501.
11. Bissell JS: Planned variation in Head Start and Follow Through. In Stanley JS (ed): *Compensatory Education for Children Ages Two to Eight: Recent Studies of Educational Intervention*. Baltimore, MD, Johns Hopkins University Press, 1973, p 63.
12. Bronfenbrenner U: Is early intervention effective? Facts and principles of early intervention: a summary. In Clarke AM, Clarke ABD (eds): *Early Experience. Myth and Evidence*. New York, The Free Press, 1976.
13. Schweinhart LJ: Comment on "intelligence research and social policy". *Phi Delta Kappan* 63: 187, 1981.
14. Brown B: Head Start. How research changed public policy. *Young Children* 40: 9–13, 1985.
15. Lazar I: Early intervention is effective. *Educ Leadership* 38: 303–305, 1981.
16. McCall RB, Hogarty PS. Hurlburt N: Transitions in infant sensorimotor development and the prediction of childhood IQ. *Am Psychol* 27: 728–748, 1972.
17. McCall RB: The development of intellectual functioning in infancy and the prediction of later IQ. In Osofsky J (ed): *Handbook of Infant Development*. New York, Academic Press, 1979, p 707.
18. Zigler E: Familial mental retardation: a continuing dilemma. *Science* 155: 292–298, 1967.
19. Moynihan DP: *Family and Nation. The Godkin Lectures*. San Diego, Harcourt Brace Jovanovich, 1986.
20. Weinberg RA: Early childhood education and intervention. Establishing an American tradition. *Am Psychol* 34: 912–916, 1979.
21. Jensen AR: How much can we boost I.Q. and scholastic achievement? *Harvard Educ Rev* 39: 1–123, 1969.
22. Eysenck H, Eysenck M: *Mindwatching*. London, Michael Joseph, 1981.
23. Bronfenbrenner U: Ecology of the family as a context for human research perspectives. *Dev Psych* 22: 723–742, 1986.
24. Bronfenbrenner U: Is 80% of intelligence genetically determined? In Bronfenbrenner U (ed): *Influences on Human Development*. Hinsdale, IL, Dryden Press, 1972, p 118.
25. Zajonc RB, Markus GB: Birth order and intellectual development. *Psychol Rev* 82: 74–88, 1975.
26. Henderson RW: Home environment and intellectual performance. In Henderson RW (ed): *Parent-Child Interaction. Theory, Research and Prospects*. New York, Academic Press, 1981, p 3.
27. Altus WD: Birth order and its sequelae. *Science* 151: 44–49, 1966.
28. Schooler C: Birth order effects: Not here, not now. *Psychol Bull* 78: 161–175, 1972.
29. Clarke AM, Clarke ADB (eds): *Early Experience. Myth and Evidence*. New York, Free Press, 1976.
30. Birch HG: Health and the education of socially disadvantaged children. In Bronfenbrenner U (ed): *Influences on Human Development*. Hinsdale, IL, Dryden Press, 1972, p 131.
31. Brim OG, Kagan J (eds) *Constancy and Change in Human Development*. Cambridge, MA, Harvard University Press, 1980.
32. Kagan J: Emergent themes in human development. *Am Scient* 64: 186–196, 1976.
33. Fagan JF: The intelligent infant: theoretical implications. *Intelligence* 8: 1–9, 1984.

34. Vygotsky LS: *Mind in Society: The Development of Higher Psychological Processes.* Cambridge, MA, Harvard University Press, 1978.
35. Palisano RJ, Lydic JS: The Peabody Developmental Motor Scales: an analysis. *Phys Occup Ther Pediatr* 4: 69–75, 1984.
36. Cattell RB: Theory of fluid and crystallized intelligence. *J Educat Psychol* 54: 1–22, 1963.
37. Horn JL, Cattell RB: Age differences in fluid and crystallized intelligence. *Acta Psychol* 26: 107–129, 1967.
38. Schaie KW, Labouvie-Vief G: Generational versus ontogenetic components of change in adult cognitive behavior: a fourteen-year cross-sequential study. *Dev Psychol* 10: 305–320, 1974.
39. Maccoby EE: Sex differences in intellectual functioning. In Maccoby EE (ed): *The Development of Sex Differences.* Stanford, CA, Stanford University Press, 1966.
40. Gardner H: *Frames of Mind. The Theory of Multiple Intelligences.* New York, Basic Books, 1983.
41. Sternberg RJ: A framework for understanding conceptions of intelligence. In Sternberg RJ, Detterman DK (eds): *What is Intelligence? Contemporary Viewpoints on its Nature and Definition.* Norwood, NJ, Ablex Publishing, 1986.
42. *Webster's Seventh New Collegiate Dictionary.* Springfield, MA, Merriam, 1972.
43. Gardner H: *The Mind's New Science. A History of the Cognitive Revolution.* New York, Basic, 1985.
44. Kagan J: *The Nature of the Child.* New York, Basic Books, 1984.
45. Baldwin AL: *Theories of Child Development.* New York, John Wiley & Sons, 1967.
46. Phillips JL: *The Origins of Intellect. Piaget's Theory,* ed 2. San Francisco, WH Freeman, 1975.
47. Fischer KW: A theory of cognitive development: The control and construction of hierarchies of skills. *Psychol Rev* 87: 477–531, 1980.
48. Piaget J: *The Origins of Intelligence in Children.* New York, WW Norton, 1952.
49. Gardner H: *Developmental Psychology. An Introduction.* Boston, Little Brown, 1978.
50. Gratch G: Recent studies based on Piaget's view of object concept development. In Cohen LB, Salapatek P (eds): *Infant Perception: From Sensation to Cognition. Vol. II. Perception of Space, Speech, and Sound.* New York, Academic Press, 1975, p 51.
51. Looft WR: Children's judgments of age. *Child Dev* 42: 1282–1284, 1971.
52. Kratchowill TR, Goldman JA: Developmental changes in children's judgments of age. *Dev Psychol* 9: 358–362, 1973.
53. Scarr S, Weinberg RA, Levine A: *Understanding Development.* New York, Harcourt Brace, 1986.
54. Bower TGR: *Development in Infancy.* San Francisco, WH Freeman, 1974.
55. Feldman DH: *Beyond Universals in Cognitive Development.* Norwood, NJ, Ablex, 1980.
56. Gelman R: Preschool thought. *Am Psychol* 34: 900–905, 1979.
57. Zaporozhets AV, Zinchenko VP, Elkonin DB: Development of thinking. In Zaporozhets AV, Elkonin DB (eds): *The Psychology of Preschool Children.* Cambridge, MA, MIT Press, 1971.
58. Harris FA: The brain is a distributed information processor. *Am J Occup Ther* 24: 264–268, 1970.
59. Hampden-Turner C: *Maps of the Mind.* New York, MacMillan, 1981.
60. Crick FHC: Thinking about the Brain. *Scient Am* 241: 219–233, 1979.
61. Kail R, Bisanz J: Information processing and cognitive development. In Reese HW (ed): *Advances in Child Development and Behavior.* New York, Academic Press, 1982, vol 17, p 45.
62. Mussen PH, Conger JJ, Kagan J: *Child Development and Personality,* ed 5. New York, Harper & Row, 1979.
63. Cohen LB, Strauss MS: Concept acquisition in the human infant. *Child Dev* 50: 419–424, 1979.
64. Paris SG, Lindauer BK: The development of cognitive skills during childhood. In Wolman B (ed): *Handbook of Developmental Psychology.* Englewood Cliffs, NJ, Prentice-Hall, 1982, p 333.
65. White SH: Evidence for a hierarchical arrangement of learning processes. In Lipsitt LP,

Spiker CC (eds): *Advances in Child Development and Behavior*. New York, Academic Press, 1965, vol 2, p 187.

66. Tanner JM: *Fetus into Man. Physical Growth from Conception to Maturity*. Cambridge, MA, Harvard University Press, 1978.

67. Kail R, Hagen JW: Memory in childhood. In Wolman B (ed): *Handbook of Developmental Psychology*. Englewood Cliffs, NJ, Prentice-Hall, 1982, p 350.

68. Fabricius WV, Wellman MH: Memory development. In Frank M (ed): *A Child's Brain. The Impact of Advanced Research on Cognitive and Social Behaviors*. New York, Haworth, 1984, p 171.

69. Perlmutter M: A life-span view of memory. In Baltes PB, Featherman DL, Lerner RM (eds): *Life-Span Development and Behavior*. Hillsdale, NJ, Lawrence Erlbaum, 1986.

70. Flavell JH: Metacognition and cognitive monitoring. A new area of cognitive developmental inquiry. *Am Psychol* 34: 906–911, 1979.

71. Glaser R: The future of testing. A research agenda for cognitive psychology and psychometrics. *Am Psychol* 36: 923–936, 1981.

# 12

# The Developing Person in a Social World

Societies are composed of individuals who not only share common traits and behavior patterns with other members of their species but also have their own, unique characteristics that distinguish them from others. To a great extent the person each of us becomes is affected by what we learn about ourselves through interactions with others in our own particular society and culture. *Socialization* is the process whereby individuals learn the appropriate behaviors, beliefs, attitudes and values of their particular culture (Fig. 12.1). In contrast, *personality* refers to the relatively enduring patterns of traits that make individuals unique. The term "relatively enduring" is significant because it differentiates personality from fleeting, temporary states such as fatigue, emotional reactions, or drug-induced behavioral changes.

*Emotions* are feelings such as joy, sadness, or envy. They are typically temporary reactions to people or events in the environment and are accompanied by changes in internal physiology as well as external facial expressions (Fig. 12.2) (1). While emotions and personality are often studied separately, they are interwoven. For example, people adopt specific strategies for handling their emotions and are often described according to the characteristic way they present themselves to the world, e.g., cheerful, angry, "flat," anxious, or sullen. In addition, emotions may become fully integrated into one's personality, such as when an individual experiences an affective disorder like chronic depression. Since emotions are often responses to a social world and emerge within a social context, these two domains are often combined into an integrated topic called "socioemotional" (2) or social-emotional development (3).

Figure 12.1. Socialization is the process whereby individuals acquire the values and behaviors of their culture.

Traditionally, the scientific study of human development tended to concentrate on domains in relative isolation from one another. Recently, however, there is a growing tendency to combine study across developmental domains. Thus not only do we now speak of "sensorimotor," "cognitive-perceptual-motor," and "socioemotional" development but there are also areas such as "social learning," "social cognition," "cognitive social learning," and others. The study of social development covers many topics and necessitates looking at sensory, perceptual, motor, and cognitive functions within a social context. Thus, consistent with the rest of this text, this chapter will take an ecological approach and recognize the interaction among developmental domains that make up the whole person. The ecological approach looks at the person, the environment, and the interrelation-

Figure 12.2.   Emotions are feelings accompanied by facial expressions and internal physiological changes.

ships between them. Social beings cannot be fully understood without also looking at their relationships with other people and objects in their environment.

This chapter will examine personality formation, language acquisition, and other behaviors that occur within a social context such as the development of sex roles and sexual behavior, and the changing roles and interactions with the social group as the individual moves from infancy to childhood and adolescence. It should be noted that entire books have been written about many of the topics covered in this text. This chapter will highlight only certain relevant issues from many diverse fields. The comprehensive list of references cited in this chapter should provide direction for continued study for the reader who wants to explore individual topics in more detail.

# SOCIALIZATION

Socialization is the process by which individuals acquire the appropriate behaviors of the society in which they live. The study of social development looks at how individuals become socialized over their lifetime. As a starting point in understanding how people develop and interact with the various social forces in their lives, some of the more common theories of personality formation and socialization will be examined.

# Theories of Social Development

How is it we progress from helpless, nonverbal, demanding newborns to expressive, communicative, sometimes cooperative, and certainly very social and sexual adolescents and adults? Different theories have been proposed to explain the socialization process (Table 12.1). These theories are expansions of those already examined in detail in chapters 5 and 6, thus the reader may want to review those chapters if unfamiliar with Freud's work or with the principles of learning and conditioning that make up learning theory.

## Identification

Most important in the development of a social self is the simultaneous formation of an individual identity. Freud's psychoanalytic theory of personality development details how an individual progresses through separate psychosexual stages, as sexual energy is transferred from significant parts of the body until it is eventually centered maturely on the genitals. According to Freud (chapter 6), during the third psychosexual stage, the child's sexual feelings toward the opposite sex parent (i.e., the boy's feelings toward his mother) emerge. He wants to be close to his father, yet he also perceives his father as a rival for his mother's affections. He recognizes his father's power and central position in his mother's life, desires what his father

Table 12.1
Socialization Theories

---

Psychoanalytic Theory
  Freud's psychosexual stages and the concept of identification
  Erikson's eight stages of psychosocial development
Learning Theory (focus on the role of environment)
  Traditional learning theory—based on trial and error performance and
    rewards or punishment for behavior
  Social learning theory—an amalgam of psychoanalytic and learning
    approaches
    Observational learning (vicarious learning, modeling)—based upon
      imitation of previously observed actions
    Cognitive social learning—based on expectancies for feedback for certain
      behaviors
Cognitive-Developmental Theory (focus on the role of the child)
  Piaget's cognitive theory of development
  Traditional maturational view of biologically controlled social behavior
    emerging gradually over time
  Contemporary cognitive/perceptual structural view of social concepts that
    direct behavior

---

possesses, and yet is afraid of his father's power and potential threat. Thus, knowing he cannot directly and successfully compete with his father, he does the next best thing—he becomes like his father. As Baldwin (4, p. 366) describes, "[O]ne of the ways that people adjust to the continual presence of a feared and powerful person is to accept his values, become like him, and repress the hostility they feel toward him."

This process of taking on the characteristics of someone who is perceived to be powerful and important is called identification. As Freud proposed, it is an unconscious process which involves actually internalizing aspects of another's personality. In so doing, the child begins to construct his or her own personality after someone else.

Certain factors seem to facilitate identification, and certain consequences occur as a result of this process. Children are most likely to identify with people whom they perceive to be powerful and to whom they consider themselves similar. Thus, the identification process may be facilitated by such remarks from others as, "What a good little boy! You are so grown up and just like your daddy" or "When your daddy isn't here, you are the 'man of the house.'"

Freud claimed that by identifying with his father, the boy resolves the Oedipal complex and becomes able to move on with his own psychosexual development. While Freud did not clearly describe the identification process for females, it is assumed that they exhibit similar responses toward their mothers. In addition, by identifying with the same-sex parent, the child actually becomes socialized by incorporating adult behaviors and taking on specific masculine or feminine characteristics as well as some behaviors indicative of adult social conscience. As an example, consider the preschooler who has been playing in the kitchen corner at school. She has been cleaning up with a broom when some other children bump into the table and knock some cups on the floor. At this moment, she places one hand on her hip and shakes her finger and her head at the children and says, in the same words and intonations as her mother: "Now you two behave or you will have to go to your rooms!" According to identification theory, by taking on some of her mother's behaviors, this little girl has naturally internalized certain adult, social standards as well as sex-typed behaviors.

## Social Learning Theory

Social learning theory is an amalgam of Freudian and learning theory. Baldwin (4) states that it takes some of the concepts and hypotheses of psychoanalytic theory and translates them into stimulus-response (S-R) terminology to describe the socialization process. In contrast, traditional learning theory (Table 12.1) excludes Freudian concepts and contends that social behavior, like other aspects of development, can be explained through the principles of conditioning and learning. Social learning theory recognizes that the infant exhibits a fundamental dependency and relies on caretakers to meet such physical and primary needs as hunger, thirst, and safety. Caretakers therefore become associated with primary reinforcers as they provide food and warmth, eliminate pain, and provide comfort and security.

According to traditional learning theory, objects repeatedly associated with primary reinforcers eventually share their properties and themselves become reinforcers, though secondary in nature. Thus, caretakers become secondary reinforcers, taking on reinforcing value themselves. At the same time that caregivers learn they are able to shape behavior by withdrawing or providing support and approval, infants are learning about their environments. They experience pleasure in the presence of caretakers and distress when separation occurs. Thus they learn to express themselves in order to bring attention and sustained contact with parents and to inhibit behaviors that provoke separation. With time, the youngster's behavior is actually shaped as infants and later children and adolescents continually learn what behaviors bring approval and what behaviors are punished or result in attention being withheld from them. Initially, parents, grandparents, and other caretakers and later teachers, siblings, and peers all become important socialization forces, as the child learns to inhibit behaviors that have brought punishment and to behave for smiles, hugs, a pat on the back, "thank you's," or other forms of social approval (see Fig. 5.3).

Social learning theory also accounts for children who always seem to be getting into trouble. Such children may have been reared in circumstances where the only time they receive attention is when they do something wrong. Thus, while negative attention seeking may result in disapproval, it at least provides some form of recognition (4). Negative attention is better than none at all.

Social stimuli, while not primary reinforcers like food or drink, are important shapers of behavior. Social reinforcers are associated with feelings

Figure 12.3. Researchers have long debated the significance of the mother-infant relationship and the role of nursing on the child's subsequent development.

of approval or rejection, and according to social learning theory, their significance is traced to the initial caretaking relationship. With continued interaction, infants and children learn the characteristics of different social

Table 12.2
Processes Involved in Observational Learning*

| |
|---|
| Attention |
|     Selective observation of the behavior to be modeled |
| Retention |
|     Remembering and retaining the observed behavior over time |
| Motor Reproduction |
|     Ability to physically perform observed behavior |
| Incentive or Motivation |
|     Ability to carry out observed behavior |

*ª* Adapted from Bandura A: *Principles of Behavior Modification.* New York, Holt, Rinehart & Winston, 1969.

and nonsocial objects in their environments. They associate certain feelings with specific people and can discriminate familiar people from strangers. Eventually they generalize those initial feelings for their caretakers to other people. Thus, through the processes of reinforcement, discrimination, and generalization (chapter 5), such complex phenomena as the attachment relationship, love, friendship, feelings of affection, and other social behaviors are explained. People become important and powerful shapers of the behavior of others, and during development, complex social behaviors are learned either directly through reinforcement or indirectly through observation and imitation (Fig. 12.1).

## Imitation or Observational Learning

The process of imitation is a powerful mechanism for learning what is, and what is not, appropriate social behavior as the child observes others, notices what behaviors are rewarded or punished, and behaves accordingly. While there is agreement that imitation accounts for much social learning, the mechanics of the imitative process are controversial. The controversy revolves primarily around two points: the degree to which modeled actions must actually be performed by the observer and the role of reinforcement in maintaining that performance.

### Reinforcement and Action in Imitation

Traditional learning theorists contend that only through trial and error activity does the child learn. Thus, the child must observe another person (called a "model"), see the model reinforced, actually perform the behavior, and then, if the behavior is to be sustained, receive reinforcement. A modification of this approach acknowledges the significance of reinforcement but recognizes that it may be intrinsic rather than overt. For example, children may observe a model and later perform the action in the absence of overt praise from someone else. In this case, some social learning theorists contend, children actually provide their own reinforcement by monitoring their performance and telling themselves they did a good job or that what they did was important (4–6).

Psychologist Albert Bandura (5) points out that numerous circumstances exist in which direct performance and overt reinforcement do not occur. He notes that it would be unimaginable for cultures to exist where customs, language, avocations and vocations, and sociopolitical practices must all be shaped gradually by trial-and-error learning. In the real world (as distinct from the laboratory, where much of learning theory originates) it would be fatal to learn so slowly. Thus, Bandura and others have proposed an account of socialization that short-circuits the more cumbersome and potentially life-threatening, slow process of trial-and-error acquisition of behavior typical of traditional learning theory.

### Imitation and Identification

According to Bandura, a form of socialization occurs in which overt reinforcement and trial-and-error shaping are not required. This form of learning is known by a variety of synonymously used terms: vicarious learning, observational learning, imitation, identification, and modeling. (In this text, the terms "imitation" and "observational learning" will be used synonymously. "Identification" will be used in reference to the process described by Freud.) Bandura claims that imitation and identification are the same processes, although other theorists recognize them as separate processes associated with two diverse theories. Identification (which may or may not be coupled with the Freudian concept of "unconscious internalization") is often viewed as a very generalized form of imitation that involves a much more subtle acquisition of another's personality by taking on their mannerisms, values, personal style, and beliefs. In contrast, imitation is believed to be a much more specific copying of someone else's overt behavior (4).

### Criteria for Observational Learning

There are certain prerequisites for observational learning (Table 12.2). Bandura notes that for children to later imitate some behavior they may have observed days or weeks earlier, they must be able to symbolically code that information, cognitively represent it and store it in memory, and then later retrieve and transform it from a mental structure to motor performance. Observational learning, therefore, requires a certain degree of sensorimotor and cognitive sophistication, which may, in turn, account for the immature socialization observed in some children who experience severe sensorimotor and cognitive delays. Some of these children may be less able to pay attention, to remember, or to motorically reproduce behaviors they have observed.

Observational learning is an important socialization force that can account for the acquisition of many social behaviors such as cultural customs and religious rituals or more mundane skills such as driving a car. In addition, imitation may, in part, account for the acquisition of language and of gender-related behaviors (sex-typing), both of which are examined in greater detail in later sections of this chapter.

### Cognitive Social Learning

Traditional learning theory focuses on trial-and-error performance and the reinforcement of overt behaviors. In contrast, Bandura's (5) theory of observational learning includes internal mental processes, such as attention and mental representation in memory, that mediate learning and behavior (6). Therefore, this area is referred to as "cognitive learning" or "cognitive social learning" because it integrates cognitive mediators within an S-R framework, or a social learning theory approach, respectively. It is important to note that the terms "observational learning" and "cognitive social learning" may be used to represent identical processes. If a difference between them is implied, it relates to the emphasis on cognitive events, such as attention and memory or the formation of mental expectancies about the likelihood of some future consequence if an observed behavior is later performed. The following example illustrates how these cognitive events are integrated into a learning theory framework.

Case Example of Observational and Cognitive Social Learning

Katie and Matthew's parents are both teachers who pride themselves on raising two thoughtful and gentle children. Katie, who is 5 years old, attends kindergarten, where she often participates with and observes her classmates at play. One day, one of the more aggressive classmates, Doug, goes over to another child, shoves him, and takes the toy with which he is playing. Katie observes this behavior and sees that it worked in obtaining for Doug her favorite toy and the one with which she wishes she could play.

At home later in the week, Katie is playing with her 8-year-old brother, Matthew. Matthew has Katie's toy. Katie wants it, but Matthew refuses to share and persists in teasing her. Katie then runs behind Matthew, shoves him, knocks him over, and grabs the toy. Her parents, watching this, are appalled that their quiet, sweet little girl has displayed an uncharacteristic and unapproved behavior. How, they ask, could this happen?

According to observational learning, Katie observed Doug at school and recognized that his behavior was successful in getting what he wanted. All the necessary requisites for observational learning were present. Katie was capable of paying attention, remembering, and physically producing the behavior she observed. Furthermore, the motivation existed. Katie wanted that toy!

A cognitive social learning approach would claim that Katie had established a certain expectancy that the act of shoving another child works in getting what you want. It worked for Doug. He got the toy he wanted (and one that Katie liked too), so it ought to work for her. That expectation of reinforcement (or of getting what you want) is a cognitive mediator and can be incorporated within an observational learning framework to describe the acquisition of social behavior.

## Comparison of Identification, Learning Theory, Observational Learning, and Cognitive Social Learning

According to cognitive social learning, children (and adults) acquire certain expectancies that if they behave in certain ways, then they will be reinforced for that behavior. Huston (6) points out that this offers one explanation for how gender-associated behaviors emerge. While Freudian identification theory would assume that the little boy begins to display masculine traits and behaviors by internalizing those aspects of the father with whom he has identified, learning theory claims that the little boy learns to behave a certain way because he receives punishment or reinforcement for his behaviors. According to observational learning, the child learns by observing other children who may be punished or rewarded for behaving in certain ways. The little girl, for example, observes her older sister who gets punished for soiling her dress and is cuddled for being "a good little girl and staying out of the mud puddle." Thus, through observation of others, the child learns what is and what is not appropriate behavior. Through this same process, but emphasizing the role of cognitive mediators, cognitive social learning theorists explain that the little girl learns to expect reinforcement for certain feminine behaviors and can, in fact, monitor her own behavior through these expectations of reinforcement or punishment. Thus, though the wording and emphasis are different, varied theoretical approaches explain the development of social behavior.

## Cognitive-Developmental Theory

Another approach to socialization is based on cognitive or maturational influences. Piaget's theory provides an example of this view. His description of imitation during the sensorimotor stage and the concept of egocentrism in preoperational thought (chapter 11) are excellent examples of the interrelationship between cognitive and social behavior. Piaget believed that early social ability was actually preceded by, and contingent upon, cognition.

Piaget's theory, like other cognitive-developmental approaches to socialization, emphasizes the role of mental or maturational factors that reside in the child (6). Thus, as Piaget contended, cognitive structures actually provide a framework for the emergence of social functions. Another example is the acquisition of sex-typed behavior, which cognitive-developmentalists claim occurs because the young child initially develops a mental concept of masculinity and femininity, and then behaves accordingly. In this way, a mental structure (a concept of sex-appropriate behavior) which resides within the child is viewed as the primary guide for behavior. This is in contrast to learning theories that emphasize environmental factors, not internal cognitive structures, as the primary determinants of behavior.

Cognitive-structural approaches are becoming increasingly popular. As chapter 9 discusses, contemporary studies of infant abilities have led to a growing recognition of the existence of sophisticated perceptual/cognitive capacities early in life. Similar conclusions have been drawn in the field of early social development by researchers who propose that innate or early emerging social processes may exist and that these may function to initiate or sustain social interactions. Condon and Sander (7) report that early infant motor responses may actually be synchronized to fluctuations in adult speech (see chapter 9). Similarly, Meltzoff and Moore (8) point out that neonates are able to imitate different adult facial gestures, and this ability may serve as a basis for maintaining a mutual rhythm in early social interactions. While these imitative abilities are not fully formed and require continued elaboration, they do appear earlier than Piaget had recognized and, as Meltzoff and Moore note, may serve as a channel for early communication and as a basis for subsequent language and social development.

The cognitive-developmental approach to socialization reasons that since these sensorimotor behaviors which act to sustain social interaction are present so early in development, they may be based on innate structures residing within the child and are therefore quite different from socialization processes that are only gradually acquired through trial:and:error learning.

## Conclusion About Socialization Theories

In the study of social development, as in other domains, the nature-nurture controversy (see chapter 2) continues to guide research. Environmental, learning approaches to social development emphasize the role of nurture, while cognitive-developmental theories focus on innate structures residing in the child. The premise of this chapter, as has been throughout the text, is that despite theories that may direct us otherwise, development is neither completely learned nor completely innate, but a mixture of both. Social development, like other domains, is a result of a rich blend of natural and learned behaviors. Thus, as students of human development, we are challenged not to adopt one theory of socialization but instead to integrate several in flexible and evolving approaches to social behavior.

## ATTACHMENT

Attachment is one of the most highly researched areas in the scientific study of human development. Different theories of attachment exist, and each draws implications regarding the subsequent social-emotional, personality, and sometimes perceptual-motor and cognitive development of the child. While it is not always clearly defined in research, the term "attachment" refers to an emotional bond between individuals and their most intimate companion(s) (9) or between organisms and some inanimate object (10). In the scientific study of human development, attachment has traditionally referred to the special relationship between infants and their mothers. The research of the 1940s and 1950s tended to focus exclusively on the attachment responses of infants (see chapter 9). This does not mean that maternal responses were considered unimportant, only that the research orientation and scope at that time was directed at the child.

In 1960, in his book *Attachment* (11), psychoanalytic theorist John Bowlby used the term "caretaking," to refer to parents' behavior that is reciprocal to attachment. Later, in the 1970s Klaus and Kennell (12) introduced the term "bonding" (see chapter 7) to refer to the emotional tie formed by parents early in their relationship with their infant. Once it was recognized that infants affected their parents and that the attachment relationship was a reciprocal, not a one-way, process then developmentalists were forced to regard the infant as an active, contributing partner in a mutually inter-active and engaging relationship. Thus, as notions of the attachment process changed over the decades, so also did existing notions about the nature of the developing child. This evolution of a contemporary view of attachment is important in the science of development and will be described in detail here.

# Theories of Attachment

Theories regarding the origin and causes of attachment vary (Table 12.1). Consistent with the theory of psychosexual development, psychoanalytic theorists emphasized the importance of the mother's breast, close physical contact, and the oral gratification received during nursing for forming early attachment responses. Psychodynamic theory recognized the mother as the object to which the infant directed affections and therefore the fundamental source of infant love. In contrast, social learning theorists contended that, by meeting the infant's needs for food and warmth, caretakers become secondary reinforcers (Fig. 12.3). Attachment is explained in terms of the infant's learning to associate certain people with the gratification of basic needs. A third approach, ethological theory, emphasizes the continuity of close maternal-infant relationships across wide varieties of species. The phenomenon of imprinting is used as a model for the human attachment process.

## Bowlby's Psychoanalytic-Ethological View

In *Attachment*, Bowlby presents a combined psychoanalytic-ethological theory which has served as a guiding influence in traditional attachment research. Bowlby described how infants of different species engage in what he calls "proximity-seeking behaviors" which include sucking, clinging, following, crying, and smiling. Not unlike imprinting described by the ethologists, the adaptive function of proximity-seeking behaviors is to bring young animals into contact with their mothers (13). Crying and smiling bring the "mother to the infant" and keep her close, whereas following and clinging behaviors bring the "infant to the mother" and keep the infant close. Following with the eyes is another common proximity-seeking behavior that functions to reassure the infant of the mother's presence (11).

Bowlby claimed that attachment follows a maturational course throughout life. While initial attachment responses are unpredictable and variable during the first year, they are strong and consistent during the second and third years of life. Bowlby claimed that the behavioral systems that maintain proximity to mother are activated by her departure and terminated when she comes within sight, sound, or touch. Thus, the infant may cry and attempt to follow mother when she leaves and will greet and approach her when she returns. Attachment behaviors are fairly strong and predictable until the end of the third year when the child undergoes developmental changes, and the need to be close to mother becomes less urgent. During adolescence, attachment gradually grows weaker; during this time and in adulthood, attachment relationships are formed with other people as well as directed toward groups or institutions. In old age, when an elderly person's parents and peers may no longer be alive, attachment behaviors are often directed toward younger generations (11).

## Ainsworth's Approach

The research of psychologist Mary D. Salter Ainsworth (14–16) supported and extended Bowlby's conclusions. Ainsworth (14) conducted a short-term longitudinal study of infants and mothers in the African country of Uganda; from this and other research, she developed general conclusions about attachment as an orderly, maturational process. She claimed that this progression is characterized by the appearance of both *separation* and *stranger anxiety*, which infants experience when their mothers leave them or when they are confronted by unfamiliar persons, respectively. Although her work supported Bowlby's view about the maturational nature of this system, Ainsworth observed that within and across cultures, there are wide individual differences in the timing and expression of attachment behaviors. This observation led her to conclude that infant care practices are able to alter (speed up or delay) attachment responses. Ainsworth notes that some scientists attach almost exclusive importance to the child's genetic and constitutional forces, whereas others have equated development with learning. In contrast, she concludes (14, p. 388), "Because these two divergent views are widely held, it seems necessary to assume a stance and to assert that my position rests on a thorough-going belief that development is an interactional process—an interaction between the constitutional elements and environmental circumstances."

Ainsworth's continued research has pointed to potential long-term effects of the quality of the early mother-infant relationship on the infant's development. As a result of another longitudinal study conducted in the United States, Ainsworth concluded that infants form different kinds of attachment relationships with their mothers. Secure attachments result when mothers are consistently responsive to their infants' signals and communications. In contrast, if mothers are inconsistent in their responsiveness or avoid contact with their infants, then anxious attachment may result (16).

Ainsworth observed that infants do not continually cling to their mothers but instead make brief excursions to explore other objects and people in the environment. The mother, she claims, serves as a secure base from which infants can direct these excursions. If, however, the infants are frightened or placed in an unfamiliar (strange) situation, then they will flee to the mother as a "haven of safety." The "strange situation" is a seminaturalistic setting where mother-infant interaction can be experimentally investigated (15).

Securely attached infants will use the mother as a base of security and will gradually begin to explore familiar and strange situations. The mother will often encourage these brief separations, and through them, the child encounters inanimate objects as well as other people in the environment. As Sroufe (17) notes, these excursions from a secure base provide a foundation for a child's sense of independence and mastery. Rheingold and Eckerman (18) report that the advantages of enhanced exploratory behavior include increased perceptual stimulation, new opportunities to manipulate and learn about objects, and the acquisition of new techniques for the child to control external events in the environment. Furthermore, Ainsworth has observed that, when compared with anxiously attached infants, securely attached youngsters will tend to be more cooperative and less aggressive toward their mothers as well as more competent and sympathetic in their interactions with peers (15). Thus, secure attachment may enhance exploratory behavior, thereby facilitating locomotor, perceptual, cognitive, and diverse forms of social-emotional development and competence.

## Spitz and Bowlby: Institutionalization and Maternal Deprivation

In the 1930s and 1940s reports detailed the effects of institutional life on child development. Psychoanalytic theorists René Spitz and Bowlby described an entire syndrome of depression and withdrawal that was seen in many of the children who had been separated from their mothers and reared in institutional settings, where stimulation was often inadequate and inconsistently provided by multiple caretakers (19). Bowlby synthesized these findings, and in 1951 reported to the World Health Organization (WHO) his concern about the "deplorable patterns of institutional upbringing" (20).

Although Spitz recognized that restricted environmental stimulation contributed to the institutionalization syndrome (21), both Spitz and Bowlby have (in some cases, erroneously) become associated with a dogmatic position regarding the deleterious effects of interrupted attachment. It has since been assumed that the profound negative effects of institutionalization, such as mental and physical retardation, delinquency, depression, and other forms of psychopathology, were actually caused by maternal deprivation. Bowlby's often-quoted statement before the WHO illustrates that view. He claimed that mother love early in life is as essential for the child's mental health as are vitamins and proteins for physical health (20). He also reported that the infant requires an intimate and continuous relationship with a mother or a permanent mother substitute, and that the threat of loss or actual loss of such an individual creates anxiety, sorrow, anger, or possibly even psychopathology (19).

## Critical Discussion of Psychoanalytic-Ethological View

Certain characteristics of attachment (Table 12.3) are associated primarily with the work of Bowlby and Ainsworth and comprise the commonly accepted view of attachment and maternal deprivation during the 1960s.

**Table 12.3**
**Characteristics of Psychoanalytic-Ethological View of Attachment**

Attachment serves an adaptive, evolutionary function:
  It can be compared to similar responses (e.g., *imprinting*) in other species.
  Infants naturally respond to their mothers.
  Adults are naturally programmed to respond to infants' signals for protection.
A single, consistent caretaker (preferably the mother) is essential.
There is a critical period for formation of attachment bond.
Absence of mother (maternal deprivation) may lead to psychopathology.
Once attachment to the mother has occurred, then similar responses may be
  directed at mother substitutes.
Attachment responses go through a certain maturational sequence, of which
  separation anxiety and stranger anxiety are a part.
Attachment serves as a basis for subsequent development in many different
  domains.
Separation and stranger anxiety measured by the "strange situation" reflect the
  quality of the attachment relationship.

This view, however, was not universally accepted. Ainsworth (14) described how anthropologist Margaret Mead disagreed with Bowlby's claims about maternal deprivation and the need for a consistent, single caretaker. Reporting to the WHO at the same time as Bowlby, Mead described her observations of effective multiple caretaking in other cultures. Learning theorist Harry Harlow was also interested in testing some of the general assumptions about attachment. Theoretical discussions of attachment provided a focus for controversy in the scientific study of development. Bowlby had attacked the fundamentals of learning theory, and he rejected the view shared by psychoanalytic and learning theorists that nursing was an important part of the attachment process. Bowlby (11, p. 180) claimed, "Food and eating are held to play no more than a minor part [in attachment]." Harlow decided to experimentally investigate this matter by measuring the extent to which feeding affected infants' attachment responses (Fig. 12.3) (22). In a long series of now considered classic studies by Harlow and various colleagues, the affectional responses of infant rhesus monkeys, reared under a variety of conditions, were examined. This vast body of research is briefly summarized next.

## Classic Studies by Harlow and Colleagues

Harlow assumed that if psychoanalytic and learning theories were valid, then researchers should be able to experimentally alter attachment responses by manipulating mothers' delivery of food. In one series of studies, infant monkeys were reared with two kinds of surrogate mothers. Both "mothers" were of the same size and shape and provided warmth, but one was constructed of wire and the other covered with terry cloth. In different experimental conditions, the terry cloth or the wire mother was rigged with a bottle for nursing. The results of these studies indicated that, contrary to the expectations of learning and psychoanalytic theories, the infants spent more time with the cloth mother, whether or not she offered food. In addition, when the infants were frightened or placed in a strange situation, they consistently fled, not to the mother that fed them, but to the mother that was covered with cloth (22, 23).

In keeping with the spirit of learning theory, Harlow interpreted this research as evidence for the existence of an additional primary need, one for contact comfort. As Harlow (23, p. 677) explained, "We were not surprised to discover that contact comfort was an important basic affectional or love variable, but we did not expect it to overshadow so completely the variable of nursing." Harlow (23, p. 677) suggested that the primary function of nursing as an affectional variable ".... is that of ensuring frequent and intimate body contact of the infant with the mother."

## Maternal Deprivation

Continued research with these monkeys had important implications in the field of social development. Longitudinal studies were conducted with the infants raised with surrogates as well as others raised in total social isolation for the first 3, 6, 9, or 12 months of their lives. With the exception of the 3-month group, these infants displayed abnormal social and personal behaviors, excessively fearing other members of their own species and sometimes engaging in self-aggression and self-abuse. When they reached sexual maturity, neither males nor females engaged in normal sexual behavior, and those females who were impregnated ended up either neglecting or physically abusing their infants (24, 25). These negative effects, however, were not seen with infants maternally deprived for less than 3 months (26, 27).

These findings led Seay and Harlow to conclude that a critical period for attachment exists (27) and that their findings, "appear to be in general accord with expectations based upon the human separation syndrome described by Bowlby" (28, p. 130). Thus, in a rare circumstance in the science of human development, the major theoretical approaches were in agreement. Representatives of ethological, psychoanalytic, and learning theories concurred that maternal deprivation during a critical period could have permanent damaging effects on the social and emotional development of the child. [Harlow had demonstrated that these deleterious effects may not generalize to cognitive/learning capacities (29)].

As I stated elsewhere about this topic, "Such widely accepted conclusions resist change" (30, p. 137). Bowlby was thought to have implied that damage might occur to children if they did not receive continuous care by one person, preferably the mother or mother substitute (20). In a review of Bowlby's work, Dinnage (31, p. 57) states, "Readers old enough to have been adult in the 1950s—especially if they were parents—are more likely to remember Bowlby as the psychologist of maternal deprivation, the man who (it was said) declared children grievously harmed if their mothers left them with grandma or the cleaning lady for 10 minutes. In fact, he said no such thing." Yet existing views of maternal deprivation were severe at this time, and now the work of Harlow and other widely respected, cross-disciplinary research seemed to support this pessimistic perspective. Imagine the profound implications of these views for practical matters in which infants or children are separated from their mothers such as day-care, adoption, or prolonged hospitalization.

### Social Rehabilitation

The theoretical consensus and pessimistic view of maternal deprivation did not last long. Throughout the 1970s Novak and Harlow (32) as well as Harlow's colleague, Stephen Suomi (26, 33) began to study ways of ameliorating the effects of social isolation. In a series of studies, infant monkeys who had been deprived for various durations (6 to 12 months) were subsequently socially rehabilitated by young members of their species. Age mates, it was found, were not effective, but juveniles younger than the deprived monkeys facilitated social contact and interaction. Although the isolates did not regain all normal social skills, their gains indicated that the effects of social isolation early in life, even for as long as 1 year (32), could be reversed. These findings forced a reconceptualization of assumptions about institutionalization, critical periods, and maternal deprivation.

# Reconceptualizing Maternal Deprivation

Other reports by various researchers provided evidence that contradicted the traditional view of maternal deprivation. Children reared in institutions or exposed to various forms of social isolation still displayed normal social and intellectual abilities (see chapter 11, "Intelligence and Environment"). Clarke and Clarke (34) reassessed and presented a much more optimistic view of the early-experience literature. They assembled evidence of children who were severely neglected, socially isolated, or institutionalized, who subsequently made up their various deficits and developed normal social, emotional, and intellectual skills. Other cases were found (21, 35) in which children, who were raised without adults, instead formed strong bonds with peers. These data were integrated with Harlow's social rehabilitation work and led to conclusions about the existence of different and compensating maternal and peer affectional systems (25, 26).

Literature reviews in the 1960s by Casler (36) and by Yarrow (37) led to additional empirical reassessment of the institutionalization and maternal deprivation research. It was proposed that the harmful effects thought to be caused by disruption of the attachment bond and multiple caretaking during institutionalization were instead due to general conditions of environmental deprivation (see chapter 9, "The Science of Early Experience"). Reviewing other early experience research that demonstrated positive effects of environmental stimulation, Casler proposed a perceptual deprivation hypothesis to account for institutionalization effects. He claimed that in many cases institutional rearing may be preferable to family rearing. Casler (36, p. 612) concluded: "The human organism does not need maternal love in order to function normally [, and] there is no evidence that social stimulation is best administered by a loving mother or mother figure."

In 1972, in his reassessment of Bowlby's work, Michael Rutter (20) pointed out that while the concept of maternal deprivation was initially important, it had since lost its significance. Initially, the concept concentrated attention on children and their environments and led to improvements in the quality of residential care for children; but now, Rutter concludes:

> The experiences included under the term 'maternal deprivation' are too heterogenous and the effects too varied for it to continue to have any usefulness. It has served its purpose and should now be abandoned. That 'bad' care of children in early life can have 'bad' effects, both short-term and long term, can be accepted as proven. What is now needed is a more precise delineation of the different aspects of 'badness,' together with an analysis of their separate effects and of the reasons why children differ in their responses. (20, p. 128)

The critical period hypothesis (see chapter 6, "Maturation and the Concepts of Readiness and Critical Periods; chapter 9, "Critical Periods") was also scrutinized. Its usefulness is still upheld by some researchers such as Scott (38, 39) and others, but Harlow's work and numerous other long-term studies of institutionalized or deprived children indicated that the damaging effects of social isolation could be reversed (21, 36, 40). As Jerome Kagan (41, p. 121) concludes in Clarke and Clarke's book:

> "The total corpus of information implies that the young animal retains an enormous capacity for change in early patterns of behaviour and cognitive competence, especially if the initial environment is seriously altered. The data offer no firm support for the popular belief that certain events during the first year can produce irreversible consequences in either human or infrahuman infants. . . . The first messages written on the 'tabula rasa' may not necessarily be the most difficult to erase." (41, p. 121)

Research conclusions concurred with Rutter's reassessment (20). Maternal deprivation is not a unitary process; it is complex and multivariable. Thus, researchers began to separately call for continued investigation of what Yarrow termed "nonmaternal variables," which affect children's social and emotional attachments and subsequent development. This involved examining other parental variables as well as characteristics of children and the many interacting systems of social behavior. Clearly, the time had come for an ecological approach to studying attachment.

# ECOLOGICAL APPROACH TO SOCIAL DEVELOPMENT

An ecological approach, which is summarized in Table 12.4, will be used to look at social development research. This approach looks at research according to whether it focuses primarily on (a) the characteristics of the individual's social and perceptual environment, (b) the characteristics and sensitivities of the individual, or (c) on the interrelationships between them. Current research in the field of social development is extensive and inconclusive. Thus, the purpose here is not to make sweeping generalizations but

Table 12.4
**Ecological Approach to Social Development Research**

• *Environmental Characteristics*
Nature of Social Stimulation (distant vs. somatosensory)
Nature of Social Deprivation
  Form, extent, intensity, duration
  Characteristics prior to and after deprivation
Child's Social Environment
  Characteristics of Caregivers
    Caregiving style, efficacy, consistency, competence
    Maternal vs. paternal roles
    Single vs. multiple caretakers
    Parents attitudes, marital relationship, employment, education
    Biological vs. adoptive, foster parents
  Number of siblings and birth order
  Number of peers
  Presence of extended family
Characteristics of Child's Physical Environment
  Location and quality of care (home, institution, day care)
  Presence and use of toys, television, other objects
Socioeconomic status of family
Culture, levels of community systems in which family is involved
• *Child-Related Characteristics*
Age, Sex, Temperament
Developmental trends, norms, milestones
  Health, physical status, genetic background, birth history
Presence of sensorimotor or cognitive deficits
Physical appearance, level of sexual and physical maturity
Characteristics of abilities that comprise early social signals
• *Interaction between Environmental and Child-Related Characteristics*
Dynamics of family systems
Characteristic changes in family interactions that accompany development
Differences between mother-child and father-child interactions
Characteristics of "good fit"
Effects of interrupted relationships (when one member in family system cannot
    respond)
Characteristics of peer relationships
  Comparisons to adult-peer interactions
  Developmental characteristics throughout life
Dynamics of home environment and relations between levels of systems,
    community services, culture
Methods of describing and classifying interactions
Methods of improving parent-child relationships, goodness of fit

to point out the direction and status of study in regard to social, emotional, and personality development. Some of the major research issues that have generated a large volume of activity in the 1970s and 1980s or that are illustrative of the status of scientific inquiry in these developmental domains are highlighted here.

## Characteristics of the Environment

Environmental stimulation is of interest to students of human development

Figure 12.4.   Cultural influences are responsible for wide variations in child and adolescent social behaviors.

as well as clinicians (chapter 9). In regard to attachment and other social development research, the environment refers initially to the infant's immediate family and then gradually expands to include friends, with whom children and adolescents form special kinds of attachment relationships. Hartup (42, p. 944) notes, "The social world to which most children are exposed initially is the family, a complex unit varying widely in composition and cohesiveness from family to family as well as from culture to culture."

Researchers in social development have long recognized that these individual and cultural variations among families may have wide-ranging influences on the child's social and emotional development (Fig. 12.4). As Table 12.4 illustrates, studies have looked at the effects of numerous social and nonsocial factors on the development of infants, children, and adolescents. Baumrind's (43) exploration of parenting styles illustrates how parents may influence children's social behavior. Baumrind identified three different types of parents according to their style of child rearing. *Authoritarian* parents tended to be warm and supportive and tended to provide consistently firm discipline along with reasons for the discipline. On the other hand, *authoritative* parents were demanding and punitive, and *permissive* parents were generally indulgent. Of the three parenting styles, the first was associated with the most independent and socially interactive children. In contrast, the children of authoritative or permissive parents tended to be dependent, immature, and less socially appropriate or interactive.

More specific research has attempted to delineate the most effective forms of environmental stimulation for eliciting certain social and emotional responses. This has led to controversy over the nature of social stimulation and whether near (somatosensory) (44) or far (visual and auditory) sensory systems (45) are more important for social interaction. Building on Harlow's work, which demonstrated stronger attachment responses to terry cloth vs. wire and rocking vs. stationary surrogates, researchers have proposed varying theories regarding the effectiveness of different forms of sensory stimulation for social development (46). Along with this research are clinical evaluations of home environments that attempt to delineate the significant social factors (parents, siblings, extended family members, and their interactions) and nonsocial variables (toys, play environments, amount and kind of television viewing) that may facilitate social maturity, social interaction, and long range social and emotional responses.

With the recognition that the child's social environment includes many important "nonmaternal variables," the roles of fathers, siblings, extended family members, and peers have come under closer scrutiny, as have other nonsocial variables. Many of these will be discussed in more detail under "Individual-Environment Interactions." Attention has recently centered on the roles of fathers and their participation in the birth process (see chapter 7, "A Special Concern: Father-Infant Bonding"), the problems of single parenthood for fathers or mothers, and the ways that parents may differentially socialize male and female children. One line of research indicates that fathers and mothers may have different expectations for social interaction with their sons and daughters, such that girls are provided greater amounts of direction and structure than boys. The result is that in play, boys may be encouraged to display greater amounts of independent exploration. This, in turn, may account for gender differences in social behavior later in life (47).

Some contemporary studies continue to investigate topics associated with traditional maternal deprivation and attachment research. Cross-cultural studies compare the long-term effects of multiple vs. single caretakers on various types of social behavior such as cooperation and competition (48–50). Other studies, some of which will be discussed later in this chapter, continue to explore short-term and long-term developmental effects of institutional rearing, hospitalization, adoption, and day-care (20, 21, 34, 51). The results of such research have sometimes led to important social changes such as policies in institutions which allow siblings or parents to "room in" with young children during hospital stays or facilities that enable whole families to be together in a home-like environment near institutions. For the most part, however, most empirical social development research oriented around the issue of maternal deprivation continues to be inconclusive and, in some cases, highly controversial.

### Day-Care as Maternal Deprivation

Research exploring day-care effects on children illustrates the controversial and inconclusive nature of social development studies. The subject of day-care is important because it illustrates several points: (a) that traditional views of attachment persist (b) that procedures for measuring attachment are controversial and incomplete (c) that the behavioral components of attachment behavior are not universally accepted, and (d) that a developmental issue (as discussed in chapter 2) can have profound sociopolitical implications.

In day-care research today, the maternal deprivation assumptions of Bowlby and Ainsworth (Table 12.3) persist. The strange situation, originally used by Ainsworth, continues to be used to assess the attachment relationship. In this procedure, infants or children are placed in an unfamiliar setting, and their emotional reactions are observed as their mothers repeatedly leave and return. In some cases, a stranger may enter the room when the mother is absent, and then the infants' approach and avoidance responses to this unfamiliar person are recorded. The assumption is that the type and intensity of the infants' reactions in this procedure: (a) reflect the degree of separation anxiety and stranger anxiety, (b) reflect the quality of attachment, and (c) may predict long-term (or possibly permanent) social and emotional behaviors that result from the original mother-infant relationship.

Blehar (52) provides an example of research in which the strange-situation procedure was used to compare the emotional responses of home-reared 2- and 3-year-old children with a similar group of children reared in day-care. The results of Blehar's study indicated that the day-care children cried more during separation from their mothers and avoided a stranger more than the home-reared children did. Thus, Blehar concluded that the experiences associated with day-care may bring about anxious or ambivalent attachment reactions.

Procedures other than the strange situation have also been used. Observations have been made of children's emotional or social behaviors, as well as the quality and quantity of social interactions with peers or with caretakers. In one such study, Schwarz and colleagues (53) noted that when compared with matched subjects who had no day-care, 3- and 4-year-old children who were in day-care from infancy displayed more aggressive behaviors, were more motorically active, and were less cooperative with adults. The authors of this study concluded that their findings are subject to multiple interpretations and that, rather than provide substantive conclusions, their data illustrate the rudimentary state of knowledge about day-care and the need for continued study.

While this book was being written, a highly popularized controversy occurred in regard to day-care effects. Developmental researcher Jay Belsky has received considerable attention in the news media (54) regarding his concerns (some of which will be published in forthcoming developmental journal articles) about the potential damaging effects of day-care. What is remarkable about Belsky's conclusions is that they have dramatically reversed in recent years. In 1978, Belsky and Steinberg (55) critically reviewed day-care literature and concluded that high-quality care increases peer interaction, does not appear to disrupt attachment responses, nor does it appear to have any negligible effects (positive or negative) on intellectual development. In contrast, in 1986, after reviewing additional day-care literature, Belsky (56) has issued a cautiously worded concern that if infants are involved in such care during the first year of life, they are apt to exhibit disrupted social behaviors later in development. Belsky's concerns were based on findings that infants placed in day-care exhibit insecure attachment responses during the strange situation and also display aggressive, noncompliant, and disobediant responses toward adults.

## Ecological Approach to Day-Care

Belsky (56) and his critics (57, 58), however, point out that the issue of day-care is very complex. Developmental researchers have expressed their concern that Belsky's analyses and resulting conclusions are oversimplified. Clinician Stella Chess (58) is fearful that research and attitudes toward day-care in the 1980s may parallel institutionalization research of the 1940s and 1950s. She criticizes Belsky's claims for their oversimplification and reliance on the strange-situation procedure, which has yet to be determined as a valid measure of attachment. These researchers (53, 57–60) note that many interacting and changing variables may alter the responses of children participating in day-care, including:

Child-teacher ratios;
Age of the child when entering day-care;
Kind of setting and whether care is offered in or outside of the home;
Quality and stability of care;
Number of peers and their relationships at the setting;
Whether a close relationship is developed with adults in the day-care setting;
Duration of care;
Child and family experiences prior to and during the time care is provided;
Family relations and family stress at home, the roles of fathers in the home and in child care;
Other ecological considerations.

In a 1982 review of day-care research, Bradley (60) concluded that continued investigations need to look at long-term effects, individual differences in children's responses, the interaction of socialization at home care and at day-care, and the relationships between child characteristics and day care. Such topics, Bradley notes, require an ecological approach, such as that provided by Bronfenbrenner's model (see chapter 6, "The Infants' Ecology and Social Systems) to adequately explore the effects of interacting systems of social interaction associated with day-care.

At present, studies of day-care are contradictory and confounded. Day-care, itself, is difficult to define, and studies vary in the types of variables they look at as well as the procedures they use. Since well-controlled studies of infant day-care are rare, conclusions about its effects are tentative at best. Until data from such approaches are available, clear conclusions about the effects of day-care cannot be drawn.

In the meantime, the subject of day-care continues to be debated (54). It is an emotionally charged issue with important political implications. Chess points out that in 1987 1 of 2 women with a child under 1 year of age is employed and needs child-care arrangements. Phillips and colleagues note (58, p. 20), "While we continue to debate the merits of infant care, the realities of economic and demographic life in America tell us that infant day-care is here to stay." It is clearly too important a topic for overgeneralizations to be made. Belsky (56) himself points out that infant day-care research is complex and that for each infant in day-care, there may be a variety of setting arrangements as well as family feelings and practices associated with these arrangements. He concludes (56, p. 7): "Infant day-care refers to complex ecological niches. This means, then, that any effects associated with care are also associated with a host of other factors. Thus, it would be misguided to attribute any effects associated with nonmaternal care to the care per se, or even to the mother's employment."

### Summary Regarding Environmental Factors

These many considerations about day-care are illustrative of other topics currently being investigated in the scientific study of social development. Hartup (42, 944–945) claims: "The traditional theories of personality and social development have spawned thousands of investigations relating parent characteristics on the one hand to child characteristics and actions on the other." The results of many of these studies are inconclusive, however, because researchers have often failed to conduct longitudinal studies and have neglected to consider that development involves a dynamic system of reciprocal interactions which change over time. Thus, it appears that instead of being at a point where conclusions can be drawn, the status of research in social development is at a frontier where variables are just being identified. It is clear that studies cannot focus solely on aspects of the child's environment without also considering that environment as a system that interacts with characteristics of the child.

# Characteristics of the Developing Individual

One of the pioneers who promoted a reconceptualization of infant and child roles in social interaction was Richard Q. Bell. In 1971, Bell (61, p. 63) expressed his concern: "Most investigators have only considered the child an object on which parental actions are registered. . . . Parent and child are clearly a social system and in such systems we expect each participant's responses to be stimuli for the other. Why, then, is the child's own contribution to an interaction overlooked by social scientists, and what can be done about the problem?"

The research in the 1970s and 1980s has been addressing that problem.

Contemporary study in human development now recognizes that at every stage of life, individuals take active parts in effecting their own development. In regard to social development, each person is now seen as a partner who is embedded in a social system, where each takes responsibility for either initiating or responding to cues from others (62). While it may seem like common sense, this new perspective of infants and children as partners in their own socialization process could not have emerged in developmental study until several events occurred: (a) reconceptualizing the capabilities of infants and children, (b) reconceptualizing the attachment relationship, and (c) creating new ecological-systems approaches that enable scientists to study organism-environmental (child-parent) relationships (see chapter 6).

### Individuals as Promoters of Their Own Development

In 1979, Susan Goldberg (63, p. 214) noted: "In the last fifteen years, the study of infant development has shown that the young infant is by no means passive, inert, or helpless when we consider the environment for which he or she is adapted—that is, an environment which includes a responsive caregiver." Research since the 1960s (already examined in detail in chapters 7–11) has demonstrated that infants possess sophisticated capabilities that may actually predispose them to social interaction. At present, we continue to gather descriptive information regarding characteristics of adolescents and children (Table 12.4). Many of these studies, however, have focused on the child's characteristics and then related this information to environmental interactions. For example, Green and his colleagues (64) looked not only at the developmental changes in social and motor abilities of infants but also at how these changes altered the youngster's social encounters with their mothers. Descriptive studies of sexual behavior look not only at the physiological and hormonal responses within children and adolescents but also at the youngsters' evolving social relationships with their parents and peers. Thus many of these studies will be examined in more detail under "Individual-Environment Relationships."

### Stranger Anxiety

As with studies that focus on environmental stimulation, child-based research is often inconclusive and controversial. An example of this is provided by ongoing investigations of stranger anxiety. Traditional attachment research by Ainsworth assumed that stranger anxiety is a normal emotional response that predictably emerges and matures in infancy (14). Experimental approaches to stranger anxiety, however, indicate that it is unclear when, if at all, the reaction develops and what social situations or objects may elicit it (65). Recent descriptive studies by Rheingold and Eckerman (66) suggest that fear of the stranger is neither universal nor maturationally predictable. They suggest that rather than assume that such a phenomenon exists and that it is somehow predictive of the quality of the attachment relationship, new descriptive studies are needed to assess the varieties of affective responses infants make in reaction to strangers in different environmental circumstances. Until such data are collected, generalizations about this social-emotional process should be considered tentative.

## Emotions and Temperament

Studies of emotional and personality development have recently focused on three main issues: (a) identification and description, (b) debate about their origins, and (c) analyses of interactive effects. In regard to emotions, psychologist Carroll Izard (67) has identified and described specific facial characteristics of emotions that are common across cultures. These findings, along with his identification of similar expressions in infants, have led Izard and others to conclude that some emotions are universal and innate (68). Such conclusions have resurrected debate about the origins of affect. Historically, ethologists and others concurred with Darwin that emotions are innate responses that ensure adaptation and survival (Fig. 12.5). In contrast, consistent with Watson's demonstration of conditioning of fear responses (chapter 5), many contemporary behaviorists contend that emotions are learned.

Thus, debate regarding the nature and nurture of affect continue, although interactive views are also proposed. Lewis and Rosenblum (1, p. 10) explain: "It is clear, then, that affect and its development may best be viewed as embedded in the child's total social, psychological, and physical maturation." They conclude that a true understanding of emotional devel-

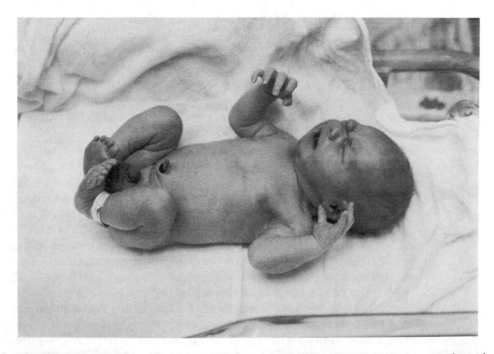

Figure 12.5. Early in life, infants display a characteristic temperament and recognizable emotions to which caregivers respond.

opment will occur only with studies that incorporate different methods, different developmental domains, and a biopsychosocial approach.

Research on temperament has progressed along similar lines. The term "temperament" was introduced by Thomas, Chess, and Birch to refer to a child's behavioral style (69). Since the 1950s, when these researchers began the New York Longitudinal Study (NYLS) from which nine different categories of temperament were distinguished, research into temperament has been oriented toward assessment as well as debate about its origins. As Worobey (70, 71) notes, over 25 instruments have been developed for identifying and assessing temperament, and attention continues to focus on describing this aspect of development at all stages of life. However, consistent with many other social, emotional, and personality variables, temperament has been studied more in children and infants than in adolescents and adults, an oversight which Thomas and Chess are now addressing (69).

As a result of the NYLS, some developmental researchers claim that since temperament has been identified very early in life, then it must be innate. This is in contrast to other common views (e.g., Freud's psychoanalytic approach or learning theory) that parent training and environment actually shape the child's personality. Thomas and Chess (69) conclude that temperament is a result of genetic-environmental interactions, and their investigations have therefore been looking at the effects of children's temperament on their own development and on their interactions with their family and friends. These investigations are clearly ecological in nature.

According to Thomas and Chess (69), children can be classified as: "easy," "difficult," or "slow to warm up." Easy children approach stimuli in their environment, are adaptable to change, and stay biologically regulated and moderate in intensity. In contrast, difficult children tend to be biologically irregular, react intensely, frequently withdraw from stimuli, adapt slowly to change, and display frequent negative moods. Slow-to-warm-up children are similar; they often withdraw and are slow to adapt, but they are mildly intense and have less biological irregularity and fewer negative mood states than difficult children.

Thomas and Chess (69, p. 241) point out that each of these temperaments interacts with the child's environment, and this interaction is responsible for the child's continued development: "At any age period, if environmental demands and expectations are consonant with the individuals capacities, abilities, motivations, and temperament, a goodness of fit will exist." Thus, in a low-stimulation environment, a slow-to-warm-up child may fit, whereas in the same setting, a difficult child will not. To fully understand the dimension of temperament, Thomas and Chess point out that we must simultaneously consider the characteristics of the environment, the characteristics of the organism, and their mutual interaction.

This approach to temperament exemplifies the state of the art in the science of human development. The work of Thomas and Chess illustrates two important types of interaction that must be explored: (a) the interaction between children and their environments and (b) the interaction among developmental domains. They claim (69, p. 253): "The role of temperament at any age period can never be considered in isolation from other organism-related characteristics or from environmental demands and expectations. It is but one factor in the constantly evolving interactional process among all these variables." Similar approaches in the study of emotional development concur. Emotional development, in general, is best understood through its interaction with other developmental domains, and the emotional development of specific individuals is understood through examination of individual:environmental relationships.

# Characteristics of Individual-Environment Relationship[a]

Hofer noted in 1975 in the introduction of *Parent-Infant Interaction* (72, p. 1),

> "Most recently, our views of the infant as a passive beneficiary, or victim, of the parents' influence has been successfully challenged by data from both animal and human studies. The infant turns out to be immensely powerful in his own special field—that of promoting and regulating the behavior of his parents. This has had the good effect of making us think, design studies, and interpret those studies in terms of an 'evolving interaction.'"

This focus on interactions which now dominates the field of social and emotional development was made possible by the emerging ecological-systems perspective that became so popular in developmental study in the 1970s. This perspective and its approach to social development has already been described in several chapters of this book: chapter 2 "Contemporary Models of Interaction in Human Development;" chapter 6 "The Infant's Ecology and Social Systems;" chapter 9, "Caretaking as Reciprocal Perceptual Stimulation."

Clearly, times had changed in developmental study. Bell (61), who in 1971 had expressed concern that research was neglecting the effects of the child, claimed in 1979 (73, p. 821): "The active infant and child have now been put back into the picture with a thinking parent [and]...a recent interest in ecology has placed the reciprocal system in the context of overlapping circles of family, neighborhood, ethnic, and larger groups." The studies that have resulted have been exploring the interrelationships between the many environmental and child-based variables that are included in Table 12.4. The studies are too numerous to review here, thus only a selection of studies will be examined to illustrate the status of research and some of the conclusions in this field.

## Measuring Social Interaction

One area of interest to many researchers is determining valid and reliable methods of assessing, classifying, and describing the qualities of social

---

[a] The term "individual" is used here to refer to the infant, child, or adolescent.

interactions (74–77). Many researchers claim that until such methods are refined, accurate descriptive and empirical studies cannot exist.

## Father-Child Relationship

The roles of fathers in the American family in the 1980s are changing (78). Fathers are increasingly participating in the birth process and in caregiving roles (Fig. 12.6). Yet, because this is a new field of study and because parenting and gender roles are in such rapid flux, it is difficult to draw conclusions about the developmental effects of father-child relationships (79, 80). Studies of fathers and infants or toddlers indicate that, like mothers, fathers form attachment relationships and that their sensitive and reciprocal interactions facilitate children's development and have important effects on the family unit (79, 81–83). In a traditional test of attachment using the strange situation, Kotelchuck (84) found that children responded similarly to both mothers and fathers. The extent of fathers' influences, however, is not understood. Studies of older children are rare, and those that do exist have tended to focus on one area, the effects of father absence. These studies indicate that paternal deprivation can have long-range negative effects, potentially interfering with boys' sex role development and adolescent females' adjustment (79).

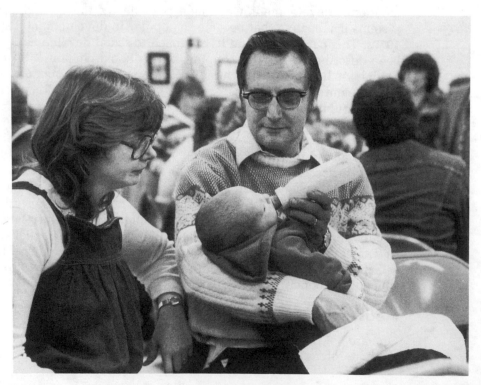

Figure 12.6.   Fathers play roles that are important to the integrity of the family system and to the development of their children.

Some studies have found differences between mother-infant and father-infant interactions with their children. These indicate that fathers tend to engage in more physical, vigorous types of play, whereas mothers tend to engage in more verbal and toy-mediated play. Other studies have found no differences between mothers' and fathers' responses to their infants, and evidence suggests that these will vary according to the family system as well as the age, sex, and reciprocal responses of the children involved (47, 80–85).

Some special studies have resulted from the increased interest in fathering. Fathers' roles with their handicapped infants and children is of special interest to clinicians (86). A series of studies by clinical researchers at the University of Washington indicates that supportive programs for such fathers may provide additional positive effects on spouses and on family units (87–89). An additional result of the recent focus on fathers is an increase in research exploring adolescent fatherhood. The general consensus is that this unique area is poorly understood and bears continued study because of the increasing number of children born to adolescent parents and the increased involvement in child care on the part of adolescent males (90). Consistent with ecological approaches, Yogman (79) suggests that our knowledge of parent-child interactions will be advanced by studies that explore the issues involved in adolescent and adult development and parenthood for both men and women.

## Interrupted Relationships

Infants normally emit cues to signal their status or needs, and caregivers respond accordingly. Infants then adjust their responses to the type of caregiving reactions they have elicited. This system of reciprocal responding can be facilitated or interrupted by either partner at any time during development (91, 92). Thus, a parent who may be particularly responsive to an infant may feel unsure and become unresponsive when that child reaches adolescence. Contemporary research is just beginning to explore the dynamics of family relationships and the subsequent results of interrupted social relations on infant, child, and adolescent development. At present, however, most descriptive research has focused on the early years of life, and only recently have these systems models been applied to adolescence (93).

An infant with sensorimotor deficits or emotional problems may not respond as parents' expect, and as a result the parent-infant relationship may break down or require adjustments to preserve the social synchrony (see chapter 7, "Interaction Coaching;" chapter 9, "Sensorimotor Deficits and Development"). Understanding the effects of different deficits such as sensory impairments (blindness, tactile hypersensitivity, deafness), physical syndromes, and severe emotional or temperament deficits has been the focus of considerable developmental and clinical research (94–96). One result of this research is the recognition that, just as parents may adjust their reactions to children with sensorimotor deficits, infants also appear to learn, at an early age, to react differently to parental modes of responding. Brazelton and colleagues (92) have demonstrated that infants respond differently to blind parents than to parents who can see and react to the infants' facial cues.

There is obvious practical value for such investigations. Understanding how social interactions develop and what effects interruptions may have will lead to methods for improving communication, preventing possible long-range social-emotional problems, and supporting the integrity of family systems. Some current approaches in mental health research suggest that inappropriate or inconsistent responses on the part of parents may contribute to long-range maladaptive social functions (96). Thus, various types of interventions have been suggested to prevent early interruptions in social interaction and to promote a goodness of fit between caregivers and their children (91, 92, 96, 97).

Broader, ecological studies have been looking at the effects of the types and availability of community services on the social systems of families with or without disabled children (59, 98). The goal is that ultimately we will be able to develop more sensitive and effective methods of family intervention so that each person's developmental potential can be maximized at every stage of life.

## Peer Interaction

Peers refer to individuals who are similar in age and who share capacities and common physical characteristics (99). Relationships with peers change with chronological age, (42), but this evolution is not fully understood. It is commonly recognized that children begin a gradual separation from their parents and become increasingly involved with peers throughout childhood, adolescence, and beyond. Yarrow (100) indicates that peers are important to all people, not just children and that people are likely to spend long portions of their time and wide varying relationships with near-age mates.

Peer relations provide important functions (Table 12.5) but the origins

Table 12.5
Functions of Peer Group Participation[a]

---

Provides support from individuals facing similar developmental tasks.
Permits increased autonomy from parents and older siblings.
Offers opportunities to experiment with different roles and values.
Provides opportunity to experience affection for non-family individuals.
Provides behavioral regulations, controls, and pressures, which can result in deviant or socially desirable behaviors.
Provides a basis for learning how to regulate emotional expressions and develop social skills.
Provides opportunity for adolescents and adults to form relationships and to develop sexual intimacy.

---

[a] Data from Lewis M, Young G, Brooks J, Michaelson L: The beginning of friendship. In Lewis M, Rosenblum LA (eds): *Friendship and Peer Relations*. New York, John Wiley & Sons, 1975, chapter 2. Newman PR: The peer group. In Wolman BB (ed): *Handbook of Developmental Psychology*. Englewood Cliffs, NJ, Prentice-Hall, 1982.

of these social interactions are not fully understood. Because mothers were at one time seen as the essential person in infants' lives and because all development was considered contingent on the attachment relationship, peer interaction was presumed to arise out of this basic maternal-based system. Contrary to this point of view, work by Harlow and others suggests that separate affectional systems may exist and that peer interaction may develop independent of, not out of, the maternal system (101). Recent studies of infant-peer interaction indicate that infants do interact with one another, directly and through object play. Eckerman and colleagues (102) indicate that 2-year-olds interact more frequently and differently with peers than they do with adults, which may be partly due to the novelty of peers and the ability of infants to imitate and react to peer responses. Nonsocial objects also play a role in facilitating peer social interaction, but this area is not well researched or understood (103).

### Trends in Peer Relations

By age 5, peers interact differently with age mates of different sexes, displaying higher levels of responding to same-sex peers (104). By middle childhood, children spend large amounts of time with peers; by ages 13 and 14, adolescents are spending increasingly long periods of time away from home (105). As peer relationships take on greater importance, changes are also seen in the reactions of parents. Parents, who spend most of their time taking care of infants and supervising toddlers, spend decreasing time supervising and intervening in peer relationships as their children grow older. With children spending their days in school, the amount of time spent at home in parental supervision and interaction diminishes, and both children and parents must make adjustments to these changes.

Dunphy (106) has described the evolution of peer relationships in late childhood and adolescence and suggests it follows certain characteristic patterns. Initially, adolescents come together in same-sex cliques, which are small groups with about 6 to 9 members. The next stage involves the combination of male and female cliques in a group activity, where interaction occurs across cliques. The cliques then gradually become heterosexual as they combine into crowds, which are large groups of many cliques and comprise about 15 to 30 members. At this time peer interaction consists of dating or forming friendships with members of the same crowd. In the final stage, the crowd disintegrates as couples are formed and are no longer dependent on the crowd for interaction.

Adolescent peer relations are often popularly misconstrued as providing social pressure that runs contrary to the values and beliefs of their family. While this is true of some adolescent interactions, stress and antagonistic interpersonal relations with parents are apparently not the norm (93). Lerner and Shea (93) indicate that even though adolescents may spend more time with peers than with parents, this does not mean that parental influences are terminated. On the contrary, many adolescents choose peers with values similar to their parents. In addition, with their increasingly sophisticated cognitive abilities, adolescents become capable of evaluating both peer and parental inputs. As Newman (105, p. 531) notes, "Individual adolescents become increasingly capable of evaluating situations and making their own choices without guidance from parents or peers."

In general, peer relations form among individuals who are similar in age, appearance, and values; status levels are associated with certain individual characteristics and with groups. Some valued physical and personal characteristics (e.g., physical appearance, early maturation, athletic ability, leadership, good grooming) are recognized as important for popularity, but these values and peer relations, themselves, vary from culture to culture and community to community (105). The ultimate influence of peers depends upon numerous factors such as the individual's sex, age, cognitive abilities, popularity, and other characteristics; the nature of the family; the school system; and many other ecological considerations.

### Value of Friendships for Disabled Children and Adolescents

A special issue concerns the friendships of handicapped individuals. Peer interaction is not only significant for normal social-emotional development, but it may also be significant for handicapped children's language, cognitive, sexual, and academic development (107). Since children tend to develop friendships with peers who are physically similar, handicapped children are often at a disadvantage in having opportunities to socialize with nonhandicapped peers. Strain (107) points out that handicapped children may have poorly developed social skills, may experience peer rejection, and have few friendships. Thus, an important goal for handicapped children is for them to experience opportunities to socialize with nonhandicapped peers. This goal has been realized in many cases due to the placement of handicapped children into the least restrictive educational settings where they are integrated with nonhandicapped peers. This process, called *mainstreaming*, provides opportunities for physically or emotionally handicapped children to not only interact with but also to imitate the actions and social behaviors of their nonhandicapped peers (107–109).

It appears that such imitation and social interaction do not occur spontaneously. They can be successfully promoted, however, through a variety of interventions that provide structured behavioral strategies for peer interaction (107–111). The goals of successful interaction among disabled and nondisabled children include the following: promoting the development of the handicapped youth, providing positive learning experiences for both groups, and positively affecting attitudes of teachers and families of the children. Ultimately, peer interaction may alter societal attitudes toward the handicapped and effectively integrate handicapped people within their communities. As Anastasiow (111, p. 216) notes, "[The goal is to facilitate] the handicapped person's social skills early in life so as to enable a larger percentage of these persons to have access to the full range of events in our society when they are adults."

Most of the research exploring peer interaction among handicapped and nonhandicapped individuals has been conducted with children. The few studies of adolescents indicate they have special needs which emerge

Figure 12.7.  Relationships with peers provide valuable experiences for children's socialization.

during this stage of development. With the onset of dating, there is an increased emphasis on physical attractiveness and social skills, which may be to the disadvantage of some disabled adolescents, who may devalue themselves or feel that others devalue them because of their disability. Those with physical or sensory limitations may have fewer opportunities for social interaction either due to physical or attitudinal barriers. As a result they may be hampered in the advancement of social skills and friendship formation. An additional issue is that adolescence is when considerable pressure is felt regarding selecting vocational and educational goals. Some handicapped adolescents may experience pressures and conflicts beyond those experienced by their peers because of disabilities or attitudinal barriers that restrict their choices. Clearly this is an area where there are wide individual differences based on the disabled adolescents' past experiences, nature and degree of their handicap, family relationships, school environment, level of intelligence, mobility, motivation, proximity to neighboring peers, and numerous other factors (112).

This special area regarding friendships of handicapped individuals illustrates that much still needs to be learned about the normal course and effects of peer interaction. Present studies recognize the difficulty of examining all the interacting variables as the developing individual must be regarded as embedded within numerous, changing social systems. Hartup (42, p. 949) has aptly concluded, "Childhood socialization is best viewed in terms of reciprocal causalities occurring within various social networks, each a social system in its own right." It will be some time before we understand how all these social systems interact and affect the development of children and adolescents.

# RELATIONSHIPS AMONG DEVELOPMENTAL DOMAINS

As described at the beginning of this chapter, traditional theories assigned primary importance to the maternal-infant relationship, assuming that attachment serves as a basis for most aspects of the child's eventual development. Current findings are from a broader, ecological perspective and somewhat modify this view. Contemporary findings emphasize the long-range social (not just maternal) environment, the responsibility of the child for his or her own development, and the resulting transactions that occur between the social environment and the child. The status and trends regarding child social development are summarized in Table 12.6.

The general conclusion is that humans are fundamentally social beings. Evidence indicates that babies are responsive to people from birth, (113, 114), and a good fit between the characteristics of the child and the caregiving environment can have effects on the child's development in all domains (Fig. 12.8). Numerous studies have found relationships between competent, sensitive, and consistent parental responding; security of attachment; and certain characteristics of the home environment on children's cognitive development (71, 82, 92, 94, 113–116), language acquisition (116,

Table 12.6
Themes in Social Development Research[a]

---

The child plays an active role in his or her own social-emotional development.

Social-emotional development has multiple causes, and multiple theories are needed to understand it.

Multiple levels of social influence are necessary in order to understand social-emotional development.

Social-emotional development varies and is influenced by different people, contexts, cultures, and time periods.

Multiple research strategies are essential in order to understand social-emotional development.

Research on social and emotional development influences and is influenced by social policy.

---

[a] Adapted from Parke RD: Emerging themes for social-emotional development. *Am Psychol* 34: 930–931, 1979.

117), persistence at problem solving (82), competence in play (118), and other social and emotional behaviors (96).

We now recognize the close relationship between developmental domains and the important role that children play in affecting their own development in each of those domains. This has been most frequently studied in regard to social and cognitive development. Bornstein (114, p. 472) notes: "How children fare in terms of their mental growth is informed by their abilities in complex continuing interplay with the early experience they provoke and those they encounter." Sroufe (2, p. 132) claims that cognitive and social-emotional aspects of development are inseparable as they mutually affect one another. Other researchers support this claim and purport that a common regulatory mechanism underlies social interaction, motor activity, and cognition (113). Lester and colleagues (113) contend that infants' early rhythmic behaviors, which are fundamental to social inter-

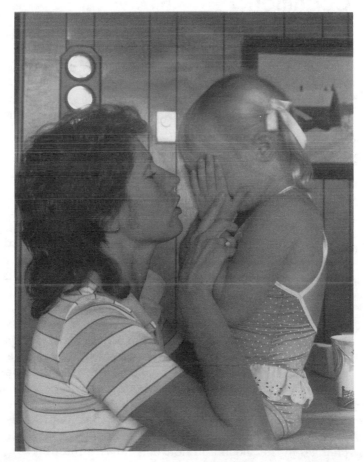

**Figure 12.8.** A nurturant environment that supports children's development is one with a good fit between the child's temperament and parental sensitivity and consistency of responses to the child's signals and needs.

change, may provide an early, if not innate, temporal structure for the organization of cognitive and affective experiences. A complex transactional model of cognitive development, as described by Olson and Lamb (115) views intelligence as a joint product of the child's enduring characteristics and environmental inputs.

The focus now appears to be not whether social or cognitive development are interrelated but whether one serves as a foundation for the other. Piaget and others have claimed that cognitive structures provide a foundation for subsequent social and emotional behavior. In contrast, other theories claim that social-emotional experiences form the basis for development in other domains. Whether one comes first, or they are contingent upon similar regulatory mechanisms, is not clear. What is clear is that development across domains is interrelated. This is exemplified by the field of social cognition.

# Social Cognition

Since social cognition is a relatively new area of research, it varies in definition and scope according to different researchers. Huston (6) explains that social cognition involves people's way of cognitively organizing their social world. Gelman and Spelke (119, p. 63) note that students of social cognition look at "... how the structures and processes that underlie cognition of the physical world apply to the social world." In their text on social cognition, Lewis and Brooks-Gunn (120) include three interrelated areas: (a) knowledge of self, (b) knowledge of others, and (c) knowledge of one's relationships to others. A form of "meta" social cognition (see chapter 11) looks at what Chandler and Boyes (121, p. 267) describe as "knowledge about the knowledge of others."

One of the questions being debated in social cognition is whether the understanding of social and physical worlds is based on the same cognitive structures or processes. Some researchers claim that social knowledge involves different features and operates according to different rules than knowledge of the physical world. Some researchers also claim that social knowledge may be essential for an understanding of inanimate objects in the physical world. Thus, person permanence, the ability to form mental representations of people, may mature sooner than object permanence (119, 122, 123). Brazelton and others have demonstrated that neonates behave differently with objects than they do with human interactants. This work as well as studies of infants' responses to their mothers' blank faces (92) indicates that very early in life infants possess a certain cognitive-perceptual "expectancy" that animate and inanimate objects in their environment respond in distinctive ways. A part of social cognition is the child's growing awareness of the relationship between his or her actions and reciprocal changes in the social and nonsocial environment (124).

## Social Interaction and Social Cognition

Social cognition is affected by social interactions. An important consequence of attachment is that infants, at some point, are able to separate themselves from their caregivers and discriminate different caregivers. Studies of peer interaction illustrate the connection between social and cognitive development. Eckerman and colleagues (102) have speculated that infants' early interest in peers may be due, in part, to their novelty. Thus, infants with a schema for adults become interested in people who are discrepant from that initial schema. Kagan (125) explains that discrepant schema attract attention, and infants' emotional states are affected by their ability to assimilate these discrepant events. Events that are easily assimilated (e.g., pictures of peers) produce excitement, whereas events that are not easily assimilated (e.g., pictures of a scary monster) produce uncertainty and anxiety. Thus, cognitive and affective development are clearly intertwined.

## Sense of Self and Identity

According to Lewis and Brooks-Gunn (120), knowledge about one's self develops independent of, and parallel to, knowledge of others. These researchers claim that in early development a biologically controlled social reflex supports social interaction between the infant and the caregiver until later social cognitions are developed. These later abilities, in turn, form the basis for mature, reciprocal social interactions. Evidence for this biologically controlled social reflex is provided by infants' abilities to imitate their caregivers without ever having looked in a mirror and practiced such actions (8). This proposed social reflex is similar to early perceptual and cognitive structures, discussed in chapter 9, which are also proposed to facilitate human social interaction.

An area not fully understood is the development of a sense of self. Self-related concepts (self awareness, self-concept, identity) vary in definition, but all relate to sensorimotor, physical, cognitive, and social-emotional feedback received from self-generated actions and from social interactants. Harter (126) points out that vision is important to self-recognition and self-awareness, and this is illustrated by the popular method of testing self-identity by putting a spot of rouge on a child's nose and then observing the reaction when the child looks into a mirror. The child who becomes embarrassed and tries to rub the rouge off his or her own nose (not the one in the mirror) is assumed to have internalized a sense of self. In addition, the child who can look into a mirror and say his or her name or, "That's me" is displaying self-recognition (Fig. 12.9).

### Erikson's Psychosocial Stage Theory

Identity probably shifts and has different meanings for infants, children, and adolescents as they change physically, cognitively, and socially and as they face different developmental tasks. Erik Erikson's (127) stages of psychosocial development illustrate this point. Although his work came before the new field of social cognition was organized, it covers the issue of identity that is now an important component of this new field. Erikson's theory is an extension and modification of Freud's psychoanalytic theory; however,

Figure 12.9. The child who can look into a mirror and say "That's me" is exhibiting social cognition, an awareness of self and the relationship between self and others in the environment.

Erikson modified Freud's sexual emphasis and recognized the significance of socialization influences.

Erikson's eight stages of psychosocial development illustrate that at different times in life, individuals face certain crises or tasks that must be resolved. Whether resolution occurs or not is determined by socialization experiences and by successful resolution of the issues of previous stages. Erikson describes these crises by labeling two poles that occur at each end of a continuum. The positive pole is associated with good socialization, resolution of the crisis, and the development of particular types of ego strength characteristic of that stage. Poor socialization is associated with the negative pole, the crisis is not resolved, and ego strength does not mature. The tasks and the ego strengths that are characteristic of each of the stages of life are listed in Table 12.7.

While his is not a social-cognitive theory per se, Erikson's stages incorporate aspects of development from different domains. This is illustrated by his fifth stage, identity formation, during adolescence. Considered by many to be the most important task during development, identity formation is affected by social and sexual experiences, cognitive abilities,

Table 12.7
Erikson's Eight Psychosocial Stages and the Ego Strengths Associated with Them[a]

| Stage | Psychosocial Crisis | Environment Social Sphere | Strength |
|---|---|---|---|
| Infancy | trust vs. mistrust | maternal caregiver | hope |
| Early Childhood | autonomy vs. shame, doubt | parental caregivers | will |
| Play Age | initiative vs. guilt | family | purpose |
| School Age | industry vs. inferiority | school neighborhood | competence |
| Adolescence | identity vs. role confusion | peers | fidelity |
| Young Adult | intimacy vs. isolation | friends, co-workers, intimates | love |
| Adulthood | generativity vs. stagnation | new generations | care |
| Old Age | integrity vs. despair | humanity | wisdom |

[a] Adapted from Erikson E: *The Life Cycle Completed. A Review.* New York, WW Norton, 1982.

self-knowledge, and all other past and ongoing socialization experiences. Gardner (128, p. 529) has aptly described this process: "In inventing an ultimate sense of identity, individuals who have negotiated the identity crisis achieve a sense of inner continuity with what they were before and what they will become. At the same time, they reconcile their own self-conceptions with the expectations and norms of the surrounding community, and particularly with the desires of significant persons in their lives."

## Identity, Language, and Gender

Two important aspects of development that relate to both identity and social cognition are language and gender acquisition. Language, as already described, is important in providing a vehicle for self-reference as well as a system of communicating one's needs and thoughts. With an individual's growing self-identity and language, his or her needs become more fully understood and more easily communicated with words. In addition, words enable children to categorize their social world, e.g., by using personal pronouns such as "me" or "I," by correctly naming people or using pronouns such as "you" or "they," and by using gender labels denoting masculine or feminine. *Gender constancy* is a social-cognitive structure that involves a stable perception of one's self and others as masculine or feminine (6). Thus, children not only develop cognitive schemas for gender-related concepts, they also begin to appropriately use words such as "boy" and "girl" to refer to themselves and others. Clearly, social, linguistic, and cognitive abilities are interrelated.

# Language

Language is an important process involving social, cognitive, cultural, sensorimotor, and affective domains. Language is a system of communication that involves sending and receiving information. In humans, this information is conveyed through different sensory modes and often includes feelings, thoughts, and other aspects of culture.

An excellent example of how language, cognition, and culture are intertwined was recently described in an article in *The New Yorker* (129) about U.S.-Soviet relations. The article gave an illustration of the controversial Whorfian hypothesis, the notion that language shapes thought: In the 1970s, the Russians had no word for "detente." It was therefore erroneously assumed that, up until that time, the Russian temperament was simply not oriented toward compromise. In 1987, President Reagan (129, p. 21) suggested (according to *The New Yorker* article) that the Russian language had no word for "freedom" (which it does) and that without the word, the Russians ". . . are trapped within the confines of their language. . . . If the Russians don't have a word for 'detente,' they can't have the concept; if they don't have a word for 'freedom,' they can never be free." The article points out the fallacy of this kind of reasoning and notes that "detente" and "freedom" represent concepts available to all of us, regardless of the words we use to describe them. Culture, politics, and social and national relations are often shaped and dependent upon successful communication through language.

## Language Acquisition

The topic of language development is obviously vast and can only be briefly summarized here. Definitions of language vary depending upon which theory of language acquisition is espoused. As a result of studies of human and nonhuman systems of communication, controversies regarding the nature of language development abound. The controversies revolve around a number of issues, two of which are the nature-nurture debate and whether or not humans possess a unique communication system.

## Nature and Nurture of Language

One theory contends that language is shaped through environmental influence and social learning. Thus, infants initially imitate the sounds and language they hear. Caregivers, in turn, provide feedback for the gradual formation of words. Through imitation and reinforcement, infants eventually put words together in longer combinations until sentences are formed. Sentence construction and written language are similarly learned.

This social learning approach is contradicted by evidence from the field of developmental psycholinguistics that supports a biological, maturational view of language acquisition. The study of language systems and processes is known as *linguistics*. This field looks at four different aspects of language:

sound, grammar, meaning, and interaction. The study of how to put sounds together to form words is *phonology*. Grammar, the system of rules for putting words together in sentences, is known as *syntax; semantics* refers to the meaning of words and sentences. The rules regarding participation in conversation by sequencing sentences and by anticipating responses from others is known as *pragmatics* (130). Developmentalists are interested in how each of these aspects of language emerge and how they work with other psychological processes to result in changes in language acquisition throughout life. This field, which studies psychological processes and language and looks at changes over time, is known as *developmental psycholinguistics*.

Developmental psycholinguists, such as Noam Chomsky and Eric Lenneberg, claim that humans are innately (and uniquely) predisposed for language. Some of the evidence to support this point of view includes humans' highly specialized vocal-respiratory apparatus and a corresponding brain organization specialized for speech and language (131, 132). Additional evidence, some of which is listed in Table 12.8, is provided by the predictable maturation of language seen in children across cultures.

### Language Development in Children

Language acquisition follows a certain predictable course that is common across cultures. Despite these commonalities, however, children only acquire the language(s) to which they are exposed. During the first 6 months, infants make noncrying sounds of cooing and babbling. That this may be biologically determined is provided by evidence that deaf babies will also make these sounds but only during the first half-year of life. Children's first words emerge around their first birthday, although there are wide individual differences. Single words, then, form the first sentences, and language comprehension at this time far exceeds production. Initially, vocabulary is limited and comprises only a few verbs and nouns, but meaning is often altered by variations in inflection. Thus, infants may say "Daddy?" if they hear someone approaching them, or they may say "Daddy!" to call for attention. Inflection is also used at the two-word sentence stage of about 24 months, when the infant may ask questions or make demands or statements with only two-word combinations.

Telegraphic speech, which follows, is the combination of nouns and words in abbreviated sentences. Function words (e.g., "is") are only gradually introduced into sentences such as in the following sequence: "Doggie go," "Doggie go out," "Doggie is going out." Eventually, by about age 4, "wh" questions (what, where, when, why) are introduced, and nouns and verbs are reversed. Thus, instead of the question, "Why you do that?" the child can ask, "Why *do* you do that?"

Much to the consternation of some parents, children, who initially do not make errors, begin to use inappropriate words or word combinations that do not seem to make sense in terms of the language to which the children have been exposed. Thus, they may say "foots" or "goed." This is

Table 12.8
Evidence for the Biological Basis of Human Language

Common maturation pattern of children across cultures.
  All children begin to speak at about the same age.
  Sound and speech patterns follow the same sequence.
  Deaf children initially go through phonology patterns similar to children who
    hear normally.
  Most children overgeneralize linguistic rules.
  Grammatical rules are applied similarly across cultures.
Speed with which language is acquired early in life.
Brain lateralization specialized for language reception and speech.
Language is closely tied with human culture, transmission of history, and
    cognition.

a common occurrence and is evidence that children go through a predictable sequence in learning rules of their language. Initially, once a rule is learned, children overgeneralize it until they learn all the exceptions. Thus, they may learn the rule: To make plural words, add "s." As a result, they over-apply this rule when it is inaccurate and produce such words as "foots" or "deers."

What is remarkable about language acquisition is its speed and universality across human cultures. By about the age of 5, most children have acquired a system of communication that contains the basic elements of adult speech. While they do not have the cognitive abilities to communicate about abstract processes and to understand some vocabulary, they do understand the basic structure (syntax and pragmatics) of the language to which they have been exposed. The remarkability of this process has been imaginatively described by Arthur Koestler:

> Take a simple example: the farmer's little boy of about three, leaning out of the window, sees the dog snapping at the postman and the postman retaliating with a vicious kick. All this happens in a flash, so fast that his vocal chords have not even had the time to get innervated; yet he knows quite clearly what happened and feels the urgent need to communicate this as yet unverbalised event, image, idea, thought, or what-have-you, to his mum. So he bursts into the kitchen and shouts breathlessly: 'The postman kicked the dog.' Now the first remarkable fact about this is that he does *not* say, 'The dog kicked the postman,' though he *might* say, 'Doggy *was* kicked *by* the postman;' and again, he will *not* say, 'Was the dog kicked by the postman?', and least of all 'Dog the by was the kicked postman'.
>
> This was an example of a very simple sentence consisting of four words only ('the' being used twice). Yet a change of the order of two words gave a totally different meaning; a more radical reshuffling, with two new words added, left the meaning unaltered; and most of the ninety-five possible permutations of the original words give no meaning at all. The problem is how a child ever learns the several thousand abstract rules and corollaries necessary to generate and comprehend meaningful sentences—rules which his parents would be unable to name and define; which you and I are equally unable to define; and which nevertheless unfalteringly guide our speech.... These recipes, or formulae, the child somehow discovers by intuitive processes ... by the time it has reached the age of four. (133, p. 28)

## Animal Communication

While this area cannot be fully examined here, the study of animal communication needs to be briefly reviewed because it has contributed general information about the acquisition of communication abilities and because it addresses psycholinguists' contentions about the uniqueness of the human system of language.

Since the 1930s and 1940s, when the Kelloggs and the Hayes (134, 135) raised chimpanzees in their homes, numerous attempts have been made to teach language to primates. Initially, attempts to teach chimpanzees a vocal language failed because of the structure of their vocal apparatus; however, new systems of communication involving American Sign Language (ASL) (which is used by people with hearing impairments) or other symbolic systems have been under study (136–139). Current investigations are looking at whether primates trained in symbolic language will use these methods to communicate with offspring or with other members of their species. In addition, the special communication systems used by many nonhuman primates and nonprimate species, such as whales and dolphins, insects, and birds, are under investigation and provide information about linguistics and the cognitive and social capabilities of other species (140–144). This information is, predictably, controversial and incomplete (132, 140, 141, 145, 146).

The question of the uniqueness of human language is fundamental to religious and philosophical views about human reasoning and thought processes and is therefore beyond the scope of this book (141, 142). The studies of animal communication, however, have informed us about how methods of communication evolve for each species depending upon the environment for which they are adapted. For example, some nocturnal animals or whales, who cannot see long distances under water, must rely heavily on nonvisual forms of communication. Other species transfer information through other modes. Many nonhuman primates, similar to humans, communicate with different facial expressions, body postures, hand signals, and vocal tones (140, 144). Understanding how these emerge in other social species will help us learn about the evolution and the function of communication among humans.

## Child and Environmental Interaction

Ecological studies of infants' early abilities, caregiver responses, and their social interaction have recently informed us about human language acquisition. It appears that infants are specially equipped for verbal social interaction and that caregivers adapt their socialization to accommodate infant and children's early language needs. This interaction between infants and caregivers results in language being acquired very early in development.

### Infant Capabilities

Studies of different infant abilities discussed throughout this text have shown us that early human communication is multisensory: kinesthetic, olfactory, visual, tactile, and auditory. These early rhythmic sensorimotor abilities may be specially adapted for the human social environment, thus possessing a timing characteristic of human language and interaction (7, 113, 147). Dowd and Tronick (147) conclude that purposeful limb and ocular movements and facial changes are used as signals to which caregivers, in turn, respond, creating an interdependence and coordination basic to human interaction and fundamental for language development. Newson (148, pp. 91–92) explains:

> The actions which babies emit have characteristic forms which are apparently pan-cultural. They move their heads, waver their arm and legs, breathe heavily, vocalize, blow raspberries, screw up their faces and shift their visual regard, etc. in ways which are clearly biologically preprogrammed. Individual movements are also integrated into coordinated action-patterns which 'run-off' in sequences of activity interspersed with natural breaks in a typical 'burst-pause' pattern. This pattern may serve as a basis for the pragmatics of language—an early turn-taking, which is a basis for reciprocal social interaction.

Thus, some theorists speculate that given the social environment to which humans are adapted, the developmental capacities discussed throughout this text (i.e., cognitive, perceptual, sensorimotor) are basic predispositions for infant human language and reciprocal interaction (Fig. 12.10). Just as some scientists have suggested that humans possess perceptual feature detectors that are sensitive to aspects of the human environment, Eimas (149, 150) proposes that linguistic feature detectors are an inherent part of the human character.

### The Child's Socially Interactive Environment

Parents appear to be particularly sensitive to infant's social signals, and this sensitivity may facilitate their youngsters' language development. Infants as young as 2 weeks produce a variety of vocalizations that signal affective states, and parents, in turn, respond to these different vocalizations with a diverse pattern of reactions (151). Language acquisition is not a passive experience. Moskowitz (130) describes a young boy with normal hearing who was confined to home because of illness. His deaf parents, who communicated only with ASL, exposed him daily to television so that he would learn English. While the child was fluent in sign language by the age of 3, he did not acquire spoken language from his passive exposure to it on television.

Most caregivers are sensitive to their infants' language skills, deliberately adjusting their reactions to facilitate understanding. Caregivers talk differently to adults than they do to infants, and they can easily shift back and forth between adult-adult and adult-infant interaction. "Baby talk," an aspect of caregiver speech, is a simplification of speech to accommodate an infant's level of cognitive and language abilities; when talking with children, parents often simplify words, sentences, and vocabulary and talk about the here and now rather than abstract or displaced events (130). Papousek and Papousek (152, p. 150) explain that parents modulate their speech to support

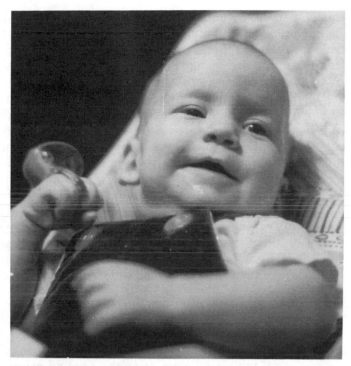

Figure 12.10.   Very early in life, infants discriminate and respond differently to inanimate and animate objects in their environments.

and facilitate language acquisition: "Thus different movements displayed by parents--facial expressions, play activities with toys or parts of the infant's body, and the first games--are modulated to simple repetitive patterns. . .The proportion of repetition and modification is balanced as if to allow the slowly learning infant to perceive and integrate the stable stimuli on one hand and to arouse the baby's attention with a change in stimulation if need be on the other." Language acquisition is clearly a product of social transactions between humans adapted to a social world.

# Sex and Gender Development

A crucial aspect of social interaction involves sexual maturity and the formation of gender identity. Sexual development is significant because without it the species would not survive. From a purely evolutionary perspective, our goal as humans is for successful mating and child-rearing to result in the reproduction of more members of the species.

The term "sex" is used here to refer to the biological traits of male or female. The term "gender" refers to the collective biological, social, and cultural influences that result in an individual's being feminine or masculine. Money and Ehrhardt (153, p. 4) indicate that gender is divided into two parts: *Gender identity* involves the persistence of one's individuality as male, female, or ambivalent as it is experienced in self-awareness and behavior. Gender identity is the private experience of gender role. *Gender role* involves the things that a person says and does to indicate to others or to the self the degree that one is male, female, or ambivalent. "Gender role is the public expression of gender identity."

Sexual development and gender role formation are complex. They involve biological, social, cognitive, perceptual, emotional, cultural, and physical forces, all in different types of interaction at different times during life. The degree to which these domains affect development is controversial and varies according to theorists' speculations. At present, one of the major controversies in the field of gender differences is the degree to which biological factors affect development. Proponents of equality among the sexes discount biological differences; to acknowledge their existence gives credibility to biological determinism and the belief that one sex is inherently superior to the other (154–156). While many recent analyses indicate that the similarities between the sexes are far greater than the differences, biological, genetic, and hormonal differences do exist.

## Sexual Development

Sexual development starts at fertilization with an XX or an XY gene combination (see chapter 7). During prenatal development some significant events occur which will determine the sex and gender of the individual in later life. The basic rule is that, to masculinize, something must be added (153). Thus, if the fetus is XX, then the reproductive anatomy will become that of a female. However, with the XY combination, an entire sequence of events must take place to ultimately result in male anatomy and masculinization. These additions to the XY embryo include: (a) a substance that inhibits the mullerian ducts which would normally develop into female reproductive anatomy and (b) the presence of testosterone, a male sex hormone. In early embryonic life, testosterone has two masculinizing influences: First it acts locally to promote the development of the wolffian ducts that form the internal male reproductive structure. Second, it circulates in the bloodstream to affect the prenatal tissue that will eventually become the external male sex organs (153). The differences between male and female sex organs are referred to as "primary sex differences."

### Potential Long-Range Psychological Influences

Not only are peripheral structures affected by the presence of testosterone, the central nervous system is also influenced. The degree to which this early influence alters gender identity is not completely understood, but work by Money and Ehrhardt and others indicates that prenatal influences may have long-range physiological *and* psychological effects. This information has been provided by clinical studies of females and males: Females may be inadvertently exposed prenatally to male sex hormones due to exogenously administered hormones or overproduction of male hormones by the adrenal glands of the fetus or mother (153). Males may be exposed to female sex hormones (157), produce low levels of testosterone, or prove insensitive to prenatal hormones (158). Work by Money and Ehrhardt (153) indicates that if the XX embryo was exposed to male hormones during the prenatal sensitive period for genital formation, masculinization occurred. These females were often born with external genitalia resembling that of a male. In addition, at adolescence when compared with females without prenatal male hormone exposure, these females tended to display more masculine or "tomboyish" traits including greater athletic behaviors, greater self-assertiveness, more functional and less adorned dress, less doll play in childhood, more focus on achievement rather than marital goals, and more masculine-oriented fantasies regarding sexual intimacy. Similarly, when compared with controls, males who were exposed prenatally to female sex hormones exhibited reduced assertiveness and athletic abilities (157).

While the results of such studies are compelling, they are also controversial. Males who are insensitive to prenatal testosterone are often raised as females, since they are not masculinized at birth (158). The degree to which these individuals display masculine or feminine behaviors at adulthood is clearly variable, depending upon whether they were reared as males or females and on the types of hormonal therapy administered during development. Many researchers point out that continued investigations are needed to explore the degree to which adolescent and adult sexual behaviors are a result of early hormone exposure, long-term socialization influences, or an interaction of both (159).

### Brain Dimorphism

Physiological sex differences in the central nervous system have been documented in many species and are assumed to occur in humans as well (160). These differences occur in the neural tissues that will ultimately control the release of male or female hormones from the gonads. The male gonads (testes) produce testosterone, which is responsible for the male secondary sex characteristics and which may also be partly responsible for some behaviors (such as aggression). In females, the female sex hormones, estrogen and progesterone, cycle rhythmically and are responsible for the formation and release of the egg and preparation of the uterus for implantation of the zygote.

The gonads do not operate autonomously. They are controlled by gonadotrophic hormones released by the anterior pituitary, and the pituitary is, in turn, regulated by the portion of brain that sits above it. These neural regions of the hypothalamus and the medial preoptic area instruct the pituitary when and to what degree hormones should be released. In some species, other neural regions respond to seasonal changes and many other environmental factors; these, in turn, interact with still other regions to affect hormone release and behavior. Thus, the actual control of sex hormones is complicated but resides, ultimately, in the brain. This means, then, that male and female brains differ, if only in their regulation of concentrations and rhythmicity of male and female sex hormones. Although male hormones may show some rhythmic alterations, they do not display the shifts characteristic of the female menstrual cycle. The brains of males and females have therefore been labeled "sexually dimorphic," meaning that some fundamental structural difference must exist to account for the variation of male and female regulation of pituitary and gonadal hormone release (161–163).

## Puberty

Genetic and hormonal forces that begin prenatally, in a sense, stay dormant until sexual maturity, known as *puberty*. At this time, many of the secondary sex characteristics, such as males' facial hair, women's breasts, and males' deeper voices, begin to emerge; and sex differences become apparent (Fig. 12.11). In females, reproductive maturity is signaled by *menarche*, the

Figure 12.11. Sexual development and maturity are important, for they ultimately lead to reproduction and perpetuation of the species.

onset of the menstrual cycle. In males, a number of gauges are used: the production of sperm, growth of testes, onset of a physical growth spurt characteristic of adolescence, or secondary sex characteristics. Other sex differences apparent during this period of rapid growth are listed in Table 12.9. The age of onset of puberty varies widely across individuals, ranging, for various cultures, from 12.4 to 18.8 years. A commonly recognized sex difference is that females tend to reach puberty about 2 years ahead of males (163, 164). In addition to these overt sex differences, other subtle gender differences may be initiated as early as infancy and certainly during childhood as a result of child-rearing, peer, and other socialization influences.

## Gender Identity and Role

As already discussed in this chapter, parents have been found to respond differently to male and female infants, and these differences may change over time (6, 47, 83, 165–167). From the time the baby is labeled "male" or "female" at birth, the youngster may be treated differently by the various people with whom he or she interacts. These sometimes subtle, and sometimes overt, influences interact with genetic, biological, hormonal, and psychological factors and result in gender identity and gender role formation. The degree to which one becomes masculine or feminine is known as *gender-typing* (or *sex-typing*), a process that is clearly culturally affected (6).

Studies of these various gender differences have virtually exploded in the past 20 years. With the influences of Women's Liberation Movement, the formation of women's studies programs, and a greater awareness of ecological effects on development, the focus on the potential causes and the apparent effects of gender influences has been a priority of many social and educational researchers. The implications of this research are enormous, as evidenced by changes that have already occurred in regard to women's changing roles in U. S. society, modifications of the English language, and other sociopolitical alterations. The general consensus has been that cultural messages can affect gender-typing. Thus, studies have been exploring not

Table 12.9
**Sex Differences Characteristic of the Adolescent Growth Spurt**[a]

Growth spurt occurs 2 years earlier in females.
Growth spurt is more pronounced in males; this is partly due to testosterone.
Males have a greater increase in size of muscles.
Males have larger heart and lungs body, greater capacity for oxygen in blood, and lower resting heartbeat.
Males recover faster from physical exhaustion.
Females' hips widen more than males'.
Males' shoulders widen more than females'.
With respect to upper arms or total body height, males' forearms are longer than females' and males tend to be taller than females.

[a] Data from Tanner JM: Growing up. *Scient Am* 229: 34–43, 1973.

only parental influences but also the effects of toys, books, television, teachers, role models, personal characteristics, and numerous other social forces on child and adolescent gender identity formation (6, 165–168).

An example of such research is found in studies of gender differences in cognitive and social functions. It has been widely recognized that males tend to perform differently than females on spatial tasks, outperforming females on tasks such as embedded figures, where they must identify shapes obscured in a background. Other widely recognized differences are males' greater aggressiveness, females' superior verbal skills, and female's tendencies to be more socially dependent and less assertive and achievement oriented than men (169). The purported causes for these differences vary.

### Biological Causes

Those researchers who favor biological interpretations point to potential brain or physiological differences between the sexes as causes for these differences. Thus, males are assumed to be more aggressive (assertive, achievement-oriented) because of their higher testosterone levels (Fig. 12.12). Males' superiority over females on spatial tasks has also been linked to hormonal differences (170) as well as differences in the male and female

Figure 12.12.   The male sex hormone, testosterone, has been linked to aggressive behavior and is claimed to be responsible for many differences between the sexes.

brains (171, 172). Other researchers, who favor social explanations, point out the various potential social and cultural causes for these differences. Thus, while it makes sense that males might be more aggressive because of their higher concentrations of testosterone, other explanations are equally valid. Males may be socialized to be more aggressive.

### Socialization Explanation

Parents have been found to react differently to male and female children. Despite females' higher survival rates, daughters are rated as softer and less alert than sons, and sons are rated as stronger and hardier than daughters. Parents tend to engage in more physical stimulation and gross motor play with male than with female infants, tend to leave male children to play alone more often than females, and encourage females to be dependent and to engage in social play. Gender stereotypes with doll and toy play are reinforced as children develop, and fathers, more than mothers, tend to perceive and respond to sex differences among their children. Both parents tend to interact more with children who are the same sex they are, and fathers tend to treat female children more gently than male children (6, 165, 168).

These same behaviors may also be reinforced in day-care and school settings, so that children gradually learn "gender-appropriate" or stereotyped behaviors at or away from home. In addition, television cartoons, sports programs, commercials, and books may continue to reinforce gender messages such as the notion that aggression is an appropriate masculine behavior. Thus, male activity, exploration, and aggression may be socially encouraged, accounting for greater opportunity for males to experiment with the world and to receive a variety of experiences, including opportunities for spatial stimulation and challenges. For females, for whom language and social skills are reinforced, aggression and exploration become less and less socially acceptable options. Instead, social interaction, dependency, and a reliance on language and communication skills become the norm.

### Interaction Among Biological and Social Influences

Those who favor biological interpretations of gender differences point out that parents interact differently with male and female children because the children, themselves, display fundamental differences in their behavior. Thus, they claim, society just reinforces what biology has already designed. Opponents of biological determinism claim that socialization is responsible for gender differences, and fundamental, biological, behavioral differences between the sexes do not exist. The nature-nurture debate continues.

An interactionist position described by Archer and Lloyd (173, p. 176) assumes that over time, biology and culture interact, provide feedback, and thereby mutually affect one another to result in gender differences. These researchers conclude: "A two-way process is involved which can be explained neither by socialization acting upon the child nor by the innate expression of the child's nature regardless of the environmental consequences." Ecological views of family and cultural systems are showing us that simple, linear accounts of gender differences are too limited, and that multiple, interactional models of development are essential (Fig. 12.13) (165).

### Contemporary Approaches to Gender Differences

Investigations of sex differences are becoming less popular as new approaches are forthcoming. Instead of looking at sex differences per se, these approaches are focusing on issues important to both males and females, on

Figure 12.13.   New approaches to gender differences suggest we look not at whether behaviors are masculine or feminine but instead at the circumstances in which certain behaviors are necessary or appropriate.

theory-testing based on evidence regarding the development of females, and on *androgeny*, the incorporation of masculine and feminine traits in both males and females (155, 174). Thus, rather than trying to separate the sexes based on certain traits and the assumption that one sex is superior to another, studies are directed at determining the circumstances in which certain traits (assertiveness, nurturance, aggression, dependency) are appropriate (Fig. 12.13). For example, nurturance, once considered a feminine trait, is currently considered an important characteristic of both males and females, especially considering the increased role of males in child care. Shepherd-Look (155, p. 428) suggests: "Rather than trying to socialize a child into having an equal number of masculine and feminine traits, it might be more profitable to alter the system of values, such that all sex-typed behaviors would be equally valued whether they were performed by a male or a female. In this way each child could develop his/her potential without regard for which traits would be more socially acceptable than others."

## Moral Development

An area of study that illustrates the contemporary approach to gender differences and also poses a problem for the science of human development is moral development. Moral development, which involves the gradual understanding of what one's culture considers right and wrong, incorporates both cognitive and social influences. Theories of how moral development emerges are diverse.

### Theories of Moral Development

Psychoanalytic theory assumes that morality is a result of the process of identification, where the child actually takes on some of the values (super-ego) of a model. In contrast, recent sociobiological approaches (see chapter 6) assumes that behaviors that support the preservation of the species are instinctive. Social learning theorists assume, like psychoanalytic theorists, that moral reasoning is acquired through socialization. Considerable research on moral development was prompted by Piaget's theory, which assumed that moral reasoning (like language and emotions) grows out of cognition. Piaget's theory was extended by psychologist Lawrence Kohlberg, whose theory of moral development is most often associated with the field. Kohlberg developed a major model of moral acquisition that has been the subject of research in the field for the past several decades.

### Kohlberg's Theory

Like cognitive development, moral development was assumed by Kohlberg to follow a certain maturational sequence. Based on a 20-year longitudinal study of middle-class and lower-class boys, ages 10, 13, 16, Kohlberg came up with six stages of moral reasoning, which he has subsequently revised to five. This sequence starts at a premoral level, where the child obeys rules in order to avoid punishment. Gradually, the child progresses and through maturation and socialization, develops an internalized sense of right and wrong. In the third stage, moral decisions are based on approval and maintaining relationships with others; but in the fourth stage, the significance of relationships diminishes while "law and order" decisions are made based upon the concepts of obedience and responsibility to the social order. In the final stage, the individual internalizes a universal sense of order or justice which guides an independent sense of morality (175, 176).

### Problems with Kohlberg's Theory

Although Kohlberg's theory has dominated the research and field of moral development, it has come under considerable criticism lately. The criticism has been directed at his methods as well as his conclusions. Cross-cultural studies do not support Kohlberg's notion of maturational sequences, and evidence indicates that some types of moral reasoning may be learned and that people may skip around and not follow any particular sequence in life (176, 177).

The most vocal critic lately has been researcher Carol Gilligan. Using a different method than Kohlberg used, Gilligan has been exploring the differences in moral reasoning of males and females (178). Some of her research was prompted by Kohlberg's findings that while males tended to progress through the stages of moral development, females tended to stay at stage 3, which is governed by the interpersonal skills of helping and pleasing others (178). Gilligan reports that Kohlberg indicated that this was the typical level of morality for housewives, mothers, or women in nurturing professions and that if these women were to enter traditional arenas of male activity, then they would advance to higher stages of morality.

### Gilligan's Hypothesis

After examining women's moral reasoning, Gilligan claims that developmental pressures differ for males and females and that these differences account for Kohlberg's findings. Gilligan claims that because of the nature of socialization experiences, males and females end up confronting separate socialization issues. Consistent with Erikson's adolescent stage of identity formation (Table 12.7), males are encouraged to separate from their mothers, to become independent, and to form separate, distinctive identities. In doing so, they become autonomous and they learn to individuate, i.e., exist as an individual separate from others and to define themselves as distinct (e.g., "I am Joe, a teacher, an athlete, a student").

In contrast, Gilligan claims that female identity is not, like males, dependent upon separation but instead is based on attachment. Females are encouraged to maintain close interpersonal relationships and therefore define themselves primarily in terms of their relationships with others (e.g., "I am a mother, a wife, a friend, a daughter, a colleague"). The issues in adulthood for males and females therefore become focused on separate but related concerns: Females, who are comfortable with intimacy, become threatened by loneliness and separation. In contrast, males, who have been taught to separate and individuate, are threatened by intimacy. Thus, tests of their moral development and reasoning, such as Kohlberg's, will find different approaches by males and females, based on their different approaches to individual vs. group needs. Such tests that assume that the approach of one sex is the correct one will, of necessity, find the other sex inferior. Gilligan (178, pp. 8–9) notes in regard to Kohlberg's scale, which was based solely on male respondents: "Women's failure to separate then becomes by definition a failure to develop."

Other studies of gender differences in social and cognitive development tend to support Gilligan's hypothesis. Hoffman (177) suggests that females are actually more morally internalized than males and display more humanistic moral values. Mahon (179) reports that female adolescents experience high levels of loneliness during the phase of detachment and separation from their parents, and Alishio and Schilling (180) note that males and females may approach interpersonal relationships differently, with males focusing on achievement and autonomy, and with females focusing on issues of trust and continuity with others. These researchers conclude that continued studies need to look not at which sex is more or less developed but at how each sex integrates their own personality through their varying social and emotional experiences in life.

### Challenging Developmental Theory

Gilligan notes that the theories of human development have been based primarily on work with males and that, therefore, psychologists have neglected an important aspect of development by not considering the issues of social interaction and intimacy that are central to females. She challenges scientists to shift the lens of developmental observation from concerns about ego identity, autonomy, and individuality toward an understanding of relationships, contexts, family issues, and interdependence. With greater understanding of these issues, the science of development will incorporate a different voice, a female voice, to round out developmental theory. Based on Gilligan's challenge, the science and theories of human development, as discussed throughout this text, may be subject to considerable modification and expansion in the years to come.

# CHAPTER SUMMARY

Research examining social and emotional development is characterized by controversy and challenges. So many factors are involved in social and emotional development, and these are difficult to study with the massive social and cultural changes in our society in the past several decades. These diverse cultural influences have affected the lives of the people growing up during these times as well as altering the approaches and perspectives of researchers. Clearly, it is difficult to understand the normal processes involved in the social development of children and adolescents as families, gender roles, occupations, the economy, attitudes, technology, schools,

Figure 12.14.   Humans seem to have a special interest in other human beings. This interest is fundamental to their subsequent linguistic, social, emotional, personality, cognitive, perceptual, sexual, and moral development.

communication systems, and other diverse aspects of our culture are in such rapid flux. The research approaches that have recently emerged, however, may enable us to accommodate these changing influences as well as incorporate the numerous interacting systems that work together to affect development. With these recent ecological and systems approaches, we may begin to more clearly understand the interaction among nature and nurture and the interactions of the numerous domains that make up human development (Fig. 12.14).

## REFERENCES

1. Lewis M, Rosenblum LA: Introduction: Issues in Affect Development. In Lewis M, Rosenblum LA (eds): *The Development of Affect.* New York, Plenum Press, 1978.
2. Sroufe AL: Socioemotional development. In Osofksy JD (ed): *Handbook of Infant Development.* New York, John Wiley & Sons, 1979, p 462.
3. Parke RD: Emerging themes for social-emotional development. *Am Psychologist* 34: 930–931, 1979.
4. Baldwin AL: *Theories of Child Development.* New York, John Wiley & Sons, 1967.
5. Bandura A: *Principles of Behavior Modification.* New York, Holt, Rinehart, & Winston, 1969.

6. Huston AC: Sex-typing. In Mussen PH (ed): *Handbook of Child Psychology*, vol IV, Hetherington EM (ed): *Socialization, Personality, and Social Development*. New York, John Wiley, 1983.
7. Condon WS and Sander L: Neonate movement is synchronized with adult speech: interactional participation in langauge acquisition. *Science* 183: 99–101, 1974.
8. Meltzoff AN, Moore MK: The origins of imitation in infancy: Paradigm, phenomena, and theories. In Lipsitt LP, Rovee-Collier CK (eds): *Advances in Infancy Research*. Norwood, NJ, ABLEX, 1983, vol II, p 265.
9. Cohen LJ: The operational definition of human attachment. *Psychol Bull* 81: 207–217, 1974.
10. Gubernick DJ: Parent and infant attachment in mammals. In Gubernick DJ, Klopfer PH (ed): *Parental Care in Mammals*. New York, Plenum Press, 1981.
11. Bowlby J: *Attachment and Loss. vol I. Attachment*. New York, Basic Books, 1969.
12. Klaus MH, Kennell JH: *Maternal-infant Bonding*. St. Louis, CV Mosby, 1976.
13. Rajecki DW, Lamb ME, Obmascher P: Toward a general theory of infantile attachment: a comparative review of aspects of the social bond. *Behav Brain Sci* 3: 417–464, 1978.
14. Ainsworth MDS: *Infancy in Uganda. Infant Care and Growth of Love*. Baltimore, Johns Hopkins Press, 1967.
15. Ainsworth MDS: Infant-mother attachment. *Am Psychologist* 34: 932–937, 1979.
16. Ainsworth MDS: Patterns of infant-mother attachment as related to maternal care: their early history and their contribution to continuity. In Magnusson D, Allen VT (eds): *Human Development. An Interactional Perspective*. New York, Academic Press, 1983.
17. Sroufe LA: Attachment and the roots of competence. *Hum Nature* October: 92–96, 1978.
18. Rheingold HL, Eckerman CO: The infant separates himself from his mother. *Science* 168: 78–83, 1970.
19. Bowlby J: *Attachment and Loss. Vol II. Separation*. New York, Basic Books, 1973.
20. Rutter M: *Maternal Deprivation Reassessed*. Baltimore, Penguin Books, 1972.
21. Bronfenbrenner U: Early deprivation in monkey and man. In Bronfenbrenner U (ed): *Influences on Human Development*. Hinsdale, IL, Dryden Press, 1972.
22. Harlow HF, Mears C: *The Human Model: Primate Perspectives*. New York, VH Winston & Sons, 1979.
23. Harlow HF: The nature of love. *Am Psychologist* 13: 673–685, 1958.
24. Harlow HF, Harlow MK, Dodsworth RO, Arling GL: Maternal behavior of rhesus monkeys deprived of mothering and peer association in infancy. *Proc Am Philosoph Soc* 110: 58–66.
25. Harlow HF, Harlow MK, Suomi SJ: From thought to therapy: lessons from a primate laboratory. *Am Scient* 59: 538–549, 1971.
26. Suomi SJ, Harlow HF: Social rehabilitation of isolate-reared monkeys. *Dev Psychol* 6: 487–496, 1972.
27. Harlow HF, Seay B: Affectional systems in rhesus monkeys. *J Arkansas Med Soc* 61: 107–110, 1964.
28. Seay B, Hansen E, Harlow HF: Mother-infant separation in monkeys. *J Child Psychol Psychiat* 3: 123–132, 1962.
29. Harlow HF, Schlitz KA, Harlow MK: Effects of social isolation on the learning performance of rhesus monkeys. *Proc 2nd Int Congr Primat Atlanta GA*. New York, Karger, Basel, 1969, vol 1, pp 178–185.
30. Short MA: Vestibular stimulation as early experience: Historical perspectives and research implications. In Ottenbacher K, Short MA (eds): *Vestibular Processing Dysfunction in Children*. New York, Haworth Press, 1985.
31. Dinnage R: Understanding loss: the Bowlby canon. *Psychol Today* May: 56–60, 1980.
32. Novak MA, Harlow HF: Social recovery of monkeys isolated for the first year of life: 1. Rehabilitation and therapy. *Dev Psychol* 11: 453–465, 1975.
33. Suomi SJ: Surrogate rehabilitation of monkeys reared in total social isolation. *J Child Psychol Psychiat* 14: 71–77, 1973.
34. Clarke AM, Clarke ADB (eds): *Early Experience: Myth and Evidence*. New York, Free Press, 1976.

35. Freud A, Dann S: An experiment in group upbringing. In Bronfenbrenner U (ed): *Influences on Human Development*. Hinsdale, IL, Dryden Press, 1972.
36. Casler L: Perceptual deprivation in institutional settings. In Newton G, Levine S (eds): *Early Experience and Behavior*. Springfield, CT, Charles C Thomas, 1968.
37. Yarrow LJ: Maternal deprivation: toward an empirical and conceptual re-evaluation. *Psychol Bull* 58: 459–490, 1961.
38. Scott JP: Critical periods in behavioral development. *Science* 138: 949–958, 1962.
39. Scott JP, Stewart JM, DeGhett VJ: Critical periods in the organization of systems. *Dev Psychobiol* 7: 489–513, 1974.
40. Tizard B: Early experience and later social behaviour. In Schaeffer D, Dunn J (eds): *The First Year of Life. Psychological and Medical Implications of Early Experience*. New York, John Wiley & Sons, 1979.
41. Kagan J: Resilience and continuity in psychological development. In Clarke AM, Clarke ABD (eds): *Early Experience: Myth and Evidence*. New York, Free Press, 1976.
42. Hartup WW: The social worlds of childhood. *Am Psychologist* 34: 944–950, 1979.
43. Baumrind D: The development of instrument competence through socialization. In Pick AD (ed): *Minnesota Symposium on Child Psychology*. Minneapolis, MN, University of Minnesota Press, vol 7, 1973.
44. Prescott JW: Somatosensory deprivation and its relationship to the blind. In Jastrzembska ZS (ed): *The Effects of Blindness and Other Impairments on Early Development*. New York, American Foundation for the Blind, 1976.
45. Walters RH, Parke RD: The role of the distant receptors in the development of social responsiveness. In Kopp CB (ed): *Readings in Early Development for Occupational and Physical Therapy Students*. New York, Charles C Thomas, 1971.
46. Harlow HF, Suomi SJ: Nature of love--simplified. *Am Psychologist* 25: 161–168, 1970.
47. Power TG: Mother- and father-infant play: a developmental analysis. *Child Dev* 56: 1514–1524, 1985.
48. Cook H, Stingle S: Cooperative behavior in children. *Psychol Rev* 81: 918–933, 1974.
49. Shapira A, Madsen MC: Cooperative and competitive behavior of kibbutz and urban children in Israel. *Child Devel* 40: 609–617, 1969.
50. Rabin AI, Beit-Hallahmi B: *Twenty Years Later Kibbutz Children Grow Up*. New York, Springer Publishing Co, 1982.
51. Teichman T, Ben Rafael M, Lerman M: Anxiety reaction of hospitalized children. *Br J Med Psychol* 59: 375–382, 1986.
52. Blehar MC: Anxious attachment and defensive reactions associated with day care. *Child Dev* 45: 683–692, 1974.
53. Schwarz JC, Strickland RG, Krolick G: Infant day care: Behavioral effects at preschool age. *Dev Psychol* 10: 502–506, 1974.
54. Wallis C: The child care dilemma. *Time* June 22: 54–60, 1987.
55. Belsky J, Stenberg LD: The effects of day care: a critical review. *Child Dev* 49: 929–949, 1978.
56. Belsky J: Infant day care: A cause for concern? *Zero to Three* 6: 1–7, 1986.
57. Phillips D, McCartney K, Scarr S, Howes C: Selective review of infant day care research: A cause for concern! *Zero to Three* 7: 18–21, 1987.
58. Chess S: Comments: "Infant day care: a cause for concern" *Zero to Three* 7: 24–25, 1987.
59. Bronfenbrenner U: Ecology of the family as a context for human development: research perspectives. *Dev Psychol* 22: 723–742, 1986.
60. Bradley RH: Day care: a brief review. *Phys Occup Ther Pediatr* 2: 73–82, 1982.
61. Bell RQ: Stimulus control of parent or caretaker behavior by offspring. *Dev Psychol* 4: 63–72, 1971.
62. Lerner RM, Busch-Rossnagel NA (eds): *Individuals as Producers of their Own Development. A Life-Span Perspective*. New York, Academic Press, 1981.
63. Goldberg S: Premature birth: consequences for the parent-infant relationship. *Am Scientist* 67: 214–220, 1979.
64. Green JA, Gustafson GE, West MJ: Effects of infant development on mother-infant interactions. *Child Dev* 51: 199–207, 1980.

65. Greenberg DJ, Hillman D, Grice D: Infant and stranger variables related to stranger anxiety in the first year of life. *Dev Psychol* 9: 207–212, 1973.

66. Rheingold HL, Eckerman CO: Fear of the stranger. A critical examination. In Reese HW (ed): *Advances in Child Development and Behavior*. New York, Academic Press, 1973, vol 8, p 186.

67. Izard CE: On the ontogenesis of emotions and emotion-cognition relationships in infancy. In Lewis M, Rosenblum LA (eds): *The Development of Affect*. New York, Plenum Press, 1978.

68. Trotter RJ: Baby Face. *Psychol Today* Aug: 15–20, 1983.

69. Thomas A, Chess S: The role of temperament in the contributions of individuals to their development. In Lerner RM, Busch-Rossnagel NA (eds): *Individuals as Producers of their Own Development. A Life-Span Perspective*. New York, Academic Press, 1981.

70. Worobey J: Convergence among assessments of temperament in the first month. *Child Dev* 57: 47–55, 1986.

71. Ahmad A, Worobey J: Attachment and cognition in a naturalistic context. *Child Study J* 14: 185–203, 1984.

72. Hofer MA: Introduction. *Parent-Infant Interaction*. New York, Elsevier, 1975.

73. Bell RQ: Parent, child, and reciprocal influences. *Am Psychol* 34: 821–826, 1979.

74. Rosenberg SA: Measures of parent-infant interaction: An overview. *Topics in Early Childhood Special Education* 6: 32–43, 1986.

75. Thomas EAC, Martin JA: Analyses of parent-infant interaction. *Psychol Rev* 83: 141–156, 1976.

76. Lytton H: Observation studies of parent-child interaction: A methodological review. *Child Dev* 42: 651–684, 1971.

77. Rheingold HL: The measurement of maternal care. *Child Dev* 31: 565–575, 1960.

78. A new kind of life with father. *Newsweek* Nov. 30: 93–97, 1981.

79. Yogman MW: Development of the father-infant relationship. In Fitzgerald HE, Lester BM, Yogman MW (eds): *Theory and Research in Behavioral Pediatrics*. New York, Plenum, 1982.

80. Lamb ME: Paternal influences and the father's role. A personal perspective. *Am Psychol* 34: 938–943, 1979.

81. Parke RD: Perspectives on father-infant interaction. In Osofksy JD (ed): *Handbook of Infant Development*. New York, John Wiley & Sons, 1979, p 549.

82. Yarrow LJ, MacTurk RH, Vietze PM, McCarthy ME, Klein RP, McQuiston S: Developmental course of parental stimulation and its relationship to mastery motivation during infancy. *Dev Psychol* 20: 492–503, 1984.

83. Easterbrooks MA, Goldberg WA: Toddler development in the family: Impact of father involvement and parenting characteristics. *Child Dev* 55: 740–752, 1984.

84. Kotelchuck M: The infant's relationship to the father: Experimental evidence. In Gardner JK (ed): *Readings in Developmental Psychology*. Boston, Little, Brown and Co, 1978, pp 61–69.

85. Frodi AM, Lamb ME, Leavitt LA, Donovan WL: Fathers' and mothers' responses to infant smiles and cries. *Infant Behav Dev* 1: 187–198, 1978.

86. Crowe TK: Father involvement in early intervention programs. *Phys Occup Ther Pediatr* 1: 35–46, 1981.

87. Meyer DJ, Vadasy PF, Fewell RR, Schell G: Involving fathers of handicapped infants: translating research into program goals. *J Div Early Childhood* 5: 64–72, 1982.

88. Vadasy PF, Fewell RR, Meyer DJ, Greenberg MT: Supporting fathers of handicapped young children: preliminary findings of program effects. *Anal Intervent Dev Disabil* 5: 151–163, 1985.

89. Vadasy PF, Fewell RR, Greenberg MT, Dermond NL, Meyer DJ: Follow-up evaluation of the effects of involvement in the fathers program. *Topics Early Childhood Spec Educ* 6: 16–31, 1986.

90. Elster AB, Lamb ME (eds): *Adolescent Fatherhood*. Hillsdale, NJ, Lawrence Erlbaum Associates, 1986.

91. Thoman E: How a rejecting baby may affect mother-infant synchrony. *Parent-Infant Interaction*. New York, Elsevier, 1975.
92. Brazelton TB, Tronick E, Adamson L, Asl H, Weise S: Early mother infant reciprocity. *Parent-Infant Interaction*. New York, Elsevier, 1975.
93. Lerner RM, Shea JA: Social behavior in adolescence. In Wolman BB (ed): *Handbook of Developmental Psychology*. Englewood Cliffs, NJ, Prentice-Hall, 1982.
94. Dunn JF: Consistency and change in styles of mothering. *Parent-Infant Interaction*. New York, Elsevier, 1975.
95. Asher J: Born to be shy? *Psychology Today* April: 56–64, 1987.
96. Crittenden A: New insights into infancy. *The New York Times Magazine* Nov. 13: 84–96, 1983.
97. Demos V: The role of affect in early childhood. In Tronick EZ (ed): *Social Interchange in Infancy. Affect, Cognition, and Communication*. Baltimore, University Park Press, 1982.
98. Short-DeGraff MA, Kologinsky E: Respite care: Roles for therapists in support of families with handicapped children. *Phys Occup Ther Pediatr*, in press, 1987.
99. Lewis M, Rosenblum LA: Introduction. In Lewis M, Rosenblum LA (eds): *Friendship and Peer Relations*. New York, John Wiley & Sons, 1975.
100. Yarrow MR: Some perspectives on research on peer relations. In Lewis M, Rosenblum LA (eds): *Friendship and Peer Relations*. New York, John Wiley & Sons, 1975.
101. Rheingold HL, Eckerman CO: Some proposals for unifying the study of social development. In Lewis M, Rosenblum LA (eds): *Friendship and Peer Relations*. New York, John Wiley & Sons, 1975.
102. Eckerman CO, Whatley JL, Kutz SL: Growth of social play with peers during the second year. *Dev Psychol* 11: 42–49, 1975.
103. Mueller EC, Vandell D: Infant-infant interaction. In Osofksy JD (ed): *Handbook of Infant Development*. New York, John Wiley & Sons, 1979, p 519.
104. Langlois JH, Gottfried NW, Seay B: The influence of sex of peer on the social behavior of preschool children. *Dev Psychol* 8: 93–98, 1973.
105. Newman PR, The peer group. In Wolman BB (ed): *Handbook of Developmental Psychology*. Englewood Cliffs, NJ, Prentice-Hall, 1982.
106. Dunphy DC: The social structure of urban adolescent peer groups. *Sociometry* 26: 230–246, 1975.
107. Strain PS: Social interactions of handicapped preschoolers in developmentally integrated and segregated settings: A study of generalization effects. In Field T, Roopnarine JL, Segal M (eds): *Friendships in Normal and Handicapped Children*. Norwood NJ, Ablex Publishing, 1984.
108. Greenwood CR: Settings or setting events as treatment in special education? A review of mainstreaming. In Wolraich M, Routh DK (eds): *Advances in Developmental and Behavioral Pediatrics*. Greenwich, CT, JAI Press, 1985, vol 6.
109. Guralnick MJ (ed): *Early Intervention and the Integration of Handicapped and Non-Handicapped Children*. Baltimore, University Park Press, 1978.
110. Strain PS, Fox JJ: Peer social imitation and the modification of social withdrawal: a review and future perspectives. *J Pediatr Psychol* 6: 417–433, 1981.
111. Anastasiow NJ: Strategies and models for early childhood intervention programs in integrated settings. In Guralnick MJ (ed): *Early Intervention and the Integration of Handicapped and NonHandicapped Children*. Baltimore, University Park Press, 1978.
112. Busch-Rossnagel NA: Where is the handicap in disability?: the contextual impact of physical disability. In Lerner RM, Busch-Rossnagel NA (eds): *Individuals as Producers of their Own Development. A Life-Span Perspective*. New York, Academic Press, 1981.
113. Lester BM, Hoffman J, Brazelton TB: The rhythmic structure of mother-infant interaction in term and preterm infants. *Child Dev* 56: 15–27, 1985.
114. Bornstein MH: How infant and mother jointly contribute to developing cognitive competence in the child. *Proc Natl Acad Sci USA* 82: 7470–7473, 1985.
115. Olson GM, Lamb ME: Premature infants. Cognitive and social development in the first year of life. In Stack JM (ed): *The Special Infant. An Interdisciplinary Approach to the Optimal Development of Infants*. New York, Human Sciences Press, 1982.

116. Olson SL, Bates JE, Bayles K: Mother-infant interaction and the development of individual differences in children's cognitive competence. *Dev Psychol* 20: 166–179, 1984.
117. Hall F, Pawlby SJ, Wolkind S: Early life experiences and later mothering behaviour: a study of mothers and their 20-week old babies. In Schaeffer D, Dunn J (eds): *The First Year of Life. Psychological and Medical Implications of Early Experiences*. New York, John Wiley & Sons, 1979.
118. Belsky J, Garduque L, Hrncir E: Assessing performance, competence, and executive capacity in infant play: Relations to home environment and security of attachment. *Dev Psychol* 20: 406–417, 1984.
119. Gelman R, Spelke E: The development of thoughts about animate and inanimate objects: implications for research on social cognition. In Flavell JH, Ross L (eds): *Social Cognitive Development*. Cambridge, Cambridge University Press, 1981.
120. Lewis M, Brooks-Gunn J: *Social Cognition and the Acquisition of Self*. New York, Plenum Press, 1979.
121. Chandler M, Boyes M: Social-cognitive development. In Wolman BB (ed): *Handbook of Developmental Psychology*, Englewood Cliffs, Prentice-Hall, 1982.
122. Hoffman ML: Perspectives of the difference between understanding people and understanding things: the role of affect. In Flavell JH, Ross L (eds): *Social Cognitive Development*. Cambridge, Cambridge University Press, 1981.
123. Damon W: Exploring children's social cognition on two fronts. In Flavell JH, Ross L (eds): *Social Cognitive Development*. Cambridge, Cambridge University Press, 1981.
124. Yarrow LJ: Emotional development. *Am Psychol* 34: 951–957, 1979.
125. Kagan J: *The Nature of the Child*. New York, Basic Books, 1984.
126. Harter S: Developmental perspectives on the self-system. In Mussen PH (ed): *Handbook of Child Psychology, vol IV*, Hetherington EM (ed): *Socialization, Personality, and Social Development*. New York, John Wiley & Sons, 1983.
127. Erikson EH: *The Life Cycle Completed*. New York, WW Norton, 1982.
128. Gardner H: *Developmental Psychology. An Introduction*. Boston, Little, Brown, & Co, 1978.
129. Notes and comments. *The New Yorker* March 30, 1987.
130. Moskowitz BA: The acquisition of language. *Scient Am* 239: 92–108, 1978.
131. Geschwind N: The organization of language and the brain. *Science* 170: 940–944, 1970.
132. Lenneberg EH: On explaining language. *Science* 164: 635–643, 1969.
133. Koestler A: *The Ghost in the Machine*. New York, MacMillan, 1967.
134. Gardner RA, Gardner BT: Teaching sign language to a chimpanzee. *Science* 165: 664–672, 1969.
135. Kellogg WN: Communication and language in the home-raised chimpanzee. *Science* 162: 423–427, 1968.
136. Premack AJ, Premack D: Teaching language to an ape. *Prog Psychobiol* San Francisco, WH Freeman, 1976, 333–340.
137. Patterson F: Conversations with a gorilla. *Natl Geographic* October: 438–466, 1978.
138. Fouts RS: Acquisition and testing of gestural signs in four young chimpanzees. *Science* 180: 978–980, 1973.
139. Linden E: *Apes, Men, and Language*. New York, EP Dutton & Co, 1970.
140. Marler P: Animal communication signals. *Science* 157: 769–774, 1967.
141. Griffin DR: *The Question of Animal Awareness. Evolutionary Continuity of Mental Experience*. New York, Rockefeller University Press, 1981.
142. Griffin DR: *Animal Thinking*. Cambridge, MA, Cambridge University Press, 1984.
143. Morgan S: The sagacious dolphin. *Natural History* 77: 32–39, 1968.
144. Wilson EO: Animal communication. *Prog Psychobiol* San Francisco, WH Freeman, 1976, 324–332.
145. Terrace HS: How Nim Chimsky changed my mind. *Psychology Today* Nov.: 65–76, 1979.
146. Bronowski J, Bellugi U: Language, name, and concept. *Science* 168: 669–673, 1970.
147. Dowd JM, Tronick EZ: Temporal coordination of arm movements in early infancy: Do infants move in synchrony with adult speech? *Child Develop* 57: 762–776, 1986.
148. Newson J: Intentional behavior in the young infant. In Schaeffer D, Dunn J (eds): *The First Year of Life. Psychological and Medical Implications of Early Experience*. New York, John Wiley & Sons, 1979.

149. Eimas PD, Siqueland ER, Jisczyk P, Vigorito J: Speech perception in infants. *Science* 171: 303–305, 1971.
150. Eimas PD: Speech perception in early infancy. In Cohen LB, Salapatek P (eds): *Infant Perception. From Sensation to Cognition*, vol II, *Perception of Space, Speech, and Sound*. New York, Academic Press, 1975.
151. Keller H, Scholmerich A: Infant vocalizations and parental reactions during the first 4 months of life. *Dev Psychol* 23: 62–67, 1987.
152. Papousek H, Papousek M: How human is the human newborn, and what else is to be done? In Bloom K (ed): *Prospective Issues in Infancy Research*. Hillsdale, NJ, Lawrence Erlbaum Associates, 1981.
153. Money J, Ehrhardt AA: *Man & Women. Boy & Girl*. Baltimore, MD, Johns Hopkins University Press, 1972.
154. Birke L: *Women, Feminism, and Biology*. New York, Metheun, 1986.
155. Shepherd-Looke DL: Sex differentiation and the development of sex roles. In Wolman BB (ed): *Handbook of Developmental Psychology*. Englewood Cliffs, NJ, Prentice-Hall, 1982.
156. Fausto-Sterling A: *Biological Theories about Women and Men*. New York, Basic Books, 1985.
157. Yalom ID, Green R, Fisk N: Prenatal exposure to female hormones. *Arch Gen Psychiatry* 28: 554–561, 1973.
158. Money J: Effects of prenatal androgenization and deandrogenization on behavior in human beings. In Ganong WF, Martini L (eds): *Frontiers in Neuroendocrinology*. New York, Oxford University Press, 1973.
159. Ehrhardt AA, Meyer-Bahlburg FL, Rosen LR, Feldman JF, Veridiano NP, Zimmerman I, McEwan BS: Sexual orientation after prenatal exposure to exogenous estrogen. *Arch Sexual Behavior* 14: 57–75, 1985.
160. Money J, Ehrhardt AA: Fetal hormones and the brain: Effect on sexual dimorphism of behavior: a review. *Arch Sex Behav* 3: 241–262, 1971.
161. Bermant G, Davidson JM: *Biological Bases of Sexual Behavior*. New York, Harper & Row, 1974.
162. Short MA: An examination of the roles of the medial preoptic nucleus and the anterior hypothalamic nucleus in the mediation of sexual behavior in the female rat. *Diss Abstracts Int (Sci)* 38, Jan-Feb 1978, University of Texas at Arlington, 1977, #77–27, 791, 95 pages.
163. Tanner JM: Growing up. *Scient Am* 229: 34–43, 1973.
164. Smart MS, Smart RC: *Adolescents. Development and Relationships*. New York, Mac-Millan, 1973.
165. Maccoby EE, Martin JA: Socialization in the context of the family: Parent-child interaction. In Mussen PH (ed): *Handbook of Child Psychology*, vol IV, Hetherington EM (ed): *Socialization, Personality, and Social Development*. New York, John Wiley, 1983.
166. MacDonald K, Parke RD: Parent-child physical play: The effects of sex and age of children and parents. *Sex Roles* 15: 367–378, 1986.
167. Weitzman N, Birns B, Friend R: Traditional and nontraditional mothers' communication with their daughters and sons. *Child Develop* 56: 894–898, 1985.
168. Schwartz LA, Markham WT: Sex stereotyping in children's toy advertisements. *Sex Roles* 12: 157–170, 1985.
169. Maccoby EE, Jacklin CN: *The Psychology of Sex Differences*. Stanford, CA, Stanford University Press, 1974.
170. Hier DB, Crowley WF: Spatial ability in androgen-deficient men. *N Engl J Med* 306: 1202–1205, 1982.
171. Goleman D: Special abilities of the sexes: Do they begin in the brain? *Psychol Today* 12: 48–59, 1978.
172. Kimura D: Male brain, female brain: the hidden difference. *Psychol Today* 19: 52–58, 1985.
173. Archer J, Lloyd B: Sex differences: biological and social interaction. In Lewin R (ed): *Child Alive*. Garden City, NJ, Anchor Books, 1975.
174. Parlee MB: The sexes under scrutiny: from old biases to new theories. *Psychol Today* 12: 62–69, 1978.

175. Kohlberg L: Revisions in the theory and practice of moral development. *New Dir Child Dev* 2: 83–87, 1978.
176. Kurtines W, Greif EB: The development of moral thought: review and evaluation of Kohlberg's approach. *Psychol Bull* 81: 453–470, 1974.
177. Hoffman ML: Development of moral thought, feeling, and behavior. *Am Psychol* 34: 958–966, 1979.
178. Gilligan C: *In a Different Voice. Psychological Theory and Women's Development.* Cambridge, MA, Harvard University Press, 1982.
179. Mahon NE: Developmental changes and loneliness during adolescence. *Topics Clin Nursing* 5: 66–76, 1983.
180. Alishio KP, Schilling KM: Sex differences in intellectual and ego development in late adolescence. *J Youth Adolesc* 13: 213–224, 1984.

# INDEX

Page numbers in *italics* denote figures; those followed by "t" denote tables